Important Physical Constants

Constant	Symbol	Value
Velocity of light (vac.)	c	$2.997930 \times 10^{10}\,\text{cm sec}^{-1}$
Planck's constant	h	$6.6256 \times 10^{-27}\,\text{erg sec}$
Avogadro's number	N	$6.0226 \times 10^{23}\,\text{mole}^{-1}$
Faraday's constant	F	$96{,}487\,\text{coulomb equiv}^{-1}$
Gas constant	R	$8.3143 \times 10^{7}\,\text{erg deg}^{-1}\,\text{mole}^{-1}$
		$1.9872\,\text{cal deg}^{-1}\,\text{mole}^{-1}$
		$8.2054 \times 10^{-2}\,\text{liter atm deg}^{-1}\,\text{mole}^{-1}$
Boltzmann constant	k	$1.3805 \times 10^{-16}\,\text{erg deg}^{-1}$
Electronic rest mass	m_c	$9.1090 \times 10^{-28}\,\text{g}$
Electronic charge	e	$-\ 4.8030 \times 10^{-10}\,\text{esu}$
		$-\ 1.6021 \times 10^{-19}\,\text{coulomb}$
Gravitational constant	g	$980.665\,\text{cm sec}^{-2}$

Energy Conversion Factors

	erg	joule	liter atm	electron volt	calorie
1 erg =	1	10^{-7}	9.8687×10^{-10}	6.242×10^{11}	2.389×10^{-8}
1 joule =	10^{7}	1	9.8687×10^{-3}	6.242×10^{18}	2.389×10^{-1}
1 liter atm =	1.0133×10^{9}	1.0133×10^{2}	1	6.3248×10^{20}	24.218
1 electron volt =	1.602×10^{-12}	1.602×10^{19}	1.5811×10^{-21}	1	3.829×10^{-20}
1 calorie =	4.184×10^{7}	4.184	4.1291×10^{-2}	2.612×10^{19}	1

Principles
of Instrumental
Analysis

Principles of Instrumental Analysis

Douglas A. Skoog
Stanford University

Donald M. West
San Jose State College

HOLT, RINEHART AND WINSTON, INC.

New York   Chicago   San Francisco   Atlanta
Dallas   Montreal   Toronto   London   Sydney

Preface

Instrumental methods of analysis have become the backbone of experimental chemistry. As a consequence, the practicing chemist has need for a broad understanding of the principles, the applications, and the limitations of these techniques. *Principles of Instrumental Analysis* has been prepared in an effort to fill this need.

We have chosen to stress principles throughout the text; information concerned with applications and with instrumentation has been selected with the aim of illustrating these principles. We agree with Sandell and Elving,[1] who have written:

> It is important in any discussion of instrumentation and so-called instrumental methods of analysis to distinguish between instrumentation as an approach to the implementation of analytical techniques and methods, and as the manipulation of instruments. The latter is a minor result of the development of physical methods of analysis; the former is an extremely important aspect of analytical chemistry. Unfortunately, many writers on analytical chemistry, as well as many chemists in general, tend to think of instrumental analysis as being essentially the manipulation of black boxes.
>
> Basically, in order to be able to apply instrumentation most efficiently to his problems, the analytical chemist must understand the fundamental relations of chemical species to their physical and chemical properties; he must know the scope, applicability, and limitations of physical property measurement in respect to qualitative and quantitative analysis. Knowing this, he can then call on the instrumentation expert to design an apparatus for the measurement—continuous or intermittent—of the desired properties with the needed precision. The analytical chemist does not necessarily need to be an electronics expert.

[1]E. B. Sandell and P. J. Elving in I. M. Kolthoff and P. J. Elving, Eds., *Treatise on Analytical Chemistry*, Part I, Vol. 1 (New York: Interscience Publishers, Inc., 1959), p. 17.

The subject matter of this text is organized into two major and several minor sections. Chapters 2 through 14 are concerned with major analytical applications based upon the absorption or emission of electromagnetic radiation. Mass spectrometry is considered in Chapter 15; a survey of radiochemical methods is presented in Chapter 16. Chapters 17 through 22 cover the principal aspects of electroanalytical chemistry. A final section (Chapters 23 and 24) is devoted to separation techniques. Topic development in each section is sufficiently independent of other sections to provide the user with flexibility in selecting the order of presentation.

In the interests of imparting a reasonable size to the work, no attempt has been made to incorporate specific laboratory directions. The manuals by Meloan and Kiser[2] as well as by Reilley and Sawyer[3] are recommended as sources for experimental work.

We wish to acknowledge the contributions of Professors Alfred Armstrong, William and Mary College, and John DeVries, California State College at Hayward, who have read the entire manuscript and have offered numerous valuable suggestions. Thanks are also extended to Professor R. M. Silverstein, State University of Forestry at Syracuse, who has reviewed the chapter concerned with nuclear magnetic resonance spectroscopy, to Professor Paul Kruger of Stanford University, who has offered comments with regard to the chapter on radiochemical methods, to Professor Richard H. Eastman of Stanford University, who has made suggestions concerning the material on absorption spectroscopy, and to Mr. Claude F. Mears of Stanford University who has provided answers to all problems and has verified the calculations that appear within the text.

April, 1971   Douglas A. Skoog
Stanford, California   Donald M. West
San Jose, California

---

[2]C. E. Meloan and R. W. Kiser, *Problems and Experiments in Instrumental Analysis* (Columbus, Ohio: Charles E. Merrill Books, Inc., 1963).

[3]C. N. Reilley and D. T. Sawyer, *Experiments for Instrumental Methods* (New York: McGraw-Hill Book Company, Inc., 1961).

Contents

Principles
of Instrumental
Analysis

1 Introduction

The goal of a chemical analysis is to provide information about the composition of a sample of matter. In some instances qualitative information concerning the presence or absence of one or more components of the sample suffices; in other instances quantitative data are sought. Regardless of the need, however, the required information is finally obtained by measuring some physical property that is characteristically related to the component or components of interest.

Classification of Analytical Methods

Analytical methods are ordinarily classified according to the property that is observed in the final measurement process. Table 1-1 lists the more important of these properties as well as the names of the methods that are based upon them. It is striking to note that until about 1920 nearly all analyses were founded on the first two properties listed — namely mass and volume. As a consequence, gravimetric and volumetric procedures have come to be known as *classical* methods of analysis. The remainder of the list comprises *instrumental* methods; it is with these that this book is concerned.

Few features clearly distinguish instrumental methods from the classical ones beyond the chronology of their development. Some instrumental techniques are more sensitive than classical techniques, but others are not. With certain combinations of elements or compounds an instrumental method may be more specific; with others a gravimetric or a volumetric procedure is less subject to interference. Generalizations on the basis of accuracy, convenience, or expenditure of time are equally difficult to draw. Nor is it necessarily true that instrumental procedures employ more

1

Table 1-1 Physical Properties Employed for Analysis

Physical Property Measured	*Analytical Methods Based on Measurement of Property*
Mass	Gravimetric
Volume	Volumetric
Absorption of radiation	Spectrophotometry (x-ray, UV, visible, IR); colorimetry; atomic absorption, nuclear magnetic resonance, and electron spin resonance spectroscopy
Emission of radiation	Emission spectroscopy (x-ray, UV, visible), flame photometry, fluorescence (x-ray, UV, visible), radiochemical methods
Scattering of radiation	Turbidimetry, nephelometry, Raman spectroscopy
Refraction of radiation	Refractometry, interferometry
Diffraction of radiation	X-ray, electron diffraction methods
Rotation of radiation	Polarimetry, optical rotatory dispersion, and circular dichroism
Electrical potential	Potentiometry, chronopotentiometry
Electrical conductance	Conductimetry
Electrical current·	Polarography, amperometric titrations
Quantity of electricity	Coulometry
Mass-to-charge ratio	Mass spectrometry
Thermal properties	Thermal conductivity and enthalpy methods

sophisticated or more costly apparatus; indeed, use of a modern automatic balance in a gravimetric analysis involves more complex and refined instrumentation than is required for many of the other methods listed in Table 1-1.

Separation Methods

More often than not, the analysis of a sample of matter requires one or more of the following steps preliminary to the final physical measurement: (1) sampling to provide a homogeneous specimen whose composition is representative of the bulk of the material; (2) preparation and solution of a measured quantity of the sample; and (3) separation of the species to be measured from components that will interfere with the final measurement. These steps are frequently more troublesome and cause greater errors than the final measurement itself.

Separation procedures are needed because the physical and chemical properties suited for the determination of concentration are ordinarily shared by more than one element or compound. In dealing with closely related substances the problem of separation becomes of major importance and requires such techniques as chromatography, fractional distillation, countercurrent extraction, or controlled potential electrolysis. These specialized separation methods also fall within the purview of this text.

Choice of Methods for an Analytical Problem

Table 1-1 suggests that the chemist who is faced with an analytical problem often has a bewildering array of methods from which to choose. The amount of time he must spend on the analytical work and the quality of his results are critically dependent on this choice. In making his decision the chemist must take into account the complexity of the materials to be analyzed, the concentration of the species of interest, the number of samples he must analyze, and the accuracy required. His choice will then depend upon a knowledge of the basic principles underlying the various methods available to him, and thus their strengths and limitations. The development of this type of knowledge represents a major goal of this book.

Instrumentation in Analysis

In the broadest sense an instrument for chemical analysis does not generate quantitative data but instead simply converts chemical information to a form that is more readily observable. Thus, the instrument can be viewed as a communication device. It accomplishes this purpose in several steps that include (1) generation of a signal, (2) transformation of the signal to one of a different nature (called transduction), (3) amplification of the transformed signal, and (4) presentation of the signal as a displacement on a scale or on the chart of a recorder. It is not necessary that all of these steps be incorporated in every instrument.

The signal employed in an instrument may be generated from the sample itself. Thus, the yellow radiation emitted by heated sodium atoms constitutes the source of the signal in a flame photometer. The gravitational force exerted by the species of interest (or a compound of that species) serves as the signal in a gravimetric method. With many instruments, however, the original signal is formed independently of the sample; it is the modification of this signal by the species of interest that is related to concentration. Thus, in a polarimeter the signal is a beam of plane-polarized light; it is the change in orientation of this plane by the sample that provides the desired information.

Many instruments employ a transducer, which serves to convert the original analytical signal to one that is more conveniently measured. For example, the photocell, the thermocouple, and the photomultiplier tube are transducers that convert radiant energy into electrical signals. A chemical coulometer is a transducer that transforms electrical signals into a related amount of a gas or a solid (which is subsequently measured).

The sensitivity of many instruments is increased by amplification of the original signal or its transduced form. Amplification is most commonly accomplished electronically, although mechanical amplification is also encountered as in, for example, the pointer of an analytical balance.

The transduced and amplified signal from an instrument is generally presented as a linear or angular displacement along a scale. The presentation will ordinarily consist of the deflection of a needle of a voltmeter, the deflection of a beam of light reflected from the mirror of a galvanometer, or a displacement of the meniscus in a buret.

The growth of instrumental analysis has closely paralleled developments of the field of electronics because the generation, transduction, amplification, and display of a signal can be rapidly and conveniently accomplished with electronic circuitry. Numerous transducers have been developed for the conversion of signals to electrical form, and enormous amplification of the resulting electrical signals is possible. In turn, the electrical signals are readily presented on meters, on recorders, or in digital form.

As a result of the appearance of so much electronic circuitry in the laboratory the modern chemist finds himself faced with the question of how much electronics he should know in order to make most efficient use of the equipment available to him for analysis. Is it more important for him to concentrate his attention on the instrument itself or should he devote the majority of his efforts to the chemical principles of the measurement and its inherent limitations and strengths? In this book we have chosen the latter course, and we discuss instrument details only to the extent that these play a direct, major role in the outcome of a measurement. Undoubtedly, a strong knowledge of electronics is advantageous to the experimental chemist, but we feel that a fundamental knowledge of the chemical aspects of chemical instrumentation is more important.

2 Electromagnetic Radiation and Its Interactions with Matter

Fundamental concepts and properties of electromagnetic radiation and its interaction with matter are reviewed in this chapter. The material serves as preparation for the instrumental methods discussed in Chapters 3 through 14.

PROPERTIES OF ELECTROMAGNETIC RADIATION

Electromagnetic radiation is a type of energy that is transmitted through space at enormous velocities. It takes many forms, the most easily recognizable being light and radiant heat. Less obvious manifestations include x-ray, ultraviolet, microwave, and radio radiations.

In order to characterize many of the properties of electromagnetic radiation it is convenient to ascribe a wave nature to its propagation and to portray these waves by such parameters as velocity, frequency, wavelength, and amplitude. However, in contrast to other wave phenomena such as sound, electromagnetic radiation requires no supporting medium for its transmission, and it readily passes through a vacuum.

5

The wave model for radiation fails completely to account for phenomena associated with the absorption or the emission of radiant energy; for these processes it is necessary to view electromagnetic radiation as discrete particles of energy called *photons*. The energy of a photon is proportional to the frequency of the radiation. These dual views of radiation as particles and as waves are not mutually exclusive. Indeed, the apparent duality is rationalized by wave mechanics and is found to apply to other phenomena, such as the behavior of streams of electrons or other elementary particles.

Wave Properties

Electromagnetic radiation is conveniently treated as an alternating electrical field in space; associated with the electrical force field, but at right angles to it, is a magnetic force field. The wave properties for the radiation can be represented by electrical and magnetic vectors as shown in Figure 2-1. The two vectors are sinusoidal and are perpendicular to the direction of propagation of the wave.

It is the electrical field of electromagnetic radiation that interacts with the electrons in matter; as a result, representation of radiation by the electrical vector alone suffices for most purposes.

Wave parameters. The time interval required for the passage of successive maxima past a fixed point in space is called the *period p* of the radiation. The *frequency ν* is the number of oscillations of the field that occur per second, and is equal to $1/p$.[1] It is of importance to realize that *the frequency*

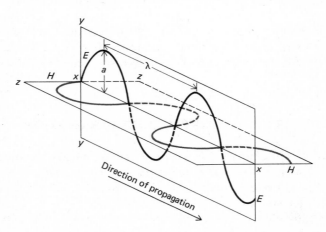

Figure 2-1. Electrical vector *E* and magnetic vector *H* of a plane-polarized electromagnetic wave of wavelength λ and amplitude *a*.

---

[1]The common unit of frequency is the *hertz* Hz, which is equal to one cycle per second. The *fresnel* is 10^{12} Hz.

is determined by the source and remains invariant regardless of the media through which the radiation travels. In contrast, the *velocity* of propagation of a wave v_i is the rate at which the wave front moves through a medium, and *is dependent upon* both the medium and the frequency; the subscript i is employed to indicate this frequency dependence. Another parameter of interest is the *wavelength* λ_i, which is the linear distance between successive maxima or minima of a wave.[2] Multiplication of the frequency (in cycles per second) by the wavelength (in centimeters per cycle) gives the velocity of the radiation (in centimeters per second); that is,

$$v_i = \nu\lambda_i \tag{2-1}$$

In a vacuum the velocity of propagation of radiation becomes independent of frequency and is at its maximum; this velocity is given the symbol c. The value of c has been accurately determined to be 2.99792×10^{10} cm/sec. Thus, for a vacuum,

$$c = \nu\lambda \cong 3 \times 10^{10} \text{ cm/sec} \tag{2-2}$$

In any other medium the rate of propagation is less because of interaction between the electromagnetic field of the radiation and the bound electrons of the medium. Since the radiant frequency is invariant and fixed by the source, the *wavelength must decrease* as radiation passes from a vacuum to a medium containing matter (equation 2-1).

The *wave number* σ is defined as the number of waves per centimeter, and is yet another way of describing electromagnetic radiation. When the wavelength *in vacuo* is expressed in centimeters, the wave number is equal to $1/\lambda$.

The *power P* of a beam of radiation is the energy of the beam reaching a given area per second; the *intensity I* is the power per unit solid angle. These quantities are related to the square of the amplitude a of the beam (see Figure 2-1). Although it is not strictly correct, power and intensity are often used synonomously.

Interference. As with other wave phenomena, electromagnetic waves can under certain conditions interact with one another when superimposed to produce a resultant wave whose intensity is either amplified or diminished depending upon the phases of the component waves. The amplitude of the resultant wave can be obtained by vector addition as shown by the solid line in Figure 2-2. Maximum destructive interference occurs when two waves are 180 deg out of phase, and maximum constructive interference occurs when the waves are exactly in phase.

[2]The units commonly used for describing the wavelength of radiation differ considerably in the various spectral regions. For example, the Ångstrom unit Å (10^{-10} m) is convenient for x-ray and short ultraviolet radiation; the nanometer nm or the synonomous millimicron $m\mu$ (10^{-9} m) is employed with visible and ultraviolet radiation; the micron μ (10^{-6} m) is commonly employed for infrared radiation.

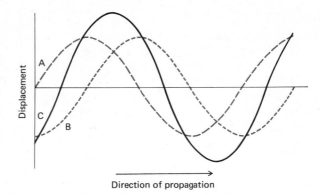

Displacement

Direction of propagation

Figure 2-2. Illustration of interference. Curve B is 70 deg out of phase with Curve A. Curve C is the resultant wave.

Diffraction of radiation. Electromagnetic radiation normally travels in straight paths. However, when a beam passes a sharp edge or travels through a narrow opening, a part appears in areas that should be in the shadow of the object in its path of travel. This bending of radiation is a form of *diffraction,* and is a direct consequence of interference.

Figure 2-3(a) is a schematic drawing of a device that was used in 1800 by Thomas Young to demonstrate that light waves could interfere with one another in the same way as could other types of waves. Radiation is propagated from slit S_1, illuminates slits S_2 and S_3 more or less equally, and then strikes a screen. If the radiation is monochromatic, a series of dark and light images or lines perpendicular to the plane of the paper is observed. If polychromatic radiation is employed, a central white band surrounded by colored fringes results. Note that the light band at E is in the shadow of the opaque material containing slits S_2 and S_3, and that the paths BE and AE are identical in length. Thus, the waves arrive at this point completely in phase and reinforce one another.

With the aid of Figure 2-3(b) the conditions for maximum constructive interference that cause the other light bands are readily derived. The lines BD and AD represent the light paths from slits S_2 and S_3 to the point on the screen of maximum intensity D. Line BC is then drawn perpendicular to AD so that triangles ABC and BCD are similar. Thus, angle ABC is equal to the angle of diffraction θ and we may write

$$\overline{AC} = \overline{AB} \sin \theta$$

The relative dimensions in Figure 2-3 have been greatly distorted for the sake of clarity. In a real system the slit spacing $\overline{AB}$ is extremely small when compared with the distance between the slit and the screen ($\overline{OE}$); that is, $\overline{OE} >> \overline{AB}$. Under these conditions $\overline{CD}$ closely approximates $\overline{BD}$, and the distance $\overline{AC}$ then is a good measure of the difference in path lengths of beams BD and AB. For the two beams to be in phase at D it is necessary that $\overline{AC}$ correspond to the wavelength of the radiation; that is,

$$\lambda = \overline{AC} = \overline{AB} \sin \theta$$

Reinforcement would also occur when the additional path length corresponds to 2λ, 3λ, and so forth. Thus, a more general expression for the light bands surrounding the central band is

$$n\lambda = \overline{AB} \sin \theta \qquad (2\text{-}3)$$

where **n** is an integer called the *order* of interference.

The linear displacement of the diffracted beam along the plane of the

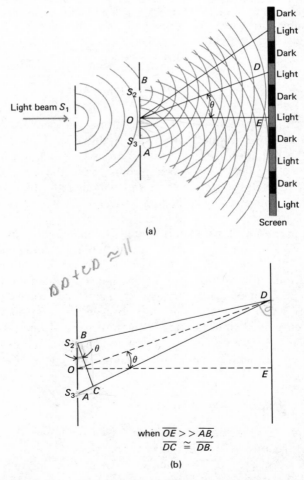

(a)

when $\overline{OE} >> \overline{AB}$,
$\overline{DC} \cong \overline{DB}$.

(b)

Figure 2-3. Diffraction of radiation by a pair of slits, S_2 and S_3.

screen DE is a function of the distance $\overline{OE}$ between the screen and the plane of the slits, and the spacing between the slits $\overline{AB}$; that is,

$$\overline{DE} = \overline{OE} \sin \theta$$

and substituting into equation (2-3) gives

$$\mathbf{n\lambda} = \frac{\overline{AB} \cdot \overline{DE}}{\overline{OE}} \tag{2-4}$$

Equation (2-4) permits the calculation of the wavelength from the three measurable quantities.

> **Example.** Suppose that the screen in Figure 2-3 is at a distance of 2.00 m and that the slit spacing is 0.300 mm. What is the wavelength of radiation if the fourth band is located 15.4 mm from the central band?
> Substituting into equation (2-4),
>
> $$4\lambda = \frac{0.300 \times 15.4}{2 \times 1000} \text{ mm}$$
>
> $$= 5.78 \times 10^{-4} \text{ mm} = 578 \text{ nm}$$

Coherent radiation.   In order for interference to occur between two wave trains such as those produced from slits S_2 and S_3 in Figure 2-3, it is necessary that the radiation from the two sources be *coherent*. The conditions for coherence are (1) the two beams must have the same frequency and wavelength (or sets of frequencies and wavelengths for a polychromatic beam) and (2) the phase relationship between the waves or sets of waves must be constant at all distances from the source. The necessity for these requirements can be shown experimentally by the illumination of each of the two slits S_2 and S_3 in Figure 2-3 with tungsten filaments; no interference patterns are then seen because the two beams are *incoherent*. That is, the emission of radiation by the atoms of the one source is not only random but is also independent of the other; as a consequence, constant phase differences at any given wavelength are not observed. In contrast, if two optical *lasers* are employed, interference is again observed because the emitting atoms of a laser act in concert to give a coherent beam that has a constant phase relationship with respect to the beam from a second laser.

Particle Properties of Radiation

Energy of electromagnetic radiation.   Certain interactions of radiation with matter require that the radiation be treated as packets of energy called *photons* or *quanta*. The energy of the photon depends upon the frequency of the radiation, and is given by

$$\checkmark \quad E = h\nu \tag{2-5}$$

where h is Planck's constant, which has a numerical value of 6.63×10^{-27} erg sec. In terms of wavelength,

$$E = \frac{hc}{\lambda} \qquad (2\text{-}6)$$

Thus, the energy associated with an x-ray photon ($\lambda \sim 10^{-8}$ cm) is of the order of ten thousand times greater than the photon emitted by a hot tungsten wire ($\lambda \sim 10^{-4}$ cm).

The photoelectric effect. The need for a particle model to describe the behavior of electromagnetic radiation can be seen by consideration of the *photoelectric effect.* When sufficiently energetic radiation impinges on a metallic surface, electrons are emitted. The energy of the emitted electron is found to be related to the frequency of the incident radiation by the equation

$$E = h\nu - w \qquad (2\text{-}7)$$

where w, the *work function,* is the work required to remove the electron from the metal to a vacuum. While E is directly dependent upon the frequency, it is found to be totally independent of the intensity of the beam, an increase in intensity merely causing an increase in the *number of electrons emitted* with energy E.

Calculations indicate that no single electron could acquire sufficient energy for ejection if the radiation striking the metal were uniformly distributed over the surface; nor could any electron accumulate enough energy for its removal in a reasonable length of time. Thus, it is necessary to assume that the energy is not uniformly distributed over the beam front, but rather is concentrated at certain points or in particles of energy.

The work w required to cause emission of electrons is characteristic of the metal. The alkali metals possess low work functions and emit electrons when exposed to radiation in the visible region. Heavier metals (such as cadmium) have larger work functions and require more energetic ultraviolet radiation to exhibit the photoelectric effect. As we shall point out in later chapters, the photoelectric effect has great practical importance in the detection of radiation with phototubes.

Energy units. The energy of a photon that is absorbed or emitted by a sample of matter can be related to an energy separation between two molecular or atomic states or to the frequency of a molecular motion of a constituent of the matter. For this reason it is often convenient to describe radiation in energy units, or in terms of frequency (Hz) or wave number (cm^{-1}) which are directly proportional to energy. On the other hand, the experimental measurement of radiation is most often expressed in terms of the reciprocally related wavelength units. The chemist must become adept at interconversion of the various units employed in spectroscopy. Table 2-1 provides conversion factors for some of the common transformations.

Table 2-1 Conversion Factors for Electromagnetic Radiation

(To convert data in units of x shown in the first column to the units indicated in the remaining columns, multiply or divide as shown.)

Units of x	Frequency Hz	Wave number cm^{-1}	Energy $kcal/mol$	Energy eV	Wavelength cm	Wavelength nm
Hz	$1.00\ x$	$3.33 \times 10^{-11}\ x$	$9.54 \times 10^{-14}\ x$	$4.14 \times 10^{-15}\ x$	$\dfrac{3.00 \times 10^{10}}{x}$	$\dfrac{3.00 \times 10^{17}}{x}$
cm⁻¹	$3.00 \times 10^{10}\ x$	$1.00\ x$	$2.86 \times 10^{-3}\ x$	$1.24 \times 10^{-4}\ x$	$\dfrac{1.00}{x}$	$\dfrac{1.00 \times 10^{7}}{x}$
kcal/mol	$1.05 \times 10^{13}\ x$	$3.50 \times 10^{2}\ x$	$1.00\ x$	$4.34 \times 10^{-2}\ x$	$\dfrac{2.86 \times 10^{-3}}{x}$	$\dfrac{2.86 \times 10^{4}}{x}$
eV	$2.42 \times 10^{14}\ x$	$8.07 \times 10^{3}\ x$	$2.31 \times 10^{1}\ x$	$1.00\ x$	$\dfrac{1.24 \times 10^{-4}}{x}$	$\dfrac{1.24 \times 10^{3}}{x}$
cm	$\dfrac{3.00 \times 10^{10}}{x}$	$\dfrac{1.00}{x}$	$\dfrac{2.86 \times 10^{-3}}{x}$	$\dfrac{1.24 \times 10^{-4}}{x}$	$1.00\ x$	$1.00 \times 10^{7}\ x$
nm	$\dfrac{3.00 \times 10^{17}}{x}$	$\dfrac{1.00 \times 10^{7}}{x}$	$\dfrac{2.86 \times 10^{4}}{x}$	$\dfrac{1.24 \times 10^{3}}{x}$	$1.00 \times 10^{-7}\ x$	$1.00\ x$

3,300 cm⁻¹   1/3,300 cm⁻¹

The *electron volt* (eV) is the energy unit ordinarily employed to describe the more energetic types of radiation, such as x-ray or ultraviolet. The electron volt is the energy acquired by an electron in falling through a potential of 1 volt. Radiant energy can also be expressed in terms of energy per mole of photons (that is, Avogadro's number of photons). For this purpose units of kcal/mol or cal/mol are often employed.

The Electromagnetic Spectrum

The electromagnetic spectrum covers an immense range of wavelengths, or energies. Figure 2-4 depicts qualitatively its major divisions. A

Energy		Wave number, σ	Wavelength, λ	Frequency, ν	Type radiation	Type spectroscopy	Type quantum transition
kcal/mol	Electron volts, eV	cm⁻¹	cm	Hz			
9.4×10^7	4.1×10^6	3.3×10^{10}	3×10^{-11}	10^{21}	Gamma ray	Gamma ray Emission	Nuclear
9.4×10^5	4.1×10^4	3.3×10^8	3×10^{-9}	10^{19}	X-ray	X-ray absorption, emission	Electronic (inner shell)
9.4×10^3	4.1×10^2	3.3×10^6	3×10^{-7}	10^{17}			
9.4×10^1	4.1×10^0	3.3×10^4	3×10^{-5}	10^{15}	Ultraviolet	Vac. UV absorption UV Vis. absorption emission, fluorescene	Electronic (outer shell)
					Visible		
9.4×10^{-1}	4.1×10^{-2}	3.3×10^2	3×10^{-3}	10^{13}	Infrared	IR absorption, Raman	Molecular vibration
9.4×10^{-3}	4.1×10^{-4}	3.3×10^0	3×10^{-1}	10^{11}	Microwave	Microwave absorption	Molecular rotation
9.4×10^{-5}	4.1×10^{-6}	3.3×10^{-2}	3×10^1	10^9		Electron paramagnetic resonance	Magnetically induced spin states
9.4×10^{-7}	4.1×10^{-8}	3.3×10^{-4}	3×10^3	10^7	Radio	Nuclear magnetic resonance	

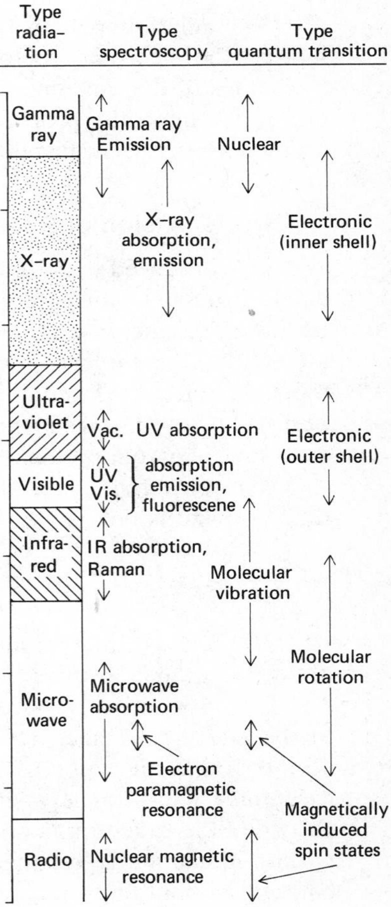

Figure 2-4. Spectral properties, applications, and interactions of electromagnetic radiation.

752-7511

logarithmic scale has been employed in this representation; note that the portion to which the human eye is perceptive is very small. Such diverse radiations as gamma rays and radio waves are also electromagnetic radiations, differing from visible light only in the matter of frequency, and hence energy.

Figure 2-4 shows the regions of the spectrum that are employed for analytical purposes, and the names of the spectroscopic methods associated with these applications. The molecular or atomic transitions responsible for absorption or emission of radiation in each region are also indicated.

THE INTERACTION OF RADIATION WITH MATTER

As radiation passes from a vacuum to the surface of a portion of matter, the electrical vector of the radiation interacts with the atoms and molecules of the medium. The nature of the interaction can vary, however, depending upon the properties of the matter. As a consequence, the radiation may be either transmitted, absorbed, reflected, or scattered.

Transmission of Radiation

It is observed experimentally that the rate at which radiation is propagated through a transparent substance is less than its velocity in a vacuum; furthermore, the rate is dependent upon the kinds and concentrations of atoms, ions, or molecules present. It follows from these observations that the radiation must interact in some way with the matter. As we have mentioned, however, a frequency change is not observed; thus, the interaction does not involve a permanent energy transfer.

The *refractive index* of a medium is a measure of its interaction with radiation and is defined by

$$n_i = \frac{c}{v_i} \qquad (2\text{-}8)$$

where n_i is the refractive index at a specified frequency i, v_i is the velocity of the radiation in the medium, and c is its velocity *in vacuo*.

The interaction involved in the transmission process can be ascribed to the alternating electrical field of the radiation, which causes oscillation of the bound electrons of the particles with respect to their heavy (and essentially fixed) nuclei; periodic polarization of the particles thus results. Provided the radiation is not absorbed, the energy required for polarization is only momentarily retained (10^{-14} to 10^{-15} sec) by the species and is re-emitted without alteration when the substance returns to its original state. Since there is no net energy change in this process, the frequency of the emitted radiation is unchanged, but the rate of its propagation has been slowed by the time required for the process to occur. Thus, transmission in a particulate medium can be viewed as a stepwise process involving oscillating atoms, ions, or molecules as intermediates.

One would expect the radiation from each polarized particle in a medium to be emitted in all directions. If the particles are small, however, it can be shown that destructive interference prevents the propagation of significant amounts of radiation in any direction other than that of the original path of the beam. On the other hand, if the medium contains large particles (such as polymer molecules or colloidal particles), this destructive effect is incomplete and a portion of the beam is *scattered* as a consequence of the interaction step. We consider scattering in a later section of this chapter.

Dispersion. As we have noted, the velocity of radiation in matter is frequency-dependent; since c in equation (2-8) is independent of this parameter, the refractive index of a substance must also change with frequency. The variation in refractive index of a substance with frequency or wavelength is called its *dispersion,* and is summarized by a plot such as Figure 2-5. Clearly, the relationship is complex; typically, however, dispersion plots exhibit two types of regions. In the *normal dispersion* region there is a gradual increase in refractive index with increasing frequency or decreasing wavelength. *Anomalous dispersion* regions are those frequency ranges in which a sharp change in refractive index is observed. Anomalous dispersion always occurs at radiation frequencies that correspond to the natural harmonic frequency associated with some part of the molecule, atom, or ion of the substance. At such a frequency permanent energy transfer from the radiation to the substance occurs and *absorption* of the beam is observed. Absorption is discussed in the next section.

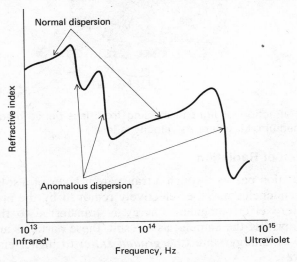

Figure 2-5. Typical dispersion curve.

Dispersion curves are of importance in the choice of materials for the optical components of instruments. A substance that exhibits normal dispersion over the wavelength region of interest is most suitable for the manufacture of lenses, for which a high and relatively constant refractive index is desirable. Chromatic aberration is minimized through the choice of such a material. In contrast, a substance with a refractive index that is not only large but also highly frequency-dependent is selected for the fabrication of prisms. The applicable wavelength region for the prism then approaches the anomalous dispersion region for the material.

Refraction of radiation. When radiation passes from one medium to another of differing physical density, an abrupt change in direction of the beam is observed as a consequence of differences in the velocity of the radiation in the two media. This *refraction* of a beam is illustrated in Figure 2-6. The extent of refraction is given by the relationship

$$\frac{\sin \theta_1}{\sin \theta_2} = \frac{n_2}{n_1} = \frac{v_i}{v_2} \tag{2-9}$$

If the medium M_1 is vacuum, v_1 becomes c and thus the refractive index n_2 of M_2 is simply the ratio of the sines of the two angles.

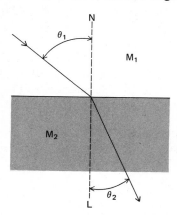

Figure 2-6. Refraction of light in passing from less-dense medium M_1 into more-dense medium, M_2 where its velocity is less.

Absorption of Radiation

When radiation passes through a transparent layer of a solid, liquid, or gas, certain frequencies may be selectively removed by the process called *absorption*. Here electromagnetic energy is transferred to the atoms or molecules comprising the sample; as a result, these particles are promoted from their lowest-energy state (the *ground state*) to higher-energy states, or *excited states*.

Excited atoms or molecules are relatively short-lived, and tend to return to their ground states after approximately 10^{-8} sec. Most commonly, the energy released in this process appears in the system as heat. In some cases, however, the excited species may undergo a chemical change that absorbs the energy (*a photochemical reaction*). In others, radiation is reemitted (usually at longer wavelengths) in the form of fluorescence or phosphorescence.

Atoms, molecules, or ions have a limited number of discrete, quantized energy levels; for absorption of radiation to occur the energy of the exciting photon must match the energy difference between the ground state and one of the excited states of the absorbing species. Since these energy differences are unique for each species, a study of the frequencies of absorbed radiation provides a means of characterizing the constituents of a sample of matter. For this purpose a plot of the decrease in radiant power (absorbance) as a function of wavelength or frequency is experimentally derived. Typical plots of this kind, called *absorption spectra,* are shown in Figure 2-7.

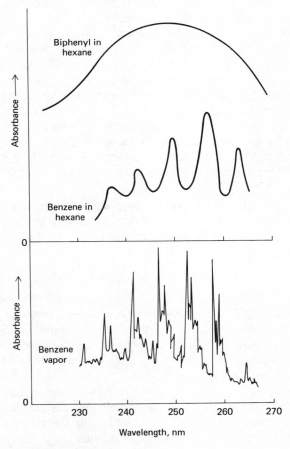

Figure 2-7. Absorption spectra.

The general appearance of absorption spectra varies widely depending upon the complexity, the physical state, and the environment of the absorbing species. It is convenient to recognize two types of spectra, namely those associated with atomic absorption and those resulting from molecular absorption.

Atomic absorption.   When polychromatic ultraviolet and visible radiation is passed through a medium containing monatomic particles (such as gaseous mercury or sodium) only a relatively few well-defined frequencies are removed by absorption because of the small number of possible energy states of the particles. Excitation can only occur by an electronic process in which one or more of the electrons of the atom is raised to a higher energy level. Thus with sodium, excitation of the $3s$ electron to the $3p$ state requires an energy corresponding to a wave number of 1.697×10^4 cm^{-1}. As a result, sodium vapor exhibits a sharp absorption peak at 589.3 nm (yellow light). Several other narrow absorption lines corresponding to other permitted electronic transitions are possible.

Ultraviolet and visible radiation have sufficient energy to cause transitions of the outermost or bonding electrons only. X-ray frequencies, on the other hand, are several orders of magnitude more energetic and capable of interacting with electrons closest to the nuclei of atoms. Absorption peaks corresponding to electronic transitions of the innermost electrons are thus observed in the x-ray region.

Regardless of the wavelength region involved, atomic absorption spectra typically consist of a limited number of very narrow peaks. Spectra of this type are discussed later in connection with x-ray absorption and atomic absorption spectroscopy.

Molecular absorption.   Absorption by polyatomic molecules, particularly in the condensed state, is a considerably more complex process because the number of energy states is greatly enhanced. Here the total energy of a molecule is given by

$$E = E_{\text{electronic}} + E_{\text{vibrational}} + E_{\text{rotational}} \tag{2-10}$$

where $E_{\text{electronic}}$ describes the electronic energy of the molecule, while $E_{\text{vibrational}}$ refers to the energy of the molecule as a whole due to interatomic vibrations of various energies. The third term in equation (2-10) accounts for the energy associated with the rotation of the molecule around its center of gravity. For each electronic energy state of the molecule there normally exist several possible vibrational states and for each of these, in turn, numerous rotational states. As a consequence, the number of possible energy levels for a molecule is much greater than for an atomic particle.

The first term in equation (2-10) is ordinarily larger than the other two, and as we have mentioned, electronic transitions generally involve energies corresponding to ultraviolet or visible radiation. Pure vibrational transitions, on the other hand, are brought about by the less energetic infrared radiation in the 1 to 15 μ range; rotational changes require even less energy, radiation in the 10 to 100 μ region being sufficient.

In contrast to atomic absorption spectra, which consist of a series of sharp, well-defined lines, molecular spectra are often characterized by absorption bands that may encompass a wide range of wavelengths (see Figure 2-7). As before, absorption by molecules in the ultraviolet and visible regions is the result of electronic transitions involving bonding electrons. Molecular spectra are much more complex, however, because there exists a large number of vibrational and rotational states for each electronic state. Thus, for a given value of $E_{\text{electronic}}$ in equation (2-10), there is a series of values for E that differs only slightly due to variations in $E_{\text{vibrational}}$ and/or $E_{\text{rotational}}$. As a consequence, the spectrum for a molecule often consists of a series of closely spaced absorption bands, such as that shown for benzene vapor in Figure 2-7. Unless a high resolution instrument is employed, the individual bands may not be detected, and the spectra appear as smooth curves. Furthermore, in the condensed state and in the presence of solvent molecules the individual bands tend to broaden to give the type of spectra shown in the upper two curves of Figure 2-7. Solvent effects are considered in later chapters.

Vibrational absorption in the absence of electronic absorption can be observed in the infrared region where the energy of radiation is insufficient for electronic transitions. Here, spectra exhibit narrow, closely spaced absorption peaks resulting from transitions among the various vibrational quantum levels. Variations in rotational levels may give rise to a series of peaks for each vibrational state; with liquid or solid samples, however, rotation is often hindered or prevented, and the effects of these small energy differences are not detected.

Pure rotational spectra for gases can be observed in the microwave region.

Absorption induced by a magnetic field. When electrons or the nuclei of certain elements are subjected to a strong magnetic field, additional quantized energy levels are produced as a consequence of magnetic properties of these elementary particles. The difference in energy between the induced states is small, and transitions between them are brought about only by absorption of radiation of long wavelengths, or low frequency. With nuclei, radio waves ranging from 10 to 200 MHz are generally employed, while for electrons microwaves with frequencies of 1000 to 25,000 MHz are absorbed.

Absorption by nuclei or by electrons in magnetic fields is studied by *nuclear magnetic resonance* (nmr) and *electron spin resonance* (esr) techniques, respectively; these methods are considered in Chapter 7.

Reflection and Scattering of Radiation

Reflection.   When radiation crosses an interface between media of different refractive index, reflection occurs, the fraction reflected becoming larger with increasing differences in refractive index. For a beam traveling normal to the interface, the fraction reflected is given by

$$\frac{I_r}{I_0} = \frac{(n_2 - n_1)^2}{(n_2 + n_1)^2} \tag{2-11}$$

where I_0 is the intensity of the incident beam and I_r is the reflected intensity; n_1 and n_2 are the refractive indexes of the two media. As radiation passes through glass ($n_2 \sim 1.5$), equation (2-11) predicts reflection losses amounting to about 4 percent at each glass-air interface. In later chapters we see that such losses assume considerable importance in various optical instruments, particularly those containing several such interfaces.

Reflective losses at a glass or quartz surface increase only slightly as the angle of the incident beam increases up to about 60 deg. Beyond this figure, however, the percentage of radiation that is reflected increases rapidly and approaches 100 percent at 90 deg.

Scattering.   As noted earlier, the transmission of radiation in matter can be pictured as the momentary retention of the radiant energy causing the polarization of the ions, atoms, or molecules, followed by its reemission in all directions as the particles return to their original state. When the particles are small with respect to the wavelength of radiation, destructive interference removes nearly all of the reemitted radiation except that traveling in the original direction of the beam; the path of the beam appears to be unaltered as a consequence of the interaction. Careful observation, however, reveals that a very small fraction of the radiation (the *scattered radiation*) is transmitted at all angles from the original path and that the intensity of the scattered radiation increases with particle size. With particles of colloidal dimensions, scattering becomes sufficiently intense to be seen by the naked eye (the Tyndall effect).

Scattering by molecules or aggregates of molecules with dimensions significantly smaller than the wavelength of the radiation is called *Rayleigh scattering*; its intensity is readily related to wavelength (an inverse fourth power effect) and the particle dimensions and their polarizability. An everyday manifestation of scattering is the blueness of the sky, which results from the greater scattering of the shorter wavelengths of the visible spectrum.

Scattering by larger particles, as in a colloidal suspension, is much more difficult to treat theoretically; the intensity varies roughly as the inverse square of wavelength.

Measurement of scattered radiation can be used to determine the size and shape of polymer molecules and colloidal particles. The phenomenon is also utilized in *nephelometry,* an analytical method that is considered in Chapter 9.

Raman scattering.   This type differs from ordinary scattering in that part of the scattered radiation suffers quantized frequency changes. These changes arise from vibrational energy level transitions occurring in the molecule as a consequence of the polarization process. Raman spectroscopy is also discussed in Chapter 9.

Polarization of Radiation

Plane polarization.   Ordinary radiation can be visualized as a bundle of electromagnetic waves in which the amplitude of vibrations is equally distributed among a series of planes centered along the path of the beam. Viewed end-on, the electrical vectors would then appear as shown in Figure 2-8(a). The vector in any one plane, say *XY*, can be resolved into two mutually perpendicular components *AB* and *CD* as shown in Figure 8(b); if the two components for each plane are combined, the resultant has the appearance shown in Figure 8(c); note that Figure 8(c) has a different scale from Figure 8(a) or 8(b) to keep its size within reason.

Removal of one of the two resultant vibration planes in Figure 2-8(c) produces a beam that is *plane polarized.* The vibration of the electrical vector of a plane-polarized beam, then, occupies a single plane in space.

Plane-polarized electromagnetic radiation is produced by certain radiant energy sources. For example, the radio waves emanating from an antenna commonly have this characteristic. Presumably, the radiation from a single

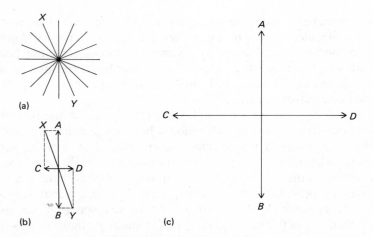

Figure 2-8.   (a) A few of the electrical vectors of a beam traveling perpendicular to the paper. (b) The resolution of a vector in plane *XY* into two mutually perpendicular components. (c) The resultant when all vectors are resolved (not to scale).

atom or molecule is also polarized; however, since common light sources contain large numbers of these particles in all orientations, the resultant is a beam that vibrates in all directions around the axis of travel.

The absorption of radiation by certain types of matter is dependent upon the plane of polarization of the radiation. For example, when properly oriented to a beam, anisotropic crystals selectively absorb radiations vibrating in one plane but not the other. Thus, a properly oriented polarizing sheet of these crystals absorbs all of the components of, say, CD and transmits nearly completely component AB. A polarizing sheet, regardless of orientation, removes approximately 50 percent of the radiation from an unpolarized beam, and transmits the other half as a plane-polarized beam. The plane of polarization of the transmitted beam is dependent upon the orientation of the sheet with respect to the incident beam. When two polarizing sheets, oriented at 90 deg to one another, are placed perpendicular to the beam path, essentially no radiation is transmitted. Rotation of one results in a continuous increase in transmission until a maximum is reached when molecules of the two sheets have the same alignment.

The way in which radiation is reflected, scattered, transmitted, and refracted by certain substances is also dependent upon direction of polarization. As a consequence, a group of analytical methods whose selectivity is based upon the interaction of these compounds with light of a particular polarization have been developed. These methods are considered in detail in Chapter 13.

EMISSION OF RADIATION

Electromagnetic radiation often results when excited particles (ions, atoms, or molecules) return to lower-energy levels or to their ground state. Excitation can be brought about by a variety of means, including bombardment with electrons or other elementary particles, exposure to a high-potential alternating current spark, heat treatment in an arc or a flame, or absorption of electromagnetic radiation.

If the radiating particles are well separated from one another, as in the gaseous state, they behave as independent bodies and often produce radiation containing relatively few specific wavelengths. The resulting spectrum is then said to be discontinuous and it is termed a *line spectrum*. A *continuous spectrum,* on the other hand, is one in which all wavelengths are represented over an appreciable range, or one in which the individual wavelengths are so closely spaced that separation is not feasible by ordinary means. Continuous spectra result from excitation of (1) solids or liquids, in which the atoms are so closely packed as to be incapable of independent behavior or (2) complicated molecules possessing many closely related energy states. Continuous spectra also occur when the energy changes involve particles with unquantized kinetic energies.

Both continuous spectra and line spectra are of importance in analytical chemistry. The former are frequently employed as sources in methods based on the interaction of radiation with matter, such as spectrophotometry. Line spectra, on the other hand, are important because they permit the identification and the determination of the emitting species.

Thermal Radiation

When solids are heated to incandescence, continuous radiation is emitted that is more characteristic of the temperature of the emitting surface than of the material of which it is composed. Radiation of this kind (called *black-body radiation*) is produced by the innumerable atomic and molecular oscillations excited in the condensed solid by the thermal energy. Theoretical treatment of black-body radiation leads to the following conclusions: (1) the radiation exhibits a maximum emission at a wavelength that varies inversely with the absolute temperature, (2) the total energy emitted by a black body (per unit of time and area) varies as the fourth power of temperature, and (3) the emissive power at a given temperature varies inversely as the fifth power of wavelength. These relationships are reflected in the behavior of several experimental radiation sources shown in Figure 2-9. The emission from these sources approaches the emission of the ideal black body. Note that the energy distribution peaks in Figure 2-9 shift to shorter wavelengths with increasing temperature; thus, very high temperatures are needed to cause a thermally excited source to emit a substantial fraction of its energy as ultraviolet radiation.

Heated solids are used to produce infrared radiation for analytical

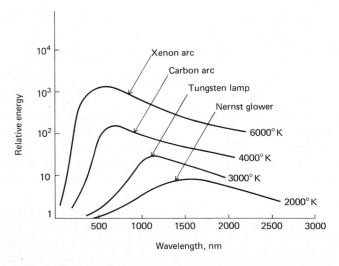

Figure 2-9. Black-body radiation curves.

instruments. Such sources are also useful as radiation sources for the visible and longer wavelength ultraviolet region.

Emission by Gases

Atoms or molecules in the gaseous state can often be excited by electrical discharge or by heat to produce radiation in the ultraviolet and visible regions. Ordinarily, such excitation causes transitions involving the outer electrons of the species; radiation is then emitted as the excited atoms return to the ground state. For simple atoms, emission spectra consist of a series of discrete lines whose energies correspond to the energy differences between the various electronic states. For molecules, emission spectra may become more complicated because there may exist several vibrational and rotational states for each possible electronic energy level; instead of a single line for each electronic transition, numerous closely spaced lines that form an emission band may be observed.

A true continuous spectrum is sometimes produced from excitation of gaseous molecules. For example, when hydrogen at low pressure is subjected to an electric discharge, an excited molecule is formed which then dissociates to give two hydrogen atoms and an ultraviolet photon. While the energy of the excited species is quantized, the radiant energy emitted by this process is not because the atoms produced can have an entire range of kinetic energies; the larger these energies are, the smaller is the energy of the photon. Thus, a continuous spectrum from about 400 to 200 nm is produced that is useful as a source for absorption spectrophotometry.

Emission of X-ray Radiation

Radiation in the x-ray region is normally produced by the bombardment of a metal target with a stream of high-speed electrons. The accelerated beam of electrons causes the innermost electrons in the atoms of the target material to be raised to higher energy levels or to be ejected entirely. The excited atoms or ions then return to the ground state by various stepwise electronic transitions that are accompanied by the emission of photons, each having an energy $h\nu$. The consequence is the production of an x-ray spectrum consisting of a series of discrete lines characteristic of the target material. The discrete spectrum is superimposed on a continuous spectrum resulting from the nonquantized radiation given off when some of the high-speed electrons are partially decelerated in passing through the target material.

Fluorescence

When a species absorbs radiant energy, its excited state typically has a lifetime of 10^{-7} to 10^{-9} sec. Under usual circumstances the excited particle loses its energy and returns to the ground state via a series of collisions with other particles in the system; the net effect of this process is a conversion of the absorbed energy to heat.

If the absorbing system consists of atoms in the gaseous state at low pressure, the excited atom may return directly to its ground state by emitting radiation of the same frequency that was responsible for excitation. This process is known as *resonance fluorescence*.

At normal pressures or in solution, collisions occur in less than 10^{-9} sec, and the probability of transferring energy from the excited species by collision becomes much greater than by the direct reemission of energy. In some instances a molecule that has been excited to a high vibrational level of an upper electronic state rapidly loses vibrational energy by collision. If the still-excited molecule is slow to lose its excess electronic energy by collision, however, it may return to the ground state by emitting radiation that is lower in frequency than the absorbed radiation. This type of fluorescence is much more common than the resonance variety.

An energy level diagram for a typical fluorescence process is shown in Figure 2-10. Note that in normal fluorescence the frequency emitted is lower than the absorbed frequency, while in resonance fluorescence the two are identical.

Fluorescence measurements provide a convenient method for the determination of a variety of molecular species.

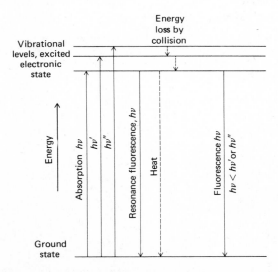

Figure 2-10. Energy changes in absorption and fluorescence.

Problems

1. Calculate the frequency in hertz and the wave number in cm^{-1} for
    (a) a monochromatic x-ray beam with a wavelength of 9.0 Å,

    (b)   the sodium D line at 589.0 nm,

    (c)   an infrared absorption peak at 12.6 μ,

    (d)   a microwave beam of wavelength 200 cm.

2. Calculate the energy of each of the photons in Problem 1 in ergs per photon, kcal per mol, and eV.

3. Calculate the wavelength and frequency (Hz) of the sodium D line (5890 Å in a vacuum) when it is transmitted by

    (a)   air ($n_D = 1.00027$),

    (b)   a solution having a refractive index of 1.340,

    (c)   a solid having a refractive index of 2.070.

4. The radiation from a sodium lamp was passed through a glass cell containing an aqueous solution with a refractive index of 1.3400. Assume that the two cell walls were normal to the beam and that the refractive index of the glass was 1.7000 and of air was 1.00027. Calculate the percent loss in intensity by reflection at each interface. Calculate the total percent reflection loss.

3 Techniques and Tools for the Measurement of Absorption of Ultraviolet and Visible Radiation

The measurement of absorption of ultraviolet and visible radiation provides a convenient means for the analysis of numerous inorganic and organic species.[1] As noted in Chapter 2, radiation in these regions is of sufficient energy to cause electronic transitions of outer, valence electrons. If the sample is composed of atoms or simple molecules in the gaseous state, its absorption spectrum usually consists of a series of sharp, well-defined lines corresponding to the limited number of permitted electronic

[1] For a more complete discussion of absorption measurements see: W. West in A. Weissberger, Ed., *Physical Methods of Organic Chemistry*, Vol. I, Part III (New York: Interscience Publishers, Inc., 1960), Chaps. 28 and 30; R. P. Bauman, *Absorption Spectroscopy* (New York: John Wiley & Sons, Inc., 1962), Chaps. 1-4; E. J. Meehan in I. M. Kolthoff and P. J. Elving, Eds., *Treatise on Analytical Chemistry*, Part 1, Vol. 5 (New York: Interscience Publishers, Inc., 1964), Chap. 55.

transitions; the discrete nature of the absorption process lends a high degree of selectivity to analyses based on such measurements. In contrast, the spectra of ions or molecules in solution ordinarily contain broad bands arising in part from the superposition of vibrational and sometimes rotational energy changes upon the electronic transitions. As a consequence, associated with each electronic absorption is a series of lines that are so closely spaced as to appear continuous. In addition, broadening of lines occurs as a result of intermolecular forces operating between the closely packed molecules or ions in the condensed medium. Selectivity is less with spectra of this type; on the other hand, the broad bands are ideally suited for accurate quantitative measurements.

In this and the following chapter we are concerned exclusively with absorption of ultraviolet and visible radiation by solutions. Absorption measurements of gaseous species are considered in Chapter 5.

QUANTITATIVE ASPECTS OF ABSORPTION MEASUREMENTS

The absorption of electromagnetic radiation by some species M can be considered to be an irreversible two-step process, the first step of which can be represented by

$$M + h\nu \rightarrow M^* \tag{3-1}$$

where M^* represents the atomic or molecular particle in the excited state resulting from absorption of the photon $h\nu$. The lifetime of the excited state is brief (10^{-8} to 10^{-9} sec), its existence being terminated by any of several *relaxation* processes. The most common type of relaxation involves conversion of the excitation energy to heat; that is,

$$M^* \rightarrow M + heat \tag{3-2}$$

Relaxation may also result from decomposition of M^* to form new species; such a process is called a *photochemical reaction*. Alternatively, relaxation may result in the fluorescent or phosphorescent reemission of radiation (p. 24). It is important to note that the lifetime of M^* is so very short that its concentration at any instant is negligible under ordinary conditions. Furthermore, the amount of thermal energy created is usually not detectable. Thus, absorption measurements have the advantage of creating a minimal disturbance of the system under study.

Beer's Law

In his very early studies on the absorption of radiation Beer[2] postulated that the decrease in radiant energy of a beam of monochromatic radiation

---

[2]A. Beer, *Ann. Physik. Chem.* (2), **86**, 78 (1852).

was proportional to the intensity or power of the beam and the amount of absorbing substance in its path. These hypotheses lead directly to the statement of Beer's law, given in equation (3-7); it is of interest, however, to rationalize these hypotheses in terms of the modern picture of radiant energy.[3]

Consider the block of absorbing matter (solid, liquid, or gas) shown in Figure 3-1. A beam of parallel radiation of power P_0 strikes the block perpendicular to a surface; after passing through a length b of the material its power is reduced to P by absorption. Consider now a cross section of the block having an area S and an infinitesimal thickness dx. Within this section there are dn absorbing particles (molecules or ions); associated with each particle we can imagine a surface at which photon capture can occur. That is, if a photon arrives by chance at one of these areas, absorption will follow immediately. The total projected area of these capture surfaces within the section is designated as dS and the probability for capture of a single photon that passes through the section is the ratio of the capture area to the total area, dS/S. On a statistical average this ratio represents the probability of capture of photons within the section.

The power of the beam entering the section P_x is proportional to the numbers of photons per cm^2 per sec, and dP_x represents the quantity removed within the section; the fraction captured is then $-dP_x/P_x$ and this ratio on the average also equals the probability of capture. The term is given a minus sign to indicate that P undergoes a decrease. Thus,

$$-\frac{dP_x}{P_x} = \frac{dS}{S} \tag{3-3}$$

Recall, now, that dS is the sum of the capture areas for each particle within the section; it must therefore be proportional to the number of particles or

$$dS = \alpha \, dn \tag{3-4}$$

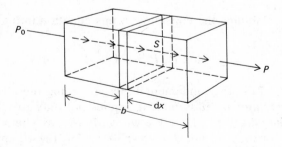

Figure 3-1. Passage of radiation through an absorbing body of length b.

[3]The discussion that follows is based on a paper by F. C. Strong, *Anal. Chem.*, **24**, 338 (1952).

where dn is the number of particles and α is a proportionality constant that can be called the capture cross section. Combining equations (3-3) and (3-4) and summing over the interval of zero to n we obtain

$$-\int_{P_0}^{P} \frac{dP_x}{P_x} = \int_{0}^{n} \frac{\alpha\, dn}{S}$$

Integration gives

$$-\ln P/P_0 = \alpha n/S$$

Upon converting to base-10 logarithms and inverting the fraction to change the sign we have

$$\log P_0/P = \frac{\alpha n}{2.303\ S} \tag{3-5}$$

where n is the total number of particles within the block shown in Figure 3-1. The cross-sectional area S can be expressed in terms of the volume of the block V and its length b. Thus,

$$S = \frac{V}{b}\ \text{cm}^2$$

Substitution of this quantity into (3-5) yields

$$\log P_0/P = \frac{\alpha n b}{2.303\ V} \tag{3-6}$$

We can readily convert n/V to concentration in moles per liter; that is,

$$c = \frac{1000\ n}{6.02 \times 10^{23}\ V}\ \text{mole/liter}$$

Substitution into (3-6) yields

$$\log P_0/P = \frac{6.02 \times 10^{23}\ \alpha b c}{2.303 \times 1000}$$

Finally, upon combining the constants in this equation into a single term ϵ, we obtain

$$\log P_0/P = \epsilon b c = A \tag{3-7}$$

Equation (3-7) is the fundamental law governing the absorption of *all types of electromagnetic radiation*; it applies not only to solutions but to gases and solids as well. It is known variously as the Lambert-Beer, Bouguer-Beer, or most commonly, *Beer's law*. The logarithmic term on the left side of the equation is called the *absorbance*, and is given the symbol A. The constant ϵ is called the *molar absorptivity* when the concentration c is expressed in terms of moles of absorber per liter, and the path length b is given in centimeters; it is simply called the *absorptivity*, and is given the symbol a when other units are used for concentration or path length. Equation (3-7) shows that the absorbance of a solution is directly proportional to the concentration of absorbing species when the length of the light path is fixed, and directly proportional to the light path when the concentration is fixed; a quantitative analysis based upon the absorption of radiation makes use of one or the other of these relationships.

Beer's law applies to a solution containing more than one kind of absorbing substance provided there is no interaction among the various species. Thus, for a multicomponent system, we may write

$$A_{\text{total}} = A_1 + A_2 + \cdots + A_n$$
$$= \epsilon_1 bc_1 + \epsilon_2 bc + \cdots + \epsilon_n bc_n \qquad (3\text{-}8)$$

where the subscripts refer to absorbing components $1, 2, \cdots n$.

Measurement of Absorption

Beer's law, as given by equation (3-7), is not directly applicable to chemical analysis. Neither P nor P_0, as defined, can be measured easily in the laboratory because the solution to be studied must be held in some sort of container. Interaction between the radiation and the walls is inevitable, producing a loss in power at each interface as a consequence of reflection (p. 20) or possibly absorption. In addition, the beam may suffer a diminution in power during its passage through the solution as a result of scattering by large molecules or inhomogeneities. These phenomena are depicted in Figure 3-2.

In order to correct for these effects the power of the beam transmitted

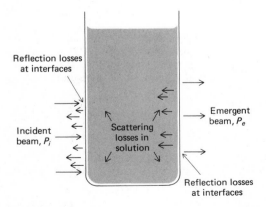

Figure 3-2. Processes that attenuate a beam of radiation upon passage through a solution. The power of the emergent beam P_e is less than that of the incident beam P_i because of (1) reflection at air-wall, wall-solution, solution-wall, and wall-air interfaces; (2) absorption by the two walls; (3) scattering within the solution; (4) absorption by components of the solution.

through the solution of interest is compared with the power of a beam passing through an identical cell containing the solvent for the sample. An experimental absorbance that closely approximates the true absorbance of the solution can then be calculated; that is,

$$A \cong \log \frac{P_{\text{solvent}}}{P_{\text{solution}}} \cong \log \frac{P_0}{P} \tag{3-9}$$

The term P_0, when used henceforth, refers to the power of a beam of radiation after it has passed through a cell containing the solvent for the component of interest.

Terminology Associated with Absorption Measurements

In recent years the attempt has been made to develop a standard nomenclature for the various quantities related to the absorption of radiation. The recommendations of the American Society for Testing Materials are given in Table 3-1 along with some of the alternative names and symbols frequently encountered. An important term in this table is the *transmittance* T, which is defined as

$$T = \frac{P}{P_0}$$

Table 3-1   Important Terms and Symbols Employed in Absorption Measurement

Term and Symbol[a]	Definition	Alternative Name and Symbol
Radiant power, P, P_0	Energy of radiation reaching a given area of a detector per second	Radiation intensity, I, I_0
Absorbance, A	$\log \frac{P_0}{P}$	Optical density, D; extinction, E
Transmittance, T	$\frac{P}{P_0}$	Transmission, T
Path length of radiation, in cm, b	—	l, d
Molar absorptivity,[b] ϵ	$\frac{A}{bc}$	Molar extinction coefficient
Absorptivity,[c] a	$\frac{A}{bc}$	Extinction coefficient, k

[a]Reprinted from *Analytical Chemistry*, **24**, 1349 (1952). With permission, A.C.S.
[b]c expressed in units of mole/liter.
[c]c may be expressed in g/liter or other specified concentration units; b may be expressed in cm or in other units of length.

The transmittance is the fraction of incident radiation transmitted by the solution; it is often expressed as a percentage. The transmittance is related to the absorbance as follows:

$$-\log T = A$$

Limitations to the Applicability of Beer's Law

The linear relationship between absorbance and path length at a fixed concentration of absorbing substances is a generalization for which no exceptions are known. On the other hand, deviations from the direct proportionality between measured absorbance and concentration when b is constant are frequently encountered. Some of these deviations are of such a fundamental nature that they represent real limitations of the law; others occur as a consequence of the manner in which the absorbance measurements are made or as a result of chemical changes associated with concentration changes; the latter two are sometimes known, respectively, as *instrumental deviations* and *chemical deviations*.

Real limitations to Beer's law. Beer's law is successful in describing the absorption behavior of dilute solutions only; in this sense it is a limiting law. At high concentrations (usually $> 0.01\ F$) the average distance between the species responsible for absorption is diminished to the point where each affects the charge distribution of its neighbors. This interaction, in turn, can alter their ability to absorb a given wavelength of radiation. Because the degree of interaction is dependent upon concentration, the occurrence of this phenomenon causes deviations from the linear relationship between absorbance and concentration.

Deviations from Beer's law also arise because ϵ is dependent upon the refractive index of the solution.[4] Thus, if concentration changes cause significant alterations in the refractive index n of a solution, departures from Beer's law are observed. A correction for this effect can be made by substitution of the quantity $\epsilon n/(n^2 + 2)^2$ for ϵ in equation (3-7). In general, this correction is not significant at concentrations of less than $0.01\ F$.

Chemical deviations. Apparent deviations from Beer's law are frequently encountered as a consequence of association, dissociation, or reaction of the absorbing species with the solvent. A classic example of a chemical deviation is observed with unbuffered potassium dichromate solutions, in which the following equilibria exist:

$$Cr_2O_7^{2-} + H_2O \rightleftharpoons 2HCrO_4^- \rightleftharpoons 2H^+ + 2CrO_4^{2-}.$$

At most wavelengths the molar absorptivities of the dichromate ion and the two chromate species are quite different. Thus, the total absorbance of any

[4]G. Kortum and M. Seiler, *Angew. Chem.*, **52**, 687 (1939).

solution is dependent upon the concentration ratio between the dimeric and the monomeric forms. This ratio, however, changes markedly with dilution, and causes a pronounced deviation from linearity between the absorbance and the total concentration of chromium. Nevertheless, the absorbance due to the dichromate ion remains directly proportional to its molar concentration; the same is true for the chromate ion. This fact is easily demonstrated by making measurements in strongly acidic or strongly basic solutions where one or the other of these species predominates. Thus, deviations in the absorbance of this system from Beer's law are more apparent than real, because they result from shifts in chemical equilibria. These deviations can, in fact, be readily predicted from the equilibrium constants for the reactions and the molar absorptivities of the dichromate and chromate ions.

Instrumental deviation.   Strict adherence of an absorbing system to Beer's law is observed only when the radiation employed is monochromatic. This observation is another manifestation of the limiting character of the relationship. Use of a truly monochromatic beam for absorbance measurements is seldom practical, and polychromatic radiation may lead to departures from Beer's law.

Consider a beam comprised of radiation of two wavelengths λ' and λ''. Assuming that Beer's law applies strictly for each of these individually, we may write for radiation λ'

$$A' = \log \frac{P_0'}{P'} = \epsilon' bc$$

or
Similarly, we can write for λ''

$$\frac{P_0'}{P'} = 10^{\epsilon' bc}$$

$$\frac{P_0''}{P''} = 10^{\epsilon'' bc}$$

When an absorbance measurement is made with radiation composed of both wavelengths, the power of the beam emerging from the solution is given by $(P' + P'')$, and that of the beam from the solvent, by $(P_0' + P_0'')$. Therefore, the measured absorbance is

$$A_M = \log \frac{(P_0' + P_0'')}{(P' + P'')}$$

which can be rewritten as

$$A_M = \log \frac{(P_0' + P_0'')}{(P_0' \, 10^{-\epsilon' bc} + P_0'' \, 10^{-\epsilon'' bc})}$$

or

$$A_M = \log(P_0' + P_0'') - \log(P_0' \, 10^{-\epsilon' bc} + P_0'' \, 10^{-\epsilon'' bc})$$

Now, when $\epsilon' = \epsilon''$, this equation simplifies to

$$A_M = \epsilon'bc$$

and Beer's law is followed. If the two absorptivities differ, however, the relationship between A_M and concentration will no longer be linear, and greater departures from linearity can be expected with increasing differences between ϵ' and ϵ''.

Experiments show that deviations from Beer's law resulting from the use of a polychromatic beam are not appreciable provided the radiation used does not encompass a spectral region in which the absorber exhibits large changes in absorbance as a function of wavelength. This observation is illustrated in Figure 3-3.

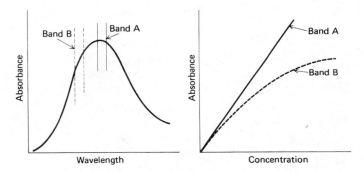

Figure 3-3. The effect of polychromatic radiation upon the Beer's law relationship. Band A shows little deviation since ϵ does not change greatly throughout the band. Band B shows marked deviations since ϵ undergoes significant changes in this region.

Spectral Curves

Spectral data are presented in a variety of ways. The frequency, the wave number, or the wavelength is commonly employed for the abscissa. The ordinate is usually expressed in units of transmittance (or percent transmittance), absorbance, or the logarithm of absorbance.

Physical chemists prefer frequency or wave number for the abscissa because of the linear relationship between these functions and energy. Other chemists, on the other hand, often express the wavelength in nanometers or Ångstrom units for ultraviolet and visible radiation.

Spectral data for three permanganate solutions are presented in three different ways in Figure 3-4. The concentrations stand in the ratio of 1:2:3. Note that employment of absorbance provides the greatest differentiation in the region where the absorbance is high (0.8 to 1.3) and the transmittance is low (< 20 percent). In contrast, greater differences occur in the trans-

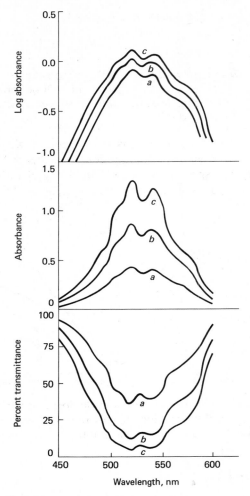

Figure 3-4. Methods for plotting spectral data. Concentration relationship: $a:b:c = 1:2:3$.

mittance curves when the transmittances lie in the range of 20 to 60 percent. With a plot of log A spectral detail tends to be lost; on the other hand, this type of plot is particularly convenient for comparing curves for solutions of different concentration since the curves are displaced an equal amount on the ordinate scale regardless of wavelength.

COMPONENTS OF INSTRUMENTS FOR ABSORPTION MEASUREMENTS

Instruments that measure the transmittance or absorbance of solutions contain five basic components: (1) a stable source of radiant energy that can be varied in intensity; (2) a device that permits employment of a restricted wavelength region; (3) transparent containers for sample and solvent; (4) a radiation detector or transducer that converts the radiant energy to a measurable signal (usually electrical), and (5) a signal indicator. The block diagram in Figure 3-5 shows the usual arrangement of these components.

The signal indicator for most instruments for absorption measurements is equipped with a linear scale that covers a range from 0 to 100 units. Direct percent transmittance readings can then be obtained by first adjusting the indicator to read zero when radiation is blocked from the detector by a shutter. The indicator is then brought to 100 with the beam passing through the solvent and impinging on the detector; this adjustment is accomplished by varying either the intensity of the source or the sensitivity of the detector. When the sample container is placed in the beam, the indicator gives percent transmittance directly, provided that the detector responds linearly to changes in radiant power. Clearly, a logarithmic scale can be scribed on the indicator to permit direct absorbance readings as well.

The nature and the complexity of the various components of absorption instruments vary enormously depending upon the wavelength region involved and how the data are to be used; regardless of the degree of sophistication, however, the function of each component is the same.

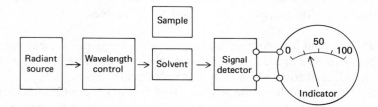

Figure 3-5. Components of instruments for measuring absorption of radiation.

Radiation Sources

In order to be suitable for absorption measurements the source of radiation must meet certain requirements. First, it must generate a beam with sufficient power for ready detection and measurement. Second, the source should provide continuous radiation; that is, its spectrum should contain all wavelengths over the region in which it is to be used. Finally, the source should be stable; the power of the radiant beam must remain constant for the period needed to measure both P and P_0. Only under these conditions are absorbance measurements reproducible. Some instruments are designed to measure P and P_0 simultaneously; here, fluctuations in the output of the source present no problem.

Sources of visible radiation. The most common source of visible radiation is the tungsten filament lamp. The energy distribution of this source approximates that of a black body and is thus temperature-dependent. Figure 2-9 illustrates the behavior of the tungsten filament lamp at 3000° K. In most absorption instruments the operating filament temperature is about 2870° K; the bulk of the energy is thus emitted in the infrared region. A tungsten filament lamp is useful for the wavelength region between 320 and 2500 nm.

In the visible region the energy output of a tungsten lamp varies approximately as the fourth power of the operating voltage. As a consequence, close voltage control is required for a stable radiation source. Constant voltage transformers or electronic voltage regulators are often employed for this purpose. As an alternative, the lamp can be operated from a 6-V storage battery, which provides a remarkably stable voltage source if it is maintained in good condition.

Sources for ultraviolet radiation. As noted in Chapter 2, a continuous spectrum in the ultraviolet region is conveniently produced by the excitation of hydrogen at low pressure with an electric discharge. Two types of hydrogen lamps are encountered. The high-voltage variety employs potentials of 2000 to 6000 V to cause a discharge between aluminum electrodes; water cooling of the lamp is required if high-radiation intensities are to be produced. In low-voltage lamps an arc is formed between a heated, oxide-coated filament and a metal electrode. About 40 V dc are required to maintain the arc.

Both high- and low-voltage lamps produce a continuous spectrum in the region between 180 and 375 nm. Quartz windows must be employed in the tubes since glass absorbs strongly in this wavelength region.

Deuterium lamps have the advantage of producing continuous radiation of higher intensities under the same operating conditions as hydrogen lamps.

Band Selection with Filters

Radiation that is restricted to a limited wavelength region is employed in quantitative techniques for three reasons. (1) The probability of adherence of the absorbing system to Beer's law is greatly enhanced (see p. 35). (2) A greater selectivity is assured, since substances that absorb in other wavelength regions are less likely to interfere. (3) A greater change in absorbance per increment of concentration will be observed if only wavelengths that are strongly absorbed are employed; thus a greater sensitivity is attained.

Of the various devices that produce limited bands of radiation, filters are the simplest and least expensive. Two types, *absorption filters* and *interference filters*, are available. The characteristics of a filter can be described in terms of the wavelength of maximum transmission and the *effective band width*. The latter defines the wavelength range over which the transmittance decreases to one-half of its maximum value (see Figure 3-6).

Absorption filters. Absorption filters limit radiation by absorbing certain portions of the spectra. The most common type consists of colored glass or of a dye suspended in gelatin and sandwiched between glass plates. The former have the advantage of greater thermal stability.

Absorption filters have effective band widths that vary from perhaps 30 to 250 nm (see Figure 3-6). Filters that provide the narrowest band widths also absorb a significant fraction of the desired radiation, and may have a transmittance of 0.1 or less at their band peaks. Glass filters with transmittance maxima throughout the entire visible region are available commercially and are relatively inexpensive.

Cut-off filters have transmittance of nearly 100 percent over a portion of the visible spectrum but then rapidly decrease to zero transmittance over the remainder. A narrow spectral band can be isolated by coupling a cut-off filter with a second filter [see Figure 3-6(b)].

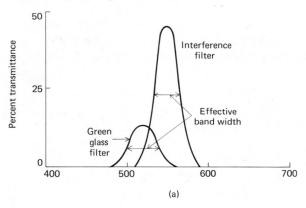

(a)

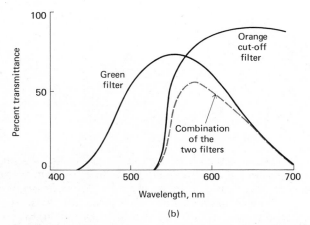

(b)

Figure 3-6. Comparison of transmittance characteristics of two filters and a filter combination.

Interference filters. As the name implies, interference filters rely on optical interference to produce relatively narrow bands of radiation [see Figure 3-6 (a)]. An interference filter consists of a transparent dielectric (frequently calcium fluoride or magnesium fluoride) that occupies the space between two semitransparent metallic films coated on the inside surfaces of two glass plates. The thickness of the dielectric layer is carefully controlled, and determines the wavelength of the transmitted radiation. When a perpendicular beam of collimated radiation strikes this array, a fraction passes through the first metallic layer while the remainder is reflected. The portion that is passed undergoes a similar partition upon striking the second metallic film. If the reflected portion from this second interaction is of the proper wavelength, it is partially reflected from the inner side of the first layer in phase with incoming light of the same wavelength. The result is that this particular wavelength is reinforced, while most others, being out of phase, suffer destructive interference.

Figure 3-7 illustrates an interference filter. For purposes of clarity the incident beam is shown as arriving at an angle θ from the perpendicular. In ordinary use θ approaches zero so that the equation accompanying Figure 3-7 simplifies to

$$n\lambda' = 2t$$

where λ' is the wavelength of radiation *in the dielectric* and t is its thickness. The corresponding wavelength in air is given by

$$\lambda = \lambda' n$$

where n is the refractive index of the dielectric medium. Thus, the wavelengths of radiation transmitted by the filter are

$$\lambda = \frac{2\,tn}{n}$$

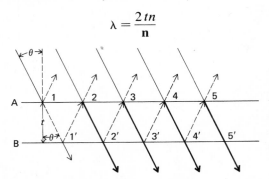

Figure 3-7. The interference filter. At point 1, light strikes the semitransparent film at an angle θ from the perpendicular, is partially reflected, partially passed. The same process occurs at 1′, 2, 2′, and so forth. For reinforcement to occur at point 2, the distance traveled by the beam reflected at 1′ must be some multiple of its wavelength in the medium λ'. Since the path length between surfaces can be expressed as $t/\cos \theta$, the condition for reinforcement is that $n\lambda' = 2t/\cos \theta$ where **n** is a small whole number.

where the integer **n** is the *order* of interference. The glass layers of the filter are often selected to absorb all but one of the reinforced bands; transmission is thus restricted to a single order.

Interference filters generally provide significantly narrower band widths (as low as 10 nm) and greater transmittances of the desired wavelength than do absorption-type filters (Figure 3-6). Interference filters that provide radiation bands from the ultraviolet up to about 6 μ in the infrared can be purchased.

Monochromators

A monochromator is a device that resolves radiation into its component wavelengths and permits the isolation of any desired portion of the spectrum from the remainder. In general, a monochromator contains a system of slits and lenses, and a dispersing element that may be either a prism or a diffraction grating.

Figure 3-8 is a schematic representation of a prism monochromator. Light is admitted through an entrance slit, is collimated by a lens, and then strikes the surface of the prism at an angle. Refraction occurs at both faces of the prism; the dispersed radiation is then focused on a slightly curved surface containing the exit slit. Radiation of the desired wavelength can be caused to pass through this slit by rotation of the prism.

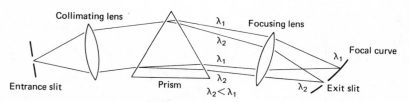

Figure 3-8. A prism monochromator.

Dispersion with prisms. The spectral purity of the radiation emitted from a prism monochromator is determined in part by the dispersion characteristics of the prism. Referring to Figure 3-9, the angular dispersion of a prism is defined as dθ/dλ, which may be broken into two parts

$$d\theta/d\lambda = \frac{d\theta}{dn} \cdot \frac{dn}{d\lambda} \tag{3-10}$$

where dθ/dn represents the change in θ as a function of the refractive index of the prism material and dn/dλ expresses the variation of the refractive index with wavelength.

The magnitude of dθ/dn in equation (3-10) is determined by the geometry of the prism and the angle of incidence I. In order to avoid problems of astigmatism it is desirable to employ an angle of incidence such that the

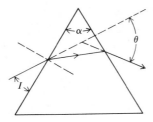

Figure 3-9. Dispersion by a prism.

path of the beam through the prism is approximately parallel to the base. Under these conditions $d\theta/dn$ depends only upon the prism angle α, and increases rapidly with this quantity. Reflection losses, however, impose an upper limit of about 60 deg on α. For such a prism it can be shown that

$$d\theta/dn = (1 - n^2/4)^{-1/2} \qquad (3\text{-}11)$$

The term $dn/d\lambda$ is related to the dispersion of the substance from which the prism is constructed. We noted in Chapter 2 that the greatest dispersion for a given material is in the region near its anomalous dispersion, which in turn is close to a region of absorption. The dispersion of substances employed for construction of prisms is shown in Figure 3-10. Note that the rapid rise in the refractive index for glass below 400 nm corresponds to the sharp increase in absorbance that prevents its use below 350 nm. In the

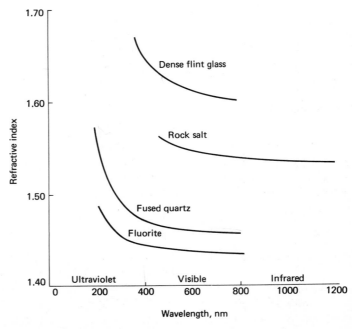

Figure 3-10. Dispersion of several optical materials.

region of 350 to 800 nm, however, glass is greatly superior to quartz for prism construction because of its greater change in refractive index with wavelength.

The complex relationship associated with the dispersion of a prism monochromator results in a nonlinear distribution of wavelengths along the plane of the exit slit (see Figure 3-8). Typical dispersion patterns for glass and quartz prism monochromators are shown in Figure 3-11; note the difference in dispersion characteristics in the visible region. For comparison, the dispersion of a grating monochromator, which exhibits a linear dispersion, is also shown.

Two types of prism monochromators are commonly used in instruments for measuring absorption of radiation. The *Bunsen* monochromator employs a 60-deg prism arranged as shown in Figure 3-8. The entrance slit is at the focal point of the first lens, which collimates the beam before it strikes the prism surface. The second lens then focuses the dispersed radiation upon the surface containing the exit slit.

When crystalline quartz is used for a monochromator of the above type, a *Cornu* prism must be employed because of the optical activity of quartz. A beam of radiation passing into quartz is divided into two circularly polarized beams that have different velocities. This problem is overcome by forming a 60-deg Cornu prism from two 30-deg prisms, one of left-hand quartz and the other of right-hand quartz. In this way the different rates of travel are compensated for.

The *Littrow* monochromator contains a 30-deg prism which is silvered along its vertical axis. Thus, the radiation enters and leaves the same surface but is refracted to the same degree as with a 60-deg prism. An example of a Littrow monochromator is shown in Figure 3-23. Littrow prisms are widely used because they provide compactness in instrument construction. Furthermore, the reversal of path in such a prism cancels out any effects of optical activity; the need for two-piece construction is thus eliminated.

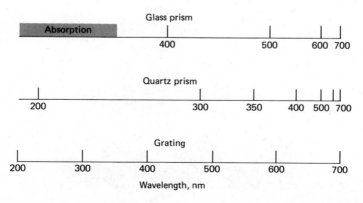

Figure 3-11.   Dispersion characteristics of three types of monochromators.

Dispersion with Diffraction Grating

Dispersion of ultraviolet, visible, and infrared radiation can be brought about by passage of a beam through a *transmission grating* or by reflection from a *reflection grating*. A transmission grating consists of a series of parallel and closely spaced grooves ruled on a piece of glass or other transparent material. A grating suitable for use in the ultraviolet and visible region has about 15,000 lines per inch. It is vital that these lines be equally spaced throughout the several inches in length of the typical grating. Such gratings require elaborate apparatus for their production, and are consequently expensive. Replica gratings are less costly. They are manufactured by employing a master grating as a mold for the production of numerous plastic replicas; the products of this process, while inferior in performance to an original grating, suffice for many applications.

When a transmission grating is illuminated by radiation from a slit. each of the grooves acts as a new light source; interference among the multitude of beams results in dispersion of the radiation into its component wavelengths (see Figure 2-3). If the dispersed radiation is focused on a plane surface, a spectrum consisting of a series of images of the entrance slit is produced.

Reflection gratings are produced by ruling a polished metal surface or by evaporating a thin film of aluminum onto the surface of a replica grating. As shown in Figure 3-12, the radiation is reflected from each of the unruled portions, and interference among the reflected beams produces dispersion. Thus, the path of beam 1 differs from that of beam 2 by the amount ($\overline{AB} - \overline{CD}$); for constructive interference to occur it is necessary that

$$n\lambda = (\overline{AB} - \overline{CD})$$

But it is readily seen that

$$\overline{AB} = d \sin i$$

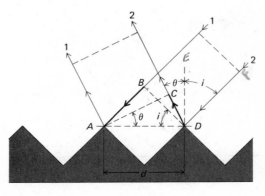

Figure 3-12. Diffraction at a reflection grating.

where d is the spacing of the rulings and i is the angle of the incident beam to the normal. It is also seen that

$$\overline{CD} = - d \sin \theta$$

where θ is the angle of reflection, and the minus sign arises because the angle of reflection, by convention, is opposite in sign to the angle of incidence. The conditions for constructive interference then are

$$n\lambda = d(\sin i + \sin \theta) \tag{3-12}$$

Equation (3-12) shows that several values of λ exist for a given diffraction angle θ. Thus, if a first-order line ($n = 1$) of 800 nm is found at θ, second-order (400 nm) and third-order (267 nm) lines also appear at this angle. Ordinarily, the first-order line is the most intense; by variation in the shapes of the grooves it is possible to concentrate as much as 90 percent of the incident intensity in this order. The higher-order lines can be generally removed with suitable filters. For example, glass, which absorbs radiation below 350 nm, eliminates the high-order spectra associated with first-order radiation in most of the visible region.

A concave grating can be produced by ruling a spherical reflecting surface. Such a diffracting element serves also to focus the radiation on the exit slit and eliminates the need for a lens.

The angular dispersion of a grating can be obtained by differentiating equation (3-12) holding i constant; thus, at any given angle of incidence

$$\frac{d\theta}{d\lambda} = \frac{n}{d \cos \theta}$$

Note that the dispersion increases as the distance d between rulings decreases or as the number of lines per inch increases. Over short wavelength ranges $\cos \theta$ does not change greatly with λ, so that the dispersion of a grating is nearly linear. By proper design of the optics of a grating monochromator, it is possible to produce an instrument that for all practical purposes has a linear dispersion of radiation along the focal plane of the exit slit. Figure 3-11 shows the contrast between a grating and a prism monochromator in this regard.

Figure 3-13 illustrates the *Ebert mounting* of a reflection grating in a monochromator. A spherical concave mirror is employed to collimate the radiation from the entrance slit and also to focus the diffracted radiation on the exit slit. Rotation of the grating permits the desired wavelength to be emitted from the monochromator.

Slits.   The slits of a monochromator play an important part in determining its quality. Slit jaws are formed by carefully machining two pieces of metal to give sharp edges, as shown in Figure 3-14. Care is taken to assure that the edges of the slit are exactly parallel to one another and that they lie on the same plane.

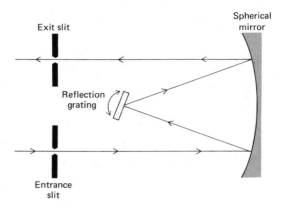

Figure 3-13.   A simple grating monochromator.

In some monochromators the openings of the two slits is fixed; more commonly, the spacing between the jaws can be adjusted with a micrometer mechanism.

The entrance slit (see Figures 3-8 and 3-13) serves as the radiation source; its image is focused on the surface containing the exit slit. If the radiation source consists of a few discrete wavelengths, a series of images appears on this surface as bright lines, each corresponding to a given wavelength. A particular line can be brought to focus on the exit slit by rotation of the dispersing element. If the entrance and exit slits are of the same size (as is often the case), the image of the entrance slit will in theory just fill the exit-slit opening when the setting of the monochromator corresponds to the wavelength of the radiation. Movement of the monochromator mount in one or the other direction results in a continuous decrease in emitted intensity, zero being reached when the entrance slit image has been displaced by its full width.

Figure 3-15 illustrates the situation in which monochromatic radiation of wavelength λ_2 strikes the exit slit. Here, the monochromator is set for λ_2 and the two slits are identical in width. The image of the entrance slit just fills the exit slit. Movement of the monochromator to a setting of λ_1 or λ_3 results in the image being moved completely out of the slit. The lower half of the figure shows a plot of the radiant power emitted as a function of mono-

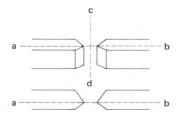

Figure 3-14.   Construction of slits.

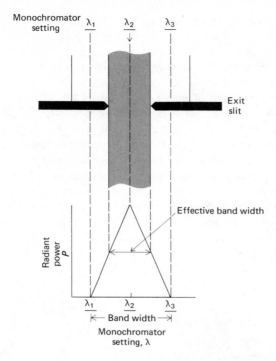

Monochromator setting λ_1 λ_2 λ_3

Exit slit

Effective band width

Radiant power *P*

λ_1 λ_2 λ_3

$\longleftarrow$ Band width $\longrightarrow$

Monochromator setting, λ

Figure 3-15. Illumination of an exit slit by monochromatic radiation λ_2 at various monochromator settings. Exit and entrance slits are identical.

chromator setting. Note that the *band width* is defined as the span of mono-chromator settings (in units of wavelength) needed to move the image of the entrance slit across the exit slit. If polychromatic radiation were employed it would also represent the span of wavelengths emitted from the exit slit at a given monochromator setting.

The *effective band width* for a monochromator is also defined in Figure 3-15 as one-half the wavelength range transmitted by the instrument. Figure 3-16 illustrates the relationship between the effective band width of an instrument and its effectiveness in resolving spectral peaks. Here, the exit slit of a monochromator is illuminated with a beam composed of just three wavelengths, λ_1, λ_2, and λ_3; each beam has the same intensity. In the top figure the effective band width of the instrument is exactly equal to the difference in wavelength between λ_1 and λ_2 or λ_2 and λ_3. When the monochromator is set at λ_2, radiation of this wavelength just fills the slit. Movement of the monochromator in either direction reduces the transmitted intensity of λ_2 but increases the intensity of one of the other lines by an equivalent amount. As shown to the right of the figure, no spectral resolution of the three wavelengths is achieved.

In the middle drawing of Figure 3-16 the effective band width of the instrument has been reduced by narrowing the openings of the exit and entrance slits to three-quarters that of the original. The figure on the right shows that partial resolution of the three lines results. When the effective band width is reduced to one-half the difference in wavelengths of the three beams, complete resolution is obtained as shown in the bottom drawing.

The effective band width of a monochromator is dependent upon both the dispersion of the prism of grating as well as the width of the entrance and exit slits. Most monochromators are equipped with variable slits so that the effective band width can be changed. The use of minimal slit widths is desirable where the resolution of narrow absorption bands is needed. On the other hand, as the slits are narrowed there is a marked decrease in the radiant power emitted, and accurate measurement of this power becomes more difficult. Thus, wider slit widths may be used for quantitative analysis than for qualitative work, where spectral detail is important.

As noted previously, the dispersion of a prism is not linear (see Figure 3-10). Much narrower slits must thus be employed at long wavelengths than

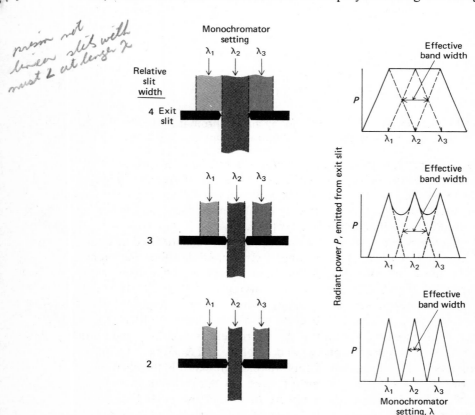

Figure 3-16. The effect of the slit width on spectra. The entrance slit is illuminated with λ_1, λ_2, and λ_3 only. Entrance and exit slits are identical. Plots on the right show changes in emitted power as the setting of monochromator is varied.

at short wavelengths in order to obtain radiation of a given effective band width. The slit width required to maintain an effective band width of one nanometer for a monochromator equipped with a quartz Littrow prism is shown in Figure 3-17. One of the advantages of a grating monochromator is that a fixed slit width produces radiation of nearly constant band width regardless of wavelength.

Spurious radiation. The exit beam of a monochromator is usually contaminated with small amounts of radiation which may be of wavelengths far different from the instrument setting. There are several sources of this unwanted radiation. Reflections of the beam from various optical parts and the monochromator housing contribute. Scattering by dust particles in the atmosphere or on the surfaces of optical parts also causes stray radiation to reach the exit slit. Generally, the effects of spurious radiation are minimized by introducing baffles in appropriate spots in the monochromator and by coating interior surfaces with flat black paint. In addition, the monochromator is sealed with windows over the slits to prevent entrance of dust and fumes. Despite these precautions some spurious radiation is still emitted; as pointed out later, its presence can have serious effects on absorption measurements under certain conditions.

Double monochromators. Many modern monochromators contain two dispersing elements; that is, two prisms, two gratings, or a prism and a

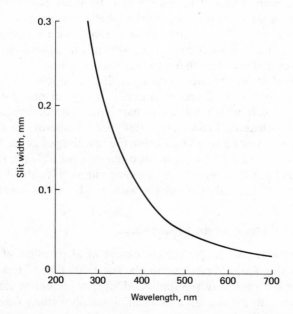

Figure 3-17. The slit width needed to maintain a constant effective band width of 1 nm from a quartz Littrow prism monochromator.

grating. This arrangement markedly reduces the amount of stray radiation and also provides greater dispersion and spectral resolution. Furthermore, if one of the elements is a grating, higher-order wavelengths are removed by the second element. Figures 3-25 and 3-26 show instruments with double monochromators.

Sample Containers

In common with the optical elements of monochromators the *cells* or *cuvettes* that hold the samples must be made of material that passes radiation in the spectral region of interest. Thus, quartz or fused silica is required for work in the ultraviolet region below 350 nm; these materials are also transparent in the visible region and to about 3 μ in the infrared region. Silicate glasses can be employed in the region between 350 nm and 2 μ. Plastic containers have also found application in the visible region.

In general, the best cells have windows that are perfectly normal to the direction of the beam in order to minimize reflection losses. Most instruments are provided with a pair of cells that have been matched with respect to light path and transmission characteristics to permit an accurate comparison of the power transmitted through the sample and the solvent. The most common cell length is 1 cm; matched, calibrated cells of this size are available from several commercial sources. Other path lengths from 0.1 cm and shorter to 10 cm can also be purchased. Transparent spacers for shortening the path length of 1-cm cells to 0.1 cm are also available.

For reasons of economy, cylindrical cells are sometimes employed. The curved surfaces of such cells require that care be taken to duplicate the position of the cell with respect to the beam; otherwise variations in reflection losses and path length lead to erroneous data.

The quality of absorbance data is critically dependent upon the way the matched cells are used and maintained. Fingerprints, grease, or other deposits on the cell wall alter its transmission characteristics markedly. Thus, thorough cleaning before and after use is imperative; the surface of the windows must not be touched during the handling of cells. Matched cells should never be dried by heating in an oven or over a flame, for such treatment may cause physical damage or a change in path length. They should be calibrated against each other regularly with an absorbing solution.

Radiation Detection and Indicators

Early instruments for the measurement of absorption of radiation required visual or photographic methods for detection. These procedures have been almost completely supplanted by photoelectric devices, which convert radiant energy into an electrical signal; our discussion is confined to detectors of this kind.

To be useful the detector must respond to radiant energy over a broad

1) Broad λ range
2) low levels of radiant power
3) produce an electric signal that can be amplified
4) signal ∝ P_0

Components of Instruments for Absorption Measurements **51**

wavelength range. It should, in addition, be sensitive to low levels of radiant power, respond rapidly to the radiation, produce an electrical signal that can be readily amplified, and have a relatively low noise level (for stability). Finally, it is essential that the signal produced be directly proportional to the power of the beam striking it; that is,

$$G = k'P + k'' \qquad (3\text{-}13)$$

where G is the electrical response of the detector in units of current, resistance, or emf. The constant k' measures the sensitivity of the detector in terms of electrical response per unit of radiant power. Many detectors exhibit a small constant response, known as a *dark current k''*, when no radiation impinges on their surfaces. Instruments with detectors that have a dark-current response are ordinarily equipped with a compensating circuit that permits application of a countersignal to reduce k'' to zero. Thus, under ordinary circumstances, we may write *electrical response*

$$P = G/k' \sim \text{constant sensitivity (of detector)} \qquad (3\text{-}14)$$

and

$$P_0 = G_0/k' \qquad (3\text{-}15)$$

where G and G_0 represent the electrical response of the detector to radiation passing through the solution and the solvent, respectively. Thus, the absorbance is given by

$$\log \frac{P_0}{P} = \log \frac{G_0/k'}{G/k'} = \log \frac{G_0}{G} = A \qquad (3\text{-}16)$$

Three basic photoelectric devices are employed for radiation detection: (1) *photovoltaic cells,* in which the radiant energy generates a current at the interface of a semiconductor and a metal; (2) *phototubes,* in which the radiation causes the photoemission of electrons from a solid surface; and (3) *photoconductive cells,* in which absorption of radiation by a semiconductor results in a change in electrical resistance. Photoconductive cells are employed primarily for detection of radiation in the region of 750 to 3500 nm, and are discussed in Chapter 6.

Photovoltaic or barrier layer cells. The photovoltaic cell is used primarily for the detection and measurement of radiation in the visible region. The typical cell has a maximum sensitivity at about 550 nm, and the response falls off to perhaps 10 percent of the maximum at 250 and 750 nm.

The photovoltaic cell consists of a flat copper or iron electrode upon which is deposited a layer of semiconducting material, such as selenium or copper(I) oxide. On the surface of the semiconductor is a transparent metallic film of gold, silver, or lead, which serves as the second or collector electrode; the entire array is protected by a transparent envelope. The interface between the selenium and the metal film serves as a barrier to the passage of electrons. Irradiation with light, however, provides some electrons in the oxide layer with sufficient energy to overcome this barrier, and electrons flow from the semiconductor to the metal film. If the film is connected via an external circuit to the plate on the other side of the semiconducting layer and if the resistance is not too great, a flow of electrons occurs. Ordinarily this current is large enough to be measured with a galvanometer or microammeter; if the resistance of the external circuit is small, the magnitude of the current is directly proportional to the power of the radiation striking the cell. Currents on the order of 10 to 100 μamp are typical.

The barrier-layer cell constitutes a rugged, low-cost means for measuring radiant power. No external source of electric energy is required. On the other hand, the low internal resistance of the cell makes difficult the amplification of its output. Thus, although the barrier-layer cell provides a readily measured response at high levels of illumination, it suffers from lack of sensitivity at low levels. Finally, a barrier-layer cell exhibits fatigue, its response falling off upon prolonged illumination; proper circuit design and choice of experimental conditions largely eliminate this source of difficulty.

Phototubes. A second type of photoelectric device is the phototube, which consists of a semicylindrical cathode and a wire anode sealed inside an evacuated transparent envelope. The concave surface of the cathode supports a layer of photoemissive material (p. 11) that tends to emit electrons upon being irradiated. When a potential is applied across the electrodes, the emitted electrons flow to the wire anode and a photocurrent results. For a given radiant intensity, the currents produced are approximately one-fourth as great as those from a photovoltaic cell. In contrast, however, amplification is easily accomplished since the phototube has a very high electrical resistance. Figure 3-18 is a schematic diagram of a typical phototube arrangement.

The number of electrons ejected from a photoemissive surface is directly proportional to the radiant power of the beam striking the surface. As the potential applied across the two electrodes of the tube is increased,

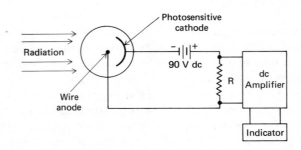

Figure 3-18. Schematic diagram of a phototube and its accessory circuit. The current induced by the radiation causes a potential drop across the resistor R; this is amplified and measured by the indicator.

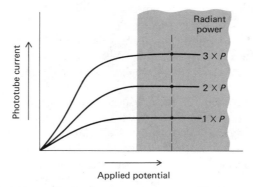

Figure 3-19. Variation of phototube response with the applied potential at three levels of illumination. The shaded portion indicates the saturation region. Note the linearity of response at the higher applied potentials.

the fraction of the emitted electrons reaching the anode rapidly increases; when the saturation potential is achieved, essentially all of the electrons are collected at the anode. At saturation, the current becomes independent of potential and directly proportional to radiant power (see Figure 3-19). Phototubes are usually operated at a potential of about 90 V, which is well into the saturation region.

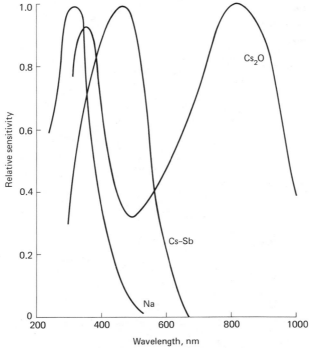

Figure 3-20. Spectral responses of some photoemissive surfaces.

The photoemissive cathode surfaces of phototubes ordinarily consist of alkali metals or alkali-metal oxides alone or combined with other metals or oxides. As shown in Figure 3-20, the coating on the cathode determines the spectral response of a phototube. When sensitivity to radiation below 350 nm is required, the tube must be supplied with an envelope or a window of quartz.

For the measurement of low radiant power the *photomultiplier* tube offers advantages over the ordinary tube. Figure 3-21 is a schematic diagram of such a device. The cathode surface is similar in composition to that of a phototube, electrons being emitted upon exposure to radiation. The tube also contains additional electrodes (nine in Figure 3-21) called *dynodes*. Dynode 1 is maintained at a potential 90 V more positive than the cathode, and electrons are accelerated toward it as a consequence. Upon striking the dynode each photoelectron causes emission of several additional electrons; these, in turn, are accelerated toward dynode 2, which is 90 V more positive than dynode 1. Again, several electrons are emitted for each electron striking the surface. By the time this process has been repeated nine times, 10^6 to 10^7 electrons have been formed for each photon; this cascade is finally collected at the anode. The resulting enhanced current is then passed through the resistor R and can be further amplified and measured.

Photomultiplier tubes are easily damaged by exposure to strong radiation and can only be used for measurement of low radiant power. To avoid irreversible changes in their performance the tubes must be mounted in light-tight housings, and care must be taken to avoid even momentary exposure to strong light.

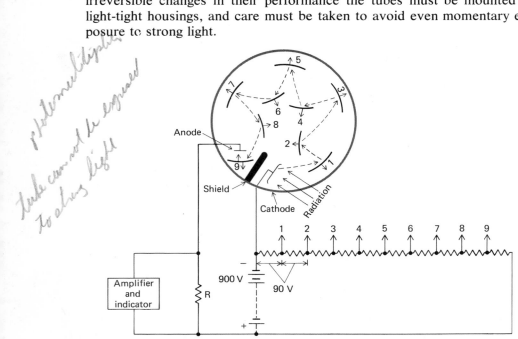

Figure 3-21. A photomultiplier tube.

INSTRUMENTS AND INSTRUMENT DESIGN

The instrumental components discussed in the previous section can be combined in various ways to produce devices for the measurement of absorption. Over the years, designs for numerous such devices have appeared in the technical literature, and a dozen or more are produced in quantity by manufacturing concerns. These commercial instruments differ greatly in complexity, performance characteristics, and cost; some of the simpler instruments can be purchased for $100 to $200, while the most sophisticated cost a hundred times this amount. No single instrument is best for all purposes; selection is governed by the type of work for which the instrument is to be used.

Some of the more important design variables encountered in absorption instruments are described in this section; in addition, a few commercial products that typify these differences are described.

Design Variables

Wavelength selection and control. Absorption instruments are categorized by the type of device employed for restricting radiation to a limited spectral region. A *spectrophotometer* has a prism or grating monochromator that permits the use of a narrow band of radiation that can be varied continuously in wavelength. Spectrophotometers are further classified on the basis of the wavelength region for which they have been designed; that is, as infrared, visible, ultraviolet, or x-ray instruments. A *photometer,* on the other hand, employs filters to provide bands of radiation that usually encompass a broader wavelength span than those obtained with a spectrophotometer. The other components are similar to those of a spectrophotometer. Photometers are largely, but not exclusively, confined to the visible region, and as a consequence are sometimes called *colorimeters* or *photoelectric colorimeters.*

We shall limit the use of the term colorimeter to those instruments that employ the human eye as the detector of radiation. Colorimetric methods represent the simplest form of absorption analysis.

Single-beam and double-beam designs. An absorption measurement with a single-beam photometer or spectrophotometer involves three steps: the indicator is first adjusted to zero transmittance (or infinite absorbance), in which no radiation strikes the detector; then the indicator is adjusted to 100 percent transmittance, or zero absorbance, with the cell filled with solvent in the beam of the instrument; finally, the solvent is replaced by the solution under study and the transmittance or absorbance is read directly from the indicator scale. Single-beam instruments produce reliable absorption data provided their electrical components are operationally stable during the period required to complete these three steps (perhaps 10 to 30 sec).

As the name implies, the dispersed radiation in a double-beam photometer is split into two components; one passes through the sample solution, the other through the solvent. In some instruments the two beams are then monitored by comparison of the outputs of twin detector-amplifier systems so that transmittance or absorbance data are obtained directly. In other instruments the split beams are mechanically chopped so that pulses of radiation pass alternately through the sample and the solvent. The resulting beams are then recombined and focused on a single detector. The pulsating electrical output from the detector is fed into an amplifier system that is programmed to compare the magnitude of the pulses and convert this information into transmittance or absorbance data.

Because comparison of the solvent with the sample is made simultaneously or nearly simultaneously, a split-beam instrument compensates for all but the most short-term electrical fluctuations, as well as for irregular performance in the source, the detector, and the amplifier. Therefore, the electrical components of a double-beam photometer need not be of as high quality as those for a single-beam instrument. Offsetting this advantage, however, is the greater number and complexity of components associated with double-beam instruments. Moreover, in photometers equipped with twin detectors and amplifiers, a close match between the components of the two systems is essential.

Single-beam instruments are particularly well-adapted to the quantitative analysis that involves an absorbance measurement at a single wavelength. Here, the simplicity of the instrument and the concomittant ease of maintenance offer real advantages. The greater speed and convenience of measurement, on the other hand, makes the double-beam instrument particularly useful for qualitative analyses, where many absorbance measurements must be made at several wavelengths. Furthermore, the double-beam device is readily adapted to continuous monitoring of absorbance; all modern recording spectrophotometers employ twin beams for this reason.

Data presentation. Another important design variable is the system of data presentation. In *direct-reading instruments* the transmittance and absorbance scales are scribed on the face of a meter (usually a milli- or microammeter), and data are indicated by the position of the pointer. In *null instruments* the signal from the detector-amplifier is fed into a bridge circuit (p. 504) or a potentiometer (p. 461), and exactly balanced against a known signal from the bridge or potentiometer. The absorbance is then established by the setting required to achieve this null point.

Direct-reading instruments are convenient and easy to use but suffer from a limit in accuracy associated with the length of the scale and the inherent uncertainties in the meter. Where moderate accuracy will suffice (of the order of $\pm$ 2 percent), the speed and ease of maintenance of these instruments is an advantage. Where greater accuracies are required, null instruments must be employed.

Colorimeters and Photometers

Colorimeters and photometers are the simplest instruments for absorption analysis. They provide entirely adequate analytical data for many purposes.

Colorimeters. The human eye as a detector and the brain as an amplifier and signal indicator are less satisfactory for absorption measurements than most photometric and spectrophotometric devices. The eye is sensitive to a limited spectral range, suffers from fatigue and slow response, and is inaccurate in judging absolute intensities. Because of the last limitation, visual colorimetric methods always involve the comparison of the sample with a set of standards until a match is found. Flat-bottomed *Nessler tubes* are frequently employed for this purpose. These tubes are calibrated so that a uniform light path is achieved. Daylight commonly serves as a radiation source. Ordinarily, no attempt is made to restrict the portion of the spectrum employed.

A somewhat more refined colorimetric procedure involves comparison of the unknown with a single standard solution in a split-beam instrument. Here, the two solutions are contained in flat-bottomed tubes; the path lengths are varied by means of adjustable transparent plungers that can be moved up and down in the solutions. After balance has been achieved visually, the path lengths are measured; the concentration of the unknown may then be calculated:

$$A_x = A_s$$

$$\epsilon b_x c_x = \epsilon b_s c_s$$

or
$$c_x = c_s \frac{b_s}{b_x}$$

where x refers to the unknown and s to the standard. A *Duboscq* colorimeter embodies these principles and is equipped with an optical system that permits the ready comparison of the beams passing through an eyepiece with a split field.

Visual colorimetric methods suffer from several disadvantages. A standard or a series of standards must always be available. Furthermore, the eye may not be capable of matching colors if a second colored substance is present in the solution. Finally, the eye is not as sensitive to small differences in absorbance as a photoelectric device; as a consequence, concentration differences smaller than about 5 percent relative cannot be detected.

Despite their limitations, visual comparison methods find wide application for routine analyses in which the requirements for accuracy are modest. For example, simple but useful colorimetric test kits are sold for determining the *p*H and chlorine content of swimming pool water; kits are also available for the analysis of soils. Water filter plants commonly employ color comparison tests for the estimation of iron, silicon, fluorine, and chlorine in city water supplies. For such an analysis a colorimetric reagent is introduced to the sample, and the resulting color is then compared with permanent standard solutions or with colored plastic disks. Accuracies of perhaps 10 to 50 percent relative are to be expected, and suffice for the purposes intended.

Photometers. The photometer provides a simple, relatively inexpensive tool for the performance of absorption analyses. Convenience, ease of maintenance, and ruggedness are properties of a filter photometer that may not be found in the more sophisticated spectrophotometer. Moreover, where spectral purity is not important to a method (and often it is not), analyses can be performed as accurately with this instrument as with more complex instrumentation.

Figure 3-22 presents schematic diagrams for two photometers. The

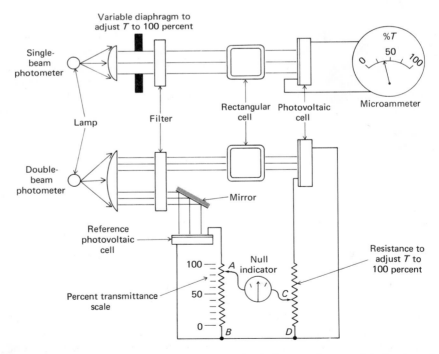

Figure 3-22. Schematic diagrams for a single-beam photometer (top) and a double-beam photometer (bottom).

first is a single-beam, direct-reading instrument consisting of a tungsten filament lamp, a lens to provide a parallel beam of light, a filter, and a photovoltaic cell. The current produced is indicated with a microammeter, the face of which is ordinarily scribed with a linear scale from 0 to 100. A cell containing an appropriate blank is first placed in the light path, and the power of the beam is regulated so that a scale reading of 100 is achieved. In some instruments this adjustment involves changing the voltage applied to the lamp; in others, the aperture size of a diaphragm located in the light path is altered. The sample is then placed in the beam; since the signal from the photovoltaic cell is linear with respect to the radiation it receives, the resultant scale reading will be the percent transmittance (that is, the percent of full scale). Clearly, a logarithmic scale could be substituted to give the absorbance of the solution directly.

Also shown in Figure 3-22 is a schematic representation of a double-beam, null-type photometer. Note that in contrast to most double-beam instruments, the solvent is not inserted in the reference half of the beam. Here, the light beam is split by a mirror, a part passing through the sample or the solvent, and thence to a photovoltaic cell. The other part is directed to a similar detector that continuously monitors the output of the lamp. The output of the working photocell is then compared with that of the reference by a suitable circuit design. In the instrument shown, the currents from the two photovoltaic cells are passed through variable resistances; one of these is calibrated as a transmittance scale in linear units from 0 to 100. A sensitive galvanometer, which serves as a null indicator, is connected across the two resistances. When the potential drop across AB is equal to that across CD, no current passes through the galvanometer; under all other circumstances a current flow is indicated. At the outset the solvent is introduced into the cell, and contact A is set at 100; contact C is then adjusted until no current is indicated. Introduction of the sample into the cell results in a decrease in radiant power reaching the working phototube and therefore a reduction in the potential drop across CD; this lack of balance is compensated for by moving A to a lower value. At balance, the percent transmittance is read directly from the scale.

Commercial photometers usually cost a few hundred dollars. The majority employ the double-beam principle.

Filter selection for photometric analysis. Photometers are generally supplied with several filters, each of which transmits a different portion of the spectrum. Selection of the proper filter for a given application is important inasmuch as the sensitivity of the measurement is directly dependent upon the filter. The color of the light absorbed is the complement of the color of the solution itself. For example, a liquid appears red because it transmits unchanged the red portion of the spectrum but absorbs in the green. It is the intensity of green radiation that varies with concentration; a green filter should thus be employed. Generally, then, the most

suitable filter for a colorimetric analysis is the one that is the color complement of the solution being analyzed. If several filters possessing the same
general hue are available, the one that causes the sample to exhibit the
greatest absorbance (or least transmittance) should be used.

Spectrophotometers

The simplest and least expensive spectrophotometers are direct-
reading, and employ a single-beam design with a grating monochromator.
At the other end of the price scale are double-beam, recording instruments
employing high-quality double monochromators. No attempt will be made
to describe all available spectrophotometers; instead, the principal features
of several typical instruments are considered.

Single-beam instruments. Figure 3-23 is a schematic diagram of a high-
quality, single-beam spectrophotometer, the Beckman DU-2. The first
versions of this instrument appeared on the market over 25 years ago, and
it is one of the most widely used spectrophotometers of its kind.

The DU-2 spectrophotometer is equipped with quartz optics and can
be operated in both the ultraviolet and visible regions of the spectrum.
The instrument is provided with interchangeable radiation sources, in-

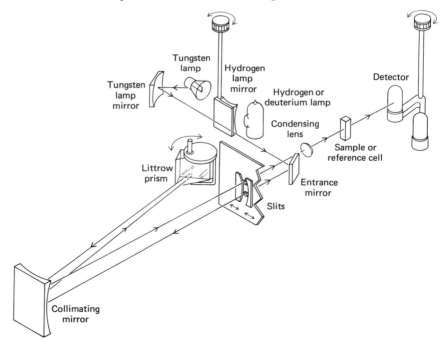

Figure 3-23. Schematic diagram of the Beckman DU-2® Spectrophotometer.
(By permission, Beckman Instruments, Inc., Fullerton, California.)

cluding a deuterium or hydrogen discharge tube for the lower wavelengths and a tungsten filament lamp for the visible region. A pair of mirrors reflect radiation through an adjustable slit into the monochromator compartment. After traversing the length of the instrument, the radiation is reflected into a Littrow prism; by adjusting the position of the prism, light of the desired wavelength can be focused on the slit. The optics are so arranged that the entrance and the exit beams are displaced from one another on the vertical axis; thus, the exit beam passes above the entrance mirror as it enters the cell compartment.

Ordinarily, the cell compartment will accommodate as many as four rectangular 1-cm cells, any one of which can be positioned in the path of the beam by movement of a carriage arrangement. Compartments are also available that will hold both cylindrical cells and cells up to 10 cm in length.

The detectors are housed in a phototube compartment; control over the incoming radiation is achieved with a manually operated shutter in the path of the beam. Interchangeable detectors are provided—a red sensitive phototube for the wavelength region beyond 625 nm and a photomultiplier tube for the range between 190 and 625 nm. The current from the phototube in the light path is passed through a fixed resistance of large magnitude; the potential drop across this resistor then gives a measure of the radiant power reaching the detector.

The Beckman instrument employs a null-point circuit to provide absorbance or transmittance data. As in Figure 3-18, the current from the phototube is passed through a resistor R (2000 megohm); the resulting potential drop is impressed upon the grid of an electrometer tube. Variations in the potential resulting from alterations in the photocurrent cause changes in the plate current of the tube. The electrometer plate current, after amplification, is detected by a simple, rugged milliammeter.

In order to counterbalance the effect of the phototube current, three potentiometer control circuits are provided, each of which can vary the grid potential of the electrometer tube to bring the needle of the milliammeter to zero. The first control (the *dark-current control*) is employed to offset the small phototube current when no radiation strikes the photoelectric device. The second control (the *sensitivity control*) serves the same purpose when radiation from the monochromator reaches the phototube after passage through the solvent. The *transmittance control* adjusts the position of the contact on a potentiometer slide wire that is calibrated in absorbance and transmittance units; it is used to null the instrument when the sample is in the radiation path. In operation, the milliammeter needle is first brought to zero with a shutter in the beam path. The solvent is then positioned in front of the phototube and the shutter is opened; a large current is observed with the milliammeter. With the transmittance scale effectively set at 100 the indicator needle is again brought to zero by adjustment of the sensitivity control and the slit opening of the monochromator. Finally, the solvent is replaced with the sample and the instrument is again zeroed, this time with the transmittance control. The length of the transmittance slide wire (about 20 cm) is such that the position of the contact at the null point can be determined with high precision.

The DU-2 design achieves photometric accuracies as good as $\pm$ 0.2 percent transmittance by employing high-quality electronic components that are operated well below their rated capacities. Narrow effective band widths (less than 0.5 nm) can be obtained throughout the spectral region by suitable variation of the slit adjustment. The instrument is particularly well-suited for research and quantitative analytical measurements that require absorbance data at a limited number of wavelengths.

The Bausch and Lomb Spectronic 20, shown schematically in Figure 3-24, may be considered as representative of instruments in which a degree of photometric accuracy is sacrificed in return for simplicity of operation and low cost. Its normal range is 350 to 650 nm, although this can be extended to 900 nm by the use of a red-sensitive phototube. The monochromator system consists of a reflection grating, lenses, and a pair of fixed slits. Because the grating produces a dispersion that is independent of wavelength, a constant band width of 20 nm is obtained throughout the entire operating region.

The Spectronic 20 is an example of a direct-reading, single-beam spectrophotometer. The photocurrent from the detector is amplified and indicated by the position of a needle on the scale of a current-indicating meter. Since the amplified current is directly proportional to radiant power, the meter scale can be scribed to read transmittance and absorbance. Stable operation of the amplifier unit is obtained by means of a Wheatstone bridge (p. 504) circuit in which the plate current of the electrometer tube that amplifies the photocurrent is balanced against the plate current of a matched amplifier tube to which no signal is applied. As a consequence of this arrangement, electrical irregularities and fluctuations in the circuit tend to be canceled out.

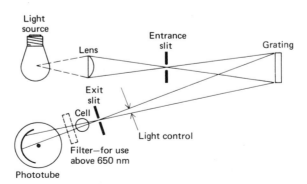

Figure 3-24. Schematic diagram of the Bausch and Lomb Spectronic 20 Spectrophotometer. (By permission, Bausch and Lomb, Inc., Rochester, N.Y.)

Double-beam instruments. Several double-beam spectrophotometers are now available on the market. All can be purchased with automatic recording capability, the double-beam design being readily adapted to this feature. Double-beam instruments are inherently more complex both optically and electronically; as a consequence, their cost is greater (from $4000 to $16,000 and more) and their maintenance is more difficult.

Figure 3-25 is a schematic diagram of the optics for the Cary Model 14 Recording Spectrophotometer,[5] an instrument with the highest quality performance characteristics. Radiation from the source (either a tungsten lamp or a hydrogen tube) enters the monochromator through slit S_1 and is dispersed by the Littrow-mounted quartz prism. After passage through slit S_2, it is further dispersed by a reflection grating, and it then leaves the monochromator via slit S_3. The double monochromator produces very narrow band widths and has freedom from stray radiation. The beam is then caused to pass alternately through the reference cell and the sample cell by means of a semicircular mirror arrangement that is rotated at 30 cycles per second. Also mounted on the synchronous motor shaft is a chopper that produces a dark interval between each half cycle. The photo-multiplier tube then receives alternate pulses of radiation, first from the reference beam and then from the sample beam. Synchronized with the alternate pulses is a system of photoelectric timing signals that permits comparison of the signals from the two beams. Any difference between these signals is corrected by automatic adjustment of a slide wire; this

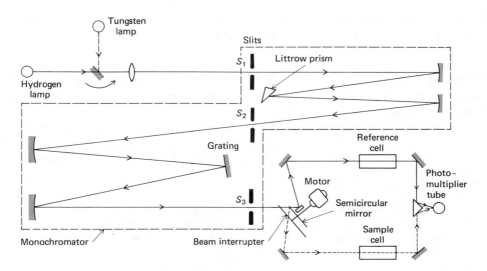

Figure 3-25. Schematic diagram of the Cary Model 14 Spectrophotometer. (By permission, Cary Instruments, Monrovia, California.)

[5]Manufactured by the Applied Physics Corporation, Monrovia, Calif.

adjustment is reflected by the position of the motor-driven pen of the recorder. The instrument is also equipped with a monochromator- and slit-drive mechanism which continuously varies the wavelength; this mechanism is synchronized with the paper drive of the recorder.

The Cary instrument has a wavelength range of 186 to 2650 nm. For the near-infrared region a special tungsten lamp and a lead sulfide cell are employed (not shown in Figure 3-25).

Figure 3-26 is a schematic diagram of the Perkin-Elmer Model 4000 Recording Spectrophotometer.[6] This instrument also uses a double-mono-

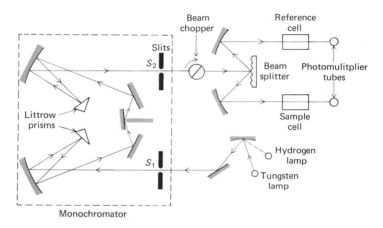

Figure 3-26. Schematic diagram of the Perkin-Elmer Model 4000 Spectrophotometer. (With permission, Perkin-Elmer Corp., Norwalk, Conn.)

chromator system, but here two Littrow-type prisms are employed. After leaving the monochromator the beam is chopped to produce signals that alternate at 50 cycles per second; it is then split by reflection from a stack of very small mirrors. In contrast to the Cary instrument the Perkin-Elmer photometer employs twin multiplier tubes for detection (or twin lead sulfide cells for detection in the near-infrared region). Any imbalance between the two signals is corrected by imposition of voltage from a potentiometer. The required correction voltage also determines the position of the pen on the recorder.

The Perkin-Elmer instrument has a range of 200 to 2800 nm, which is comparable with the Cary instrument. In terms of resolution, photometric accuracy, and absence of stray radiation, the Cary instrument is somewhat superior; on the other hand, its cost is significantly greater.

Several other less expensive double-beam instruments are available; the performance specifications of these, while not as good as the two instruments just described, are still adequate for many purposes.

[6]Manufactured by the Perkin-Elmer Corporation, Norwalk, Conn.

Problems

1. A 7.50×10^{-5} F $KMnO_4$ solution has an absorbance of 0.658 when measured in a 1.50-cm cell at 525 nm.
    (a) Calculate the molar absorptivity ϵ for $KMnO_4$ on the basis of these data; indicate the units of ϵ.
    (b) Calculate the absorptivity at this wavelength when the concentration is expressed in ppm $KMnO_4$; indicate the units of a.
    (c) Calculate the percent transmittance of the above solution at 525 nm if the measurement had been made with a 1.00-cm cell.

2. At 440 nm a 5.00×10^{-4} M solution of the complex CoX_2^- exhibits an absorbance of 0.750 when measured in a 2.50-cm cell.
    (a) Calculate the molar absorptivity ϵ for the complex at this wavelength.
    (b) Calculate the absorbance of an 8.00×10^{-5} M solution of this complex when measured at 440 nm in a 1.00-cm cell.
    (c) Calculate the transmittance of a 2.00×10^{-3} M solution of the complex when measured at 440 nm with a 0.110-cm cell.

3. An 8.0×10^{-5} M solution of the nickel chelate NiY^{2-} has an absorbance of 0.580 when measured in a 2.50-cm cell at 460 nm, the wavelength of its maximum absorption. Calculate the transmittance of a 5.0×10^{-5} M solution of the chelate at this wavelength in a 1.00-cm cell.

4. At 480 nm a 4.01×10^{-5} M solution of $FeSCN^{2+}$ has a transmittance of 50.0 percent when measured in a 1.50-cm cell. What will be the percent transmittance if a 4.00-cm cell is used for measurement?

5. At 575 nm, the wavelength of maximum absorption, solutions of the chelate CuX_2^{2+} show adherence to Beer's law over a wide concentration range. Neither component by itself absorbs at this wavelength. When contained in a 1.00-cm cell, a 3.40×10^{-5} M solution of CuX_2^{2+} has a transmittance of 18.2 percent.
    (a) Calculate the absorbance of this solution.
    (b) Calculate the absorbance of a CuX_2^{2+} solution whose transmittance at this wavelength is 36.4 percent.
    (c) Calculate the absorbance of a solution in which the concentration of the chelate is one-half that of the original solution described above.
    (d) Calculate the CuX_2^{2+} concentration of the solution described in (b).

6. At 420 nm solutions containing M^{3+} and the chelating agent X obey Beer's law over a wide range, provided the pH is held constant at 5.0 and the formal concentration of X exceeds that of M^{3+} by a factor of 5 or more. Under these optimum conditions a solution that is 6.20×10^{-5} F with respect to M^{3+} and 4.0×10^{-3} F with respect to X has an absorbance of 0.335 at this wavelength, when measured in a 1.00-cm cell. Neither M^{3+} nor X absorbs at 420 nm.
    (a) Calculate the molar concentration of the complex in a solution that has an absorbance of 0.602 when measured in a 1.50-cm cell under the conditions specified above.
    (b) What will be the percentage transmittance of the solution in (a)?
    (c) What will be the molar concentration of the complex in a solution that possesses three times the transmittance of the solution in (a) when measured under the same conditions?

7. An interference filter was constructed with a dielectric layer of exactly 0.500 μ

thickness and a refractive index of 1.400. What visible and ultraviolet wavelengths would be transmitted by this filter? What orders of interference are involved in the transmissions?

8. The acid-base indicator HIn dissociates as follows in aqueous solution:

$$HIn \rightleftharpoons In^- + H^+ \qquad K_a = 1.0 \times 10^{-5}$$

Studies in strongly alkaline and strongly acidic solutions demonstrated that absorbance of both HIn and In$^-$ at 490 nm obeyed Beer's law, with the molar absorptivity of HIn being 3000 and of In$^-$, 60.

   (a) A series of unbuffered solutions having the following total indicator concentrations were prepared: 4.00×10^{-4}, 3.00×10^{-4}, 1.00×10^{-4}, and 0.50×10^{-4}. Calculate the absorbance at 490 nm of each of these solutions and plot as a function of concentration.

   (b) Repeat the calculation in (a) but assume that the solutions were *buffered* to *p*H 4.0.

4 The Applications of Ultraviolet and Visible Absorption Measurements

In this chapter four applications of absorption of ultraviolet and visible radiation are considered: (1) qualitative analysis, in which the goal is the identification of a pure compound, the determination of the presence or absence of a particular species in a mixture, or the identification of certain functional groups in a compound under structural investigation; (2) quantitative analysis of one or more species in a mixture; (3) spectrophotometric titrations, wherein absorption measurements are employed to locate the equivalence point of a titration; and (4) the determination of equilibrium constants.

Before discussing each of these applications, the kinds of species that absorb in the ultraviolet and visible regions are considered.

ABSORBING SPECIES

As we have pointed out in Chapter 2, absorption in the wavelength region under consideration results primarily from electronic transitions in which the outermost or bonding electrons are promoted to higher-energy levels. Both organic and inorganic species exhibit this behavior, and although the fundamental absorption process is the same for both, it is convenient to treat the two separately.

Absorption by Organic Compounds

All organic compounds are capable of absorbing electromagnetic radiation because all contain valence electrons that can be excited to higher-energy levels. The excitation energies associated with electrons forming most single bonds are high; thus, absorption by this type of electron is restricted to the so-called vacuum ultraviolet region ($\lambda < 180$ nm) where components of the atmosphere also absorb strongly. The experimental difficulties associated with the vacuum ultraviolet are formidable; as a result, most spectrophotometric investigations of organic compounds have involved the wavelength region greater than 180 nm. Absorption of longer-wavelength ultraviolet and visible radiation is restricted to a limited number of functional groups (called *chromophores*) that contain valence electrons with relatively low excitation energies.

The electronic spectra of polyatomic organic molecules containing chromophores are usually complex because the superposition of vibrational transitions on the electronic transitions leads to spectra made up of an intricate combination of overlapping series of lines; the result is a broad band of continuous absorption. The complex nature of the spectra makes detailed theoretical analysis difficult or impossible; nevertheless, qualitative or semiquantitative statements concerning the types of electronic transitions responsible for a given absorption spectrum can be derived from molecular orbital considerations. We therefore need to consider briefly molecular orbital descriptions of absorbing groups.

Molecular orbital treatment.[1]   The electrons that contribute to the absorption characteristics of an organic molecule are: (1) those that participate directly in bond formation between atoms and are thus associated with more than one atom; (2) nonbonding or unshared outer electrons that are largely localized about such atoms as oxygen, the halogens, sulfur, and nitrogen.

Covalent bonding occurs because the electrons forming the bond move in the field about two atomic centers in such a manner as to minimize the repulsive coulombic forces between these centers. The nonlocalized fields between atoms occupied by bonding electrons are called *molecular orbitals* and can be considered to be the result of overlap of atomic orbitals. When two atomic orbitals combine, either a low-energy *bonding molecular orbital* or a high-energy *antibonding molecular orbital* results. In the ground state of the molecule the electrons occupy the former.

[1]For further details see C. N. R. Rao, *Ultra-Violet and Visible Spectroscopy* (New York: Plenum Press, 1967), Chaps. 2 to 6; and R. P. Bauman, *Absorption Spectroscopy* (New York: John Wiley & Sons, Inc., 1962), Chaps. 6 and 8.

The molecular orbitals associated with the single bonds in organic molecules are designated as *sigma* (σ) *orbitals*, and the corresponding electrons are σ electrons. As shown in Figure 4-1, the distribution of charge density of a sigma orbital is rotationally symmetric around the axis of the bond.

The double bond in organic molecules contains two types of molecular orbitals: a *sigma* (σ) orbital corresponding to one pair of the bonding electrons, and a *pi* (π) *molecular orbital* associated with the other. Pi orbitals are formed by the parallel overlap of atomic *p* orbitals. The charge distribution in this type of orbital is characterized by a *nodal plane* (a region of low-charge density) along the axis of the bond and a maximum density in regions above and below the plane (see Figure 4-1).

Also shown in Figure 4-1 are the charge-density distributions for the antibonding sigma and pi orbitals; these orbitals are designated by σ^* and π^*.

In addition to σ and π electrons we also need to consider the nonbonding electrons in a molecule. Unshared electrons are designated by the symbol *n*. Examples of the three types of electrons associated with organic molecules are to be found in formaldehyde.

$$\sigma \longleftarrow \overset{H}{\underset{H}{\ddots}} C :: \ddot{O} : \overset{\displaystyle n}{\underset{\pi}{\sigma}}$$

As shown in Figure 4-2, the energies for the various types of molecular orbitals differ significantly. Quite generally, the energy level of a nonbonding electron lies between those of the bonding and the antibonding orbitals. Electronic transitions among certain of the energy levels can be brought

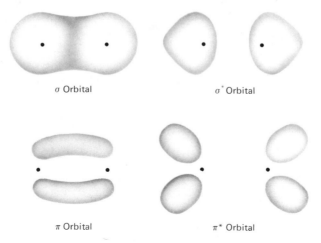

σ Orbital σ^* Orbital

π Orbital π^* Orbital

Figure 4-1.   Electron distribution in sigma and pi molecular orbitals.

about by the absorption of radiation. As shown in Figure 4-2, the commonly encountered transitions are of four types: $\sigma \rightarrow \sigma^*$, $n \rightarrow \sigma^*$, $n \rightarrow \pi^*$, and $\pi \rightarrow \pi^*$.

$\sigma \rightarrow \sigma^*$ *transitions.* Here an electron in a bonding σ orbital of a molecule is excited to the corresponding antibonding orbital by the absorption of radiation. The molecule is then described as being in the σ,σ^* excited state. Relative to the other possible transitions, the energy required to induce a $\sigma \rightarrow \sigma^*$ transition is large, corresponding to radiant frequencies in the vacuum ultraviolet region. Methane, for example, which contains only single C—H bonds and thus undergoes $\sigma \rightarrow \sigma^*$ transitions, exhibits an absorption maximum at 125 nm. Ethane has an absorption peak at 135 nm, which must also arise from a $\sigma \rightarrow \sigma^*$ transition, but here, electrons of the C—C bond appear to be involved. Because the strength of the C—C bond is less than that of the C—H bond, less energy is required for excitation. Thus, an absorption peak at a longer wavelength is observed.

Since absorption maxima due to $\sigma \rightarrow \sigma^*$ transitions are never observed in the ordinary, accessible ultraviolet region, we need not consider them further.

$n \rightarrow \sigma^*$ *transitions.* Saturated compounds containing atoms with unshared electron pairs (nonbonding electrons) are capable of $n \rightarrow \sigma^*$ transitions. Generally these transitions require less energy than the $\sigma \rightarrow \sigma^*$ type, and can be brought about by radiation in the region of 150 to 250 nm, with most absorption peaks appearing below 200 nm. Table 4-1 shows absorption data for some typical $n \rightarrow \sigma^*$ transitions. It will be seen that the energy requirements for such transitions depend primarily upon the kind of atomic bond and to a lesser extent upon the structure of the molecule. The molar absorptivities (ϵ) associated with this type of absorption are intermediate in magnitude and usually range between 100 and 3000 liter cm^{-1} mol^{-1}.

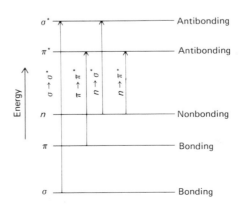

Figure 4-2. Electronic molecular energy levels.

Absorption maxima for formation of the n,σ^* state tend to shift to shorter wavelengths in the presence of polar solvents such as water or ethanol. The number of organic functional groups with $n \rightarrow \sigma^*$ peaks in the readily accessible ultraviolet region is relatively small.

$n \rightarrow \pi^*$ and $\pi \rightarrow \pi^*$ transitions. Most applications of absorption spectroscopy to organic compounds are based upon transitions of n or π electrons to the π^* excited state because the energies required for these processes bring the absorption peaks into an experimentally convenient spectral region (200 to 700 nm). Both transitions require the presence of an unsaturated functional group to provide the π orbitals. Strictly speaking, it is to these unsaturated absorbing centers that the term chromophore applies.

The molar absorptivities for absorption peaks associated with excitation to the n,π^* state are the lowest of any we will encounter, and ordinarily range from 10 to 100 liter cm^{-1} mol^{-1}. Values for $\pi \rightarrow \pi^*$ transitions, on the other hand, are 100- to 1000-fold greater. Another characteristic difference between the two types of absorption is found in the effect of solvent on the wavelength of the peaks. The peaks arising from $n \rightarrow \pi^*$ transitions are generally shifted to shorter wavelengths (*hypsochromic* or *blue shift*) as the polarity of the solvent is increased. Usually, but not always, a reverse trend (*bathochromic* or *red shift*) is observed for $\pi \rightarrow \pi^*$ transitions. The hypsochromic effect apparently arises from the increased solvation of the unbonded electron pair, which lowers the energy of the n orbital. The most dramatic effects of this kind (blue shifts of 30 nm or more) are seen with polar hydrolytic solvents, such as water or alcohols, in which hydrogen-bond formation between the solvent protons and the nonbonded electron pair is extensive. Here the energy of the n orbital is lowered by an amount that is approximately equal to the energy of the hydrogen bond. When an $n \rightarrow \pi^*$ transition occurs, however, the remaining single n electron is not adequate to maintain the hydrogen bond; thus, the energy of the n,π^* *excited* state is not affected by this type of solvent interaction. A blue shift, roughly corresponding to the energy of the hydrogen bond, is therefore observed.

A second solvent effect that undoubtedly influences both $\pi \rightarrow \pi^*$ and $n \rightarrow \pi^*$ transitions leads to a bathochromic shift with increased solvent polarity. This effect is small (usually less than 5 nm), and as a result is com-

Table 4-1 Some Examples of Absorption Due to $n \rightarrow \sigma^*$ Transitions[a]

Compound	$\lambda_{max}(nm)$	ϵ_{max}	Compound	$\lambda_{max}(nm)$	ϵ_{max}
H_2O	167	1480	$(CH_3)_2S^b$	229	140
CH_3OH	184	150	$(CH_3)_2O$	184	2520
CH_3Cl	173	200	CH_3NH_2	215	600
CH_3I	258	365	$(CH_3)_3N$	227	900

[a]Samples in vapor state.
[b]In ethanol solvent.

pletely overshadowed in $n \rightarrow \pi^*$ transitions by the hypsochromic effect just discussed. Here, attractive polarization forces between the solvent and the absorber are involved which tend to lower the energy levels of both the unexcited and the excited states. The effect on the excited state is greater, however, and the energy differences thus become smaller with increased solvent polarity; small bathochromic shifts result.

Organic chromophores. Table 4-2 lists common organic chromophores and the approximate location of their absorption maxima. The data for position and peak intensity can serve only as a rough guide for the identification of functional groups since the position of maxima is also affected by solvent and structural details. Furthermore, the peaks are ordinarily broad because of vibrational effects; the precise determination of the position of the maximum is thus difficult.

Effect of conjugation of chromophores. In the molecular orbital treatment, π electrons are considered to be further delocalized by the conjuga-

Table 4-2 Absorption Characteristics of Some Common Chromophores

Chromophore	Example	Solvent	$\lambda_{max}(nm)$	ϵ_{max}	Type of transition
Alkene	$C_6H_{13}CH{=}CH_2$	n-Heptane	177	13,000	$\pi \rightarrow \pi^*$
Alkyne	$C_5H_{11}C{\equiv}C{-}CH_3$	n-Heptane	178	10,000	$\pi \rightarrow \pi^*$
			196	2000	—
			225	160	—
Carbonyl	$CH_3\overset{\overset{O}{\|\|}}{C}CH_3$	n-Hexane	186	1000	$n \rightarrow \sigma^*$
			280	16	$n \rightarrow \pi^*$
	$CH_3\overset{\overset{O}{\|\|}}{C}H$	n-Hexane	180	large	$n \rightarrow \sigma^*$
			293	12	$n \rightarrow \pi^*$
Carboxyl	$CH_3\overset{\overset{O}{\|\|}}{C}OH$	Ethanol	204	41	$n \rightarrow \pi^*$
Amido	$CH_3\overset{\overset{O}{\|\|}}{C}NH_2$	Water	214	60	$n \rightarrow \pi^*$
Azo	$CH_3N{=}NCH_3$	Ethanol	339	5	$n \rightarrow \pi^*$
Nitro	CH_3NO_2	Isooctane	280	22	$n \rightarrow \pi^*$
Nitroso	C_4H_9NO	Ethyl ether	300	100	—
			665	20	$n \rightarrow \pi^*$
Nitrate	$C_2H_5ONO_2$	Dioxane	270	12	$n \rightarrow \pi^*$

tion process; the orbitals thus involve four (or more) atomic centers. The effect of this delocalization is to lower the energy level of the π^* orbital and give it less antibonding character. The absorption maxima are shifted to longer wavelengths as a consequence.

As seen from the data in Table 4-3, the absorptions of multichromophores in a single organic molecule are approximately additive provided the chromophores are separated from one another by more than one single bond. Conjugation of chromophores, however, has a profound effect on spectral properties. For example, it is seen in Table 4-3 that 1,3-butadiene, $CH=CHCH=CH_2$, has a strong absorption band that is displaced to a longer wavelength by 30 nm as compared with the corresponding peak for an unconjugated diene. When three double bonds are conjugated, the bathochromic effect is even larger.

Conjugation between the doubly bonded oxygen of aldehydes, ketones, and carboxylic acids and an olefinic double bond gives rise to similar behavior (see Table 4-3). Analogous effects are also observed when two carbonyl or carboxylate groups are conjugated with one another. For α-β unsaturated aldehydes and ketones the weak absorption peak due to $n \rightarrow \pi^*$ transitions is shifted to longer wavelengths by 40 nm or more. In addition, a strong absorption peak corresponding to a $\pi \rightarrow \pi^*$ transition appears. For unconjugated aldehydes or ketones this latter peak occurs only in the vacuum ultraviolet.

The wavelengths of absorption peaks for conjugated systems are sensitive to the types of groups attached to the doubly bonded atoms. Various empirical rules developed for predicting the effect of such substitutions upon

Table 4-3 Effect of Multichromophores on Absorption

Compound	Type	$\lambda_{max}(nm)$	ϵ_{max}
$CH_3CH_2CH_2CH=CH_2$	Olefin	184	~10,000
$CH_2=CHCH_2CH_2CH=CH_2$	Diolefin (unconjugated)	185	~20,000
$H_2C=CHCH=CH_2$	Diolefin (conjugated)	217	21,000
$H_2C=CHCH=CHCH=CH_2$	Triolefin (conjugated)	250	—
$CH_3CH_2CH_2CH_2\overset{O}{\overset{\|}{C}}CH_3$	Ketone	282	27
$CH_2=CHCH_2CH_2\overset{O}{\overset{\|}{C}}CH_3$	Unsaturated ketone (unconjugated)	278	30
$CH_2=CH\overset{O}{\overset{\|}{C}}CH_3$	Unsaturated ketone (conjugated)	324	24
		219	3600

the absorption maxima for a variety of compounds have proved useful for structural determinations.[2]

Absorption by aromatic systems. The ultraviolet spectra of aromatic hydrocarbons are characterized by three sets of bands that originate from $\pi \rightarrow \pi^*$ transitions. For example, benzene has a strong absorption peak at 184 nm ($\epsilon_{max} \sim 60,000$); a weaker band, called the E_2 band, at 204 nm ($\epsilon_{max} = 7900$); and a still weaker peak, termed the B band, at 256 nm ($\epsilon_{max} = 200$). The long wavelength band of benzene and many other aromatics contains a series of sharp peaks (see Figure 2-7) arising from the superposition of vibrational transitions upon the basic electronic transitions. Polar solvents tend to eliminate this fine structure, as do certain types of substitution.

All three of the characteristic bands for benzene are strongly affected by ring substitution; the effects on the two longer wavelength bands are of particular interest because they can be readily studied with ordinary spectrophotometric equipment. Table 4-4 illustrates the effects of some common ring substituents.

By definition, an *auxochrome* is a functional group that does not absorb in the ultraviolet region, but has the effect of shifting chromophore peaks to longer wavelengths as well as increasing their intensities. It is seen in Table 4-4 that —OH and —NH$_2$ have an auxochromic effect on the benzene chromophore, particularly with respect to the longer wavelength band. Auxochromic substituents have at least one pair of n electrons capable of interacting with the π electrons of the ring. This interaction apparently has the effect of stabilizing the π^* state and thus lowering its energy; a

Table 4-4 Absorption Characteristics of Aromatic Compounds

Compound		E_2 Band		B Band	
		$\lambda_{max}(nm)$	ϵ_{max}	$\lambda_{max}(nm)$	ϵ_{max}
Benzene	C_6H_6	204	7900	256	200
Toluene	$C_6H_5CH_3$	207	7000	261	300
m-Xylene	$C_6H_4(CH_3)_2$	—	—	263	300
Chlorobenzene	C_6H_5Cl	210	7600	265	240
Phenol	C_6H_5OH	211	6200	270	1450
Phenolate ion	$C_6H_5O^-$	235	9400	287	2600
Aniline	$C_6H_5NH_2$	230	8600	280	1430
Anilinium ion	$C_6H_5NH_3^+$	203	7500	254	160
Thiophenol	C_6H_5SH	236	10,000	269	700
Naphthalene	$C_{10}H_8$	286	9300	312	289
Styrene	$C_6H_5CH{=}CH_2$	244	12,000	282	450

[2]For a summary of these rules see R. M. Silverstein and G. C. Bassler, *Spectrometric Identification of Organic Compounds,* 2d ed. (New York: John Wiley & Sons, Inc., 1967), pp. 157-162.

bathochromic shift results. Note that the auxochromic effect is more pronounced for the phenolate anion than for phenol itself, probably because the anion has an extra pair of unshared electrons to contribute to the interaction. With aniline, on the other hand, the nonbonding electrons are lost by formation of the anilinium cation, and the auxochromic effect disappears as a consequence.

Absorption by Inorganic Systems

In Chapter 2 it was pointed out that the absorption spectra of atoms in the gaseous state consist of a limited number of sharp lines corresponding to electronic transitions of outer electrons. This type of absorption is treated in more detail in the next chapter; in this section our discussion is confined to absorption of inorganic species in solution. These spectra often resemble the spectra of organic compounds, with broad absorption maxima and little fine structure. An exception to this behavior is observed in the absorption spectra of several lanthanide and actinide elements.

Absorption by lanthanide and actinide ions. The ions of most of the lanthanide and actinide elements absorb in the ultraviolet and visible regions. In distinct contrast to the behavior of most inorganic and organic absorbers, their spectra consist of narrow, well-defined, and characteristic absorption peaks which are little affected by the type of ligand associated with the metal ion. A portion of a typical spectrum is shown in Figure 4-3.

In the lanthanide series the transitions responsible for absorption appear to involve the various energy levels of $4f$ electrons, while with the actinide

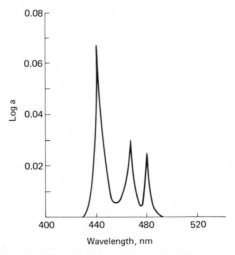

Figure 4-3. The absorption spectrum of a praseodymium chloride solution; a = absorptivity in liter cm^{-1} g^{-1}. From T. Moeller and J. C. Brantley, *Anal. Chem.*, **22**, 433 (1950). (With permission of the American Chemical Society.)

series it is the $5f$ electrons that interact with radiation. With both, these inner orbitals are largely screened from external influences by occupied orbitals with higher principal quantum numbers. As a consequence, the bands are narrow and relatively unaffected by the nature of the species bonded by the outer electrons.

Absorption by elements of the first and second transition-metal series. With very few exceptions, the ions and complexes of the 18 elements in the first two transition series are colored in one if not all of their oxidation states. In contrast to the lanthanide and actinide elements, however, the absorption bands are often broad (Figure 4-4) and are strongly influenced by environmental factors. Compare, for example, the pale blue color of the aquo copper(II) ion with the much darker blue of the complex with ammonia.

Metals of the first two series are characterized by having five partially occupied d-orbitals ($3d$ in the first series and $4d$ in the second), each capable of accommodating a pair of electrons. The electrons in these orbitals do not generally participate in bond formation; nevertheless, it is clear that the spectral characteristics of transition metals arise from electronic transitions that involve the various energy levels of these d-orbitals.

Two theories have been advanced to rationalize the colors of transition-metal ions and the profound influence of environment on these colors. The *crystal field theory*, which we shall discuss briefly, is the simpler of the two and is adequate for a qualitative understanding. The more complex molecular orbital treatment, however, provides a better quantitative treatment of the phenomenon.[3]

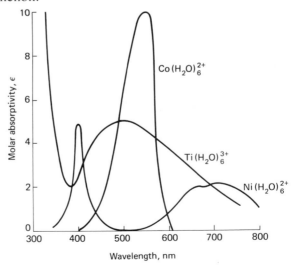

Figure 4-4.   Absorption spectra of some transition metal ions.

[3]For a nonmathematical discussion of these theories, see L. E. Orgel, *An Introduction to Transition Metal Chemistry, Ligand-Field Theory* (New York: John Wiley & Sons, Inc., 1960).

Both theories are based upon the premise that the energies of d-orbitals $d \rightarrow d^*$ of the transition-metal ions in solution are not identical, and that absorption involves the transition of electrons from a d-orbital of lower energy to one of higher energy. In the absence of an external electrical and a magnetic field (as in the dilute gaseous state), the energies of the five d-orbitals are identical, and absorption of radiation is not required for an electron to move from one orbital to another. In solution, on the other hand, complex formation occurs between the metal ion and water or some other ligand. Splitting of the energies of the d-orbitals then results from the differential electrostatic forces of repulsion between the electron pair of the donor and the electrons in the various d-orbitals of the central metal ion. In order to understand this effect we must first consider the spatial distribution of electrons in the various d-orbitals.

Figure 4-5 is a schematic representation of the electron-density distribution of the five d-orbitals around the central nucleus. Three of the orbitals, termed d_{xy}, d_{xz}, and d_{yz}, are similar in every regard except for their spatial orientation. Note that these orbitals occupy spaces *between* the three axes; consequently, they have minimum electron densities along the axes and maximum densities on the diagonals between axes. In contrast, the electron density of the $d_{x^2-y^2}$ and the d_{z^2} orbitals are directed along the axes.

Let us now consider a transition-metal ion that is coordinated to six molecules of water (or some other ligand). These groups can be imagined as being evenly distributed around the central atom, one ligand being located at each end of the three axes shown in Figure 4-5; the resulting octahedral structure is the most common orientation for transition-metal complexes.

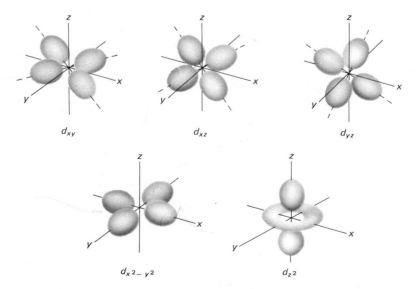

Figure 4-5. Electron density distribution in various d-orbitals.

In this arrangement, then, the negative ends of the water dipoles are pointed toward the metal ion, and the electrical fields from these dipoles tend to have a repulsive effect on all of the *d*-orbitals, thus increasing their energy; the orbitals are then said to be *destabilized*. The negative field has a greater effect on the d_{z^2} orbital than on the d_{xy}, d_{xz}, or d_{yz} orbitals, however, because the maximum charge density for the d_{z^2} orbital is along the axis on which the bonding water molecules are located. Thus, greater destabilization of the d_{z^2} orbital will occur and its energy level will be higher than the energy levels of the d_{xy}, d_{xz}, and d_{yz} orbitals. Since the latter three orbitals differ only in orientation and since we have assumed a symmetrical distribution for the water molecules, the field effect should be the same on each, and their energy levels should remain identical. The effect of the electrical field on the $d_{x^2-y^2}$ orbital is less obvious, but quantum calculations have shown that it is destabilized to the same extent as the d_{z^2} orbital. Thus, the energy-level diagram for the octahedral configuration (Figure 4-6) shows that the energies of all of the *d*-orbitals rise in the presence of a ligand field, but in addition, the *d*-orbitals are split into levels differing in energy by Δ. Also shown are energy diagrams for complexes involving four coordinated bonds. Two configurations are encountered: the *tetrahedral*, in which the four groups are symmetrically distributed around the metal ion, and the *square planar*, in which the four ligands and the metal ion lie in a single plane. Unique *d*-orbital splitting patterns for each configuration can be deduced by arguments similar to those used for the octahedral structure.

The magnitude of Δ (Figure 4-6) is dependent upon a number of factors, including the valence state of the metal ion and the position of the parent element in the periodic table. An important variable attributable to the ligand is the so-called *ligand field strength*, which is a measure of the extent to

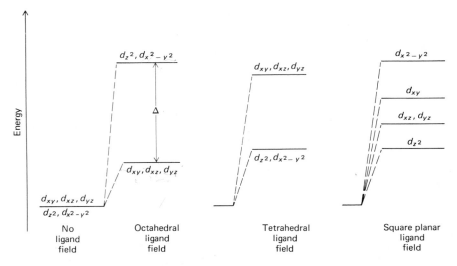

Figure 4-6.   Effect of ligand field on *d*-orbital energies.

which a complexing group will split the energies of the *d*-electrons; that is, a complexing agent with a high ligand field strength will cause Δ to be large.

It is possible to arrange the common ligands in the order of increasing ligand field strengths: $I^- < Br^- < SCN^- < Cl^- < OH^- < H_2O < NH_3 <$ ethylenediamine $< o$-phenanthroline $< NO_2^- < CN^-$. With only minor exceptions, this order of ligand field strength applies to all transition-metal ions and permits qualitative predictions as to the relative positions of absorption peaks for the various complex of a given transition-metal ion. Since Δ increases with increasing field strength, the wavelength of the absorption maxima decreases. This effect is demonstrated by the data shown in Table 4-5.

Charge-transfer absorption.[4]   For analytical purposes the most important type of absorption by inorganic species is *charge-transfer absorption* because the molar absorptivities of the band peaks are very large ($\epsilon_{max} >$ 10,000). Thus, a highly sensitive means for detecting and determining the absorbing particles is provided. Many inorganic and organic complexes exhibit charge-transfer absorption and are called, therefore, *charge-transfer complexes*. Common examples include the thiocyanate and phenolic complexes of iron(III), the *o*-phenanthroline complex of iron(II), the iodide complex of molecular iodine, and the ferro-ferricyanide complex responsible for the color of Prussian blue.

In order for a complex to exhibit a charge-transfer spectrum it is necessary that one of its components have electron-donor characteristics and the other, electron-acceptor properties. Absorption of radiation then involves transition of an electron of the donor to an orbital that is largely associated with the acceptor. As a consequence, the excited state is the product of a kind of internal oxidation-reduction process. This behavior differs from that of an organic chromophore where the electron in the excited state is in the *molecular* orbital formed by two or more atoms.

Table 4-5   Effect of Ligands on Absorption Maxima Associated with *d–d* Transitions

(λ_{max} (nm) for the indicated ligands)

Central ion	Increasing ligand field strength				
	$6Cl^-$	$6H_2O$	$6NH_3$	$3en^a$	$6CN^-$
Cr(III)	736	573	462	456	380
Co(III)	—	538	435	428	294
Co(II)	—	1345	980	909	—
Ni(II)	1370	1279	925	863	—
Cu(II)	—	794	663	610	—

[a]en = ethylenediamine, a bidentate ligand

[4]For a brief discussion of this type of absorption see C. N. R. Rao, *Ultra-Violet and Visible Spectroscopy, Chemical Applications* (New York: Plenum Press, 1967), Chap. 11.

A well-known example of charge-transfer absorption is observed in the iron(III)-thiocyanate ion complex. Absorption of a photon results in the transition of an electron from the thiocyanate ion to an orbital that is largely associated with the iron(III) ion. The product is thus an excited species involving predominately iron(II) and the thiocyanate radical SCN. As with an intramolecular excitation, the electron, under ordinary circumstances, returns to its original state after a brief period. Occasionally, however, dissociation of the excited complex may occur to produce a photochemical oxidation-reduction process.

As the tendency for electron transfer increases, less radiant energy is required for the charge-transfer process, and the resulting complexes absorb at longer wavelengths. For example, the thiocyanate ion is a better electron donor (reducing agent) than is the chloride ion; thus, the absorption of the thiocyanate complex with iron(III) is in the visible region, whereas the absorption maximum for the corresponding yellow chloride complex is in the ultraviolet region. Presumably, the iodide complex of iron(III) would absorb at still longer wavelengths, but this has not been observed because the electron-transfer process is complete, giving iron(II) and iodine as products.

In most charge-transfer complexes involving a metal ion, the metal serves as the electron acceptor. An exception is found in the *o*-phenanthroline complex of iron(II) or copper(I), where the ligand is the acceptor and the metal ion is the donor. Other examples of this type of complex are known.

Organic compounds form many interesting charge-transfer complexes. An example is quinhydrone (a 1:1 complex of quinone and hydroquinone), which exhibits strong absorption in the visible region. Other examples include the iodine complexes with amines, aromatics, and sulfides, among others.

QUALITATIVE ANALYSIS

Absorption spectroscopy provides a useful tool for qualitative analysis. Identification of a pure compound by this method involves an empirical comparison of the details of the spectrum (maxima, minima, and inflection points) of the unknown with those of pure compounds; a close match is considered good evidence of chemical identity, particularly if the spectrum of the unknown contains a number of sharp and well-defined peaks. Absorption in the infrared region, which we consider in Chapter 6, is particularly useful for qualitative purposes because of the wealth of fine structure found in the spectra of many compounds. The application of ultraviolet and visible spectrophotometry to qualitative analysis is more limited because the absorption bands tend to be broad and hence lacking in detail. Nevertheless, spectral investigations in this region frequently provide useful qualitative information concerning the presence or absence of certain functional groups (such as carbonyl, aromatic, nitro, or conjugated diene) in organic compounds. A further important application involves the detection of highly absorbing impurities in nonabsorbing media; if an absorption peak for the contaminant has a sufficiently high absorptivity, the presence of trace amounts can be readily established.

Qualitative Techniques

Solvents. In choosing a solvent, consideration must be given not only to its transparency but also to its possible effects upon the absorbing system. Quite generally, polar solvents such as water, alcohols, esters, and ketones tend to obliterate spectral fine structure arising from vibrational effects; spectra that more closely approach those of the gas phase (see Figure 4-7) are most likely to be observed in nonpolar solvents such as hydrocarbons. In addition, the positions of absorption maxima are also influenced by the nature of the solvent. Clearly, it is important to employ identical solvents when comparing absorption spectra for identification purposes.

Table 4-6 lists some common solvents and the approximate wavelength below which they cannot be used because of absorption. These minima are strongly dependent upon the purity of the solvent.[5]

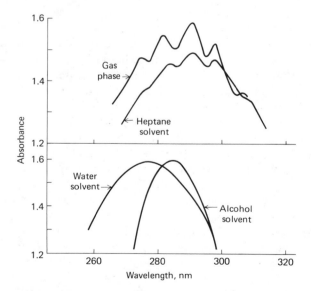

Figure 4-7. Effect of solvent on the absorption spectrum of acetaldehyde.

[5]For a brief discussion of purification of spectral solvents see A. Weissberger, Ed., *Physical Methods of Organic Chemistry,* Vol. I, Part III (New York: Interscience Publishers, Inc., 1960), p. 1854.

Table 4-6 Solvents for the Ultraviolet and the Visible Regions

Solvent	Approximate transparency minimum (nm)	Solvent	Approximate transparency minimum (nm)
Water	180	Carbon tetrachloride	260
Ethanol	220	Diethyl ether	210
Hexane	200	Acetone	330
Cyclohexane	200	Dioxane	320
Benzene	280	Cellosolve	320

Comparison of spectra. It is usually easier to compare spectra that have been plotted in terms of log A versus wavelengths. Beer's law can be written in the form

$$\log A = \log \epsilon + \log bc$$

Of the terms on the right side of this equation only ϵ varies with wavelength. Thus, if the concentrations or cell lengths for the spectra being matched are not identical, log A is displaced by exactly the same absolute amount at each wavelength (see Figure 3-4). This is in contrast to the variable displacement of an absorbance plot (shown in Figure 3-4). Plots involving log ϵ are also readily compared for the same reason.

Effect of slit width. In order to obtain an absorption spectrum useful for qualitative comparison, absorbance data should be collected with the narrowest possible band width. Otherwise, significant details of the spectrum may be lost. This effect is demonstrated in Figure 4-8.

The band width of a monochromator is determined not only by the dispersion characteristics of the prism or grating but also by the width of the exit slit (p. 45). Most spectrophotometers are equipped with a variable slit; the operator thus has some control of the band width to be employed. The descriptive brochures supplied with most instruments give the slit setting needed to achieve a given effective band width; with prism instruments this setting must be varied as a function of wavelength if a constant band width is to be maintained.

The two factors that limit the minimum width to which the slit of a spectrophotometer may be set are the intensity of the source and the sensitivity of the detecting device. At wavelengths where either becomes low, wider slits, and consequently wider band widths, must be employed in order to provide a beam of sufficient power for accurate measurement.

Effect of scattered radiation. We have noted in Chapter 3 that the radiation emitted from the exit slit of a monochromator often contains small amounts of stray radiation of wavelengths that differ considerably from the wavelength to which the instrument is set. This contamination arises from reflections off the various optical and housing surfaces within the mono-chromator. With high-quality instruments scattered radiation amounts at most to a few tenths of a percent at all wavelength settings; under usual circumstances its presence has no detectable effect on absorbance data. When measurements are being made at the wavelength extremes of an in-strument, however, scattered radiation may cause appreciable errors. For example, the typical spectrophotometer for the visible region is equipped with glass optics and cells that begin to absorb in the region between 350 and 400 nm. Furthermore, such instruments employ phototubes or photo-cells that exhibit a maximum photoelectric response between 500 and 700 nm; at 350 nm, then, the response may be only ten percent (or less) of the maximum. Finally, an instrument of this kind employs a tungsten filament source, the maximum energy of which is in the upper visible region; at 350 nm, its output is but a fraction of its maximum. These three factors, then, limit the low wavelength capability of the instrument; in order to obtain absorbance data in this region it is necessary to employ maximum amplifica-tion of the detector signal, maximum intensity of the source, and a relatively wide slit width. Only under these circumstances can the instrument be made to read 100 percent transmittance with the blank in the light path. The effect of scattered radiation of longer wavelengths now becomes greatly

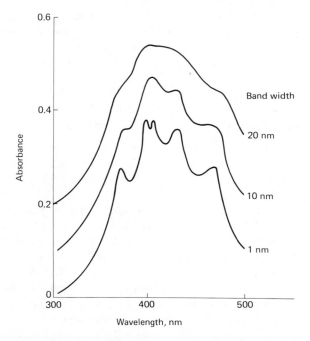

Figure 4-8. Effect of the band width on spectra of identical solutions. Each spectrum has been displaced by 0.1 absorbance unit for clarity.

magnified. First, the source generates a much greater intensity of wavelengths that can be scattered; second, the wide slit width enhances the probability of scattered radiation emerging from the monochromator; third, the scattered radiation is absorbed to a lesser extent by the cell walls; and finally, the detector is much more responsive to the wavelengths of the scattered radiation. As a consequence, at a monochromator setting of 350 nm, the absorbance measured by such an instrument may be due as much to the absorbance of longer scattered wavelengths as to 350-nm radiation.

Identification of Compounds

The certainty with which an identification can be made by absorption measurement is directly related to the number of separate spectral features (peaks, minima, and inflection points) that can be observed between the unknown and an authentic reference standard. For the ultraviolet and visible regions, the number of these characteristic features is often limited; identification based on a single pair of spectra may thus be ambiguous. Occasionally, further confirmation can be obtained from additional spectra of the unknown and the standard in a variety of solvents, at different *p*H values, and after suitable chemical treatment. Unfortunately, however, the electronic absorption peaks of many chromophores are not greatly influenced by structural features of attached nonabsorbing groups; as a consequence, only the absorbing functional groups can be identified, and other means must be used to determine the remaining features of the compound under study.

Catalogs of ultraviolet and visible absorption data for organic compounds are available in several publications. These compilations, which frequently are of help in organic qualitative work, include:

1. American Petroleum Institute, *Ultraviolet Spectral Data, A.P.I. Research Project 44*, Pittsburgh, Pennsylvania: Carnegie Institute of Technology. Ultraviolet-visible spectra on $8\frac{1}{2} \times 11$-inch loose-leaf sheets. Data are being added continually.

2. H. M. Hershenson, *Ultraviolet and Visible Absorption Spectra Index for 1930-1954*, New York: Academic Press, Inc., 1956.

3. M. J. Kamlet and H. E. Ungnade, Eds., *Organic Electronic Spectral Data*, 2 vols., New York: Interscience Publishers, Inc., 1960. These volumes contain data abstracted from the literature for the period 1946 to 1955. Each compound is listed by its empirical formula; the name, solvent, wavelength maxima, and approximate values for molar absorptivity at maxima are given. Literature references are also supplied.

4. *Sadtler Ultraviolet Spectra*, Sadtler Research Laboratories, 3314-20 Spring Garden Street, Philadelphia, Pennsylvania. The spectra are photographically reproduced, three on a page. Information includes compound

name, structural formula, melting and boiling points, source of compound, solvent, and concentration.

5. R. A. Friedel and M. Orchin, *Ultraviolet Spectra of Aromatic Compounds*, New York: John Wiley & Sons, Inc., 1951. A large number of spectra have been reduced to a common scale and published in loose-leaf form.

6. American Society for Testing Materials, Committee E-13; 1916 Race Street, Philadelphia, Pennsylvania. Ultraviolet spectra are coded on IBM cards.

Detection of Functional Groups

Even though they may not provide the unambiguous identification of an organic compound, absorption measurements in the visible and the ultra-violet regions are nevertheless useful for detecting the presence of certain functional groups that act as chromophores. For example, a weak absorption band in the region of 280 to 290 nm that is displaced toward shorter wavelengths with increased solvent polarity is a strong indication of the presence of the carbonyl group. A weak absorption band at about 260 nm with indications of vibrational fine structure constitutes evidence for the presence of an aromatic ring. Confirmation of the presence of an aromatic amine or a phenolic structure may be obtained by comparing the effects of pH on the spectra of solutions containing the sample with those shown in Table 4-4 for phenol and aniline.

QUANTITATIVE ANALYSIS BY ABSORPTION MEASUREMENTS

Absorption spectroscopy is one of the most useful tools available to the chemist for quantitative analysis. Important characteristics of spectro-photometric and photometric methods include:

1. *Wide applicability.* As we have pointed out, a wide variety of inorganic and organic species absorb in the ultraviolet and visible ranges, and are thus susceptible to quantitative determination. In addition, many nonabsorbing species can be analyzed after conversion to absorbing species by suitable chemical treatment.

2. *High sensitivity.* Molar absorptivities in the range of 10,000 to 40,000 are common, particularly for the charge-transfer complexes of inorganic species. Thus, analyses for concentrations in the range of 10^{-4} to 10^{-5} M are ordinary; by modification of the procedure, the range can often be extended to 10^{-6} or even 10^{-7}.

3. *Moderate to high selectivity.* Through the judicious choice of conditions, it may be possible to locate a wavelength region in which the only absorbing component in a sample is the substance being determined. Furthermore, where overlapping absorption bands do occur, corrections based

on additional measurements at other wavelengths are sometimes possible. As a consequence, the separation step can be omitted.

4. *Good accuracy.* For the typical spectrophotometric or photometric procedure, the relative error in concentration measurements lies in the range of 1 to 3 percent. By employing special techniques to be discussed later, errors can often be reduced to a few tenths of a percent.

5. *Ease and convenience.* Spectrophotometric or photometric measurements are easily and rapidly performed with modern instruments.

Scope

The applications of quantitative absorption methods are not only numerous, but also touch upon every area in which quantitative chemical information is required. The reader can obtain a notion of the scope of spectrophotometry by consulting a series of review articles published periodically in *Analytical Chemistry*[6] and from monographs on the subject.[7]

Applications to absorbing species. Tables 4-2, 4-3, and 4-4 list many of the common organic chromophoric groups. Spectrophotometric analysis for any organic compound containing one or more of these groups is potentially feasible; many examples of this type of analysis are found in the literature.

A number of inorganic species also absorb and are thus susceptible to direct determination; we have already mentioned the various transition metals. In addition, a number of other species also show characteristic absorption. Examples include permanganate, nitrate, and chromate ions; osmium and ruthenium tetroxides; molecular iodine; and ozone.

Applications to nonabsorbing species. Numerous reagents react with nonabsorbing species to yield products that absorb strongly in the ultraviolet or visible regions. The successful application of such reagents to quantitative analysis usually requires that the color-forming reaction be forced to near completion. It should be noted that these reagents are frequently employed for the determination of an absorbing species, such as a transition metal ion; the molar absorptivity of the product will frequently be orders of magnitude greater than that of the uncombined species.

A host of complexing agents have been employed for the determination of inorganic species. Typical inorganic reagents include thiocyanate ion

[6]For ultraviolet spectrometry see W. Crummett and R. Hummel, *Anal. Chem.*, **42**, 239R (1970); **40**, 330R (1968). For light absorption spectrometry see D. F. Boltz and M. G. Mellon, *Anal. Chem.*, **42**, 152R (1970); **40**, 255R (1968).

[7]See, for example, E. B. Sandell, *Colorimetric Determination of Traces of Metals*, 3d ed. (New York: Interscience Publishers, Inc., 1959); D. F. Boltz, Ed., *Colorimetric Determination of Nonmetals* (New York: Interscience Publishers, Inc., 1958); F. D. Snell and C. T. Snell, *Colorimetric Methods of Analysis*, 3d ed., 4 vols. (Princeton, N.J.: D. Van Nostrand Co., Inc., 1959).

for iron, cobalt, and molybdenum; the anion of hydrogen peroxide for titanium, vanadium, and chromium; and iodide ion for bismuth, palladium, and tellurium. Of even more importance are organic chelating agents which form stable, colored complexes with cations. Examples include *o*-phenanthroline for the determination of iron, dimethylglyoxime for nickel, diethyldithiocarbamate for copper, and diphenylthiocarbazone for lead.

Certain nonabsorbing organic functional groups are also determined by absorption procedures. For example, low-molecular-weight aliphatic alcohols react with cerium(IV) to produce a red 1:1 complex that can be employed for quantitative purposes.

Procedural Details

Before a photometric or spectrophotometric analysis can be undertaken it is necessary to choose a set of working conditions and to prepare a calibration curve relating concentration to absorbance.

Selection of wavelength. In a spectrophotometric analysis, absorbance measurements are ordinarily made at a wavelength corresponding to an absorption peak because the change in absorbance per unit of concentration is greatest at this point; the maximum sensitivity is thus realized. In addition, the absorption curve is often flat in this region; under these circumstances good adherence to Beer's law can be expected (p. 35). Finally, the measurements are less sensitive to uncertainties arising from failure to reproduce precisely the wavelength setting of the instrument.

The absorption spectrum, if available, aids in choosing the most suitable filter for a photometric analysis; if this information is lacking, the alternative method for selection given on page 59 may be used.

In order to avoid interference from other absorbing substances, a wavelength other than a peak may be appropriate for a particular analysis. In this event the region selected should, if possible, be one in which the change in absorptivity with wavelength is not too great.

Variables that influence absorbance. Common variables that influence the absorption spectrum of a substance include the nature of the solvent, the *p*H of the solution, the temperature, high electrolyte concentrations, and the presence of interfering substances. The effects of these variables must be known and a set of analytical conditions must be chosen such that the absorbance will not be materially influenced by small, uncontrolled variations in their magnitudes.

Determination of the relationship between absorbance and concentration. After deciding upon a set of conditions for the analysis, it is necessary to prepare a calibration curve from a series of standard solutions. These standards should approximate the over-all composition of the actual samples and should cover a reasonable concentration range of the species being determined. Seldom, if ever, is it safe to assume adherence to Beer's law and use only a single standard to determine the molar absorptivity. It is even less prudent to base the results of an analysis on a literature value for the molar absorptivity.

Analysis of mixtures of absorbing substances. The total absorbance of a solution at a given wavelength is equal to the sum of the absorbances of the individual components present. This relationship makes possible the analysis of the individual components of a mixture even if an overlap in their spectra exists. Consider, for example, the spectra of M and N, shown in Figure 4-9. There is obviously no wavelength at which the absorbance of this mixture is due simply to one of the components; thus, an analysis for either M or N is impossible by a single measurement. However, the absorbances of the mixture at the two wavelengths λ' and λ'' may be expressed as follows:

$$A' = \epsilon'_M b c_M + \epsilon'_N b c_N \qquad (\text{at } \lambda')$$
$$A'' = \epsilon''_M b c_M + \epsilon''_N b c_N \qquad (\text{at } \lambda'')$$

The four molar absorptivities ϵ'_M, ϵ'_N, ϵ''_M, and ϵ''_N can be evaluated from individual standard solutions of M and of N, or better, from the slopes of their Beer's law plots. The absorbances of the mixture, A' and A'', are experimentally determinable as is b, the cell thickness. Thus, from these two equations the concentration of the individual components in the mixture, c_M and c_N, can be readily calculated. These relationships are valid only if Beer's law is followed. The greatest accuracy in an analysis of this sort is attained by choosing wavelengths at which the differences in molar absorptivities are large.

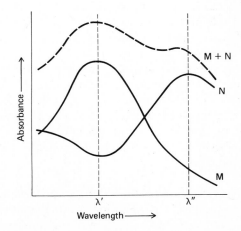

Figure 4-9. Absorption spectrum of a two-component mixture.

Mixtures containing more than two absorbing species can be analyzed, in principle at least, if a further absorbance measurement is made for each added component. The uncertainties in the resulting data become greater, however, as the number of measurements increase.

Analytical errors in absorption analysis. In many (but certainly not all) methods based upon the absorption of radiation, the major source of indeterminate error lies in the measurement of the absorbance. Determination of this quantity requires evaluation of the power of a beam that has passed through the solution, as well as its power after passage through the blank. Because of the logarithmic nature of the relationship between these quantities and concentration, however, the magnitude of the effect of uncertainties in the power measurements on the analytical result is not immediately obvious.

In many instances the measurement of radiant power involves an absolute uncertainty that is independent of the magnitude of the power. As a consequence, the ratio P/P_0 (that is, the transmittance) is affected by a constant absolute error over a considerable range. To see the effect of this constant error in transmittance upon the uncertainty of the results of an analysis it is convenient to write Beer's law in the form

$$-\log T = \epsilon bc \qquad (4\text{-}1)$$

After converting the left-hand side to a natural logarithm, the derivative of equation (4-1) is

$$-\frac{0.434}{T}\, dT = \epsilon b\, dc \qquad (4\text{-}2)$$

Dividing equation (4-2) by equation (4-1) and rearranging gives

$$\frac{dc}{c} = \frac{0.434}{T \log T}\, dT \qquad (4\text{-}3)$$

and this may be written as

$$\frac{\triangle c}{c} = \frac{0.434}{T \log T}\, \triangle T \qquad (4\text{-}4)$$

The quantity $\triangle c/c$ is a measure of the *relative error* in a concentration measurement arising from the absolute error $\triangle T$ in the measurement of T or P/P_0. The reader should note that the relative error in concentration varies as a function of the magnitude of T. This relationship is shown by the data in Table 4-7, in which an absolute error in T of 0.005 or 0.5 percent was assumed in a series of transmittance measurements;[8] the relative concentration error goes through a minimum at an absorbance of about 0.4. (By setting the derivative of equation (4-4) equal to zero, it can be shown that

[8]C. F. Hiskey, *Anal. Chem.*, **21**, 1440 (1949), recommends that $\triangle T$ be taken as twice the average deviation in transmittance obtained by repeated measurement of T for a single solution.

this minimum occurs at a transmittance of 0.368 or an absorbance of 0.434.) A plot of the data in Table 4-7 is shown by curve (a) in Figure 4-11 (p. 96).

The indeterminate error of transmittance measurements for commercial spectrophotometers will be in the range of ± 0.002 to ± 0.01, with the figure of ± 0.005 employed in Table 4-7 being typical. Concentration errors of about 1 to 2 percent relative are thus inherent in most absorption methods when absorbances measured lie in the range of 0.15 to 1.0. The higher error associated with more strongly absorbing samples can usually be avoided by suitable dilution prior to the instrumental measurement. Relative errors greater than 2 percent are inevitable, however, when absorbances are less than 0.1.

It is important to emphasize that the conclusions just described are based on the assumption that the limiting uncertainty in an analysis is a constant photometric error ΔT that is independent of transmittance. The potential for other types of errors always exists, of course, and the magnitude of many of these may indeed vary with transmittance. Examples include errors arising from sample preparation, from imperfections in the cells, from uncertainties in the wavelength setting, and from source fluctuations.

Table 4-7   Variations in the Percent Error in Concentration as a Function of Transmittance and Absorbance from Equation (4-4)[a]
(Error in transmittance measurement, ΔT, assumed to be ± 0.005, or 0.5 percent)

Transmittance, T	*Absorbance, A*	*Percent Error in Concentration,* $\dfrac{\Delta c}{c} \times 100$
0.95	0.022	± 10.2
0.90	0.046	± 4.74
0.80	0.097	± 2.80
0.70	0.155	± 2.00
0.60	0.222	± 1.63
0.50	0.301	± 1.44
0.40	0.399	± 1.36
0.30	0.523	± 1.38
0.20	0.699	± 1.55
0.10	1.000	± 2.17
0.030	1.523	± 4.75
0.020	1.699	± 6.38

[a]From *Fundamentals of Analytical Chemistry*, Second Edition, by Douglas A. Skoog and Donald M. West. Copyright © 1963, 1969 by Holt, Rinehart and Winston, Inc. Reprinted by permission of Holt, Rinehart and Winston, Inc.

Precision or Differential Absorption Methods[9]

By modifying the way in which absorbance is measured, it is often possible to reduce the relative concentration error resulting from instrumental uncertainties to a few tenths of one percent. The techniques employed are called differential or precision methods and fall into three categories: (1) the *high-absorbance method,* which is particularly applicable to more concentrated solutions; (2) the *low-absorbance method,* which is useful for very dilute solutions having absorbances so low as to lead to large analytical errors by the usual method; and (3) the *method of ultimate precision,* which is a combination of the two.

Precision is gained in all of these procedures by variation of instrumental settings so that the full transmittance (or absorbance) scale encompasses a smaller concentration range; as a consequence, a constant instrumental error results in a correspondingly smaller concentration error. This gain in accuracy is somewhat offset by the inconvenience of the more limited concentration range over which each method is applicable.

Figure 4-10 illustrates the way in which the ordinary transmittance scale is expanded in the three methods. In the high-absorbance procedure

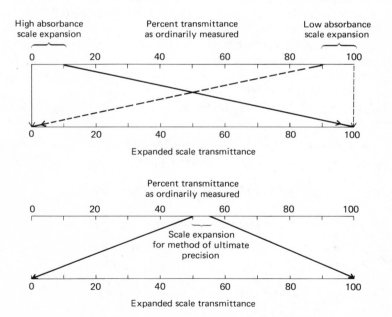

Figure 4-10. Types of scale expansion by various precision methods.

[9]For a complete analysis of these methods see C. N. Reilley and C. M. Crawford, *Anal. Chem.,* **27,** 716 (1955) and C. F. Hiskey, *Anal. Chem.,* **21,** 1440 (1949).

a percent transmittance region between zero and some percent less than 100 is expanded to cover the entire transmittance scale; for the low-absorbance method a similar expansion involves a percent transmission region between 100 and some value that is greater than zero. In the method of ultimate precision a small region lying between a percent transmittance greater than zero and smaller than 100 is expanded.

Principles of the techniques. A photometric measurement is ordinarily preceded by two adjustments which permit use of the full scale of the instrument. The first of these is performed with a shutter imposed in the beam to prevent radiation from reaching the detector; the dark current is then compensated for electrically so that the indicator exhibits a reading of zero. The second adjustment is made with the beam passing through the solvent and impinging on the detector; here the radiation intensity is varied until the indicator reads full scale (usually 100). These adjustments are made under different conditions in the differential methods. For the low-absorbance method the zero setting is made with radiation striking the detector after passing through a standard solution that is somewhat more concentrated than the sample to be measured; the full-scale adjustment is made in the usual manner. In contrast, the high-absorbance technique makes use of a standard somewhat more dilute than the sample for establishing the full-scale setting, while the zero adjustment is accomplished in the ordinary fashion. The method of ultimate precision requires two standards, one more concentrated and one more dilute than the sample; the former is employed to set the zero of the scale and the latter to adjust the full-scale reading. A comparison of these methods is shown in Table 4-8.

To develop equations for these techniques, the following terms are defined.

P_x = power of a beam of radiation after passing through the sample.

P_d = power of the same beam during adjustment of the zero of the indicator.

P_s = power of the beam during adjustment of the full-scale setting of the instrument.

P_0 = power of the beam after passage through pure solvent.

The concentrations of the standards are chosen such that

$$P_d < P_x < P_s < P_0$$

As we have pointed out (p. 51), a photoelectric detector suitable for absorption measurements exhibits a linear electrical response G to variations in power; that is,

$$G = k'P + k'' \qquad (4\text{-}5)$$

where k' and k'' are constants.

If the zero adjustment of the indicator scale ($G = 0$) is performed with a standard solution in the beam path instead of the shutter, equation (4-5) becomes

$$0 = k'P_d + k''$$

or

$$k'' = -k'P_d$$

Thus, the subsequent response of the instrument becomes

$$G = k'P - k'P_d = k'(P - P_d) \tag{4-6}$$

If, now, the full-scale adjustment is performed with a second standard that is less concentrated than the sample ($P_s > P_x$), we may express the indicator behavior as

$$G_s = k'(P_s - P_d) \tag{4-7}$$

With the sample in the beam, then

$$G_x = k'(P_x - P_d) \tag{4-8}$$

We now define the *relative transmittance* of the sample T_r as

$$T_r = \frac{G_x}{G_s} = \frac{P_x - P_d}{P_s - P_d} \tag{4-9}$$

This quantity is then employed in the differential method in place of transmittance as usually defined.

It is convenient to express the relative transmittance in terms of the

Table 4-8 Comparison of Methods for Absorption Measurements[a]

| Method | Imposed in Beam Path for Indicator Setting of | | Power Relationships[b] | | |
	Zero	Full Scale	Power, P_d, for Zero Setting ($T = 0$)	Power, P_s, for Full-Scale Setting ($T = 1$)	Power, P_x, for Sample
Ordinary	Shutter	Solvent	$P_d = 0$	$P_s = P_0$	$0 < P_x < P_0$
Low-absorbance	Standard (conc. > sample)	Solvent	$P_d > 0$	$P_s = P_0$	$P_d < P_x < P_0$
High-absorbance	Shutter	Standard (conc. < sample)	$P_d = 0$	$P_s < P_0$	$0 < P_x < P_s$
Ultimate precision	Standard (conc. > sample)	Standard (conc. < sample)	$P_d > 0$	$P_s < P_0$	$P_d < P_x < P_s$

[a]From *Fundamentals of Analytical Chemistry*, Second Edition, by Douglas A. Skoog and Donald M. West. Copyright © 1963, 1969 by Holt, Rinehart and Winston, Inc. Reprinted by permission of Holt, Rinehart and Winston, Inc.

[b]P_0 = radiant power after passage through solvent.

normal transmittances of the three solutions *with respect to the pure solvent.*
Dividing the numerator and the denominator in (4-9) through by P_0, the
power of the radiation after passage through pure solvent, gives

$$T_r = \frac{P_x/P_0 - P_d/P_0}{P_s/P_0 - P_d/P_0}$$

or
$$T_r = \frac{T_x - T_d}{T_s - T_d} \tag{4-10}$$

Variation of relative transmittance with concentration.   In differential
methods the concentrations of one or both standards are held constant so
that T_s and/or T_d in (4-10) remain unchanged, while T_x varies with concentration. To see the relationship between the experimental quantity,
T_r, and concentration, we may rewrite equation (4-10) to give

$$T_x = T_r T_s - T_r T_d + T_d$$

From Beer's law we know that

$$- \log T_x = \epsilon b c_x$$

Thus

$$- \log (T_r T_s - T_r T_d + T_d) = \epsilon b c_x \tag{4-11}$$

or, from (4-9)

$$- \log \left(\frac{G_x}{G_s} T_s - \frac{G_x}{G_s} T_d + T_d \right) = \epsilon b c_x \tag{4-12}$$

Equation (4-12) reverts to the normal Beer's law relationship when
appropriate substitutions are made for T_d and T_s. Thus, in ordinary practice, the dark current is adjusted with a shutter in front of the detector;
as a consequence, $T_d = 0$. The response of the indicator, G_s, is determined
with the pure solvent in the light path so that $T_s = 1.0$. Substituting these
values into (4-12) gives

$$- \log \frac{G_x}{G_s} = \epsilon b c_x$$

Analytical errors in differential methods.   The indeterminate error associated with a differential transmittance measurement is ordinarily similar in
nature and size to the error encountered in an ordinary absorbance measurement. The relative error in concentration resulting from the uncertainty in
the measured quantity can be found in a manner similar to that shown on
page 89. If we take the derivative of equation (4-11), remembering that T_d
and T_x are constant, we obtain

$$\frac{-0.434(T_s - T_d)\, dT_r}{T_r T_s - T_r T_d + T_d} = \epsilon b\, dc_x$$

Dividing this equation by (4-11) and expressing the quotient in terms of the finite intervals ΔT_r and Δc_x gives

$$\frac{\Delta c_x}{c_x} = \frac{0.434(T_s - T_d)}{(T_rT_s - T_rT_d + T_d)\log(T_rT_s - T_rT_d + T_d)}\,\Delta T_r \qquad (4\text{-}13)$$

It is seen that equation (4-13) reduces to (4-4) when $T_d = 0$ and $T_s = 1.00$.

This equation permits evaluation of the analytical error associated with the three types of differential methods.

High-absorbance method. In this technique the zero adjustment is made in the usual fashion with a shutter imposed between the beam and the detector. In effect, then, $T_d = 0$. The beam is then allowed to impinge on the detector after it has passed through a standard solution that is somewhat less concentrated than the sample; thus, T_s is smaller than 1 but greater than T_x. The intensity of the source or the amplification of the indicator circuit is adjusted so that the meter reads full-scale relative (usually 100). The standard is then replaced by the sample and the percent relative transmittance is read directly. Under these circumstances equations (4-9) and (4-10) become

$$T_r = G_x/G_s = P_x/P_s \qquad (4\text{-}14)$$

and

$$T_r = T_x/T_s \qquad (4\text{-}15)$$

Equation (4-11) converts to

$$-\log T_rT_s = \epsilon b c_x$$

or

$$-\log T_r = \epsilon b c_x + \log T_s \qquad (4\text{-}16)$$

Since T_s is constant, we see that the negative logarithm of the relative transmittance is linearly related to concentration.

To understand the relationship between relative transmittance measured by the high absorbance method and the transmittance measured with the pure solvent as reference it is helpful to consider a specific example. Suppose we have a set of samples with transmittances T_x in the range between 2 and 8 percent when measured against pure solvent. When these are compared with a standard solution having a transmittance T_s of 10 percent, the following relationship is obtained from equation (4-15).

Percent transmittance

Relative to solvent, $T_s \times 100$	Relative to standard, $T_r \times 100$
8	80
6	60
4	40
2	20

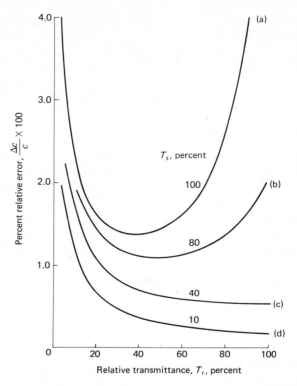

Figure 4-11. Error curves for analyses performed with reference solutions having transmittances shown. In each case uncertainty in measured transmittance is assumed to be ±0.5 percent.

Thus, the transmittance scale for this concentration range has been expanded by a factor of 10 (see Figure 4-10).

The effect of this scale expansion on the concentration error can be found by rewriting equation (4-13) for the condition $T_d = 0$:

$$\frac{\Delta c_x}{c_x} = \frac{0.434}{T_r \log T_r T_s} \Delta T_r \qquad (4\text{-}17)$$

Figure 4-11 gives several plots of the left-side of equation (4-17) against relative transmittance for several standards with transmittances T_s. As before, an absolute instrumental error (ΔT_r) of ± 0.005 was assumed. The uppermost curve is the error function when pure solvent is used as the reference (that is, a plot of the data in Table 4-7). It is apparent that the greatest gain in accuracy occurs with reference solutions that have low transmittances or high absorbances; thus, this technique is most useful for the analysis of more concentrated solutions of the absorbing species.

Example. A sample was found to have a transmittance of 2.00 percent when measured against pure solvent. What would be its relative transmittance when measured against a standard having a transmittance of 10 percent? Compare the theoretical relative concentration errors when the analysis is based (1) on measurement with pure solvent as the reference, and (2) with the standard as a reference. Assume an absolute instrumental error of ± 0.5 percent.

The relative transmittance against the standard having a T_s of 10 percent can be obtained from equation (4-15):

$$T_r = \frac{T_x}{T_s} \times 100 = \frac{2}{10} \times 100 = 20 \text{ percent}$$

To calculate the error associated with the two measurements we use equation (4-17). For the measurement against pure solvent

$$\frac{\Delta c_x}{c_x} = \frac{0.434\ \Delta T_r}{T_r \log(T_r T_s)} = \frac{0.434 \times (\pm 0.005)}{0.02 \times \log(0.02 \times 1.00)}$$

$$= \pm 0.0638 \text{ or } \pm 6.38 \text{ percent}$$

For the measurement against the standard

$$\frac{\Delta c_x}{c_x} = \frac{0.434 \times (\pm 0.005)}{0.20 \log(0.2 \times 0.1)}$$

$$= \pm 0.0064 \text{ or } \pm 0.64 \text{ percent}$$

It has been shown experimentally that the high-absorbance method is capable of analytical accuracies that rival those obtained by volumetric or gravimetric procedures; that is, errors can be diminished to approximately 0.2 percent relative.

Low-absorbance method. This technique is particularly useful for improving the accuracy of absorption methods when the concentration of the absorbing species is so low that transmittances measured in the usual way lie in the range above 80 percent. A standard solution somewhat more concentrated than the samples is employed to adjust the zero reading of the indicator scale; the indicator is then adjusted to give a full-scale response against the solvent in the usual way. Upon substitution of sample for the blank in the light path, the instrument will indicate the relative transmittance T_x.

For the low-absorbance method, the conditions imposed are that

$$P_d > 0$$
$$P_s = P_0$$
$$T_s = 1.00 > T_x > T_d > 0$$

Equations (4-9), (4-10), and (4-11) become

$$T_r = \frac{G_x}{G_s} = \frac{P_x - P_d}{P_s - P_d} \tag{4-18}$$

$$T_r = \frac{T_x - T_d}{1 - T_d} \qquad\qquad (4\text{-}19)$$

$$-\log T_x = -\log(T_r - T_r T_d + T_d) = \epsilon b c_x \qquad (4\text{-}20)$$

Let us consider the effect of this technique on the transmittance range by assuming a set of samples having transmittances ranging between 91 and 99 percent. A standard solution having a transmittance of 90 percent, when measured in the usual way, is employed to zero the instrument; that is, $P_d = 0.90$. Then the measured T_r for a sample with an ordinary transmittance of 92 percent is

$$T_r = \frac{0.92 - 0.90}{1.00 - 0.90} = 0.20$$

The transmittances of the samples measured in the usual way and by this modified technique are as follows:

T_x, percent	T_r, percent
90	0
92	20
94	40
96	60
98	80

Thus, the change in the zero adjustment has expanded the transmittance range by a factor of 10 (see Figure 4-10).

It is apparent from an examination of equation (4-20) that a plot of $-\log T_r$ versus concentration will not be linear. As a consequence, a number of standard solutions must be employed to establish with accuracy the shape of the calibration curve to be used for analysis.

The error function for the low-absorbance technique can be obtained by setting T_s in equation (4-13) at unity. Plots of the relative error in concentration as a function of T_r for three values of T_d are shown in Figure 4-12. It is apparent that significant gains in accuracy are to be expected as T_d becomes large.

Method of ultimate precision. This procedure is simply a combination of the two techniques previously described. The sample is bracketed with two standards, one with an absorbance slightly greater than the sample and the other slightly less; the former is employed to set the zero end of the instrument scale and the latter to set the 100 percent transmittance point. The extent to which the transmittance scale is expanded depends upon the closeness of the two standards. For example, if the standards have transmittances of 50 and 55 percent against pure solvent, then this 5 percent difference would be expanded by a factor of 20, with a concomittant gain in analytical accuracy (see Figure 4-10).

Equations (4-9), (4-10), and (4-11) describe the relationships among the various quantities involved in the method of ultimate precision. It is apparent that the calibration curve of concentration versus $-\log T_r$ will be nonlinear, as in the case of the low-absorbance technique.

Plots of equation (4-13), analogous to Figures 4-11 and 4-12, indicate that this method leads to smaller relative concentration errors than either of the other two precision procedures and that the error becomes smaller as T_d and T_s approach one another.

Instrumental requirements for the applications of precision methods. In order to employ the low-absorbance method it is clearly necessary that the spectrophotometer possess a dark-current compensating circuit that is capable of offsetting larger currents than are normally produced when no radiation strikes the photoelectric detector. For the high absorbance method, on the other hand, the instrument must have a sufficient reserve capacity to permit setting the indicator at 100 percent transmittance when an absorbing

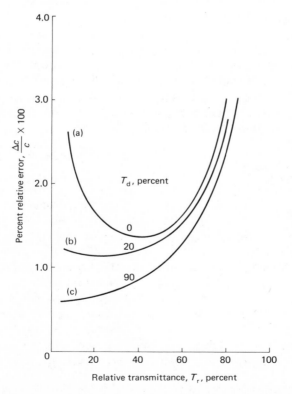

Figure 4-12. Error curves for analyses performed by low-absorbance techniques. Labels give percent transmittance of standard used to set zero on instrument. In each case uncertainty in measured transmittance is assumed to be 0.5 percent.

solvent is placed in the beam path. Here, the full-scale reading may be realized by increasing the radiation intensity (most often by widening the slits) or by increasing the amplification of the photoelectric current. The method of ultimate precision requires both of these instrumental qualities.

The capacity of a spectrophotometer to be set to full scale with an absorbing solution in the radiation path will depend both upon the quality of its monochromator and the stability of its electronic current. Furthermore, this capacity will be wavelength-dependent since the intensity of the source and the sensitivity of the detector change as the wavelength is varied. In regions where the intensity and sensitivity are low an increase in slit width may be necessary in order to realize a full-scale setting; under these circumstances scattered radiation may lead to errors unless the quality of the monochromator is high. Alternatively, a very high current amplification may be necessary; unless the electronic stability is good, significant photometric error may result.

PHOTOMETRIC TITRATIONS[10]

Photometric or spectrophotometric measurements can be employed to advantage in locating the equivalence point in a titration. The end point in a direct photometric titration is the result of a change in the concentration of a reactant or a product, or both; clearly, at least one of these species must absorb radiation. In the indirect method the absorbance of an indicator is observed as a function of added titrant.

Titration Curves

A photometric titration curve consists of a plot of absorbance versus the volume of titrant.[11] If conditions are chosen properly, the plots will consist of two straight-line portions of differing slopes, the one occurring at the outset of the titration and the other located well beyond the equivalence point. Usually a marked curvature occurs in the equivalence-point region; the end point is taken as the intersection of the extrapolated straight-line portions. Figure 4-13 shows some typical titration curves. Titration of a nonabsorbing species with a colored titrant that is decolorized by the reaction produces a horizontal line in the initial stages followed by a rapid rise in absorbance beyond the equivalence point [Figure 4-13(a)]. The formation of a colored product from colorless reactants, on the other hand, initially produces a linear rise in the absorbance followed by a region in

---

[10]For further information concerning this technique see J. B. Headridge, *Photometric Titrations* (New York: Pergamon Press, Inc., 1961).

[11]In order to obtain a satisfactory end point it is usually necessary to correct the absorbance for volume changes by multiplication by $(V + v)/V$, where V is the original volume of the solution and v is the volume of added titrant.

which the absorbance becomes independent of reagent volume [Figure 4-13(b)]. Depending upon the absorption characteristics of the reactants and the products, the other curve forms shown in Figure 4-13 are also possible.

In order to obtain a satisfactory photometric end point it is necessary that the absorbing system(s) obey Beer's law; otherwise, the titration curve will lack the linear portions needed for end-point extrapolation. Chemical deviations from Beer's law (p. 33) are to be expected if the absorbance is measured under conditions where product formation is not favored from the equilibrium standpoint. Thus, if the formation constant of the product is small or if dilute solutions are employed, the titration data must be collected well away from the equivalence point in order to take advantage of the common ion effect to force the reaction toward completion.

Instrumental deviations from Beer's law may also cause unsatisfactory titration curves. Often, for example, the absorbing species has such a large absorptivity that the absorbance at the wavelength of its absorption peak is well beyond the range of the instrument. For such systems, measurements must be made well away from the peak; appreciable deviations from Beer's law can be expected unless radiation of high spectral purity is employed (p. 35).

Instrumentation

Photometric titrations are ordinarily performed with a spectrophotometer or a photometer that has been modified to permit insertion of the titration vessel in the light path.[12] After the zero adjustment of the meter scale

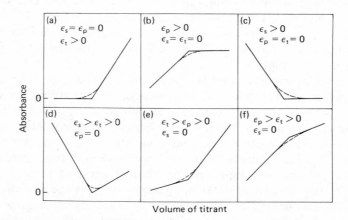

Figure 4-13. Typical photometric titration curves. Molar absorptivities of the substance titrated, the product and the titrant are given by ϵ_s, ϵ_p, and ϵ_t, respectively.

[12]See, for example, R. F. Goddu and D. N. Hume, *Anal. Chem.,* **26**, 1740 (1954); also C. Rehm, J. I. Bodin, K. A. Connors, and T. Higuchi, *ibid.,* **31**, 483 (1959).

has been made, radiation is allowed to pass through the solution to be titrated, and the instrument is adjusted by variation of the light intensity or by the sensitivity of the detector until a convenient absorbance reading is obtained. Ordinarily no attempt is made to measure the true absorbance since the relative absorbance obtained in this way is perfectly adequate for the purpose of end-point detection. Data for the titration are then collected without alteration of the instrument setting.

The power of the radiation source and the response of the detector must be reasonably constant during the period required for a photometric titration. Cylindrical containers are ordinarily used, and care must be taken to avoid any movement of the vessel that might cause changes in the length of the radiation path.

Both filter photometers and spectrophotometers have been employed for photometric titrations. The latter are preferred, however, because their narrower band widths enhance the probability of adherence to Beer's law.

Application of Photometric Titrations

Photometric titrations often provide more accurate results than a direct photometric analysis because the data from several measurements are pooled in determining the end point. Furthermore, the presence of other absorbing species may not interfere since only a change in absorbance is being measured.

The photometric end point possesses an advantage over many other commonly used end points; namely, the experimental data are taken well away from the equivalence-point region. Thus, the reactions employed need not have as favorable equilibrium constants as those required for a titration that is dependent upon observations near the equivalence point (for example, potentiometric or indicator end points). For the same reason, more dilute solutions may be employed.

The photometric end point has been applied to all types of reactions.[13] Most of the reagents used in oxidation-reduction titrations have characteristic absorption spectra and thus produce photometrically detectable end points. Acid-base indicators have been employed for photometric neutralization titrations. The photometric end point has also been used to great advantage in titrations with EDTA and other complexing agents. Figure 4-14 illustrates the application of the photometric end point to the successive titration of bismuth(III) and copper (II). At the wavelength employed (745 nm) neither the cations nor the reagent absorbs, nor does the more stable bismuth complex, which is formed in the first part of the titration; the copper complex, however, does absorb. Thus, the solution exhibits no absorbance until essentially all of the bismuth has been con-

---

[13]See, for example, the review by A. L. Underwood in C. N. Reilley, Ed., *Advances in Analytical Chemistry and Instrumentation* (New York: Interscience Publishers, 1964), Vol. 3, pp. 31–104.

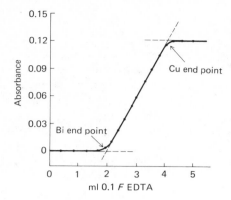

Figure 4-14. Photometric titration curve of 100 ml of a solution that was 2.0×10^{-3} F in Bi^{3+} and Cu^{2+}. Radiation of 745 nm was employed. (From A. L. Underwood, *Anal. Chem.*, **26**, 1322 (1954). With permission of the American Chemical Society.)

sumed. With the first formation of the copper complex, an increase in absorbance results. The increase continues until the copper equivalence point is reached. Further reagent additions cause no further absorbance change. Clearly, two well-defined end points result.

The photometric end point has also been adapted to precipitation titrations; here, the suspended solid product has the effect of diminishing the radiant power by scattering; titrations are carried to a condition of constant turbidity.

SPECTROPHOTOMETRIC STUDIES OF COMPLEX IONS

Spectrophotometry is one of the most useful tools for elucidating the composition of complex ions in solution and for determining their formation constants. The power of the technique lies in the fact that quantitative absorption measurements can be performed without fear of disturbing the equilibria under consideration. Although most spectrophotometric studies of complexes involve systems in which one of the reactants or the product absorbs, this condition is not a necessity provided that one of the components can be caused to participate in a competing equilibrium that does produce an absorbing species. For example, complexes of iron(II) with a nonabsorbing ligand might be studied if the effect of this ligand on the color of the iron(II)-orthophenanthroline complex (p. 80) is examined. Data on the formation constant and the composition of the nonabsorbing species can then be obtained, provided the corresponding data are available for the phenanthroline complex.

Three of the most common techniques employed for complex ion studies are (1) the method of continuous variations, (2) the mole-ratio method, and (3) the slope-ratio method. Each of these is examined briefly.

The Method of Continuous Variations[14]

In the method of continuous variations, solutions of the cation and the ligand with identical formal concentrations are mixed in varying volume ratios, but in such a way that the total volume of each mixture is the same. The absorbance of each solution is measured at a suitable wavelength and corrected for any absorbance of the mixture if no reaction has occurred. The corrected absorbance is then plotted against the volume fraction (which is equal to the mole fraction) of one of the reactants; that is, $V_M/(V_M + V_L)$ where V_M is the volume of the cation solution and V_L that of the ligand. A typical plot is shown in Figure 4-15. A maximum (or a minimum if the complex absorbs less than the reactants) occurs at a volume ratio V_M/V_L corresponding to the combining ratio of cation and ligand in the complex. In Figure 4-15 $V_M/(V_M + V_L)$ is 0.33 and $V_L(V_M + V_L)$ is 0.66; thus, V_M/V_L is 0.33/0.66, which suggests that the complex has the formula ML_2.

The curvature of the experimental lines shown in Figure 4-15 is the result of incompleteness of the complex-formation reaction. By measuring the deviations from the theoretical straight lines indicated in the figure, a formation constant for the complex can be calculated.

To determine whether more than one complex forms between the

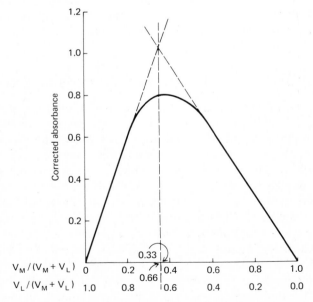

Figure 4-15. Continuous variation plot for the 1:2 complex, ML_2.

[14]See W. C. Vosburgh and G. R. Cooper, *J. Am. Chem. Soc.*, **63**, 437 (1941).

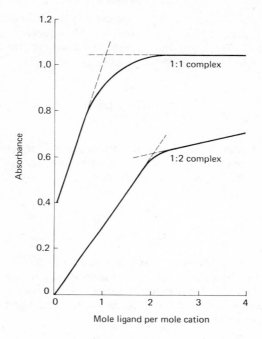

Figure 4-16. Mole-ratio plots for a 1:1 and a 1:2 complex. The 1:2 complex is the more stable.

reactants, the method is ordinarily repeated using different reactant concentrations and measurements at several wavelengths.

The Mole-ratio Method

In this method a series of solutions is prepared in which the formal concentration of one of the reactants (often the metal ion) is held constant while that of the other is varied. A plot of the absorbance versus mole ratio of the reactants is then prepared. If the formation constant is reasonably favorable, two straight lines of different slope are obtained; the intersection occurs at a mole ratio corresponding to the combining ratio in the complex. Typical mole-ratio plots are shown in Figure 4-16. Note that the ligand in the 1:2 complex absorbs at the wavelength selected; as a result, the slope beyond the equivalence point is greater than zero. The uncomplexed cation involved in the 1:1 complex absorbs, since the initial point has an absorbance greater than zero.

From the experimental plots, the formation constant can be obtained.

Example. The experimental plot for the 1:1 complex shown in Figure 4-16 can be analyzed as follows.

The known experimental variables are the formal concentration of the cation F_M, the formal concentration of the ligand F_L, and the measured absorbance A. These quantities can be related to the *equilibrium* concentration of the species present; that is,

$$F_M = [M] + [ML] \qquad (4\text{-}21)$$
$$F_L = [L] + [ML] \qquad (4\text{-}22)$$
$$A = \epsilon_M[M] + \epsilon_{ML}[ML] \qquad (4\text{-}23)$$

The fact that the right-hand side of the plot has a zero slope indicates that $\epsilon_L = 0$; therefore, the term $\epsilon_L[L]$ does not appear in the third equation

The two molar absorbances ϵ_M and ϵ_{ML} are readily obtained. Thus, when no ligand is present, $F_L = 0$ and $[ML] = 0$; then from equations (4-23) and (4-21)

$$\epsilon_M = A/[M] = A/F_M$$

The molar absorptivity for the complex ϵ_{ML} can be determined from the constant absorbance at high ligand-to-metal ratios. Here the absorbance is constant because essentially all the metal ion is present as the complex and $[M] \ll [ML]$. Under these conditions equations (4-23) and (4-21) convert to

$$A = \epsilon_{ML}[ML]$$
$$F_M = [ML]$$
or $\qquad\qquad \epsilon_{ML} = A/F_M$

With ϵ_M and ϵ_{ML} known we now have three independent equations containing the three unknowns $[M]$, $[ML]$, and $[L]$. Employing the absorbance and the formal concentration data in the region where the reaction is least complete, one can thus calculate equilibrium concentrations and from these obtain a value for the formation constant

$$K_f = \frac{[ML]}{[M][L]}$$

If the complex formation reaction is relatively incomplete, the mole-ratio plot appears as a continuous smooth curve with no straight-line portions that can be extrapolated to give the combining ratio. Such a reaction can sometimes be forced to completion by the addition of a several-hundredfold excess of the ligand until the absorbance becomes independent of further additions of the ligand. The constant absorbance can then be employed to calculate the molar absorptivity of the complex by assuming that essentially all of the metal ion has combined with the ligand and that the complex concentration is equal to the formal concentration of the metal. One can then successively assume that the complex has various compositions such as 1:1, 1:2, and so forth, and employ equations (4-21) to (4-23) to calculate equilibrium concentrations. The composition that gives constant numerical values for the formation constant over a wide concentration range can then be assumed to be the proper one for the complex.

A mole-ratio plot may reveal the stepwise formation of two or more complexes as successive slope changes, provided the complexes have different molar absorptivities, and provided also that the formation constants are not too similar.

The Slope-ratio Method

This procedure is particularly useful for weak complexes; it is applicable only to systems in which a single complex is formed. The method assumes that the complex-formation reaction can be forced to completion in the presence of a large excess of either reactant and that Beer's law is followed under these circumstances. For the reaction

$$mM + lL \rightleftharpoons M_mL_l$$

the following equation can be written, when L is present in very large excess:

$$[M_mL_l] \cong F_M/m$$

If Beer's law is obeyed

$$A_m = \epsilon b[M_mL_l] = \epsilon b F_M/m$$

and a plot of A with respect to F_M will be linear. When M is very large with respect to L

$$[M_mL_l] \cong F_L/l$$

and

$$A_l = \epsilon b[M_mL_l] = \epsilon b F_L/l$$

The slopes of the straight lines (A/F_m and A/F_l) are obtained under these conditions; the combining ratio between L and M is obtained from the ratio of the slopes

$$\frac{A_m/F_m}{A_l/F_l} = \frac{S_m}{S_l} = \frac{\epsilon b/m}{\epsilon b/l} = l/m$$

Problems

1. Starting with the equations on p. 88, derive expressions for the concentrations of two absorbing species M and N whose absorption curves overlap, given molar absorptivity data for each component at two wavelengths, λ' and λ'', and the absorbances A' and A'' for the mixture of M and N at these wavelengths.
2. A spectrophotometric study of solutions of substance A and substance B indicated that both followed Beer's law over a wide wavelength range. The following data were obtained with 1.00-cm cells.

	Absorbance	
	A	B
λ (nm)	$(5.00 \times 10^{-4} \, F)$	$(8.00 \times 10^{-5} \, F)$
415	0.626	0.171
430	0.680	0.128
440	0.683	0.106
450	0.669	0.094
460	0.633	0.090
475	0.561	0.099
490	0.477	0.132
505	0.395	0.195
520	0.320	0.283
535	0.256	0.370
550	0.204	0.426
565	0.172	0.460
580	0.146	0.469
590	0.132	0.470
600	0.118	0.466
615	0.101	0.447
630	0.086	0.410
645	0.069	0.363

(a) Draw an absorption curve for a mixed solution that is $5.00 \times 10^{-4} \, F$ in A and $8.00 \times 10^{-5} \, F$ in B.

(b) Calculate the absorbance at 440 and 590 nm of a solution that is $1.50 \times 10^{-4} \, F$ with respect to A and $1.00 \times 10^{-4} \, F$ with respect to B.

(c) Calculate the absorbance at 440 and 590 nm of a solution that is $8.00 \times 10^{-5} \, F$ with respect to A and $8.00 \times 10^{-5} \, F$ with respect to B.

(d) Calculate the absorbance at 440 and 590 nm of a solution that is $2.00 \times 10^{-4} \, F$ with respect to A and $3.00 \times 10^{-5} \, F$ with respect to B.

3. Use data from Problem 2 as well as the information given below to calculate the concentration of A and B in the following solutions. Solutions:

	A_{440}	A_{590}
(a)	1.022	0.414
(b)	0.878	0.253
(c)	1.131	0.348

4. Employing the data from Problem 2, construct an absorption curve for a mixed solution that is

	conc. A (F)	conc. B (F)
(a)	4.20×10^{-4}	2.10×10^{-5}
(b)	1.00×10^{-4}	5.80×10^{-5}
(c)	2.00×10^{-4}	6.40×10^{-5}

5. A spectral study of a series of solutions containing the pure compounds X and Y produced the following data for 1.00-cm cells.

	Solution number			
	(1)	(2)	(3)	(4)
Concentration of X	2.00×10^{-4}	0	?	?
Concentration of Y	0	3.00×10^{-4}	?	?
A at 394 nm (λ_{max} for X)	0.973	0.084	0.776	0.812
A at 502 nm (isosbestic point)	0.364	0.546	–	0.602
A at 610 nm (λ_{max} for Y)	0.102	1.076	0.934	–

(The isosbestic point for two species, in this case 502 nm, is the wavelength at which their molar absorptivities are identical.)
(a) Calculate the concentrations of X and Y in solution (3).
(b) Calculate the absorbance of solution (3) at the isosbestic point.
(c) Calculate the concentrations of X and Y in solution (4).
(d) What would be the absorbance of solution (4) at 610 nm?

6. At ordinary temperatures the acid dissociation constant for the indicator HIn has a value of 5.40×10^{-7}. Absorbance data (1.00-cm cells) for 5.00×10^{-4} F solutions of the indicator in strongly acidic and strongly alkaline media are given below.

λ (nm)	Absorbance pH 1.00	pH 13.00	λ (nm)	Absorbance pH 1.00	pH 13.00
440	0.401	0.067	570	0.303	0.515
470	0.447	0.050	585	0.263	0.648
480	0.453	0.050	600	0.226	0.764
485	0.454	0.052	615	0.195	0.816
490	0.452	0.054	625	0.176	0.823
505	0.443	0.073	635	0.160	0.816
535	0.390	0.170	650	0.137	0.763
555	0.342	0.342	680	0.097	0.588

(a) Predict the color of the acid form of the indicator HIn.
(b) What color filter would be suitable for the photometric analysis of the indicator in a strongly alkaline medium?
(c) What wavelength would be suitable for the spectrophotometric analysis of the indicator in its acidic form?
(d) What would be the absorbance of a 1.00×10^{-4} F solution of the indicator in its alkaline form when measured at 590 nm in a 2.00-cm cell?
(e) At what wavelength would the absorbance of the indicator be independent of pH?

7. A solution that is 5.00×10^{-4} F with respect to the indicator in Problem 6 has an absorbance of 0.309 at 485 nm when measured in a 1.00-cm cell.
(a) What is the pH of the solution?
(b) What will be the absorbance of this solution at 555 nm?

8. Calculate the absorbance of a solution that is 4.00×10^{-3} F with respect to Na_2HPO_4, 2.5×10^{-2} F with respect to NaH_2PO_4, and 5.00×10^{-4} F with

respect to the indicator in Problem 6 at 440 and 680 nm, the measurements being performed in 1.00-cm cells.

9. A 25.00-ml portion of a solution containing a purified weak acid of unknown composition was found to require 44.60 ml of standard 0.1050 N NaOH for neutralization. This amount of base was then precisely added to a 50.00-ml portion of the acid. Sufficient indicator was then introduced to render the solution 5.00×10^{-4} F with respect to the dyestuff described in Problem 6. The absorbance of the resulting solution, measured in a 1.00-cm cell, was found to be 0.378 at 485 nm and 0.298 at 625 nm.
    (a) What was the pH of the solution?
    (b) Calculate K_a for the weak acid.

10. Construct absorption curves for solutions in which the total formal concentration of the indicator in Problem 6 is constant at 5.00×10^{-4} F, 1.00-cm cells are used, and
    (a) the ratio between HIn and In$^-$ is 3:8,
    (b) the solution is buffered to $[H^+] = 5.40 \times 10^{-7}$,
    (c) the solution is buffered to $[H^+] = 1.44 \times 10^{-6}$.

11. The acid-base indicator In behaves as a weak base in aqueous solution. An 8.00×10^{-5} F solution of the indicator was prepared and 25.0-ml portions were pipetted into 50.0-ml volumetric flasks. To one of these was added about 10 ml of 1 N NaOH; to the second, about 10 ml of 1 N HCl; and to the third, 20 ml of a concentrated buffer of pH 8.00. After diluting to the mark, the following data were obtained with a 1.00-cm cell:

Solution	A_{420}	A_{550}	A_{680}
Strong acid	0.773	0.363	0.028
Strong base	0.064	0.363	0.596
Buffered to pH 8.00	0.314	0.363	0.396

    (a) Calculate the basic ionization constant for the indicator.
    (b) What would be the absorbance at the three wavelengths for the same concentration of indicator in a solution of pH 7.00?
    (c) A small quantity of the indicator solution was introduced into a solution of unknown pH. What was the pH of the solution if
    $A_{420} = 0.322$ and $A_{550} = 0.420$?

12. If a series of measured dilutions of the original unbuffered indicator solution in Problem 11 were carried out, would the absorbances at any of the three wavelengths be proportional to total indicator concentration? Explain your answer.

13. A 0.600-g sample of an alloy was dissolved in acid, treated with an excess of periodate, and heated to oxidize any manganese present to the 7+ state

$$5IO_4^- + 2Mn^{2+} + 3H_2O \rightarrow 2MnO_4^- + 5IO_3^- + 6H^+$$

The resulting solution was diluted to 250 ml in a volumetric flask. Color match between the diluted solution and an 8.25×10^{-4} F KMnO$_4$ solution was achieved when the light path through the standard was 6.50 cm and the light path through the test solution was 5.20 cm. Calculate the percentage of manganese in the alloy.

14. The iron content of a sample of well water was determined by treating a 25.0-ml sample with nitric acid and an excess of KSCN; the solution was then diluted to 50.0 ml. A 10.0-ml portion of a 7.30×10^{-4} F solution of Fe^{3+} was treated in an identical fashion. A color match between the two solutions was observed when the light path through the standard was 2.62 cm and the light path through the sample was 1.98 cm. Calculate the milligrams of iron per liter of sample.

15. Combination of solutions containing the ligand L and the cation M^{2+} results in formation of a complex with an absorption maximum at 520 nm. A series of solutions is prepared in which the formal concentration of M^{2+} is held constant at 1.35×10^{-4} F while the concentration of L is varied. Absorbance data for these solutions are as follows:

Concentration of L, fw/liter	Absorbance, 520 nm (1.00-cm cells)
3.00×10^{-5}	0.068
7.00×10^{-5}	0.163
1.20×10^{-4}	0.274
2.00×10^{-4}	0.419
3.00×10^{-4}	0.514
4.00×10^{-4}	0.573
5.00×10^{-4}	0.608
6.00×10^{-4}	0.626
7.00×10^{-4}	0.630
8.00×10^{-4}	0.630

(a) Evaluate the composition of the complex.

(b) Calculate a value for the formation constant for the complex.

16. The absorption spectrum for the chelate CuX_2^{2+} shows a maximum at 575 nm. Provided the formal ligand concentration exceeds that of the cation by a factor of 20 or more, absorbance at this wavelength is dependent only upon the formal concentration of Cu^{2+}. A solution that is 3.10×10^{-5} F with respect to Cu^{2+} and 2.00×10^{-3} F with respect to X has an absorbance of 0.675. Another solution, which is 5.00×10^{-5} F with respect to Cu^{2+} and 6.00×10^{-4} F with respect to X, has an absorbance of 0.366. Use this information to evaluate the equilibrium constant for the process

$$Cu^{2+} + 2X \rightleftharpoons CuX_2^{2+}$$

17. O. Menis, D. Manning, and G. Goldstein [*Anal. Chem.,* **29,** 1426 (1957)] report that thorium forms a yellow complex with quercetin (3, 3', 4', 5, 7-pentahydroxyflavone) with an absorption peak at 422 nm. Evaluate the composition of the complex on the basis of the following absorbance data obtained in 1.00-cm cells.

With Thorium Concentration Constant at 5.7×10^{-5} F		With Quercetin Concentration Constant at 5.7×10^{-5} F	
Concentration of quercetin, F	A_{422}	Concentration of Th^{4+}, F	A_{422}
6.0×10^{-6}	0.101	4.0×10^{-6}	0.134
1.1×10^{-5}	0.185	9.0×10^{-6}	0.302
1.5×10^{-5}	0.253	1.6×10^{-5}	0.537
2.0×10^{-5}	0.338	2.0×10^{-5}	0.675
2.5×10^{-5}	0.422	2.3×10^{-5}	0.778

18. The method of continuous variations was used to investigate the species responsible for an absorption peak at 480 nm when solutions of iron(III) and thiocyanate ion were mixed. The accompanying data were obtained upon mixing the indicated volumes of 1.90×10^{-3} F Fe^{3+} with sufficient 1.90×10^{-3} F KSCN to give a total volume of 20.0 ml; both stock solutions were 0.20 F with respect to HNO_3.

Volume of Fe^{3+} solution taken, ml	Absorbance, A_{480}	Volume of Fe^{3+} solution taken, ml	Absorbance, A_{480}
0.00	0.000	12.00	0.493
2.00	0.183	14.00	0.435
4.00	0.340	16.00	0.336
6.00	0.440	18.00	0.185
8.00	0.501	20.00	0.002
10.00	0.525	—	—

Elucidate the composition of the complex.

19. Manganese (II) is found to react with Q^- to give a colored complex. In order to characterize this complex a series of solutions was prepared in which the concentration of Mn^{2+} was 2.00×10^{-4} F and in which the concentration of Q^- was varied. The following data were obtained at 525 nm with a 1.00-cm cell.

Conc. NaQ, F	A_{525}	Conc. NaQ, F	A_{525}
0.250×10^{-4}	0.055	2.50×10^{-4}	0.449
0.500×10^{-4}	0.112	3.00×10^{-4}	0.463
0.750×10^{-4}	0.162	3.50×10^{-4}	0.472
1.00×10^{-4}	0.216	4.00×10^{-4}	0.468
2.00×10^{-4}	0.372	4.50×10^{-4}	0.470

(a) Plot the data.
(b) What is the molar absorptivity of the complex at 525 nm?
(c) What is the formula of the complex?
(d) Calculate the formation constant for the complex.

20. Within the limits of detection the zinc ion can be considered to be totally bound as ZnQ_2^{2-} when the formal concentration of the chelating agent exceeds

that of the cation by a factor of 40 or more. A solution in which the formal concentrations of Zn^{2+} and Q^{2-} were 8.00×10^{-4} and 4.00×10^{-2}, respectively, was found to have an absorbance of 0.364 when measured in 1.00-cm cells at 345 nm. Another solution in which the formal concentrations were again 8.00×10^{-4} for Zn^{2+}, and now 2.10×10^{-3} for Q^{2-}, had an absorbance of 0.273 when measured under the same conditions.

Calculate a value for the equilibrium constant associated with the process

$$Zn^{2+} + 2Q^{2-} \rightleftharpoons ZnQ_2^{2-}$$

21. In the presence of large excess of the complexing agent X^-, a $1.3 \times 10^{-4} \, F$ solution of nickel has an absorbance of 0.460 when measured at 385 nm in a 1.00-cm cell. Estimate the range of nickel concentrations over which this particular set of experimental conditions could be expected to provide concentration data with a relative error of less than 2 percent if the photometric uncertainty ΔT is 0.005.

22. Calculate the relative error in concentration associated with a photometric error ΔT of 0.0040 for solutions in which
    (a) $T = 0.204$,
    (b) $A = 0.195$,
    (c) $A = 0.280$,
    (d) $T = 94.4$ percent.

23. A $1.46 \times 10^{-4} \, F \, KMnO_4$ solution has an absorbance of 0.857 at 525 nm. After the instrument is adjusted to give a full-scale indication with this solution in the light path, the measured absorbance for an unknown permanganate solution is found to be 0.203. What is the concentration of $KMnO_4$ in the unknown solution?

24. A cadmium complex with a molar absorptivity of 2.00×10^4 was to be used for the analysis of a series of aqueous solutions having cadmium concentrations in the range of 0.50×10^{-4} to $1.00 \times 10^{-4} \, F$. The spectrophotometer was equipped with 1.00-cm cells only and was known to have an instrumental uncertainty of 0.004 transmittance; this uncertainty was independent of T.
    (a) What would be the expected range of absorbances and of transmittances?
    (b) Calculate the potential relative error in the analyses arising from the instrumental error for a sample that was $0.50 \times 10^{-4} \, F$ in Cd^{2+}; $1.5 \times 10^{-4} \, F$ in Cd^{2+}.
    (c) Repeat the calculations in (b) for solutions in which one volume of the sample had been diluted to exactly 5 volumes.
    (d) Repeat the calculations in (a) and (b) when the full-scale (100-percent T) reading of the instrument is obtained with a $0.45 \times 10^{-4} \, F$ solution of Cd^{2+}.

25. A routine method for the determination of iron in ground water was based upon the formation of an iron(III) complex having a molar absorptivity of 3.00×10^4. The spectrophotometer to be employed was equipped with 1.00-cm cells and was known to have an instrumental uncertainty of 0.0060 in transmittance; the uncertainty was found to be independent of T.

    The method and instrument were to be used for the analysis of samples from a region where the water was known to be 5×10^{-5} to $10 \times 10^{-5} \, F$ in

Fe(III). The procedure was such that the iron concentration in the solution to be measured would be exactly one-half of the original sample concentration.

(a) Calculate the expected absorbance and transmittance ranges of the samples.

(b) Calculate the potential relative error due to the instrument for samples that were originally $5.0 \times 10^{-5} F$ and $10.0 \times 10^{-5} F$ in Fe(III).

(c) Repeat the calculations in (b) for a modification of the method in which one volume of the sample is diluted to exactly four volumes (rather than two) before the photometric measurement.

(d) Repeat the calculations in (a) and (b) for another modification of the method in which a standard Fe(III) solution having a concentration of $4.8 \times 10^{-5} F$ was carried through the procedure (including the dilution) and the final solution, therefrom used to set the full-scale reading of the spectrophotometer.

26. A cadmium complex with a molar absorptivity of 2.00×10^4 was to be used for the analysis of Cd^{2+} in a series of solutions in which the cadmium concentration was in the range of 5.0×10^{-7} to $25 \times 10^{-7} F$. The spectrophotometer was equipped with 1.00-cm cells and was known to have an instrumental uncertainty of 0.0040 transmittance; this uncertainty was independent of T.

(a) What would be the expected absorbance and transmittance ranges?

(b) Calculate the potential relative error arising from the instrumental error for samples that were 5.0×10^{-7} and $25 \times 10^{-7} F$ in Cd^{2+}.

(c) Repeat the calculations in (a) and (b) when the zero setting of the instrument is made with a $30 \times 10^{-7} F$ solution of Cd^{2+}.

27. The method and instrument described in Problem 25 was to be used for a set of water samples in which the Fe(III) concentration was in the range of 1.0×10^{-6} to $5.0 \times 10^{-6} F$.

(a) Calculate the expected absorbance and transmittance ranges of the samples. (Remember that the samples have been diluted so that the Fe(III) concentration is half of the above.)

(b) Calculate the potential relative error due to the instrument for a sample originally $1.0 \times 10^{-6} F$ in Fe(III). Calculate the potential error for a solution that was $5.0 \times 10^{-6} F$.

(c) Repeat the calculations in (a) and (b) for a modification of the method in which a standard Fe(III) solution having a concentration of $5.6 \times 10^{-6} F$ was carried through the procedure (including the dilution) and used to set the zero reading of the spectrophotometer.

5 Atomic Absorption Spectroscopy

Atomic absorption spectroscopy involves the study of the absorption of radiant energy (usually in the ultraviolet and visible regions) by neutral atoms in the gaseous state. The principles of atomic absorption are basically the same as those already considered for the absorption of ultraviolet and visible radiation by solutions. The sample-handling techniques, the equipment, and the appearance of the spectra differ sufficiently, however, to warrant the treatment of atomic absorption as a separate topic.

The potential usefulness of atomic absorption for the analysis of metallic elements was first suggested in 1955 by Walsh and by Alkemade and Milatz.[1] Since that time, methods for the determination of some 65 elements have been developed, and numerous commercial instruments designed specifically for this type of analysis have become available.

PRINCIPLES OF ATOMIC ABSORPTION

In an atomic absorption analysis the element being determined must be reduced to the elemental state, vaporized, and imposed in the beam of radiation from the source. This process is most frequently accomplished by drawing a soluton of the sample, as a fine mist, into a suitable flame. The flame thus serves a function analogous to that of the cell and solution in conventional absorption spectroscopy (Chapters 3 and 4).

[1] A. Walsh, *Spectrochim. Acta,* **7**, 108 (1955); C. T. J. Alkemade and J. M. W. Milatz, *Appl. Sci. Research,* **B4**, 289 (1955); *J. Opt. Soc. Amer.,* **45**, 583 (1955).

Atomic Absorption Spectra

The absorption spectrum of an element in its gaseous, atomic form consists of a series of well-defined, narrow lines arising from electronic transitions of the outermost electrons. For metals the energies of many of these transitions correspond to wavelengths in the ultraviolet and visible regions. The energy-level diagram for the outer electrons of an element provides a convenient means of showing the types of transitions responsible for atomic absorption.

Energy level diagrams. The energy-level diagram for sodium shown in Figure 5-1 is typical. Note that the energy scale is linear in units of wave numbers (cm^{-1}) with the 3s-orbital being assigned a value of zero. The scale extends to 41,499 cm^{-1}, the energy necessary to remove the single 3s electron from the influence of the central atom, thus producing a sodium ion.

The energies of several atomic orbitals are indicated on the diagram by horizontal lines. Note that the p-orbitals are split into two levels that differ but slightly in energy. This difference is rationalized by assuming that an electron spins about its own axis, and that the direction of this motion may either be the same as or opposed to its orbital motion. Both the spin and the

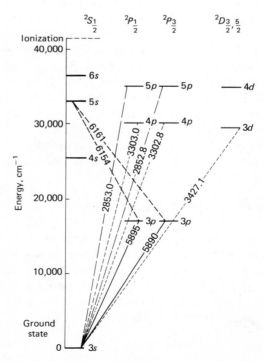

Figure 5-1. Partial energy level diagram for sodium.

orbital motions create magnetic fields owing to the rotation of the charge carried by the electron. The two fields interact in an attractive sense if these two motions are in the opposite direction; a repulsive force is generated when the motions are parallel. As a consequence, the energy of the electron whose spin opposes its orbital motion is slightly smaller than one in which the motions are alike. Similar differences exist in the *d*- and *f*-orbitals, but their magnitudes are ordinarily so slight as to be undetectable; thus, only a single energy level is indicated for *d*-orbitals in Figure 5-1.

Flame emission. At room temperature essentially all of the atoms of a sample of matter are in the ground state. Thus, the single outer electron of sodium metal occupies the 3*s*-orbital under these circumstances. Excitation of this electron to higher orbitals can be brought about by the heat of a flame. The lifetime of the excited atom is brief ($< 10^{-9}$ sec), however, and its return to the ground state is accompanied by the emission of a quantum of radiation. The vertical lines in Figure 5-1 indicate some of the common electronic transitions that follow flame excitation; the wavelength of the resulting radiation is also indicated.

The fraction of atoms excited by heat to a particular energy level is given by the Boltzmann equation,

$$\frac{N_j}{N_0} = \frac{P_j}{P_0} \exp\left(-\frac{E_j}{kT}\right) \tag{5-1}$$

where k is the Boltzmann constant (1.38×10^{-16} erg/deg), T is temperature in degrees Kelvin, and E_j is the energy difference in ergs between the excited state and the ground state. The quantities N_j and N_0 refer to the number of atoms in the excited state and the ground state, respectively, while P_j and P_0 are statistical factors that are determined by the number of states having equal energy at each quantum level. As shown by the following example, the fraction of sodium atoms excited to the 3*p* states in a typical gas flame ($T = 2500°K$) is very small.

Example. Calculate the ratio of number of sodium atoms in the 3*p* excited states to the number in the ground state at 2500°K.

In order to calculate E_j in equation (5-1) we employ an average wavelength of 5892 Å for the two sodium emission lines involving the 3*p* to 3*s* transitions.

$$\text{Wave number} = \frac{1}{5892\text{Å} \times 10^{-8} \text{ cm/Å}}$$
$$= 1.698 \times 10^4 \text{ cm}^{-1}$$
$$E_j = 1.698 \times 10^4 \text{ cm}^{-1} \times 1.986 \times 10^{-16} \text{ erg/cm}^{-1}$$
$$= 3.372 \times 10^{-12} \text{ erg}$$

There are two quantum states in the 3*s* level and six in the 3*p* level. Thus,

$$\frac{P_j}{P_0} = \frac{6}{2} = 3$$

Substituting into equation (5-1),

$$\frac{N_j}{N_0} = 3 \exp\left(-\frac{3.372 \times 10^{-12}}{1.380 \times 10^{-16} \times 2500}\right)$$

which can be rewritten as

$$\log_{10}\frac{N_j}{3N_0} = -\frac{3.372 \times 10^{-12}}{2.303 \times 1.380 \times 10^{-16} \times 2500}$$

or

$$\frac{N_j}{N_0} = 1.7 \times 10^{-4}$$

Atomic absorption in flames. The foregoing example illustrates that sodium atoms are predominately in the ground state at ordinary flame temperatures; this situation obtains for other metals as well.

In a flame medium, sodium atoms are capable of *absorbing* radiation of a wavelength that is characteristic of an electronic transition from the $3s$ state to one of the higher excited states. For example, sharp absorption peaks at 5890, 5895, 3302.8, and 3303.0 Å are observed experimentally; referring again to Figure 5-1, it is apparent that each adjacent pair of these peaks corresponds to transitions from the $3s$ level to the $3p$ and the $4p$ levels, respectively. Note that absorption due to the $3p$ to $5s$ transition is so weak as to go undetected because the number of sodium atoms in the $3p$ state is so very small in a flame. Thus typically, an atomic absorption spectrum from a flame consists predominately of *resonance lines*, which are the result of transitions from the ground state to upper levels.

Line widths. Atomic absorption peaks are much narrower than those observed for ions or molecules in solution. The natural width of an atomic absorption line can be shown to be about 10^{-4} Å. Two effects, however, tend to broaden the line to an observed dimension ranging between 0.02 and 0.05 Å. *Doppler broadening* arises from the rapid motion of the absorbing particles with respect to the source. For those atoms traveling toward the source, the radiation is effectively decreased in wavelength by the well-known Doppler effect; thus, somewhat longer wavelengths are absorbed. The reverse is true of atoms moving away from the source. *Pressure broadening* also occurs; here collisions among atoms cause small changes in the ground-state energy levels and a consequent broadening of peaks.

Relationship between Atomic Absorption and Flame Emission Spectroscopy

It is of interest to point out the fundamental difference between atomic absorption spectroscopy and flame emission spectroscopy, which is considered in Chapter 11. Superficially, the two methods resemble one another in that both are based upon the events that occur when a sample is sprayed into a flame. In flame photometry, however, it is the radiation emitted by the *excited* atoms that is related to concentration, while in atomic absorption it is the radiation absorbed by the *unexcited* atoms that is determined. Through use of equation (5-1) we have seen that the fraction of excited atoms is relatively small in a flame and is exponentially related to the temperature. Thus, while temperature variations have a profound effect on the number of excited atoms, the influence of this variable on the much larger number of unexcited atoms is negligible. Since atomic absorption depends only upon the number of unexcited atoms, the *absorption intensity* is not directly affected by the temperature of the flame; the *emission intensity,* in contrast, being dependent upon the number of excited atoms, is greatly influenced by temperature variations.

Absorption measurements are influenced indirectly by temperature fluctuations, however. The total number of atoms produced from the sample and available for absorption ordinarily increases with temperature. In addition, line broadening and a consequent decrease in peak height occurs at higher temperatures because the atomic particles travel at greater rates and enhance the Doppler effect. High concentrations of gaseous atoms also cause pressure broadening of the absorption lines. As a consequence of these indirect effects, a reasonable control of the flame temperature is required for quantitative atomic absorption measurements.

Measurement of Atomic Absorption

Because atomic absorption lines are so very narrow and because transition energies are unique for each element, analytical methods based on this type of absorption are potentially highly specific. On the other hand, the limited line widths create a measurement problem not encountered in solution absorption. Recall that although Beer's law applies only for monochromatic radiation, a linear relationship between absorbance and concentration can be expected if the band width is narrow with respect to the width of the absorption peak (p. 35). No ordinary monochromator is capable of yielding a band of radiation that is as narrow as the peak width of an atomic absorption line (0.02 to 0.05 Å). Thus, when a continuous radiation source is employed with a monochromator, only a fraction of the emerging radiation is of a wavelength that is absorbed, and the relative change in intensity of the emergent band is small in comparison to the change that occurs to the radiation corresponding to the absorption peak. Under these conditions Beer's law is not followed; in addition, the sensitivity of the method is greatly lowered.

Walsh overcame this problem in an ingenious way. He employed a source of radiation that emits a line of the same wavelength as that to be used for the absorption analysis. For example, if the 5890 Å absorption line of sodium (Figure 5-1) is chosen for the analysis of that element, a sodium vapor lamp can be employed as a source. In such a lamp, gaseous sodium atoms are excited by electrical discharge; the excited atoms then emit characteristic radiation as they return to lower energy levels. A part of the emitted radiation will have exactly the same wavelengths as the resonance absorption lines. Thus, for example, sodium atoms that have been excited to $3p$ states will emit 5890 and 5895 Å lines (Figure 5-1). Return from the $5s$ state may give two lines in the vicinity of 6160 Å corresponding to the $5s$ to $3p$ transitions; also observed will be radiation resulting from subsequent $3p$ to $3s$ transitions. With a properly designed source, the emission lines will have band widths that are significantly less than the absorption band widths. Thus, the monochromator need only have the capability of isolating a suitable emission line for the absorption measurement (see Figure 5-2). The radiation employed in the analysis is then sufficiently limited in wavelength span to permit absorbance measurements at the absorption peak. Greater sensitivity and better adherence to Beer's law results.

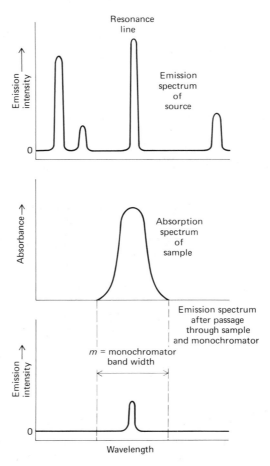

Figure 5-2. Absorption of a resonance line by atoms.

The main disadvantage of this technique is the need for a separate lamp source for each element being analyzed. In order to avoid this inconvenience, attempts have been made to employ a continuous source with a very high resolution monochromator, or alternatively to produce a line source by introducing a compound of the element to be determined in a high-temperature flame. Neither of these techniques is as satisfactory as the employment of a specific lamp for each element.

INSTRUMENTS

An instrument for atomic absorption measurements has the same basic components as a spectrophotometer for measuring the absorption of solutions (see Figure 3-5). These include a source, a monochromator, a sample container (in this case, a flame), a detector, and an amplifier-indicator. The major instrumental differences between atomic and solution absorption equipment are found in the source and in the sample container; the features of these components are discussed in detail.

Both single-beam and double-beam instruments have been designed for atomic absorption studies. The inherent advantages and disadvantages of these designs are the same as for the spectrophotometers described in Chapter 3.

Radiation Sources

Sources for atomic absorption spectrophotometers are ordinarily *hollow cathode lamps* or *gaseous discharge tubes*.

Beam modulation. The output from the radiation source, regardless of its type, must be modulated in order to eliminate interference from the flame that holds the sample. This flame emits a more-or-less continuous spectrum resulting from molecular excitation of the fuel molecules; in addition, it may contain a line spectrum attributable to excitation of the metal atoms from the sample. We have noted that the fraction of atoms undergoing thermal excitation is small at flame temperatures; nevertheless, the few atoms that are excited emit radiation corresponding to the resonant absorption line selected for analysis, and thus represent a serious source of interference.

In order to remove most of the radiation of the flame the monochromator is always located between the flame and the detector; this arrangement differs from that encountered in most spectrophotometers. Clearly, however, the monochromator transmits the emission line corresponding to the wavelength of the absorption peak. Thus, the observed power P will not be $P = P_0 - P_a$, where P_a is the power of absorbed radiation; rather, it will be $P = P_0 - P_a + P_e$, where P_e consists of the power of the line emitted in the flame plus the corresponding background power.

The effects of the emission from the flame are twofold. First, the sensitivity of the method is decreased. Second, the emitted power P_e, being dependent in part upon the number of excited particles, shows greater variation with flame temperature than does the absorbed power P_a. Therefore, the measured absorbance becomes increasingly temperature-dependent.

These difficulties are eliminated by modulation of the source. Here, the intensity of the beam is caused to fluctuate at a constant frequency. The detector then receives two types of signal: an alternating one from the source and a continuous one from the flame. These signals are converted to the corresponding types of electrical signals. A relatively simple electronic system is then employed to respond to and amplify the ac part of the signal and to ignore the unmodulated dc signal.

A simple and entirely satisfactory way of modulating the radiation source is to interpose a circular disk in the beam between the source and the flame; alternate quadrants of this disk are removed to permit passage of light. Rotation of the disk at constant speed provides an intermittent beam that is chopped to the desired frequency by control of the motor speed. As an alternative, intermittent or ac operation of the source can be provided in the design of the power supply.

Hollow cathode tube. The hollow cathode tube is the most common radiation source for atomic absorption spectroscopy. Figure 5-3 is a schematic diagram of a typical source unit. The tube, which is constructed of thick-walled glass, has a transparent window affixed to one end. Two tungsten wires are sealed into the other end of the tube. One of these serves as the anode. To the end of the other wire is attached a hollow metal cylinder 10 to 20 mm in diameter; this cylinder acts as the cathode. The cylinder is constructed of the metal whose spectrum is desired or serves to support a layer of this metal. The tube is filled with pure helium or argon at 1 to 2 mm pressure.

Ionization of the gas occurs when a potential is applied across the electrodes, and a current flows as a result of movement of the ions to the electrodes. If the potential is sufficiently large, the gaseous cations acquire sufficient kinetic energy to dislodge some of the metal atoms from the cathode surface and produce an atomic cloud; this process is called *sput-*

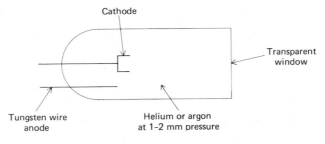

Figure 5-3. Schematic diagram of the hollow cathode tube.

tering. A portion of the sputtered metal atoms are in excited states and thus emit their characteristic radiation in the usual way. As the process goes on, the metal atoms diffuse back to the cathode surface or to the glass walls of the tube and are redeposited.

The cylindrical configuration of the cathode tends to concentrate the radiation in a limited region of the tube; this design also enhances the probability that redeposition will occur at the cathode rather than on the glass walls.

The efficiency of the hollow cathode tube is dependent upon its geometry and the operating potential. High potentials, and thus high currents, lead to greater intensities. This advantage is offset somewhat by an increase in Doppler broadening of the emission lines. Furthermore, the greater currents result in an increase in the number of unexcited atoms in the cloud; the unexcited atoms, in turn, are capable of absorbing the radiation emitted by the excited ones. This *self-absorption* leads to lowered intensities, particularly at the center of the emission band.

A variety of hollow cathode tubes is available commercially. The cathodes of some consist of a mixture of several metals; such lamps permit the analysis of more than a single element.

Gaseous discharge tubes. Gas discharge tubes produce a line spectrum as a consequence of the passage of an electrical current through a vapor of metal atoms; the familiar sodium and mercury vapor lamps are examples. Sources of this kind are particularly useful for producing spectra of the alkali metals.

Devices for Formation of an Atomic Vapor

In an atomic absorption analysis the elements in the sample must be reduced to neutral atomic particles, vaporized, and dispersed in the radiation beam in such a manner that their numbers are reproducibly related to their concentrations in the sample. This process ordinarily is the least efficient part of the method and is responsible for the largest analytical uncertainties.

A number of devices for forming atomic vapors have been investigated; included among these are (1) ovens, in which the sample is brought rapidly to a high temperature; (2) electric arcs and sparks, in which either solid or liquid samples are subjected to a high current or to a high-potential, alternating current spark; (3) sputtering devices, in which the sample, held on a cathode, is bombarded by the positive ions of a gas; and (4) flame atomization, in which the sample is sprayed into a gas flame. To date, flame atomization has proved to be the most practical technique and is the one employed in all commercial instruments.

Flame profiles. In the typical flame atomizer all or part of a solution of the sample is sprayed, as a fine mist, into a flame that is located in the path of the radiation from the source.

Important regions of the flame, from bottom to top, include the base, the inner cone, the reaction zone, and the outer mantle. The sample enters the base of the flame in the form of minute droplets. Within this region water evaporates from a substantial fraction of the droplets; some of the sample thus enters the inner cone in the form of solid particles. Here, vaporization and decomposition to the atomic state occurs; it is here, also, that the excitation and absorption processes commence. Upon entering the reaction zone the atoms are converted to oxides; these then pass into the outer mantle and are subsequently ejected from the flame. This sequence is not necessarily suffered by every drop that is aspirated into the flame; indeed, depending upon the droplet size and the flow rate, a portion of the sample may pass essentially unaltered through the flame.

The region of the flame in which the maximum emission or absorption occurs depends upon such variables as droplet size, the type of flame used, the ratio of oxidant to fuel, and the tendency for the species to enter into oxide formation. *Flame profiles* relate the absorption or emission signal-intensity to height above the burner tip. Typical flame profiles for three elements are shown in Figure 5-4. Magnesium exhibits a maximum in absorbance at about the middle of the flame because of two opposing effects. The initial increase in absorbance as the distance from the tip becomes larger results from an increased number of atomic magnesium particles produced with longer exposure to the heat of the flame. In the reaction zone, however, appreciable oxidation of the magnesium starts to occur. This process leads to an eventual decrease in absorbance because the oxide particles formed are nonabsorbing at the wavelengths used. To obtain maximum analytical sensitivity, then, the flame must be adjusted with respect to the beam until a maximum absorbance reading is obtained.

The behavior of silver, which is not readily oxidized, is quite different; here a continuous increase in atoms is observed from the base to the periphery of the flame. In contrast, chromium, which forms very stable oxides, shows a continuous decrease in absorbance beginning at the base;

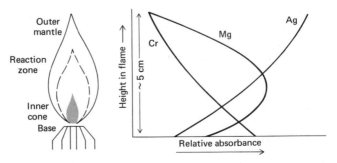

Figure 5-4. Flame absorption profiles for three elements. Reprinted from J. W. Robinson, *Atomic Absorption Spectroscopy,* p. 69, by courtesy of Marcel Dekker, Inc.

this observation suggests that oxide formation predominates from the start. Clearly, a different portion of the flame should be used for the analysis of each of these elements.

Types of burners. Two types of burners are employed in atomic absorption spectroscopy. In a *total consumption* burner the sample solution, the fuel, and the oxidizing gas are carried through separate passages and meet at an opening at the base of the flame. In a *premix* burner the sample is aspirated into a large chamber by a stream of the oxidant; here, the fine mist of the sample, the oxidant, and the fuel supply are mixed and then forced to the burner opening. Larger drops of sample collect in the bottom of the chamber and are drained off. Figure 5-5 gives schematic diagrams of both burner designs.

Table 5-1 lists the advantages and disadvantages attributed to each type of burner. Clearly, both have problems associated with their use;

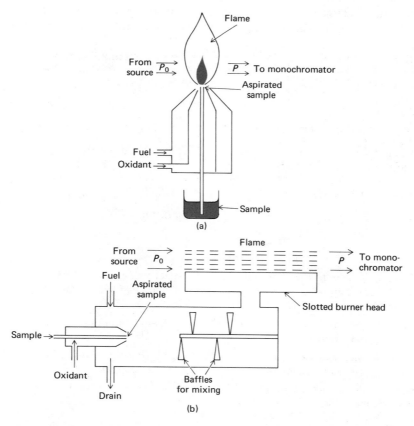

Figure 5-5. Types of burners for atomic absorption spectroscopy: (a) total consumption burner, (b) premix burner.

further effort will undoubtedly be expended toward developing better techniques for conversion of the sample to an atomic form.

Oxidants and fuels. Fuels used for flame production include natural gas, propane, butane, hydrogen, and acetylene; the last is perhaps most widely employed. The common oxidants are air, oxygen-enriched air, oxygen, and nitrous oxide. The nitrous oxide–acetylene mixture is advantageous when a hot flame is required because of its lower explosion hazard.

For elements that are readily converted to the atomic state, such as copper, lead, zinc, and cadmium, low-temperature flames (natural gas–air, for example) are used to advantage. Many elements (the alkaline earths, for example), on the other hand, form refractory oxides which require somewhat higher temperatures for decomposition; for these, an acetylene-air mixture often produces the most sensitive results. Elements such as aluminum, beryllium, and the rare earths form unusually stable oxides; a reasonable concentration of their atoms can be obtained only at the high temperatures developed in an oxygen-acetylene or a nitrous oxide-acetylene flame.

With increasing flame temperatures some ionization of the atoms can

Table 5-1 Comparison of Total Consumption and Premix Burners

Total Consumption	*Premix*
Advantages	
1. Representative sample reaches flame.	1. Long path provides sensitivity.
2. Amount of sample in flame is large, leading to enhanced sensitivity.	2. Only small drops reach flame; better reproducibility results.
3. No explosive hazard.	3. Quiet operation.
	4. Little tendency to clog.
Disadvantages	
1. Short path leads to lower sensitivity (can be partially offset by arranging several burners in a row).	1. Selective evaporation of mixed solvents can lead to analytical errors.
2. Nonuniform droplet size may cause poor reproducibility.	2. Rate of sample introduction low.
3. Large droplets not entirely decomposed; may cause clogging.	3. Possibility of explosion in mixing chamber.
4. Rate of sample introduction strongly dependent on viscosity.	4. Rate of sample introduction moderately dependent upon viscosity.
5. Noisy.	

be expected. Sufficient ionization of the alkali and some of the alkaline earth metals occurs in the hotter flames to lower significantly the number of absorbing particles. For the analysis of such elements, a lower-temperature flame is desirable.

The ratio of fuel to oxidant also influences the extent of atom formation in a flame. The effects are complex; the best mixture must be determined empirically. The flow rates of fuel and oxidant must be controlled in order to obtain reproducible results.

Monochromators or Filters

An instrument for atomic absorption analysis must be capable of providing a sufficiently narrow band width to separate the line chosen for the measurement from other lines that may interfere with or reduce the sensitivity of the analysis. For some of the alkali metals, which have only a few widely spaced resonance lines in the visible region, a glass filter suffices. An instrument employing readily interchangeable interference filters is available commercially; a separate filter (and light source) is used for each element; satisfactory results for the analysis of 22 metals is claimed. Most instruments, however, incorporate a good quality, ultraviolet and visible monochromator; several of these can be purchased from instrument manufacturers.[2]

Detectors and Indicators

The detector-indicator components for an atomic absorption spectrophotometer are fundamentally the same as for the typical ultraviolet-visible solution spectrophotometer (see Chapter 3). Generally, photomultiplier tubes (p. 54) are employed for conversion of the radiant energy signal to an electrical one. As we have pointed out, the electronic system is capable of discriminating between the modulated signal from the source and the continuous signal from the flame. Both null-point and direct-reading meters are used; these are calibrated in terms of absorbance or transmittance.

APPLICATIONS

Atomic absorption spectroscopy provides a sensitive means for the determination of more than 60 elements. Table 5-2 lists the detection limit

---

[2]For specifications see W. T. Elwell and J. A. F. Gidley, *Atomic-Absorption Spectrophotometry* (New York: Pergamon Press, 1966); R. G. Martinek, *Laboratory Management,* **6,** 24 (1968); J. Ramirez-Muñoz, *Atomic-Absorption Spectroscopy* (New York: Elsevier Publishing Company, 1968), pp. 182-202.

and the wavelength of the most sensitive absorption peaks for some of the more common elements. Details concerning the quantitative determination of these and other elements are found in several publications.[3]

Interferences

The absorption spectrum of each element is unique to that species; with a reasonably good monochromator it is possible to find an absorption peak for an element that is free from interference by other elements. Unfortunately, however, this specificity does not extend to the chemical reactions occurring in the flame; here, interferences do arise as a consequence of interactions and chemical competitions that affect the number of atoms present in the light path under a specified set of conditions. These effects are not predictable from theory and must therefore be determined by experiment.

Table 5-2 Some Elements Determined By Atomic Absorption Spectroscopy[a]

Element	Wavelength, $\mathring{A}$	Detection limit,[b] $\mu g/ml$
Aluminum	3093	1
Antimony	2176	0.5
Arsenic	1937	5
Barium	5536	8
Bismuth	2231	1
Cadmium	2288	0.03
Calcium	4227	0.08
Chromium	3579	0.05
Copper	3248	0.1
Iron	2483	0.1
Lead	2170	0.3
Magnesium	2852	0.01
Mercury	2536	10
Nickel	2320	0.13
Potassium	7665	0.03
Silicon	2516	3
Silver	3281	0.1
Sodium	5890	0.03
Tin	2863	5
Zinc	2139	0.03

[a]From W. Slavin, *Atomic Absorption Spectroscopy*, p. 61, New York, Interscience Publishers, 1968. With permission.

[b]Sensitivity is defined as the concentration in $\mu g/ml$ that absorbs 1 percent of P_0. This value corresponds to a transmittance of 99 percent or an absorbance of 0.004.

[3]W. T. Elwell and J. A. F. Gidley, *Atomic-Absorption Spectrophotometry* (New York: Pergamon Press, 1966); J. W. Robinson, *Atomic Absorption Spectroscopy* (New York: Marcel Dekker, Inc., 1966); J. Ramirez-Muñoz, *Atomic-Absorption Spectroscopy* (New York: Elsevier Publishing Company, 1968).

Cation interferences. A few examples have been found in which the absorption due to one cation is affected by the presence of a second. For example, the presence of aluminum is found to cause low results in the determination of magnesium; it has been shown that this interference results from the formation of a heat-stable, aluminum-magnesium compound and a consequent reduction in the concentration of magnesium atoms in the flame. Beryllium, aluminum, and magnesium are reported to have a similar effect on calcium analyses. Fortunately, this type of interference is relatively rare.

Anion interferences. The height of an absorption peak for a metal may be influenced by the type and the concentration of anions present in the sample solution. This effect is not surprising inasmuch as the energy required to form atomic species from compounds must vary with the strength of the attraction between anion and cation. Such effects ordinarily become smaller with increases in flame temperature, and may disappear entirely in some of the hotter flames.

Anion interference can sometimes be avoided by the addition of a complexing agent (EDTA, for example) to both the standards and the samples. In this way atom formation always results from decomposition of the complex, instead of from different compounds. Alternatively, the standards can be made to approximate the sample in anion composition to compensate for the anion effect.

Analytical Variables

We have already noted a number of variables that influence the measured absorbance. These include the rate of flow of the oxidant and the fuel gases, the position of the beam with respect to the flame, the kind of fuel and oxidant, and the nature of the anions present. Another important parameter that must be controlled is the rate of introduction of the sample into the flame. Usually there is an optimum rate that can be determined only by experiment. If the rate is too low, the number of atomic particles is small; on the other hand, an excessive rate causes an inordinate consumption of the flame energy by the evaporation process and leaves little for the formation of atoms.

The heights of atomic absorption peaks are also influenced by the viscosity of the solvent (primarily due to a change in the rate of sample introduction) and the presence of combustible organic solvents. Enhancement of absorption is frequently observed in the presence of such solvents as acetone or hydrocarbons.

Analytical Techniques

Measurement of absorbance. In a single-beam instrument the full-scale adjustment of the meter is made while pure water is aspirated into the flame (or alternatively with the flame removed from the light path). Samples or standards are then aspirated into the flame, and the percent transmittance or the absorbance is read directly from the scale. In double-beam instruments the power of the beam passing through the flame is automatically compared with the power of a beam that travels in a path out of the flame.

Calibration curves. While, in theory, absorbance should be proportional to concentration, deviations from linearity are often observed. Thus, empirical calibration curves must be employed. In addition, there are sufficient uncontrollable variables in the production of an atomic vapor to warrant the measurement of the absorbance of at least one standard solution each time the instrument is used. The deviation of the standard from the original calibration curve can then be employed to correct the analytical results.

Accuracy. Under usual conditions the relative error associated with an atomic absorption analysis is of the order of 1 to 2 percent. With special precautions this figure can be lowered to a few tenths of 1 percent.

6 Infrared Absorption Spectroscopy*

The infrared region of the spectrum encompasses radiation having wave numbers ranging from roughly 13,000 to 33 cm^{-1}, or wavelengths from 0.75 to 300 μ. The majority of applications of infrared absorption measurements, however, have been confined to the region extending from about 4000 to 667 cm^{-1} (2.5 to 15 μ).

Infrared spectrophotometry is most widely used for the identification of organic compounds because their spectra are generally complex and provide numerous maxima and minima that can be employed for comparison purposes. Indeed, the infrared absorption spectrum of an organic compound represents one of its truly unique physical properties. With the exception of optical isomers, no two compounds have identical absorption curves.

In addition to its application as a qualitative analytical tool, infrared spectrophotometry has been employed for quantitative analysis as well. Often, however, other procedures can be found that are more accurate or more convenient; as a consequence, the quantitative applications are of less significance than the qualitative ones.

*For detailed discussions of infrared spectroscopy see R. P. Bauman, *Absorption Spectroscopy* (New York: John Wiley & Sons, Inc., 1962); N. B. Colthup, L. N. Daly and S. E. Wiberley, *Introduction to Infra Red and Raman Spectroscopy* (New York: Academic Press, Inc., 1964); A. L. Smith, in I. M. Kolthoff and P. J. Elving, Eds., *Treatise on Analytical Chemistry,* Part 1, Vol. 6 (New York: Interscience Publishers, 1965), Chap. 66; and H. A. Szymanski, *Theory and Practice of Infrared Spectroscopy* (New York: Plenum Press, 1964).

THEORY OF INFRARED ABSORPTION

Introduction

A typical infrared spectrum is shown in Figure 6-1. In contrast to most ultraviolet and visible spectra, a bewildering array of maxima and minima are observed.

Spectral plots. As shown in Figure 6-1, infrared data are customarily plotted with percent transmittance rather than absorbance as the ordinate; in addition, many chemists prefer to use the unit of reciprocal centimeters (cm^{-1}) for the abscissa rather than wavelength. This preference is based upon the direct proportionality between the wave number and the energy as well as the frequency of the radiation; the frequency can, in turn, be directly related to molecular vibrational frequencies. When employed, wavelength is expressed in the convenient units of microns (μ).

Dipole changes during vibrations and rotations. Most electronic transitions require energies in the ultraviolet or visible regions; absorption of infrared radiation is thus confined largely to molecular species for which small energy differences exist between various vibrational and rotational states.

In order to absorb infrared radiation a molecule must undergo a net change in dipole moment as a consequence of its vibrational or rotational motion. Only under these circumstances can the alternating field of the radiation interact with the molecule and cause changes in its motion. For example, the charge distribution about a molecule such as carbon monoxide or nitric oxide is not symmetric, one atom having a greater electron density than the other. When the distance between the centers fluctuates, as it does in a vibration, an oscillating electrical field is established which can interact with the electric field associated with radiation. If the frequency of the radiation matches a natural vibrational frequency of the molecule, there occurs a net transfer of energy that results in a change in the amplitude of the molecular vibration; absorption of the radiation is the consequence. Similarly, the rotation of unsymmetric molecules around their centers of mass results in a periodic dipole fluctuation; again, interaction with radiation is possible. No net change in dipole moment occurs during the vibration or rotation of homonuclear species such as O_2, N_2, or Cl_2, and these compounds do not absorb in the infrared.

Rotational transitions. The energy required to cause a change in rotational level is very small and corresponds to radiation of 100 μ and greater (< 100 cm^{-1}). Because rotational levels are quantized, absorption by gases in this far-infrared region is characterized by discrete, well-defined lines. In liquids or solids, however, intramolecular collisions and interactions

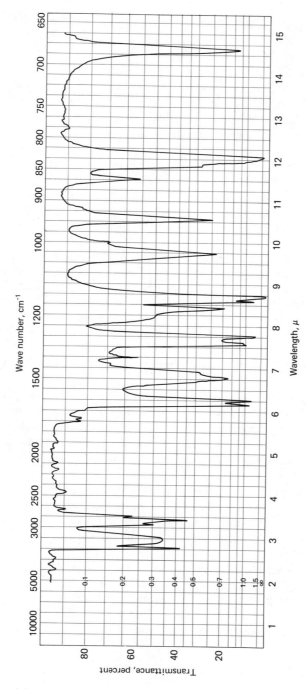

Figure 6-1. Infrared absorption spectrum of 3,5-dimethylphenol. (Spectrum courtesy of Thermodynamics Research Center Data Project, Texas A. & M. University, College Station, Texas.)

133

cause broadening of the lines into a continuum. The far-infrared region is experimentally difficult to study; pure rotational spectra are not considered further.

Vibrational-rotational transitions. Vibrational energy levels are also quantized, and the energy differences between quantum states correspond to the readily accessible regions of the infrared from about 13,000 to 675 cm^{-1} (0.75 to 15 μ). Because there are several rotational energy states for each vibrational state, the infrared spectrum of a gas usually consists of a series of closely spaced lines. In the liquid and solid state, on the other hand, rotation is highly restricted, and discrete vibrational–rotational lines disappear, leaving only somewhat broadened vibrational peaks. Our concern is primarily with the spectra of solutions and solids in which rotational effects are minimal.

Types of molecular vibrations. The relative positions of atoms in a molecule are not exactly fixed, but instead fluctuate continuously as a consequence of a multitude of different types of vibrations. In a simple diatomic or triatomic molecule it is easy to define the number and nature of such vibrations and relate these to energies of absorption. With polyatomic molecules, however, an analysis of this kind becomes difficult if not impossible, not only because of the large number of vibrating centers, but also because interactions among several centers occur and must be taken into account.

Vibrations fall into the basic categories of *stretching* and *bending*. A stretching vibration involves a continuous change in the interatomic distance along the axis of the bond between two atoms. Bending vibrations are characterized by a change in the angle between two bonds and are of four types: *scissoring, rocking, wagging,* and *twisting.* The various types of vibrations are shown schematically in Figure 6-2.

In a molecule containing more than two atoms, all of the vibration types shown in Figure 6-2 may be possible. In addition, interaction or *coupling* of vibrations can occur if the vibrations involve bonds to a single central atom. The result of coupling is a change in the characteristics of the vibrations involved.

In the treatment that follows, we first consider isolated vibrations by employing a simple mechanical model called the *harmonic oscillator.* Modifications of the theory of the mechanical oscillator needed to describe a molecular system are taken up next. Finally, the effects of vibrational interactions in molecular systems are considered.

Mechanical Model of Stretching Vibrations

The characteristics of an atomic stretching vibration can be approximated by a mechanical model consisting of two masses connected by a spring. A disturbance of one of these masses along the axis of the spring results in a vibration called a *simple harmonic motion*.

For simplicity, let us first consider the vibration of a mass attached to a spring that is hung from an immovable object. If the mass is displaced a distance x from its equilibrium position by application of a force along the axis of the spring, the restoring force, according to Hooke's law, is proportional to the displacement. That is,

$$F = -kx \qquad (6\text{-}1)$$

where F is the restoring force and k is the *force constant* that depends upon the stiffness of the spring. The negative sign indicates that F is a restoring force.

Potential energy of a harmonic oscillator. The potential energy of the mass and spring can be considered to be zero when the mass is in its rest or equilibrium position. As the spring is compressed or stretched, however, the potential energy of the system increases by an amount equal to the work required to displace the mass. If, for example, the mass is moved from some position x to $(x + dx)$, the work and hence the change in potential energy E is equal to the force F times the distance dx. Thus,

$$dE = -F\,dx \qquad (6\text{-}2)$$

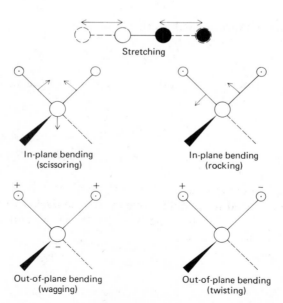

Figure 6-2. Types of molecular vibrations. *Note:* + indicates motion from plane of page toward reader; − indicates motion from plane of page away from reader.

Combining equations (6-2) and (6-1) yields

$$dE = kx dx$$

Integrating between the equilibrium position $(x = 0)$ and x gives

$$\int_0^E dE = k \int_0^x x \, dx$$

$$E = \frac{1}{2} k x^2 \qquad (6\text{-}3)$$

The potential energy curve for a simple harmonic oscillation, derived from equation (6-3), is shown in Figure 6-3(a). It is seen that the potential energy is a maximum when the spring is stretched or compressed to its maximum amplitude A and decreases parabolically to zero at the equilibrium position.

Vibrational frequency. The motion of the mass as a function of time t can be deduced as follows. Newton's law states that

$$F = ma$$

where m is the mass and a is its acceleration. But acceleration is the second derivative of distance with respect to time:

$$a = \frac{d^2x}{dt^2}$$

Substituting these expressions into (6-1) gives

$$m \frac{d^2x}{dt^2} = -kx \qquad (6\text{-}4)$$

One solution (but not the only one) to equation (6-4), as can be proved by substitution, is

$$x = A \sin\left(\sqrt{\frac{k}{m}} \, t\right) \qquad (6\text{-}5)$$

where A is the amplitude of the vibration, a constant that is equal to the maximum value of x.

One complete cycle in a sine function involves a change of 2π. Therefore, the time required to complete one cycle (the *period* τ of the motion) is given by a value of t such that the quantity in the parenthesis in equation (6-5) changes by 2π. Thus,

$$\tau = t \qquad \text{when} \qquad t = 2\pi \sqrt{\frac{m}{k}} \qquad (6\text{-}6)$$

The frequency of the vibration ν is the reciprocal of its period. Thus,

$$\nu_m = \frac{1}{\tau} = \frac{1}{2\pi} \sqrt{\frac{k}{m}} \qquad (6\text{-}7)$$

The frequency ν_m is the *natural frequency* of the mechanical oscillator. While it is dependent upon the force constant of the spring and the mass of the attached body, the natural frequency is *independent* of the energy imparted to the system; changes in energy merely result in a change in the amplitude A of the vibration.

The equations we have just developed can be readily modified to describe the behavior of a system made up of two masses m_1 and m_2 con-

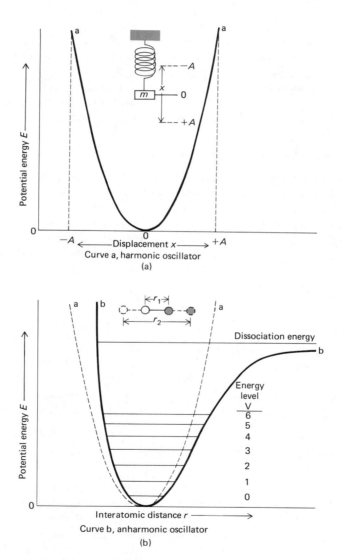

Curve a, harmonic oscillator

(a)

Curve b, anharmonic oscillator

(b)

Figure 6-3. Potential energy diagrams.

nected by a spring. Here it is only necessary to substitute the *reduced mass* μ for the single mass m where

$$\mu = \frac{m_1 m_2}{m_1 + m_2} \tag{6-8}$$

Thus, the vibrational frequency for such a system is given by

$$\nu_m = \frac{1}{2\pi} \sqrt{\frac{k}{\mu}} = \frac{1}{2\pi} \sqrt{\frac{k(m_1 + m_2)}{m_1 m_2}} \tag{6-9}$$

Molecular vibrations. The approximation is ordinarily made that the behavior of a molecular vibration is analogous to the mechanical model just described. Thus, the frequency of the molecular vibration is calculated from equation (6-9) by substituting the masses of the two atoms for m_1 and m_2; the quantity k becomes the force constant for the chemical bond, which is a measure of its stiffness (but not necessarily its strength).

Quantum Treatment of Vibrations

Harmonic oscillators. The equations of ordinary mechanics, such as we have used thus far, do not completely describe the behavior of particles with atomic dimensions. For example, the quantized nature of molecular vibrational energies (and of course other atomic and molecular energies as well) does not appear in these equations. It is possible, however, to employ the concept of the simple harmonic oscillator for the development of the wave equations of quantum mechanics. Solutions of these equations for potential energies are found when

$$E = \left(v + \frac{1}{2} \right) \frac{h}{2\pi} \sqrt{\frac{k}{\mu}} \tag{6-10}$$

where v, *the vibrational quantum number,* can take only positive integer values (including zero). Thus, in contrast to ordinary mechanics where vibrators can have any positive potential energy, quantum mechanics requires that only certain discrete energies be assumed by a vibrator.

It is of interest to note that the term $\sqrt{k/\mu}/2\pi$ appears in both the mechanical and the quantum equations; by substituting equation (6-9) into (6-10) we find

$$E = \left(v + \frac{1}{2} \right) h\nu_m \tag{6-11}$$

where ν_m is the vibrational frequency of the mechanical model.

We now assume that transitions in vibrational energy levels can be brought about by radiation, provided the energy of the radiation exactly matches the difference in energy levels ΔE between the vibrational quantum states (and provided also that the vibration causes a fluctuation in dipole).

This difference is identical between any pair of adjacent levels since v in equations (6-10) and (6-11) can assume only whole numbers; that is,

$$\Delta E = h\nu_m = \frac{h}{2\pi} \sqrt{\frac{k}{\mu}} \qquad (6\text{-}12)$$

At room temperature the majority of molecules are in the ground state (v = 0); thus $E_0 = \frac{1}{2} h\nu_m$. To move to the first excited state with energy $E_1 = \frac{3}{2} h\nu_m$ requires radiation of energy ($\frac{3}{2} h\nu_m - \frac{1}{2} h\nu_m$) = $h\nu_m$. The frequency of radiation ν that will bring about this change is *identical to the classical vibration frequency of the bond* ν_m. That is,

$$E_{\text{radiation}} = h\nu = \Delta E = h\nu_m = \frac{h}{2\pi} \sqrt{\frac{k}{\mu}}$$

or

$$\nu = \nu_m = \frac{1}{2\pi} \sqrt{\frac{k}{\mu}} \qquad (6\text{-}13)$$

If we wish to express the radiation in wave numbers,

$$\sigma = \frac{1}{2\pi c} \sqrt{\frac{k}{\mu}} = \frac{1}{2\pi c} \sqrt{\frac{k(m_1 + m_2)}{m_1 m_2}} \quad reduce \qquad (6\text{-}14)$$
more

σ = wave number of absorption peak (cm^{-1})
k = force constant of the bond (dynes/cm)
c = velocity of light (3 × 10^{10} cm/sec)
m_1, m_2 = mass of atoms 1 and 2 (g)

Equation (6-14) and infrared measurements have been employed for the determination of the force constants for various types of chemical bonds. Generally, k has been found to lie in the range of 3 × 10^5 to 8 × 10^5 dynes/cm for most single bonds, with 5 × 10^5 serving as a reasonable average value. Double and triple bonds are found by this same means to have force constant of about two and three times this value, respectively. By employing these average experimental values, equation (6-14) can be employed for the estimation of the wave number of the fundamental absorption peak (the absorption peak due to the transition from the ground state to the first excited state) for a variety of bond types. The following example demonstrates such a calculation.

Example. Calculate the approximate wave number and wavelength of the fundamental absorption peak due to the stretching vibrations of a carbonyl group $\diagdown C = O$.

The force constant for a doub⌄ bond has an approximate value of 1 × 10^6 dynes/cm. The masses of the carbon and oxygen atoms are approximately 12/6.0 × 10^{23} and 16/6.0 × 10^{23} or 2.0 × 10^{-23} and 2.6 × 10^{-23} g per atom. Thus

$$\sigma = \frac{1}{2 \times 3.14 \times 3 \times 10^{10}} \sqrt{\frac{1 \times 10^6 (2.0 + 2.6) \times 10^{-23}}{2.0 \times 2.6 \times 10^{-46}}}$$

$$= 1.6 \times 10^3 \text{ cm}^{-1}$$

$$\lambda = \frac{10^4}{1.6 \times 10^3} = 6.3 \ \mu$$

The carbonyl stretching bond is found experimentally to be in the region of 5.3 to 6.7 μ or 1500 to 1900 cm^{-1}.

Selection rules. As given by equations (6-12) and (6-13) the energy for a transition from energy level 1 to 2 or from level 2 to 3 should be identical to that for the $0 \rightarrow 1$ transition. Furthermore, quantum theory demonstrates that the only transitions that can take place are those in which the vibrational quantum number changes by unity; that is, the so-called selection rule states that $\Delta v = \pm 1$. Since the vibrational levels are equally spaced, only a single absorption peak should be observed for a given transition.

Anharmonic oscillator. Thus far we have considered the classical and quantum mechanical treatments of the harmonic oscillator. The potential energy of such a vibrator changes periodically as the distance between masses fluctuates [Figure 6-3(a)]. From qualitative considerations it is apparent, however, that this is an imperfect description of a molecular vibration. For example, as the two atoms approach one another, coulombic repulsion between the two nuclei produces a force that acts in the same direction as the restoring force of the bond; thus, the potential energy can be expected to rise more rapidly than the harmonic approximation predicts. At the other extreme of oscillation a decrease in the restoring force, and thus the potential energy, occurs as the interatomic distance approaches that at which dissociation of atoms takes place.

In theory, the wave equations of quantum mechanics permit the derivation of more correct potential energy curves for molecular vibrations. Unfortunately, however, the mathematical complexity of these equations preclude their quantitative application to all but the very simplest of systems. It is qualitatively apparent, however, that the curves must take the *anharmonic* form shown in Figure 6-3(b). These curves depart from harmonic behavior to various degrees depending upon the nature of the bond and the atoms involved. Note, however, that the harmonic and anharmonic curves are nearly alike at low potential energies. This fact accounts for the success of the approximate methods described.

Anharmonicity leads to deviations of two kinds. At higher quantum numbers ΔE becomes smaller [see Figure 6-3(b)], and the selection rule is not rigorously followed; as a result, transitions of $\Delta v = \pm 2$ or ± 3 are observed. Such transitions are responsible for the appearance of *overtone lines* at frequencies approximately twice or three times that of the fundamental line; often the intensity of overtone absorption is low and the peaks may not be observed.

Vibrational spectra are also complicated by the fact that two different vibrations in a molecule can interact to give absorption peaks with frequencies that are approximately the sums or differences of their fundamental frequencies. Again, the intensities of combination and difference peaks are generally low.

Vibrational Modes

The complexity of infrared spectra of polyatomic molecules arises from the multitude of vibrations that can occur in a molecule containing several atoms and several bonds. It is ordinarily possible to deduce the number and kinds of vibrations in simple diatomic and triatomic molecules, and whether these vibrations will lead to absorption. With complex molecules such an analysis is difficult at best, and may be impossible.

The number of possible vibrations in a polyatomic molecule can be calculated as follows. Three coordinates are needed to locate a point in space; to fix N points in space requires a set of three coordinates for each, or a total of $3N$ coordinates. In considering the motion of a polyatomic molecule containing N atoms, each coordinate corresponds to one degree of freedom of motion of one of the atoms; for this reason the molecule is said to have $3N$ *degrees of freedom*.

In defining the motion of a molecule we need to consider (1) the motion of the entire molecule through space (that is, the translational motion of its center of gravity); (2) the rotational motion of the entire molecule around its center of gravity; and (3) the motion of each of its atoms relative to the other atoms (in other words, its individual vibrations). To specify the translational motion requires three coordinates and uses up 3 degrees of freedom of the molecule. Another 3 degrees of freedom are needed to describe the rotation of the molecule as a whole. The remaining $(3N - 6)$ degrees of freedom involve interatomic motion, and hence represent the number of possible vibrations within the molecule. A linear molecule is a special case since, by definition, all of the atoms lie on a single, straight line. Rotation about the bond axis is not possible, and 2 degrees of freedom suffice to describe rotational motion. Thus, the number of vibrations for a linear molecule is given by $(3N - 5)$. Each of the $(3N - 6)$ or $(3N - 5)$ vibrations is called a *normal mode*.

For each normal mode of vibration there exists a potential energy relationship such as that shown in Figure 6-3(b). The same selection rules discussed earlier apply for each of these. In addition, to the extent that a vibration approximates harmonic behavior, the energy differences between the energy levels of a given vibration are the same; that is, a single absorption peak appears for each vibration in which there is a change in dipole.

In fact, however, the number of normal modes does not necessarily correspond exactly to the number of observed absorption peaks. Often the number of peaks is less because (1) the symmetry of the molecules is such that no change in dipole results from the vibration, (2) the energies of two or more vibrations are identical or nearly identical, (3) the absorption intensity is so low that it is not detectable by ordinary means, or (4) the vibration energy is in a wavelength region beyond the range of the instrument. As we have pointed out, additional peaks arise from overtones as well as from combination or difference frequencies.

Vibrational Coupling

The energy of a vibration, and thus the wavelength of its absorption peak, may be influenced by other vibrators in the molecule. A number of factors that influence the extent of such coupling can be identified.

1. Strong coupling between stretching vibrations occurs only when there is an atom common to the two vibrations.

2. Interaction between bending vibrations requires a common bond between the vibrating groups.

3. Coupling between a stretching and a bending vibration can occur if the stretching bond forms one side of the angle that varies in the bending vibration.

4. Interaction is greatest when the coupled groups have individual energies that are approximately equal.

5. Little or no interaction is observed between groups that are separated by two or more bonds.

6. Coupling requires that the vibrations be of the same symmetry species.[1]

As an example of coupling effects let us consider the infrared spectrum of carbon dioxide. If no coupling occurred between the two C=O bonds, an absorption peak would be expected at the same wave number as the peak for the C=O stretching vibration in an aliphatic ketone (about 1700 cm^{-1}, or 6 μ; see example, p. 139). Experimentally, carbon dioxide exhibits two absorption peaks, the one at 2330 cm^{-1} (4.3 μ) and the other at 667 cm^{-1} (15 μ).

Carbon dioxide is a linear molecule and thus has four normal modes $(3 \times 3 - 5)$. Two stretching vibrations are possible; furthermore, interaction between the two can occur since the bonds involved are associated with a common carbon atom. As shown below, one of the coupled vibrations is symmetric and the other is asymmetric.

symmetric   asymmetric

[1]For a discussion of symmetry operations and symmetry species see R. P. Bauman, *Absorption Spectroscopy* (New York: John Wiley & Sons, Inc., 1962), Chap. 10.

The symmetric vibration causes no change in dipole since the two oxygen atoms simultaneously move away from or toward the central carbon atom. Thus, the symmetric vibration is infrared-inactive. In the asymmetric vibration one oxygen approaches the carbon atom as the other moves away. As a consequence, a net change in charge distribution occurs periodically; absorption at 2330 cm^{-1} results.

The remaining two vibrational modes of carbon dioxide involve scissoring as shown below.

The two bending vibrations are the resolved components (at 90 deg to one another) of the bending motion in all possible planes around the bond axis. The two vibrations are identical in energy and thus produce but one peak at 667 cm^{-1}. (Quantum energy differences that are as identical as these are said to be *degenerate*.)

It is of interest to compare the spectrum of carbon dioxide with that of a nonlinear, triatomic molecule such as water, sulfur dioxide, or nitric oxide. These molecules have $(3 \times 3 - 6)$, or 3, vibrational modes which take the following forms:

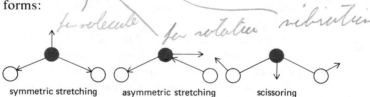

| symmetric stretching | asymmetric stretching | scissoring |

Since the central atom is not in line with the other two, a symmetric stretching vibration will produce a change in dipole and will thus be responsible for infrared absorption. For example, stretching peaks at 3650 and 3760 cm^{-1} (2.74 and 2.66 μ) are observed for the symmetric and asymmetric vibrations of the water molecule. Only one component to the scissoring vibration exists for the nonlinear molecule since motion of the plane of the molecule constitutes a rotational degree of freedom. For water, the bending vibration causes absorption at 1595 cm^{-1} (6.27 μ).

The difference in behavior of linear and nonlinear triatomic molecules (2 and 3 absorption peaks, respectively) illustrates how infrared absorption spectroscopy can sometimes be used to deduce the shape of a molecule.

Coupling of vibrations is a common phenomenon; as a result, the position of an absorption peak corresponding to a given organic functional group cannot be specified exactly. For example, the C—O stretching frequency in methanol is 1034 cm^{-1} (9.67 μ); in ethanol it is 1053 cm^{-1} (9.50 μ); and in methylethylcarbinol it is 1105 cm^{-1} (9.05 μ). These variations result from a coupling of the C—O stretching with the adjacent C—C or C—H vibrations.

While interaction effects may lead to uncertainties in the identification of functional groups contained in a compound, it is this effect that provides the unique features of an infrared absorption spectrum that are so important for the positive identification of a specific compound.

INFRARED INSTRUMENTATION

Infrared spectrophotometers have the same basic components as do instruments used for the study of absorption in the ultraviolet and visible regions of the spectrum (see Figure 3-5). In detail, however, each of these components is quite different from those we have encountered to this point.

Instrument Components

In this section the salient features of infrared sources, monochromators, and detectors are considered; discussion of cells is deferred to the section concerned with sample handling.

Sources. The common infrared source is an inert solid heated electrically to temperatures between 1500° and 2000°K. Continuous radiation approximating that of a black body results (see Figure 2-13). The maximum radiant intensity at these temperatures occurs at 1.7 to 2 μ (5000 to 6000 cm^{-1}). At longer wavelengths, the intensity falls off continuously until it is about one percent of the maximum at 15 μ (667 cm^{-1}). On the short wavelength side the decrease is much more rapid, and a similar reduction in intensity is observed at about 1 μ (10,000 cm^{-1}).

Three types of sources are used in commercial infrared instruments; namely the *Nernst glower,* the *Globar,* and incandescent wires.

The Nernst glower is composed of rare-earth oxides formed into a cylinder having a diameter of 1 to 2 mm and a length of perhaps 20 mm. Platinum leads are sealed to the ends of the cylinder to permit passage of current. This device has a large negative temperature coefficient of electrical resistance, and it must be heated externally to a dull red heat before a sufficient current passes to maintain the desired temperature. Because the resistance decreases with increasing temperature, the source circuit must be designed to limit the current; otherwise the glower rapidly becomes so hot that it is destroyed.

A Globar is a silicon carbide rod, usually about 5 cm in length and 0.5 cm in diameter. It also is electrically heated, and has the advantage of a positive coefficient of resistance. On the other hand, water cooling of the electrical contacts is required to prevent arcing. Spectral energies of the Globar and the Nernst glower are comparable except in the region below 5 μ, where the Globar provides a significantly greater output.

A source of somewhat lower intensity but longer life is a tightly wound spiral of nichrome wire heated by passage of current. A rhodium wire heater sealed in a ceramic cylinder is also employed.

Monochromators. An infrared monochromator consists of a variable entrance and exit slit system, one or more dispersing elements, and several mirrors for reflecting and focusing the beam of radiation. Lenses are not employed because of problems with chromatic aberrations.

The slit system in an infrared spectrophotometer serves the same function as in its ultraviolet and visible counterparts, and the same compromise must be made in choosing the slit width to be employed. Narrow slits provide smaller band widths, better spectral definition, and freedom from scattered radiation; wider slits, on the other hand, result in greater radiant energy reaching the detector and consequently greater photometric accuracy.

Both prisms and gratings are employed for dispersing infrared radiation; the general use of the latter is a relatively recent development.

Several materials have been used for prism construction. Quartz is employed for the near-infrared region (0.8 to 3 μ or 12,500 to 3300 cm^{-1}) even though its dispersion characteristics for this region are far from ideal. It absorbs strongly beyond about 4 μ. Crystalline sodium chloride is the most common prism material. Its dispersion is high in the region between 5 and 15 μ and is adequate to 2.5 μ. Beyond 20 μ, sodium chloride absorbs strongly and cannot be used. Crystalline potassium bromide and cesium bromide provide prism materials for the far-infrared region (15 to 40 μ or 675 to 250 cm^{-1}), while lithium fluoride is useful in the near-infrared region (1 to 5 μ or 10,000 to 2000 cm^{-1}). Many spectrophotometers are designed for convenient prism interchange. Unfortunately, all of the common infrared transmitting materials except quartz are easily scratched and are water soluble. Protection from condensation of moisture with desiccants or with heat is thus necessary.

Reflection gratings offer a number of advantages as dispersing elements for the infrared region, and appear to be replacing prisms as a consequence. Inherently better resolution is possible because there is less loss of radiant energy than in a prism system; thus, narrower slits can be employed. Other advantages of gratings include the more nearly linear dispersion and the resistance to attack by water. An infrared grating is usually constructed from glass or plastic that is coated with aluminum.

The disadvantage of a grating lies in the greater amounts of scattered radiation and the appearance of radiation of other spectral orders. In order to minimize these effects, gratings are blazed to concentrate the radiation into a single order. In addition, prisms or filters are used in conjunction with the grating.

Both transmission and interference filters have been developed for the infrared region.

Detectors. The measurement of infrared radiation is difficult as a result of the low intensity of available sources and the low energy of the infrared photon. As a consequence of these properties, the electrical signal from an infrared detector is small, and its measurement requires large amplification factors. It is usually the detector system that limits the sensitivity and the precision of an infrared spectrophotometer.

The convenient phototubes discussed in Chapter 3 are generally not applicable in the infrared because the photons in this region lack the energy to cause photoemission of electrons. Thus, thermal detectors and detection based upon photoconduction are employed. Neither of these is as satisfactory as the photocell.

Photoconductors are constructed from such semiconductor crystals as lead sulfide, lead selenide, and germanium. Absorption of radiation by these materials results in excitation of nonconducting electrons to an excited conducting state. The consequent increase in conduction or decrease in resistance can be readily measured and is directly related to the number of photons reaching the semiconductor surface. The most common photoconductor is lead sulfide, which is sensitive in the region between 0.8 and about 2 μ (12,500 to 5000 cm^{-1}). The maximum wavelength that can be detected with a photoconductor is about 5 μ (2000 cm^{-1}).

Thermal detectors, whose responses depend upon the heating effect of the radiation, can be employed for detection of all but the shorter infrared wavelengths. In these devices the radiation is absorbed by a small black body and the resultant temperature increase is measured. The radiant power level from a spectrophotometer beam is very low (10^{-7} to 10^{-9} watt) so that the heat capacity of the absorbing element must be as small as possible if a detectable temperature change is to be produced. Every effort is made to minimize the size and the thickness of the target element and to concentrate the entire infrared beam on its surface. Under the best of circumstances, temperature changes are confined to a few thousandths of a degree Celsius.

The problem of measuring infrared radiation by thermal means is compounded by thermal effects from the surroundings. In order to reduce heating of the detector by nearby extraneous objects, the absorbing element is housed in a vacuum and is carefully shielded from thermal radiation emitted by other bodies in the area.

Three types of temperature-detecting devices are used: *thermocouples, bolometers,* and *Golay cells*. Thermocouples depend upon the *Peltier effect,* in which a potential develops between two junctions of dissimilar metals or semiconductors when the junctions are at different temperatures. The receiver element is often a blackened gold or platinum foil to which is welded the fine wires comprising the thermoelectric junction. The second junction is designed to have a relatively large heat capacity and is carefully shielded from the incident radiation. A well-designed receiver responds to a temperature change of about 10^{-6}°C.

A bolometer is constructed from a metal or a semiconductor that exhibits a large change in electrical resistance as a function of temperature.

The conducting element is formed as a thin film by vacuum condensation, by rolling, or by sputtering. This element forms one arm of the Wheatstone bridge circuit used to measure its resistance. The same design factors noted for thermocouples are also requirements for bolometers.

A Golay detector is essentially a sensitive gas thermometer. The gas enclosed in the detector is warmed by the radiation reaching it; the resultant pressure increase is then converted to an electrical signal.

Commercial Instruments

Several instrument companies produce infrared spectrophotometers, and the total number of models now available exceeds three dozen. Cross[2] has summarized the important features of the various models marketed in 1964; several new instruments have appeared since that date. Our discussion is confined to some common design characteristics. A single example of a modern spectrophotometer is then presented; descriptive literature from the manufacturer should be consulted for detailed information concerning any specific instrument.

Double-beam design. Almost without exception, commercial infrared spectrophotometers employ a double-beam design, with the radiant energy passing alternately through the sample and then a reference to a single detector. As we noted earlier (p. 56), the double-beam design is not as demanding in terms of performance and stability in the source and the detector components as is the single-beam arrangement. The low energies of infrared radiation, the inherent lower stabilities of the source and the detector, and the need for large signal amplification make the double-beam design almost imperative for infrared instruments. In addition, this arrangement is more readily adapted to automatic recording, a feature that is highly desirable because of the complexity of infrared spectra.

Basic components. All of the types of sources and detectors described in the previous section are incorporated in one or more of the commercial instruments. Thermocouples appear to be favored over the other two detectors, but all three sources are widely used.

A clear trend away from prisms in favor of gratings as dispersing elements is apparent in the design of commercial instruments. Filters are frequently employed to remove stray radiation and different order lines.

Beam chopping. All commercial designs incorporate a low-frequency chopper (5 to 13 cycles per minute) to modulate the output from the source. This feature permits the detector-amplifier system to discriminate between the signal from the source and signals from extraneous radiation, such as

---

[2]A. D. Cross, *An Introduction to Practical Infra-Red Spectroscopy,* 2d ed. (Washington; Butterworths, 1964), p. 18.

infrared emission arising from various bodies surrounding the detector. Low-frequency chopping must be employed because of the relatively long response time of infrared detectors.

Beam attenuation. Infrared instruments are generally of a null type in which the power of the reference beam is reduced, or *attenuated*, to match that of the beam passing through the sample. Attenuation is accomplished by imposing a device that removes a variable fraction of the reference beam. The attenuator commonly takes the form of a fine-toothed comb, the teeth of which are tapered so that a linear relationship exists between the lateral movement of the comb and the decrease in power of the beam. Movement of the comb occurs when a difference in power of the two beams is sensed by the detector. This movement is synchronized with the recorder pen so that its position gives a measure of the relative power of the two beams and thus the transmittance of the sample.

Instrument costs. Infrared spectrophotometers vary in price from about $3000 to perhaps $30,000 depending upon their wavelength range, their resolution, their photometric accuracy, their versatility, and their convenience. For the routine identification of functional groups and most organic compounds the simpler instruments in the $3000 to $8000 range serve very nicely.

Instrument design. Figure 6-4 shows schematically the arrangement of components in a typical infrared spectrophotometer. Note that three types of systems link the components: (1) a radiation linkage indicated by dotted lines, (2) a mechanical linkage shown by thick dark lines, and (3) an electrical linkage shown by narrow lines.

Radiation from the source is split into two beams, half passing into the sample cell and the other half into the reference cell. The reference beam then passes through the attenuator and on to the chopper. The chopper consists of a motor-driven disk that alternately reflects the reference beam or transmits the sample beam into the monochromator. After dispersion by the prism or grating, the alternating beams fall on the detector and are converted to an electrical signal. The signal is amplified and passed to the synchronous rectifier, a device that is mechanically or electrically coupled to the chopper to cause the rectifier switch and the beam leaving the chopper to change simultaneously. If the two beams are identical in power, the signal from the rectifier is an unfluctuating direct current. If the two beams differ in power, a fluctuating current is produced, the polarity of which is determined by which beam is the more intense. The current from the rectifier is filtered and further amplified to drive the synchronous motor in one direction or the other, depending upon the polarity of the input current. The synchronous motor is mechanically linked to both the attenuator and the pen drive of the recorder, and causes both to move until a null is achieved. A second synchronous motor drives the chart and varies the wavelength simultaneously.

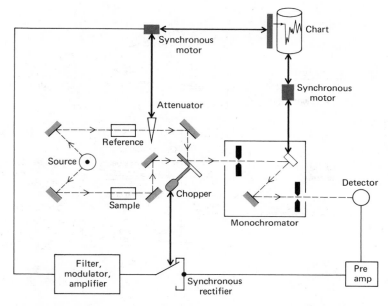

Figure 6-4. Schematic diagram of a double-beam spectrophotometer. Heavy dark line = mechanical linkage; light line = electrical linkage; dotted line = radiation path.

Often there is a mechanical linkage to the slits so that the radiant power reaching the detector is approximately constant.

Figure 6-5 shows the optical plan of a typical, medium-priced spectrophotometer, the Perkin-Elmer Model 237B. The instrument is equipped with a pair of interchangeable gratings that cover the wavelength range of 2.5 to 16 μ (4000 to 625 cm^{-1}). This company manufactures a series of instruments employing essentially the same optical design as the one shown in Figure 6-5. In some, a potassium bromide prism replaces the gratings. Others have gratings that cover a larger range (2.5 to 25 μ or 4000 to 400 cm^{-1}); still others have the capacity to generate spectra in terms of absorbance as well as transmittance. These instruments range in price from about \$6000 to \$8000.

SAMPLE HANDLING TECHNIQUES[3]

The techniques for preparing and examining samples in the infrared region are often quite different from those used in other spectral regions.

[3]For a more complete discussion see N. B. Colthup, L. H. Daly and S. E. Wiberley, *Introduction to Infra Red and Raman Spectroscopy* (New York: Academic Press, Inc., 1964), p. 60; R. P. Bauman, *Absorption Spectroscopy* (New York: John Wiley & Sons, Inc., 1962), p. 184; I. M. Kolthoff and P. J. Elving, Eds., *Treatise on Analytical Chemistry*, Part I, Vol. 6 (New York: Interscience Publishers, 1965), p. 3582.

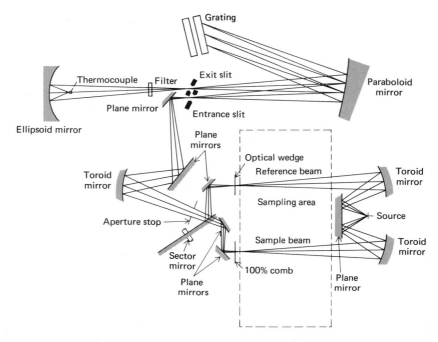

Figure 6-5. Schematic diagram of the Perkin-Elmer Model 237B Recording Infrared Spectrophotometer. (With permission, Perkin-Elmer Corp., Norwalk, Conn.)

Thus, consideration is first given to sample handling. This is followed by a discussion of qualitative and quantitative spectrophotometry.

The sample for an infrared examination may be in the form of gas, a solution, a pure liquid, or a solid suspended in a suitable matrix.

Gas Samples

The spectra of a low-boiling liquid or gas can be obtained by permitting the sample to expand into an evacuated cell. A variety of cells are available for this purpose with path lengths that range from a few centimeters to several meters. The longer path lengths are obtained in compact cells by providing reflecting internal surfaces, so that the beam makes numerous passes through the sample before exiting from the cell.

Solutions

The employment of dilute solutions for infrared studies offers many advantages; principal among these is the enhanced reproducibility of the data. Moreover, through suitable choice of concentration and of cell length, the shape and the structure of important bands can be made clearly evident. Despite these obvious advantages, it is frequently impossible to find a liquid with the required solvent capability that does not itself absorb strongly in the spectral region of interest. In addition, chemical reaction between solvent and sample may rule out use of the solution technique.

Solvents. Carbon disulfide is a common solvent for the region between 1330 and 625 cm^{-1} (7.5 to 16 μ). Carbon tetrachloride is useful in the 4000 to 1330 cm^{-1} (2.5 to 7.5 μ) region. Both of these solvents are volatile, toxic, and must be used carefully.

Polar solvents are required for the solution of many organic compounds; unfortunately, however, strong polarity is directly related to infrared absorption. Consequently, there are no polar solvents that are transparent over large ranges of the infrared. Some solvents that can be used in limited regions include chloroform, dioxane, and dimethyl formamide. The transparent regions for some common solvents are shown in Figure 6-6.

Care must be taken to dry solvents so as to eliminate the strong absorption bands of water and also to prevent attack upon the windows of the cells.

Cells. Because of the tendency for solvents to absorb, infrared cells are ordinarily much narrower (0.1 to 1 mm) than those employed in the ultraviolet and visible regions. Light paths in this range normally require sample concentrations from 0.1 to 10 percent. The cells are frequently demountable with spacers to allow variation in path length. Fixed path-length cells can be filled or emptied with a hypodermic syringe.

Sodium chloride windows are most commonly employed; even with care these eventually become fogged due to absorption of moisture. Polishing with a buffing powder returns them to their original condition.

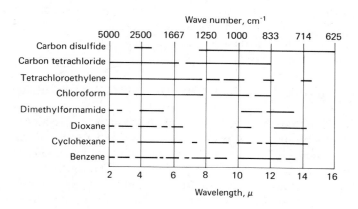

Figure 6-6. Infrared solvents. Black lines indicate useful regions.

Pure Liquids

When the amount of sample is small or when a suitable solvent is not available, it is common practice to obtain spectra on the pure (neat) liquid. Here, only a very thin film has a sufficiently short path length to produce satisfactory spectra. Commonly a drop of the neat liquid is squeezed between two rock salt plates to give a layer that is 0.01 mm or less in thickness. The two plates, held together by capillarity, are then mounted in the beam path. Clearly, such a technique does not give very reproducible transmittance data, but it is satisfactory for many qualitative investigations.

Solids

Solids that cannot be dissolved in an infrared-transparent solvent can be suspended in a suitable transparent medium to form a two-phase mixture called a *mull*. An essential condition for the acquisition of satisfactory spectra is that the particle size of the suspended solid be smaller than the wavelength of the radiation; if this condition is not realized, a significant portion of the radiation is lost to scattering.

Two techniques are employed. In one, 2 to 5 mg of the finely ground sample (particle size $< 2 \, \mu$) is further ground in the presence of one or two drops of a heavy hydrocarbon oil (Nujol). If the hydrocarbon bands interfere, Fluorolube, a halogenated polymer, can be used. In either case the resulting mull is then examined as a thin film between flat salt plates.

In the second technique a milligram or less of the finely ground sample is intimately mixed with about 100 mg of dried potassium bromide powder. Mixing can be carried out with a mortar and pestle; a small ball mill is more satisfactory, however. The mixture is then pressed in a special die at 10,000 to 15,000 pounds per square inch to yield a transparent disk. Best results are obtained if the disk is formed in a vacuum to eliminate occluded air. The disk is then held in the instrument beam for spectroscopic examination. The resulting spectra frequently exhibit bands at 2.9 and 6.1 μ (3448 and 1639 cm^{-1}) due to absorbed moisture.

QUALITATIVE APPLICATIONS OF INFRARED ABSORPTION

We have noted that the approximate frequency at which an organic functional group, such as C=O, C=C, CH$_3$, and C≡C, absorb infrared radiation can be calculated from the masses of the atoms and the force constant of the bond between them (equation 6-13). These frequencies, called *group frequencies,* are seldom totally invariant because of interactions with other vibrations associated with one or both of the atoms comprising the group. On the other hand, such interaction effects are often not large; as a result, a range of frequencies can be assigned in which it is highly probable that the absorption peak for a given functional group will be found.

Group frequencies often make it possible to establish unambiguously whether a given functional group is present or absent in a molecule.

Correlation Charts

Over the years a mass of empirical information concerned with the frequency range within which various functional groups can be expected to absorb has been accumulated. *Correlation charts* provide a concise means for summarizing this information in a form that is useful for identification purposes. A number of correlation charts have been developed,[4] one of which is in Figure 6-7.

Correlation charts permit intelligent guesses to be made as to the functional groups present and absent in a molecule on the basis of its absorption spectrum. Ordinarily, it is impossible to identify unambiguously either the sources of all of the peaks in a given spectrum or the exact identity of the molecule. Rather, correlation charts serve as a starting point in the identification process.

Important Spectral Regions in the Infrared

The chemist interested in identifying an organic compound by the infrared technique usually examines certain regions of the spectrum in a systematic way in order to obtain clues as to the presence or absence of certain group frequencies. Some of the important regions are considered briefly.

Hydrogen stretching region 3700 to 2700 cm^{-1} (2.7 to 3.7 μ). The appearance of strong absorption peaks in this region usually results from a stretching vibration between hydrogen and some other atom. The motion is largely that of the hydrogen atom since it is so much lighter than the atom with which it bonds; as a consequence, the absorption is not greatly affected by the rest of the molecule. Furthermore, the hydrogen stretching frequency is much higher than that for other chemical bonds, with the result that interaction of this vibration with others is usually small.

Absorption peaks in the region of 3700 to 3100 cm^{-1} (2.7 to 3.2 μ) are ordinarily due to various O—H and N—H stretching vibrations, with the former tending to appear at higher wave numbers. The O—H bonds are

---

[4]N. B. Colthup, *J. Opt. Soc. Am.*, **40**, 397–400 (1950); A. D. Cross, *Introduction to Practical Infra-Red Spectroscopy*, 2d ed. (Washington: Butterworths, 1964), pp. 56–62; R. N. Jones, *Infrared Spectra of Organic Compounds: Summary Charts of Principal Group Frequencies* (Ottawa: National Research Council of Canada, 1959); K. Nakanishi, *Infrared Absorption Spectroscopy* (San Francisco: Holden-Day, 1962), pp. 14–15; R. M. Silverstein and G. C. Bassler, *Spectrometric Identification of Organic Compounds*, 2d ed. (New York: John Wiley & Sons, Inc., 1967), pp. 73–77.

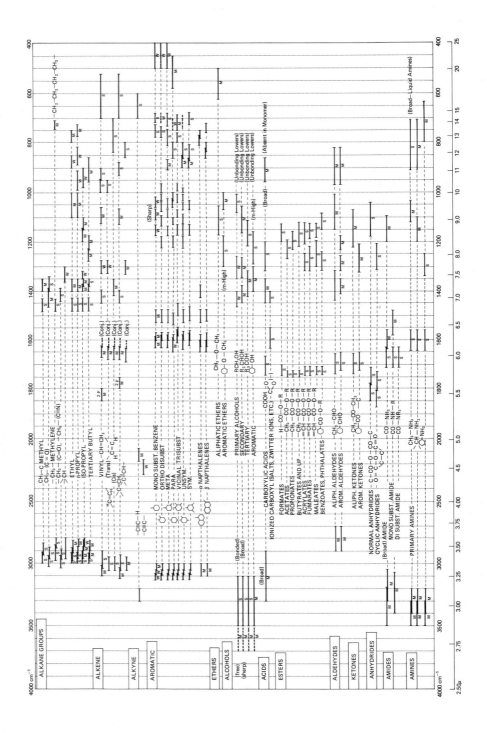

154

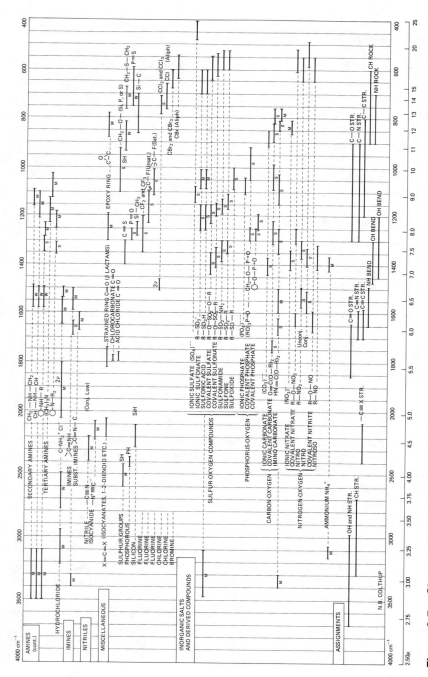

Figure 6-7. Characteristic infrared group frequencies. From N. B. Colthup, *J. Optical Soc. Am.*, **40**, 397 (1950). (With permission.)

155

often broader than N—H bonds, and appear only in dilute nonpolar solvents. Hydrogen bonding tends to broaden the peaks and move them toward lower wave numbers.

Aliphatic CH vibrations fall in the region between 3000 and 2850 cm^{-1} (3.3 to 3.5 μ). Most aliphatic compounds have a sufficient number of CH bonds to make this a prominent peak. Any structural variation that affects the CH bond strength will cause a shift in the maximum. For example, the band for Cl—C—H lies just above 3000 cm^{-1} (< 3.3 μ), as do the bands for olefinic and aromatic hydrogen. The acelytenic CH bond is strong and occurs at about 3300 cm^{-1} (3.0 μ). The hydrogen on the carbonyl group of an aldehyde usually produces a distinct peak in the region of 2745 to 2710 cm^{-1} (3.64 to 3.69 μ). Substitution of deuterium for hydrogen causes a shift to lower wave numbers by the factor of approximately $1/\sqrt{2}$, as would be predicted from equation (6-14); this effect has been employed to identify C—H stretching peaks.

The triple bond region between 2700 and 1850 cm^{-1} (3.7 to 5.4 μ). A limited number of groups absorb in this spectral region; their presence is thus often readily apparent. Triple-bond stretching results in a peak at 2250 to 2225 cm^{-1} (4.44 to 4.49 μ) for —C≡N, at 2180 to 2120 cm^{-1} (4.59 to 4.72 μ) for —N$^+$≡C$^-$, and at 2260 to 2190 cm^{-1} (4.42 to 4.57 μ) for —C≡C—. Also present in this region are peaks for SH at 2600 to 2550 cm^{-1} (3.85 to 3.92 μ), PH at 2440 to 2350 cm^{-1} (4.10 to 4.26 μ), and SiH at 2260 to 2090 cm^{-1} (4.42 to 4.78 μ).

The double bond region between 1950 and 1550 cm^{-1} (5.1 to 6.5 μ). The carbonyl stretching vibration is characterized by absorption throughout this region. Ketones, aldehydes, acids, amides, and carbonates all have absorption peaks around 1700 cm^{-1} (5.9 μ). Esters, acid chlorides, and acid anhydrides tend to absorb at slightly higher wave numbers; that is, 1770 to 1725 cm^{-1} (5.65 to 5.80 μ). Conjugation tends to lower the absorption peak by about 20 cm^{-1}. It is frequently impossible to determine which type of carbonyl is present solely on the basis of absorption in this region; however, examination of additional spectral regions may provide the supporting evidence needed for clear-cut identification. For example, esters have a strong C—O—R stretching peak at about 1200 cm^{-1} (8.3 μ), while aldehydes have a distinctive hydrogen stretching peak at just above 2700 cm^{-1} (3.7 μ), as noted previously.

Absorption peaks arising from C=C and C=N stretching vibrations are located in the 1690 to 1600 cm^{-1} (5.9 to 6.2 μ) range. Valuable information concerning the structure of olefins can be obtained from the exact position of such a peak.

The region between 1650 and 1450 cm^{-1} (6.1 to 6.9 μ) provides important information about aromatic rings. Aromatic compounds with a low degree of substitution exhibit four peaks near 1600, 1580, 1500, and 1460 cm^{-1} (6.25, 6.33, 6.67, and 6.85 μ). Variations of the spectra in this region and at lower wave numbers with the number and arrangement of substituent groups are usually consistent but independent of the type of substituent; considerable structural information can thus be gleaned from careful study of aromatic absorption in the infrared region.

The "fingerprint" region between 1500 and 700 cm^{-1} (6.7 to 14 μ).   In this region of the spectrum, small differences in the structure and the consti-tution of a molecule result in significant changes in the distribution of absorption peaks. As a consequence, a close match between two spectra in this region (as well as others), constitutes strong evidence for the identity of the compounds yielding the spectra. Most single bonds give rise to absorp-tion bands at these frequencies; because their energies are about the same, strong interaction occurs between neighboring bonds. The absorption bands are thus composites of these various interactions, and depend upon the over-all skeleton structure of the molecule. Because of the complexity, exact interpretation of spectra in this region is seldom possible; on the other hand, it is this complexity that leads to uniqueness and the consequent usefulness of the region for identification purposes.

A few important group frequencies are to be found in the fingerprint region. These include the C—O—C stretching vibration in ethers and esters at about 1200 cm^{-1} and the C—Cl stretching vibration at 700 to 800 cm^{-1}. A number of inorganic groups such as sulfate, phosphate, nitrate, and carbonate also absorb at wave numbers below 1200 cm^{-1}.

Limitations to the use of correlation charts.   The unambiguous establish-ment of the identity or the structure of a compound is seldom possible from correlation charts alone. Uncertainties frequently arise from overlapping group frequencies, spectral variations as a function of the physical state of the sample (that is, whether it is a solution, a mull, in a pelleted form, and so forth), and instrumental limitations.

In employing group frequencies it is essential that the whole spectrum, rather than a small isolated portion be considered and interrelated. Inter-pretation based on one part of the spectrum should be confirmed or rejected by study of other regions.

To summarize, then, correlation charts serve only as a guide for further and more careful study. Several excellent monographs describe the absorp-tion characteristics of functional groups in detail.[5] A study of these and the other physical properties of the sample often allows positive identification. Infrared spectroscopy, when used in conjunction with other methods such as mass spectroscopy, nuclear magnetic resonance, and elemental analysis, usually makes possible the positive identification of a species.

[5]N. B. Colthup, L. N. Daly and S. E. Wiberley, *Introduction to Infra Red and Raman Spectroscopy* (New York: Academic Press, Inc., 1964); and H. A. Szymanski, *Theory and Practice of Infrared Spectroscopy* (New York: Plenum Press, 1964).

The examples that follow illustrate how infrared spectra are employed in the identification of pure compounds. In every case, confirmatory tests would be desirable. A comparison of the experimental spectrum with that of the pure compound would suffice for this purpose.

Example. The spectrum in Figure 6-8 was obtained for a pure color-less liquid in a 0.01-mm cell; the liquid boiled at 190°C. Suggest a structure for the sample.

The four absorption peaks in the region of 1450 to 1600 cm^{-1} are characteristic of an aromatic system, and one quickly learns to recognize this grouping immediately. The peak at 3100 cm^{-1} corresponds to a hydrogen-stretching vibration, which for an aromatic system is usually above 3000 cm^{-1}. For aliphatic hydrogens this type of vibration causes absorption at or below 3000 cm^{-1} (see Figure 6-7). Thus it would appear that we are dealing with a system that is predominately if not exclusively aromatic.

We then note the sharp peak at 2250 cm^{-1} and recall that only a few groups absorb in this region. From Figure 6-7 we see that these include —C≡C—, —C≡CH, —C≡N, and SiH. The —C≡CH group can be eliminated, however, since it has a hydrogen stretching frequency of about 3250 cm^{-1}.

The pair of strong peaks at 680 and 760 cm^{-1} suggest that the aromatic ring may be singly substituted, although the pattern of peaks in the 1000 to 1300 cm^{-1} is confusing and may lead to doubt about this conclusion.

Two likely structures that fit the infrared data would appear to be $C_6H_5C≡N$ and $C_6H_5C≡C—C_6H_5$. The latter, however, is a solid at room temperature; in contrast, we find that the benzonitrile has a boiling point (191°C), similar to that of our sample. Thus, we tentatively conclude that the compound under investigation is $C_6H_5C≡N$.

Example. A colorless liquid was found to have the empirical formula $C_6H_{12}O$ and a boiling point of 130°C. Its infrared spectrum (neat in a 0.025-mm cell) is given in Figure 6-9. Suggest a structure.

The presence of an aliphatic structure is suggested in Figure 6-9 by both the strong band at 3000 cm^{-1} and the empirical formula. The intense band at 1720 cm^{-1} and the single oxygen in the formula strongly indicate the likelihood of an aldehyde or a ketone. The band at about 2800 cm^{-1} could be attributed to the shift in stretching vibration associated with hydrogen that is bonded directly to a carbonyl carbon atom; if this interpretation is correct, the substance is an aldehyde. The peak at 3450 cm^{-1} is puzzling, for absorption in this region is usually the result of an NH or an OH stretching vibration; the possibility that the substance is an aliphatic alcohol must thus be considered. The empirical formula, however, is inconsistent with a structure that contains both an aldehyde and a hydroxyl group. Examination of the lower-frequency region fails to shed further information. This dilemma can be resolved by attributing absorption at 3450 cm^{-1} to the strong OH stretching band of water that is present as a contaminant. Thus, we conclude that the sample is probably an aliphatic aldehyde. We are unable to determine the extent of chain branching from the spectrum; to be sure, the peak at 730 cm^{-1} suggests the presence of

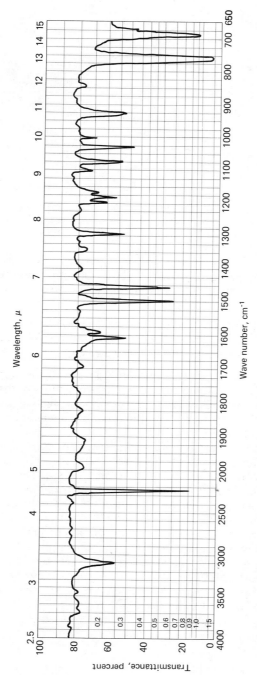

Figure 6-8. Spectrum of a colorless liquid. (Spectrum courtesy of Thermodynamics Research Center Data Project, Texas A. & M. University, College Station, Texas.)

159

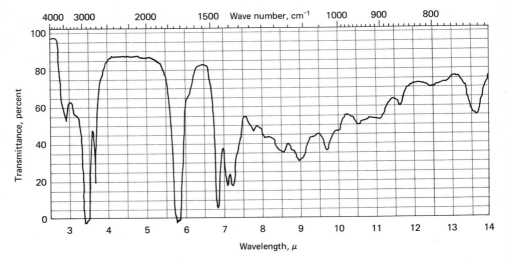

Figure 6-9. Infrared spectrum of an unknown. (Spectrum courtesy of Thermodynamics Research Center Data Project, Texas A. & M. University, College Station, Texas.)

four or more methylene groups in a row. The boiling point of *n*-hexanal is 131°C; we are thus inclined to believe that the sample is this compound. Confirmatory tests would be needed, however.

Example. The spectrum shown in Figure 6-10 was obtained on a 0.5 percent solution of a liquid sample in a 0.5-mm cell. For the region of 2 to 8 μ the solvent was CCl_4, and for the higher wavelengths CS_2 was employed. The empirical formula of the compound was found to be $C_8H_{10}O$. Suggest a probable structure.

The empirical formula indicates that the compound is probably aromatic in nature, and this supposition is borne out by the characteristic pattern of peaks in the 1450 to 1600 cm^{-1} range. The C—H stretching bands, however, are somewhat lower in wave number than those of a purely aromatic system, which suggests the existence of some aliphatic groups as well. The broad band at 3300 cm^{-1} appears important, and implies an OH or an N—H stretching vibration. From the formula we conclude that the molecule must thus contain a phenolic or an alcoholic OH group. From consideration of Figure 6-7 we conclude that the spectrum in the region of 1100 to 1400 cm^{-1} is compatible with the postulated OH group, but we are unable to decide from the spectral data whether the sample is a phenol or an alcohol.

From our observations thus far and from the empirical formula, likely structures for the unknown appear to be:

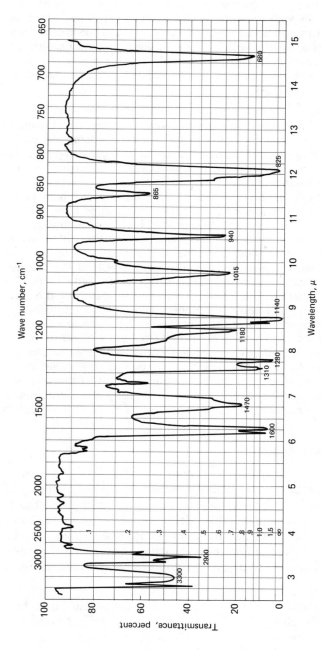

Figure 6-10. Spectrum of a sample in CCl_4 (2 to 8 μ) and CS_2 (8 to 15 μ). (Spectrum courtesy of Thermodynamics Research Center Data Project, Texas A. & M. University, College Station, Texas.)

We now turn to the low-frequency end of the spectrum where correlations of peaks with aromatic substitution patterns can often be found. The two strong peaks at 680 and 825 cm^{-1} and the weaker one at 865 cm^{-1} appear to fit best with the pattern for a symmetrically trisubstituted benzene (Figure 6-7). Thus, 3,5-dimethylphenol seems the most logical choice. Confirmatory data would be needed, however, to establish the identity unambiguously.

Example. A colorless organic liquid (boiling point 67°C) containing only carbon, hydrogen, and oxygen was examined spectroscopically. A dilute alcoholic solution of the material had no absorption peaks in the region of 220 to 400 nm. The neat liquid in a 0.01-mm cell gave the infrared spectrum shown in Figure 6-11. What structural information can be obtained from these data?

The lack of absorption peaks in the ultraviolet region immediately eliminates aromatic and carbonyl functional groups from consideration. The infrared spectrum in the 2800 to 3100 cm^{-1} region suggests that both olefin and paraffin groups may be present. We further note that the strong peaks at 925 and 1000 cm^{-1} and the weaker peak at 1650 cm^{-1} would be compatible with the presence of a vinyl group ($-CH{=}CH_2$). The low, broad band at 3500 cm^{-1} could well be due to a water contaminant. Thus, we turn to the most obvious characteristic of the spectrum, the very strong band appearing just above 1100 cm^{-1}. The position and breadth of this band immediately suggest the presence of an aliphatic ether. The series of peaks in the region of 1350 to 1000 cm^{-1} are difficult to interpret, but it seems probable that they are associated with the various hydrogen bending vibrations in the hydrocarbon part of the molecule. Without further information, then, we can only conclude that the sample is most probably an aliphatic ether having a vinyl group in its structure. (In fact, the infrared spectrum is that of allyl ethyl ether, $CH_2{=}CHCH_2OCH_2CH_3$.)

Collections of Spectra

As may be seen from the foregoing examples, correlation charts seldom suffice for the positive identification of an organic compound from its infrared spectrum. There are available, however, several catalogs of infrared spectra that assist in qualitative identification by providing comparison spectra for a large number of pure compounds. These collections have become so extensive as to require edge-punched cards or IBM cards for efficient retrieval. The principal spectral collections and indexes include the following.

1. *Sadtler Standard Spectra,* Samuel P. Sadtler and Sons, Inc., 2100 Arch Street, Philadelphia, Pa. This is probably the largest collection available. Spectra can be obtained on cards or $8\frac{1}{2} \times 11$ inch loose leaf pages (three spectra per page). A special index for locating spectra is also provided.

2. *American Petroleum Institute Research Project 44* and the *Manufacturing Chemists Association Research Project,* Chemical and Thermo-

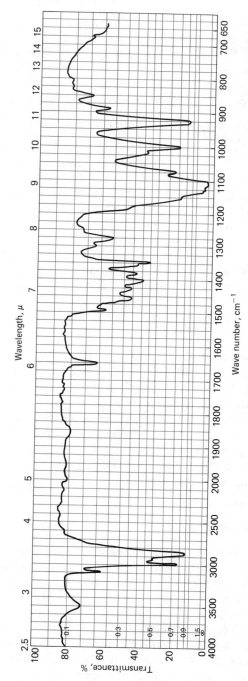

Figure 6-11. Spectrum of an organic liquid. (Spectrum courtesy of Thermodynamics Research Center Data Project, Texas A. & M. University, College Station, Texas.)

dynamics Properties Center, Texas A & M, College Station, Texas. Spectra are reproduced on $8\frac{1}{2} \times 11$ inch loose sheets.

3. *Documentation of Molecular Spectroscopy*, Butterworth Scientific Publications, London. Spectra are reproduced on edge-punched cards.

4. *National Research Council* — National Bureau of Standards, Washington, D.C. Spectra are reproduced on edge-punched cards.

Several indexes of literature references to the spectra of specific compounds are also available. These include:

1. *Molecular Formula List of Compounds, Names and References to Published Infrared Spectra*, American Society for Testing Materials, 1916 Race Street, Philadelphia, Pa. It includes listings of all of the spectra in the collections listed above plus many others. Spectral information is punched on IBM cards, which can be mechanically sorted by positions of peaks, by specific compounds, or by classes of compounds.

2. *Documentation of Molecular Spectroscopy*, Butterworth Scientific Publications, London. References to spectra are on edge-punched cards.

3. *An Index to Published Infrared Spectra*, H. M. Stationary Office, London.

QUANTITATIVE APPLICATIONS

In Chapters 3 and 4 we considered in detail the theory and the practical aspects of quantitative measurements based upon the absorption of ultraviolet and visible radiation. Most of this material applies equally well to infrared radiation; thus, the present discussion is confined to the differences in technique necessitated by the greater complexity of infrared spectra, the narrowness of infrared absorption bands, and certain instrumental limitations of infrared spectrophotometers.

Deviations from Beer's Law

With infrared radiation, instrumental deviations from Beer's law are more common than with ultraviolet and visible wavelengths because infrared absorption bands are relatively narrow. Furthermore, the low intensity of sources and low sensitivities of detectors in the infrared region require the use of relatively wide monochromator slit widths; thus, the band widths employed are frequently of the same order of magnitude as the widths of absorption peaks. As we have pointed out (see Figure 3-3), this combination of circumstances usually leads to a nonlinear relationship between absorbance and concentration. Calibration curves, determined empirically, are therefore required for quantitative work.

Absorbance Measurement

Reference absorber. In the ultraviolet and visible regions matched absorption cells for solvent and solution are commonly employed, and the measured absorbance is then found from the relation

$$A \cong \log \frac{P_{\text{solvent}}}{P_{\text{solution}}}$$

The use of the solvent in a matched cell as a reference absorber has the advantage of largely canceling out the effects of radiation losses due to reflection at the various interfaces, scattering and absorption by the solvent, and absorption by the container windows (p. 31). This technique is usually not practical for infrared radiation, however, because of the difficulty in obtaining cells whose transmission characteristics are identical. As we have noted, most infrared cells have very short path lengths that are difficult to duplicate exactly. In addition, the cell windows are readily attacked by contaminants in the atmosphere and the solvent; thus, their transmission characteristics change continually with use. For these reasons a reference absorber is often dispensed with entirely in infrared work, and the intensity of the beam passing through the sample is simply compared with that of the unobstructed beam; alternatively, a salt plate may be placed in the reference beam.

Recorder calibration. Nearly all infrared instruments are of a double-beam design and equipped for automatic recording. For quantitative work it is essential that calibration of the recorder scale for 0 and 100 percent transmittance be carefully and frequently performed. Ordinarily, the 100 percent adjustment is readily accomplished by simply removing the cell from the sample beam path (if no cell is being employed for the reference beam). On the other hand, care must be exercised in calibrating to zero transmittance because most infrared instruments are of the optical null design. As a consequence, when the sample beam is blocked from the detector, the attenuator moves to reduce the reference beam intensity to zero as well. Under these circumstances essentially no energy reaches the detector and the exact null position cannot be located with precision. In practice, the intensity of the sample beam is slowly diminished by the gradual introduction of a shutter across its path; in this way the tendency for the pen drive to overshoot the zero is avoided and a more accurate calibration is achieved. Figure 6-9 illustrates the tendency of a typical recording instrument to overshoot at zero transmittance. Fortunately, the inevitable small error arising from the uncertainty in the zero is not serious provided the transmittances being measured are not too low (< 10 percent).

Determination of P_0 and transmittance. Two methods are employed for the determination of P_0. In the so-called *cell in-cell out*, or point, method the spectra of the solvent and the sample are obtained successively with respect to the unobstructed reference beam. The same cell is used for each measurement. The transmittance of each solution versus the reference beam is then determined at an absorption maximum of the constituent of interest. We can then write

$$T_0 = P_0/P_r$$

and

$$T_s = P/P_r$$

where P_r is the power of the reference beam and T_0 and T_s are the transmittances of the solvent and sample, respectively, against this reference. If P_r remains constant during the two measurements, then the transmittance of the sample versus the solvent can be obtained by division of the two equations. That is,

$$T = T_s/T_0 = P/P_0$$

An alternative way of obtaining P_0 and T is the *base-line* method, in which the solvent transmittance is assumed to be constant or at least to change linearly between the shoulders of the absorption peak. This technique is demonstrated in Figure 6-12.

Advantages of Quantitative Infrared Spectroscopy

With the exception of homonuclear molecules, all organic and inorganic molecular species absorb in the infrared region; thus, infrared spectrophotometry offers the potential for the determination of an unusually large number of substances. Furthermore, the uniqueness of infrared spectra leads to a degree of specificity that is matched or exceeded by relatively few other analytical methods. This specificity has found particular application to analysis of mixtures of closely related organic compounds.

A typical example of the application of quantitative infrared spectroscopy involves the determination of the constitution of a mixture of C_8H_{10} isomers which includes *o*-xylene, *m*-xylene, *p*-xylene, and ethylbenzene. The infrared absorption spectra of cyclohexane solutions of each of these components in the 12 to 15 μ range is shown in Figure 6-13. Useful

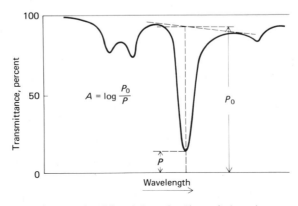

Figure 6-12. Base-line method for determination of absorbance.

absorption peaks for determination of the individual compounds occur at 13.47, 13.01, 12.58, and 14.36 μ, respectively. Unfortunately, however, the absorbance of a mixture at any one of these wavelengths is not entirely determined by the concentration of just one of the components because of overlapping absorption bands. If Beer's law is followed, the absorbance a_1 at wavelength 1 is given by the expression

$$a_1 = k_{11}c_1 + k_{12}c_2 + k_{13}c_3 + k_{14}c_4$$

where k_{11} is the absorptivity of component 1 at this wavelength and c_1 is its concentration; similarly, k_{12} is the absorptivity of component 2 at wavelength 1, c_2 is its concentration, and so forth. Three other analogous equations describe the absorbances at wavelengths 2, 3, and 4. The absorptivities of each pure component at each of the four wavelengths is determined with standard solutions of the individual compounds. The analysis of a mixture then requires the measurement of absorbance at the four wavelengths followed by the solution of a set of four simultaneous equations for c_1, c_2, c_3, and c_4. Such calculations are most easily performed with a computer.

When the relationship between absorbance and concentration is nonlinear (as frequently occurs in the infrared), the algebraic manipulations associated with an analysis of several components having overlapping absorption peaks are considerably more complex.[6]

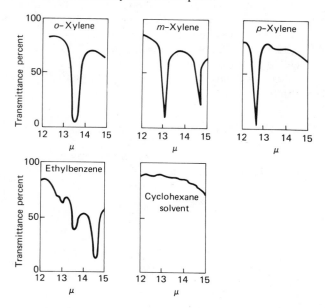

Figure 6-13. Spectra of C_8H_{10} isomers in cyclohexane solvent. From R. P. Bauman, *Absorption Spectroscopy* (New York: John Wiley & Sons, Inc., 1962), p. 406. With permission.

[6]For a discussion of the treatment of infrared data for various types of mixtures see R. P. Bauman. *Absorption Spectroscopy* (New York: John Wiley & Sons, Inc., 1962), pp. 403–419.

Disadvantages and Limitations to Quantitative Infrared Methods

Several disadvantages attend the application of infrared methods to quantitative analysis. Among these are the frequent nonadherence to Beer's law and the complexity of spectra; the latter enhances the probability of the overlap of absorption peaks. In addition, the narrowness of peaks and the effects of stray radiation make absorbance measurements critically dependent upon the slit width and the wavelength setting. Finally, the narrow cells required for many analyses are inconvenient to use and may lead to significant analytical uncertainties.

The analytical errors associated with a quantitative infrared analysis can seldom be reduced to the level associated with ultraviolet and visible methods even with considerable care and effort.

Problems

1. The infrared spectrum of CO shows a vibrational absorption peak at 2170 cm^{-1}.
   - (a) What is the force constant for the CO bond?
   - (b) At what wave number would the corresponding peak for ^{14}CO occur?
2. The force constant for the bond in HF is about 9×10^5 dynes/cm.
   - (a) Calculate the vibrational absorption peak for HF.
   - (b) Calculate the absorption peak for DF.
3. Indicate whether the following vibrations will be active or inactive in the infrared spectrum.

Molecule	Motion
(a) $CH_3—CH_3$	C—C stretching
(b) $CH_3—CCl_3$	C—C stretching
(c) SO_2	Symmetric stretching
(d) $CH_2{=}CH_2$	C—H stretching:

| (e) $CH_2{=}CH_2$ | C—H stretching: |

(f) CH_2=CH_2

CH_2 wag:

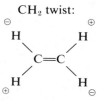

(g) CH_2=CH_2

CH_2 twist:

4. Calculate the absorption frequency corresponding to the —C—H stretching vibration treating the group as a simple diatomic C—H molecule. Compare the calculated value with the range found in correlation charts. Repeat the calculation for the deuterated bond.

5. The spectrum in Figure 6-14 (page 170) was obtained for a liquid with an empirical formula of C_3H_6O. Identify the compound.

6. The spectrum in Figure 6-15 (page 171) is that of a high-boiling liquid having an empirical formula $C_9H_{10}O$. Identify the compound as closely as possible.

7. The spectrum in Figure 6-16 (page 172) is for an acrid-smelling liquid that boils at 52°C. What is the compound? What impurity is clearly present?

8. The spectrum in Figure 6-17 (page 173) is that of a nitrogen-containing substance that boils at 97°C. What is the compound?

9. The spectrum in Figure 6-18 (page 174) was obtained from CCl_4 and CS_2 solutions of a white crystalline compound having a melting point of 54°C and an empirical formula of $C_{12}H_{11}N$. Identify the compound.

10. The spectra in Figure 6-19 (page 175) are for a pure liquid compound in cells of a different length. What conclusion can be drawn as to the probable nature of the compound?

11. The spectrum in Figure 6-20 (page 176) is for a pure liquid compound containing only C, H, and O. The boiling point of the liquid is about 179°C and its molecular weight is 106. Identify the compound.

12. The empirical formula for a liquid compound was found to be $C_8H_{12}O_2$. Its spectrum is shown in Figure 6-21 (page 177). What conclusion can be drawn regarding the identity of the compound?

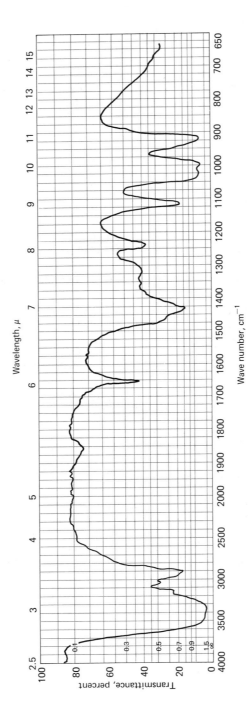

Figure 6-14. (Spectrum courtesy of Thermodynamics Research Center Data Project, Texas A. & M. University, College Station, Texas.) See Problem 5.

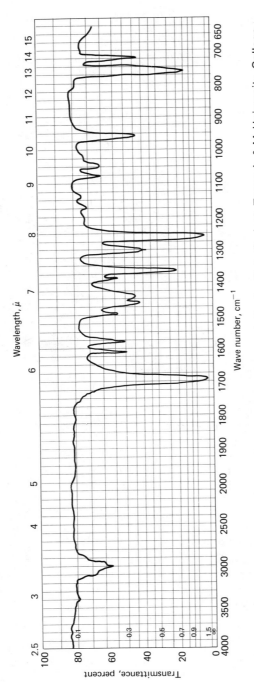

Figure 6-15. (Spectrum courtesy of Thermodynamics Research Center Data Project, Texas A. & M. University, College Station, Texas.) See Problem 6.

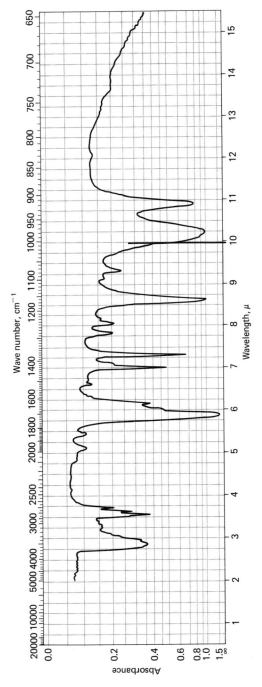

Figure 6-16. (Spectrum courtesy of Thermodynamics Research Center Data Project, Texas A. & M. University, College Station, Texas.) See Problem 7.

172

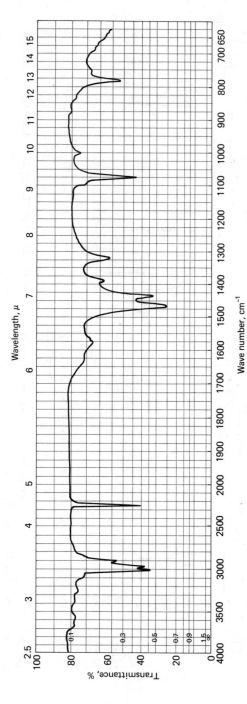

Figure 6-17. (Spectrum courtesy of Thermodynamics Research Center Data Project, Texas A. & M. University, College Station, Texas.) See Problem 8.

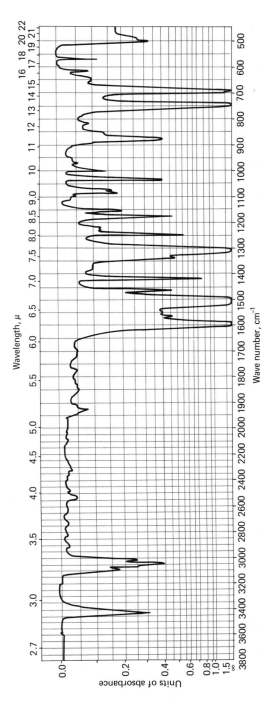

Figure 6-18. (Spectrum courtesy of Thermodynamics Research Center Data Project, Texas A. & M. University, College Station, Texas.) See Problem 9.

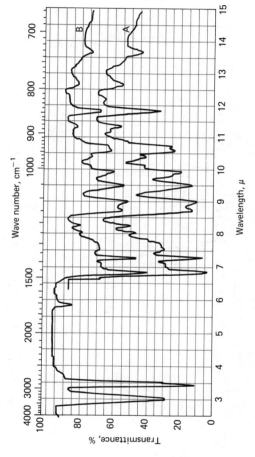

Figure 6-19. (Spectrum courtesy of Thermodynamics Research Center Data Project, Texas A. & M. University, College Station, Texas.) See Problem 10.

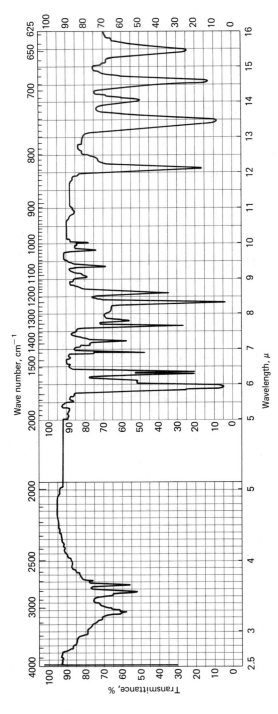

Figure 6-20. (Spectrum courtesy of Thermodynamics Research Center Data Project, Texas A. & M. University, College Station, Texas.) See Problem 11.

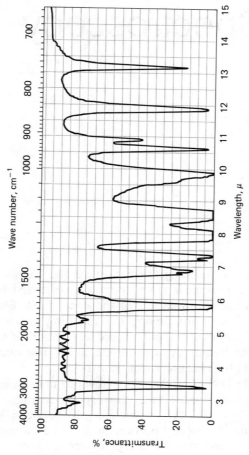

Figure 6-21. (Spectrum courtesy of Thermodynamics Research Center Data Project, Texas A. & M. University, College Station, Texas.) See Problem 12.

7 Nuclear Magnetic Resonance Spectroscopy

In the presence of a strong magnetic field the energies of the nuclei of certain elements are split into two or more quantized levels as a consequence of the magnetic properties of these particles. Electrons behave in a similar way. Transitions among the resulting magnetically induced energy levels can be brought about by the absorption of electromagnetic radiation of suitable frequency, just as electronic transitions are caused by the absorption of ultraviolet or visible radiation.

The energy differences between the magnetic quantum levels for atomic nuclei are of such magnitudes as to correspond to radiation in the frequency range of 0.1 to 100 MHz[1] (wavelengths between 3000 and 3 m), which is in the radio frequency portion of the electromagnetic spectrum (see Figure 2-4). For electrons the energy differences are much larger than for nuclei. Here the corresponding electromagnetic energies are in the frequency range of 10,000 to 80,000 MHz (wavelengths 3 to 0.375 cm), which falls in the microwave spectral region.

The study of absorption of radio-frequency radiation by nuclei is called *nuclear magnetic resonance*[2] (often abbreviated nmr or NMR); it

[1] MHz $= 10^6$ Hz $= 10^6$ cycles per second.

[2] The following references are recommended for additional study: E. D. Becker, *High Resolution NMR* (New York: Academic Press, Inc., 1969); F. A. Bovey, *Nuclear Magnetic Resonance Spectroscopy* (New York: Academic Press, Inc., 1969); J. W. Emsley, *High Resolution Nuclear Magnetic Resonance Spectroscopy*, 2 vols. (New York: Pergamon Press,

has proved to be one of the most powerful tools available for the determination of the structure of both organic and inorganic species. Electron spin resonance (esr or ESR) refers to the absorption of microwave radiation by electrons in a magnetic field; it also provides useful structural information that is of a less general nature than nuclear magnetic resonance.

In this chapter we are mainly concerned with the theory, the instrumentation, and the applications of nmr spectroscopy. From time to time, however, it will be convenient to mention the analogous features of esr spectroscopy.

THEORY OF NUCLEAR MAGNETIC RESONANCE

As early as 1924 Pauli suggested that certain atomic nuclei might have the properties of spin and of magnetic moment, and that as a consequence, exposure to a magnetic field would lead to splitting of their energy levels. During the next decade experimental verification of these postulates was obtained both for nuclei and for the electron. It was not until 1946, however, that Bloch at Stanford and Purcell at Harvard, working independently, were able to demonstrate the absorption of electromagnetic radiation as a consequence of energy-level transitions of nuclei in a strong magnetic field. In 1952 the two physicists were awarded the Nobel prize for their work. In the first five years following the discovery of nuclear magnetic resonance, chemists became aware that the molecular environment influences the absorption by a nucleus in a magnetic field and that this effect can be correlated with molecular structure. Since then the growth of nuclear magnetic resonance spectroscopy has been explosive, and the technique has had profound effect on the development of organic chemistry, inorganic chemistry, and biochemistry. It is doubtful that there has ever been as short a delay between an initial discovery and its widespread application and acceptance.

Angular Momentum of Elementary Particles

In order to account for some of the properties of elementary particles, such as the electron or a nucleus, it is necessary to assume that they rotate about an axis and thus have the property of *spin*. Furthermore, it is necessary to assume that the angular momentum associated with the spin of the

Inc., 1965); L. M. Jackman and S. Sternhall, *Nuclear Magnetic Resonance Spectroscopy*, 2d ed. (New York: Pergamon Press, Inc., 1969); J. A. Pople, W. G. Schneider and H. J. Bernstein, *High-resolution Nuclear Magnetic Resonance* (New York: McGraw-Hill Book Company, Inc., 1959); J. D. Roberts, *Nuclear Magnetic Resonance* (New York: McGraw-Hill Book Company, Inc., 1959); R. M. Silverstein and G. C. Bassler, *Spectrometric Identification of Organic Compounds*, 2d ed. (New York: John Wiley & Sons, Inc., 1967), Chap. 4.

Table 7-1 Spin Quantum Number for Various Nuclei

Number of Protons	Number of Neutrons	Spin Quantum Number I	Examples
Even	Even	0	^{12}C, ^{16}O, ^{32}S
Odd	Even	$\frac{1}{2}$	^{1}H, ^{19}F, ^{31}P
		$\frac{3}{2}$	^{11}B, ^{79}Br
Even	Odd	$\frac{1}{2}$	^{13}C
		$\frac{3}{2}$	^{127}I
Odd	Odd	1	^{2}H, ^{14}N

particle must be an integral or a half-integral multiple of $h/2\pi$, where h is Planck's constant. The maximum spin component for a particular particle is its *spin quantum number I*; it is found that the particle will then have $(2I + 1)$ discrete states. The component of angular momentum for these states in any chosen direction will have values of $I, I - 1, I - 2, \cdots, -I$. In the absence of an external field the various states have identical energies.

The spin number for both the electron and the proton is $\frac{1}{2}$; thus each has two spin states, corresponding to $I = +\frac{1}{2}$ and $I = -\frac{1}{2}$. Heavier nuclei, being assemblages of various elementary particles, have spin numbers that range from zero (no net spin component) to at least 9/2. As shown in Table 7-1, the spin number of a nucleus is related to the relative number of protons and neutrons it contains.

Magnetic Properties of Elementary Particles

Since a nucleus (or an electron) bears a charge, its spin gives rise to a magnetic field that is analogous to the field produced when an electric current is passed through a coil of wire. The resulting magnetic dipole μ is oriented along the axis of spin and has a value that is characteristic for each kind of particle.

Magnetic quantum number m. The interrelation between particle spin and magnetic moment leads to a set of observable magnetic quantum states given by

$$m = I, I - 1, I - 2, \cdots, -I \tag{7-1}$$

Energy levels in a magnetic field. When brought into the influence of an external magnetic field, a particle that possesses a magnetic moment tends to become oriented such that its magnetic dipole, and hence its spin axis, is parallel to the field. The behavior of the particle is somewhat like that of a small bar magnet when introduced into such a field, the potential energy of either being dependent upon the orientation of the dipole with respect to the field. With the magnet, this energy can assume an infinite number of values depending upon its alignment; in contrast, however, the energy of the atomic particle is limited to $(2I + 1)$ discrete values (that is, the alignment is limited to $2I + 1$ positions). In either the quantized or the nonquantized case the potential energy is given by the relationship

$$E = -\mu_H H_0 \tag{7-2}$$

where μ_H is the *component* of magnetic moment in the direction of the field and H_0 is the strength of the external field.

The quantum character of atomic particles limits to a few the number of possible energy levels. Thus, for a particle with a spin number of I and a magnetic quantum number of m the energy of a quantum level is given by

$$E = -\frac{m\mu}{I} \beta H_0 \tag{7-3}$$

where H_0 is the strength of the external field in gauss and β is a constant called the nuclear magneton, 5.049×10^{-24} erg-gauss^{-1}; μ is the magnetic moment of the particle expressed in units of nuclear magnetons. The value of μ for the proton is 2.7927 nuclear magnetons, while for the electron it is -1836.

Turning now to the proton, for which $I = \frac{1}{2}$, we see from equation (7-1) that this particle has magnetic quantum numbers of $+\frac{1}{2}$ and $-\frac{1}{2}$. The energies of these states in a magnetic field (equation 7-3) take the following values:

$$m = +\frac{1}{2} \qquad E = -\frac{1/2 \, (\mu \, \beta \, H_0)}{1/2}$$

$$= -\mu \, \beta \, H_0$$

$$m = -\frac{1}{2} \qquad E = -\frac{-1/2 \, (\mu \, \beta \, H_0)}{1/2}$$

$$= +\mu \, \beta \, H_0$$

These two quantum levels correspond to the two possible orientations of the spin axis with respect to the magnetic field; as shown in Figure 7-1, for the lower-energy state ($m = \frac{1}{2}$) the vector of the magnetic moment is aligned with the field and for the higher-energy state ($m = -\frac{1}{2}$) the alignment is reversed. The energy difference between the two levels is given by

$$\Delta E = 2\mu \, \beta \, H_0$$

Also shown in Figure 7-1 are the orientations and energy levels for a nucleus such as ^{14}N, which has a spin number of 1. Here three energy levels ($m = 1, 0$, and -1) are found, and the difference in energy between each is $\mu \, \beta \, H_0$. In general, the energy differences are given by

$$\Delta E = \mu \, \beta \, \frac{H_0}{I} \tag{7-4}$$

As with other types of quantum states, excitation to a higher nuclear magnetic quantum level can be brought about by absorption of a photon with energy $h\nu$ that just equals ΔE. Thus, equation (7-4) can be written as

$$h\nu = \mu \beta \frac{H_0}{I} \qquad (7\text{-}5)$$

In nuclear magnetic resonance studies, field strengths of about 10^4 gauss are employed. Thus, for the proton to absorb, the radiation frequency must be

$$\nu \cong \frac{2.79 \times 5.05 \times 10^{-24} \times 10^4}{6.6 \times 10^{-27} \times 1/2}$$

$$\cong 4 \times 10^7 \text{ Hz}$$

which lies in the radio-frequency range. Excitation by such radiation in-

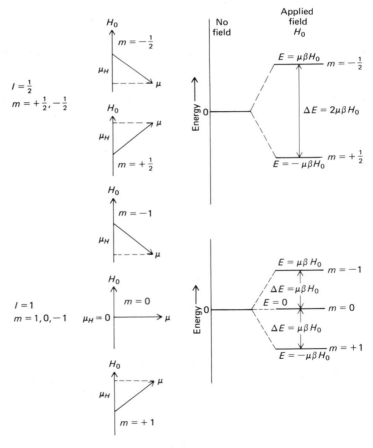

Figure 7-1. Orientations of magnetic moments and energy levels for particles in a magnetic field H_0.

volves a change in alignment of the magnetic moment of the proton from a direction that parallels the field to a direction that opposes the field.

Distribution of particles between magnetic quantum states. In the absence of a magnetic field the energies of the magnetic quantum states are identical. Consequently, under this circumstance a large assemblage of protons contains an identical number of nuclei with $m = +\frac{1}{2}$ and $m = -\frac{1}{2}$. When placed in a field, however, the nuclei tend to orient themselves so that the lower-energy state ($m = +\frac{1}{2}$) predominates. Because thermal energies at room temperatures are several orders of magnitude greater than these magnetic energy differences, thermal agitation tends to offset the magnetic effects, and only a small (< 10 ppm) excess of nuclei persists in the lower-energy state.

The success of nuclear magnetic resonance depends upon the slight excess of nuclei in the lower-energy state. If the number of protons in the two states were identical, the probability for absorption of radiation would be the same as the probability for reemission by particles passing from the higher-energy state to the lower. Under these circumstances the net absorption would be nil.

Absorption of Radiation

Thus far our discussion of the theory of magnetic resonance has been largely based upon quantum mechanical considerations. To understand the absorption process, and in particular the measurement of absorption, a more classical picture of the behavior of a charged particle in a magnetic field is helpful.

Precession of particles in a field. Let us first consider the behavior of a nonrotating magnetic body, such as a compass needle, in an external magnetic field. The needle, if not aligned, will swing in a plane about its pivot as a consequence of the force exerted by the field on its two ends; in the absence of friction the ends of the needle then would fluctuate back and forth about the axis of the field. A quite different motion occurs because of the gyroscopic effect, however, if the magnet is spinning rapidly around its north-south axis. Here the force applied by the field to the axis of rotation causes movement not in the plane of the force but at right angles to the plane; the axis of the rotating particle, therefore, moves in a circular path (or *precesses*) around the magnetic field. This motion, illustrated in Figure 7-2, is similar to the motion of a gyroscope when it is displaced from the vertical by application of a force.

From classical mechanics it is known that the angular velocity of precession is directly proportional to the applied force and inversely proportional to the angular momentum of the spinning body to which the force is applied. The force on a spinning nucleus in a magnetic field is the product of the field strength H_0 and the magnetic moment of the particle μ β; as noted earlier, the angular momentum is given by $I(h/2\pi)$. Therefore, the precessional velocity is

$$\omega_0 = \frac{2\pi\ \mu\ \beta}{I\ h} \cdot H_0 = \gamma\ H_0 \qquad (7\text{-}6)$$

where γ is a constant called the *magnetogyric ratio* (or less appropriately the gyromagnetic ratio). The magnetogyric ratio expresses the relation between the magnetic moment and the angular momentum of a rotating particle; that is,

$$\gamma = \frac{\mu\ \beta}{I(h/2\pi)} \qquad (7\text{-}7)$$

The magnetogyric ratio has a characteristic value for each type of nucleus.

Equation (7-6) can be converted to a frequency of precession ν_0 (the *Larmor frequency*) by division by 2π. Thus,

$$\nu_0 = \frac{\omega_0}{2\pi} = \frac{\gamma\ H_0}{2\pi} \qquad (7\text{-}8)$$

Equations (7-8) and (7-7) can also be combined to give

$$h\ \nu_0 = \frac{\mu\ \beta}{I}\ H_0 \qquad (7\text{-}9)$$

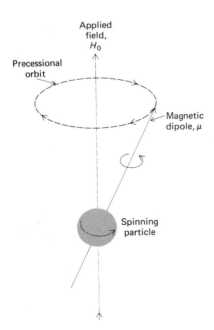

Figure 7-2.   Precession of a rotating particle in a magnetic field.

A comparison of equations (7-9) and (7-5) suggests that the precessional frequency of the particle derived from classical mechanics is identical to the quantum mechanical frequency of radiant energy required to bring about the transition of a rotating particle from one spin state to another; that is $\nu_0 = \nu$. Substituting this equality into equation (7-8) gives a useful relationship between the frequency of absorbed radiation and the strength of the magnetic field:

$$\nu = \frac{\gamma H_0}{2\pi} \tag{7-10}$$

Absorption process. The classical model of the precessing magnetic dipole can be extended to provide a picture of the mechanism by which absorption takes place. We imagine the absorption process as involving a flipping of the magnetic moment that is oriented in the field direction to a state in which the moment is in the opposite direction. The process is pictured in Figure 7-3. In order for the dipole to flip there must be a magnetic force at right angles to the fixed field and one that has a circular component that can move in phase with the precessing dipole. Circularly polarized radiation (p. 319) of a suitable frequency has these necessary properties; that is, its magnetic vector has a circular component as represented by the dotted line in Figure 7-3. If the rotational frequency of the magnetic vector of the radiation is the same as the precessing frequency, absorption and flipping can occur. The process is reversible, and the excited particle can thus return to the ground state by reemission of the radiation.

As is shown in Chapter 13, plane-polarized radiation can be considered to be composed of two circularly polarized beams rotating in opposite directions, in phase, and in a plane at 90 deg to the plane of linear polarization. Thus, by irradiating nuclear particles with a beam polarized at 90 deg to the direction of the fixed magnetic field, circularly polarized radiation is introduced in the proper plane for absorption. Only that component of the beam that rotates in the precessional direction is absorbed; the other half of the beam, being out of phase, passes through the sample unchanged. The process is depicted in Figure 7-4.

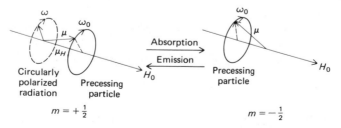

Figure 7-3. Model of the absorption of radiation by a precessing particle.

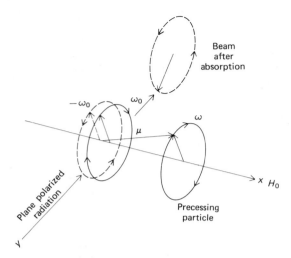

Figure 7-4. Absorption of one circular component of a beam that is polarized in the *xy* plane.

Relaxation Processes

We must now consider mechanisms by which a nucleus in an upper-energy or excited spin state can return to its lower-energy state. One obvious path would involve emission of radiation of a frequency corresponding to the energy difference between the states. Radiation theory, however, predicts a low probability of occurrence for this process; it is necessary, therefore, to postulate radiationless pathways by which energy can be lost by nuclei in the higher spin states. The various mechanisms for radiationless energy transfers are termed nuclear *relaxation processes*.

The rates at which relaxation processes occur affect the nature and the quality of an nmr signal; to some extent these rates are subject to experimental control because they do depend upon the physical state of the sample. It is clear that relaxation processes are needed for the observation of a steady nmr absorption signal. Recall that such a signal depends upon the few parts per million excess of lower-energy nuclei that exist in a strong magnetic field. Since absorption depletes this excess, the signal would rapidly fall to zero if additional low-energy particles were not being produced at a sufficient rate by some radiationless energy-transfer processes. These processes must also be responsible for the creation of the slight excess of lower-energy particles when the sample is first introduced into the magnetic field.

To produce a readily detectable absorption signal, then, the relaxation process should be as rapid as possible; that is, the lifetime of the excited state should be small. A second factor, namely, the inverse relationship between the lifetime of an excited state and the width of its absorption line, negates the advantage of very short lifetimes. Thus, when relaxation rates are high or the lifetimes low, line broadening is observed which prevents high-resolution measurements. As a consequence of these two opposing factors, the optimum half-life for an excited species ranges from perhaps 0.1 to 1 second.

Two types of nuclear relaxation are recognized. The first is called *longitudinal* or *spin-lattice* relaxation; the second is termed *transverse* or *spin-spin* relaxation.

Spin-lattice relaxation. The absorbing nuclei in an nmr experiment are part of a larger assemblage of atoms that constitutes the sample. The entire assemblage is termed the lattice regardless of whether the sample is a solid, a liquid, or a gas. In the latter two states particularly, the various nuclei comprising the lattice are in violent vibrational and rotational motion which creates a complex magnetic field about each magnetic nucleus. The resulting lattice field thus contains an infinite number of magnetic components, at least some of which must correspond in frequency and phase with the precessional frequency of the magnetic nuclei of interest. These vibrationally and rotationally developed components are capable of interacting with and converting nuclei from a higher to a lower spin state; the absorbed energy then simply increases the amplitude of the thermal vibration or rotation. This change corresponds to a minuscule temperature rise for the sample.

Spin-lattice relaxation is a first-order process that can be characterized by a time T_1, which is a measure of the average lifetime of the nuclei in the higher-energy state. In addition to being dependent upon the magnetogyric ratio of the absorbing nuclei, T_1 is strongly affected by the mobility of the lattice. In crystalline solids and viscous liquids where mobilities are low, T_1 is large. As the mobility increases (at higher temperatures, for example), the vibrational and rotational frequencies increase and the probability for existence of a magnetic fluctuation of the proper magnitude for a relaxation transition is enhanced; thus T_1 becomes shorter. At very high mobilities, on the other hand, the fluctuation frequencies are further increased and spread over such a broad range that the probability of a suitable frequency for a spin-lattice transition again decreases. Thus, there is a minimum in the relationship between T_1 and lattice mobility.

The spin-lattice relaxation time is greatly shortened in the presence of an element with an unpaired electron which, because of its spin, creates strong fluctuating magnetic fields. A similar effect is caused by nuclei that have spin numbers greater than $\frac{1}{2}$. These particles are characterized by a nonsymmetrical charge distribution, and their rotation also produces a strong fluctuating field that provides yet another pathway for an excited nucleus to give up its energy to the lattice. Because of the marked shortening of T_1 in the presence of species such as these, line broadening is observed. An example is found in the nmr spectrum for the proton attached to a nitrogen atom (for ^{14}N, $I = 1$).

Spin-spin relaxation and line broadening. Several other effects tend to diminish relaxation times and thereby broaden nmr lines. These effects are normally lumped together and described by a spin-spin relaxation time T_2. Values for T_2 are generally so small for crystalline solids or viscous liquids (as low as 10^{-4} sec) as to prohibit the use of samples of these kinds for high-resolution spectra.

When two neighboring nuclei of the same kind have identical precession rates but are in different magnetic quantum states, the magnetic fields of each can interact to cause an interchange of states. That is, a nucleus in the lower-spin state can be excited while the excited nucleus relaxes to the lower-energy state. Clearly, no net change in the relative spin state population results, but the average lifetime of a particular excited nucleus is shortened. Line broadening is the result.

Two other causes of line broadening should be noted. Both arise if H_0 in equation (7-10) differs slightly from nucleus to nucleus; under these circumstances a corresponding band of frequencies rather than a single frequency is absorbed. One cause for this variation in the static field is the presence in the sample of other magnetic nuclei whose spins create local fields which may act to enhance or diminish the external field acting on the nucleus of interest. In a mobile lattice these local fields tend to cancel because of the rapid motions of the nuclei causing them. In a solid or a viscous liquid, however, the local fields may persist long enough to produce a range of field strengths and thus a range of absorption frequencies. Variations in the static field also result from small inhomogeneities in the field source itself. This effect can be largely offset by rapidly spinning the entire sample in the magnetic field.

Measurement of Absorption

Absorption signal. In all of the types of absorption spectroscopy that we have discussed thus far, the measurement has consisted of determining the decrease in the power (attenuation) of radiation caused by the absorbing sample. While this same technique has been applied in nmr spectroscopy, it suffers from the disadvantage that the excess of absorbing particles is so small that the resulting beam attenuation is difficult to measure accurately. Consequently, nearly all nmr spectrometers employ a method by which the magnitude of a positive, absorption related, signal is determined.

Figure 7-5 is a schematic representation of the three major components of an nmr spectrometer; these are described in greater detail in a subsequent section. The radiation source is a coil that is part of a radio-frequency oscillator circuit. Electromagnetic radiation from such a coil is plane-polarized (in the xz plane in the figure). The detector is a second coil located at right angles to the source (on the y axis in the figure) and is part of a radio-receiver circuit. The magnetic field used in nmr experiments has a direction along the z axis perpendicular with respect to both the source and the detector.

As shown in Figure 7-5(c), the plane-polarized radiation can be resolved into two circularly polarized vectors that rotate in opposite directions to one another (see also pp. 319 to 322). Vector addition of these components, regardless of their angular position, indicates that there is no net component along the y axis. Thus, no signal is received by a detector located on this axis.

Figure 7-5(b) and (d) show a sample positioned at the origin of the three axes. If the source has a frequency that is absorbed by one of the types of nuclei in the sample, the power of one of the two circular components of the beam is reduced. Vector addition of the resulting components indicates that the radiation now has a fluctuating component in the y direction, which causes the detector to respond. Thus, the sample serves to couple the detector to the receiver, provided the frequency of radiation corresponds to the precessing frequency of the nuclei in the magnetic field. The extent of coupling and thus the signal strength depends upon the number of absorbing nuclei.

Absorption spectra. In theory, an nmr spectrum could be obtained in the same way as an infrared or an ultraviolet spectrum. That is, the sample (held in a fixed field H_0) would be scanned with radiation of continuously varying frequency and the strength of the resulting signal would be measured. In practice, this technique is not practical because of the difficulty of

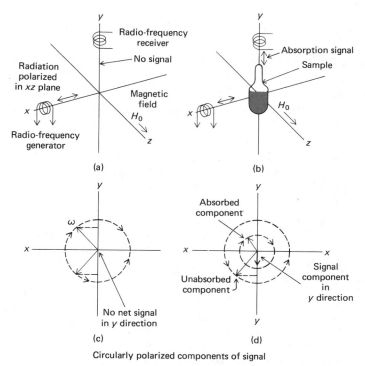

Circularly polarized components of signal

Figure 7-5. Absorption signal in nmr.

constructing a highly stable oscillator source whose frequency can be varied continuously; nor are there dispersing elements, analogous to a prism or a grating, for radio-frequency radiation. It is, however, feasible to hold the oscillator frequency constant and vary the field H_0 continuously. Since for a given nucleus, frequency and field strength are directly proportional (equation 7-10), H_0 can be used equally well as the abscissa for an absorption spectrum. Figure 7-6 shows such a plot. Table 7-2 presents nmr data for a number of the common nuclei.

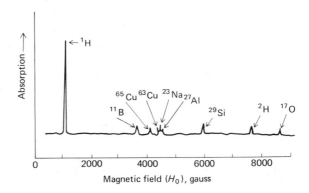

Figure 7-6. nmr Spectrum of water in glass container. Frequency = 5MHz; low resolution. (Courtesy of Varian Associates, Palo Alto, Calif.)

EXPERIMENTAL METHODS OF nmr SPECTROSCOPY

Nuclear magnetic resonance spectrometers are of two types: namely, *wide-line* instruments and *high-resolution* instruments. The latter are capable of resolving the fine-structure that is associated with the absorption peak for a given kind of nucleus; the nature of this fine-structure is determined by the chemical environment of the nucleus. Wide-line instruments cannot detect such detail; they are useful, however, for quantitative elemental analysis and for the study of the physical environment of a nucleus. Wide-line spectrometers are much simpler and less expensive than their high-resolution counterparts.

From an instrumental standpoint, the equipment required for high-resolution nmr spectroscopy is the most elaborate needed for any absorption method. Thus, most chemists are forced to accept only a general understanding of the operating principles of the nmr spectrometer, and leave its design and maintenance in the hands of those skilled in electronics. On the other hand, the sample handling techniques, the interpretation of the spectra, and an appreciation of the effects of variables are no more complex than for other types of absorption spectroscopy; it is upon these areas that the chemist focuses his attention.

Table 7-2 Magnetic Properties of Some Common Nuclei[a]

Nucleus	Absorption Frequency ν in MHz for a 10^4 Gauss Field	Magnetic Moment μ in Nuclear Magnetons	Isotopic Abundance, Percent	Sensitivity Relative to Equal Number of Protons	
				Constant H_0	Constant ν
^{1}H	42.577	2.7927	99.98	1.00	1.000
^{2}H	6.536	0.8574	1.56×10^{-2}	9.64×10^{-3}	0.409
^{13}C	10.705	0.7022	1.108	1.59×10^{-2}	0.251
^{14}N	3.076	0.4036	99.64	1.01×10^{-3}	0.193
^{17}O	5.722	−1.8930	3.7×10^{-2}	2.91×10^{-2}	1.58
^{19}F	40.055	2.6273	100	0.834	0.941
^{31}P	17.235	1.1305	100	6.64×10^{-2}	0.401
^{35}Cl	4.172	0.8209	75.4	4.71×10^{-3}	0.490
Electron	27,794	−1836	—	2.85×10^8	658

[a]Data taken from H. S. Gutowsky, in A. Weissberger, Ed., *Physical Methods of Organic Chemistry*, Vol 1, Part IV (New York: Interscience Publishers, Inc., 1960), p. 2674. With permission.

Instrumentation

A schematic diagram showing the important components of an nmr spectrometer is given in Figure 7-7. A brief description of each of these components follows.

1. The magnet. Both permanent and electromagnets of large dimensions are used in commercial nmr spectrometers. Typically, a field of about 14,000 gauss is provided between pole pieces that have diameters of 12 or more inches. The performance specifications for the magnet are stringent, especially for high-resolution work. The field produced must be homogeneous to 1 part in 10^8 within the sample area and must be stable to a similar degree for short periods of time. Elaborate instrumentation with feedback devices to correct for fluctuation is required to meet these specifications.

2. The magnetic field sweep. A pair of coils located parallel to the magnet faces permit alteration of the applied field over a small range. By varying a direct current through these coils, the effective field can be changed by a few hundred milligauss without loss of field homogeneity.

Ordinarily, the field strength is changed automatically and linearly with time and this change is synchronized with the linear drive of a chart recorder. For a 60-MHz instrument the sweep range is 1000 Hz (235 milligauss) or some integral fraction thereof.

3. The radio-frequency source. The signal from a radio-frequency oscillator (transmitter) is fed into a pair of coils mounted at 90 deg to the path of

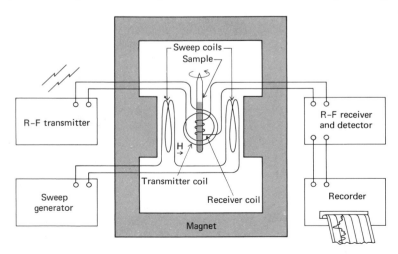

Figure 7-7. Schematic diagram of an nmr spectrometer. (Courtesy of Varian Associates, Palo Alto, Calif.)

the field. A plane-polarized beam of radiation results. A fixed oscillator of exactly 60 MHz is ordinarily employed; for high-resolution work the frequency must be constant to about 1 part in 10^8. The power output of this source is less than a watt and should be constant to perhaps 1 percent over a period of several minutes.

4. The signal detector and recorder system. The radio-frequency signal produced by the resonating nuclei is detected by means of a coil that surrounds the sample and is at right angles to the source coil. The electrical signal generated in the coils is small and must be amplified by a factor of 10^5 or more before it can be recorded.

5. The sample holder and sample probe. The usual nmr sample cell consists of a 5-mm O.D. glass tube that contains about 0.4 ml of liquid. Microtubes for smaller sample volumes are also available.

The sample probe is a device for holding the sample tube in a fixed spot in the field. As will be seen from Figure 7-8, the probe contains not only the sample holder but the sweep source and detector coils as well to assure reproducible positioning of the sample with respect to these components. The detector and receiver coils are oriented at right angles to one another to minimize the transfer of power in the absence of sample. Even with this arrangement some leakage of power between the source and receiver does occur; the so-called paddles, shown in Figure 7-8, reduce this leakage to a tolerable level.

The sample probe is also provided with an air-driven turbine for rotating the sample tube along its longitudinal axis at several hundred rpm. This rotation serves to average out the effects of inhomogeneities in the field; sharper lines and better resolution are obtained as a consequence.

Cost of nmr instruments. The foregoing brief discussion suggests that nmr instruments are complex. As might be expected, their cost is high,

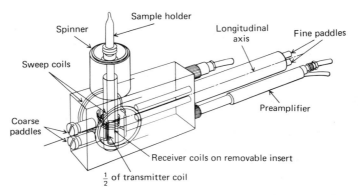

Figure 7-8. nmr probe. (Courtesy of Varian Associates, Palo Alto, Calif.)

ranging typically from $25,000 to $100,000. Less expensive instruments are, however, beginning to appear on the market.

Determination of Peak Areas

The area under an nmr absorption peak is directly proportional to the number of nuclei responsible for that peak. In contrast, peak heights are not entirely satisfactory measures of concentrations; a number of variables, including the rate of sample spinning, impurities in the sample, and chemical exchange reactions, can cause the broadening of peaks and a consequent lowering of their heights.

Area determinations permit the estimation of the relative number of absorbing nuclei in each chemical environment. This information is vital to the deduction of chemical structure. In addition, of course, such data can be employed for quantitative analytical purposes.

Area determinations for peaks on a chart paper can be made with a planimeter or by cutting out the peaks and weighing. Most nmr recorders are equipped with an electronic integrator which presents directly the relative peak areas on the ordinate scale of the chart (see Figure 7-9).

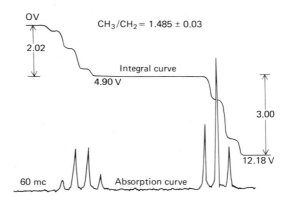

Figure 7-9. Absorption and integral curve for a dilute ethylbenzene solution (aliphatic region). (Courtesy of Varian Associates, Palo Alto, Calif.)

Sample Handling

For high-resolution work, samples must be in a nonviscous, liquid state. Most commonly solutions of the sample (2 to 15 percent) are employed. To be sure, the sample can be examined neat if it has suitable physical properties.

The best solvents for proton nmr spectroscopy contain no protons, and from this standpoint carbon tetrachloride is ideal. The low solubility of many compounds in carbon tetrachloride limits its value, however, and a variety of deuterated solvents are used instead. Deuterated chloroform, $CDCl_3$, and deuterated benzene are commonly encountered.

APPLICATION OF PROTON nmr SPECTROSCOPY TO STRUCTURAL STUDIES

Figure 7-6 is an example of a low-resolution nmr spectrum, in which each kind of nucleus is characterized by a single absorption peak, the location of which appears to be independent of the chemical state of the atom. If, however, the spectral region around one of the nuclear absorption peaks is examined in detail by means of an instrument that permits the determination of absorption over very much smaller increments of the abscissa (H_0 or ν), the single peak is usually found to be composed of several peaks (see for example the proton spectra in Figure 7-10); moreover, the position

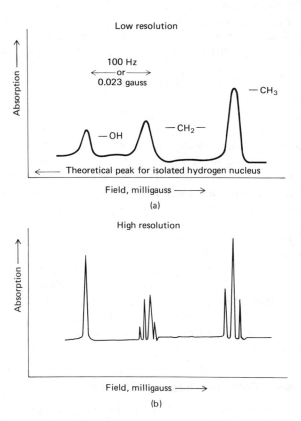

Figure 7-10. nmr spectra of ethanol at a frequency of 60 MHz. (a) Resolution $\sim 1/10^6$; (b) resolution $\sim 1/10^7$.

and the intensity of the component peaks are critically dependent upon the chemical environment of the nucleus responsible for the absorption. It is this dependence of the fine-structure spectrum on the environment of a nucleus that makes nmr spectroscopy such a valuable tool.

The effect of chemical environment on nuclear magnetic absorption spectra (Figure 7-10) is general; as a consequence, nmr represents a profitable approach to the study of atoms that possess magnetic properties. The fact that the proton appears in most organic compounds and in many inorganic ones as well has resulted in a concentration of effort upon this particular nucleus; our discussion follows this trend.

Types of Environmental Effects

The spectra for ethyl alcohol, shown in Figure 7-10, illustrate two types of environmental effects. The curve in Figure 7-10(a) was obtained with a lower-resolution instrument and shows three proton peaks with areas in the ratio 1:2:3 (left to right). On the basis of this ratio it appears logical to attribute the peaks to the hydroxyl, the methylene, and the methyl protons, respectively. Other evidence confirms this conclusion; for example, if the hydrogen atom of the hydroxyl group is replaced by deuterium, the first peak disappears from this part of the spectrum. Thus, small differences occur in the absorption frequency of the proton; such differences depend upon the group to which the hydrogen atom is bonded. This effect is called the *chemical shift*.

The higher-resolution spectrum of ethanol, shown in Figure 7-10(b), reveals that two of the three proton peaks are split into additional peaks. This secondary environmental effect, which is superimposed upon the chemical shift, has a different cause; it is termed *spin–spin splitting*.

Both the chemical shift and spin–spin splitting are important in structural analysis.

Experimentally, the two types of nmr peaks are readily distinguished, for it is found that the peak separations (in units of ν or H_0) resulting from a chemical shift are directly proportional to the field strength or to the oscillator frequency. Thus, if the spectrum in Figure 7-10(a) were to be obtained at 100 MHz rather than at 60 MHz, the horizontal distance between any pair of the peaks would be increased by $\frac{2}{3}$. In contrast, the distance between the fine-structure peaks within a group (lower spectrum) would not be altered by this frequency variation.

Measurement of the Chemical Shift and Spin-spin Splitting

Source of the chemical shift. The chemical shift arises from a circulation of the electrons surrounding the nucleus under the influence of the applied magnetic field. This phenomenon is discussed later in greater detail; for the present it suffices to state that this movement of electrons creates a small magnetic field that ordinarily opposes the applied field. As a consequence, the nucleus is exposed to an effective field that is somewhat smaller (but in some instances larger) than the external field. The magnitude of the field developed internally is directly proportional to the applied external field so that we may write

$$H_0 = H_{appl} - \sigma H_{appl} = H_{appl} (1 - \sigma) \qquad (7\text{-}11)$$

where H_{appl} is the applied field and H_0 *is the resultant field which determines the resonance behavior of the nucleus.* The quantity σ is the *shielding parameter,* which is determined by the electron density around the nucleus and is dependent upon the structure of the compound containing the nucleus.

The shielding parameter for the protons in a methyl group is larger than σ for methylene protons; this parameter is even smaller for the proton in an —OH group. It is of course zero for an isolated hydrogen nucleus. Thus, in order to bring any of the protons in ethanol into resonance at a given oscillator frequency ν it is necessary to employ a field H_{appl} that is greater than H_0 (equation 7-11), the resonance value for the isolated proton. Since σ differs for protons in various functional groups, the required applied field differs from group to group. This effect is shown in the spectrum of Figure 7-10(a) where the hydroxyl proton appears at the lowest applied field, the methylene protons next, and finally the methyl protons. Note that all of these peaks occur at an applied field greater than the theoretical one for the isolated hydrogen nucleus. Note also that if the applied field is held constant at a level necessary to excite the methyl proton, an increase in frequency would be needed to bring the methylene protons into resonance.

Source of spin-spin splitting. The splitting of chemical shift peaks can be explained by assuming that the effective field around one nucleus is further enhanced or reduced by local fields generated by *the hydrogen nuclei bonded to an adjacent atom.* Thus, the fine structure of the methylene peak shown in Figure 7-10 can be attributed to the effect of the local fields associated with the adjacent methyl protons. Conversely, the three methyl peaks arise from the adjacent methylene protons. These effects are independent of the applied field and are superimposed on the effects of the chemical shift.

Magnitude of the chemical shift and spin-spin splitting. The term σH_{appl} in equation (7-11) is infinitesimal in comparison with H_0; as a result, the measurement of the chemical shift is a major instrumental problem. To illustrate, let us calculate the effective field required to produce a resonance for an isolated proton when a fixed oscillator frequency of 60 MHz is employed. The data in Table 7-2 indicate that the absorption frequency for the proton for a 10^4 gauss field is 42.577 MHz. Substituting these values into equation (7-10) gives

$$\gamma/2\pi = \frac{42.577}{10^4} \text{ MHz/gauss}$$

The field strength to produce a resonance in an isolated proton at 60 MHz is then

$$H_0 = \frac{60}{42.577 \times 10^{-4}} = 14,092.1 \text{ gauss}$$

For the typical proton, σH_{appl} is 100 milligauss or less at a frequency of 60 MHz. Thus, the entire chemical shift spectrum of most compounds are found at an applied field lying between 14,092.1 and 14,092.2 gauss. Therefore, in order to detect the spectral detail shown in Figure 7-7 it is necessary to have an instrument in which the magnetic field can be varied over a minute range relative to the total field strength. For example, the separation between the chemical shift peaks in the upper spectrum is roughly 0.025 gauss; to discriminate between these peaks it is necessary to vary the field in increments smaller than this figure—that is, by less than 0.025/14,000, or by about 2 ppm. To detect the spin–spin peaks shown in the lower spectrum requires an instrument with even better resolution (about 1 part in 10^7). The development of instruments that are routinely capable of this order of resolution is a truly remarkable engineering accomplishment; it is not surprising that nmr spectrometers are expensive. As a consequence of these severe instrumental requirements, most high-resolution nmr spectrometers are restricted to the examination of but one kind of a nucleus. The most widely used instruments thus yield proton spectra alone and require modification if other nuclei are to be studied.

Abscissa scales for nmr spectra. The typical proton nmr spectrometer employs an oscillator of fixed frequency (usually 40, 60, or 100 MHz) and a magnet that produces a constant field of the appropriate size (that is, 14,092 gauss for a 60-MHz oscillator). The field can then be increased continuously over a small milligauss range by means of the pair of sweep coils wound around the magnet pole pieces or located within the pole gap (see Figure 7-7). The determination of the absolute field strength to an accuracy of 1 part in 10^7 is difficult or impossible; on the other hand, it is entirely feasible to determine to a few milligauss the *change* in field strength caused by the subsidiary sweep coils. Thus, it is expedient to report the position of resonance absorption peaks relative to the resonance peak for a standard substance that can be measured at essentially the same time. In this way the effect of fluctuations in the fixed magnetic field are minimized. The use of an internal standard is also advantageous in that chemical shifts can be reported in terms that are independent of the oscillator frequency.

A variety of internal standards have been employed, but the compound that is now most generally accepted is tetramethylsilane (TMS), $(CH_3)_4Si$. All of the protons in this compound are identical, and for reasons to be considered later, the shielding parameter for TMS is larger than for most other protons. Thus, the compound provides a single sharp peak at a high applied field that is well-separated from most of the peaks of interest in a spectrum.

In addition, TMS is inert, readily soluble in most organic liquids, and easily removed from samples by distillation (b. p. = 27°C). Unfortunately, TMS is not water-soluble; in aqueous media the sodium salt of 2,2-dimethyl-2-silapentane-5-sulfonate, $(CH_3)_3SiCH_2CH_2CH_2SO_3Na$, is used in its stead. The methyl protons of this compound produce a peak analogous to TMS; the methylene protons give a series of small peaks that are readily identified and can thus be ignored.

The field strength H_{ref} required to produce the TMS resonance line at frequency ν is given by equation (7-11)

$$H_0 = H_{ref} (1 - \sigma_{ref})$$

which can be rewritten as

$$\sigma_{ref} = \frac{H_{ref} - H_0}{H_{ref}} \qquad (7\text{-}12)$$

Similarly, for a given absorption peak of the sample we may write

$$\sigma_{sple} = \frac{H_{sple} - H_0}{H_{sple}} \qquad (7\text{-}13)$$

where H_{sple} is the field necessary to produce the peak. We then define a *chemical shift parameter* δ as

$$\delta = (\sigma_{ref} - \sigma_{sple}) \times 10^6 \qquad (7\text{-}14)$$

and substituting (7-12) and (7-13) we obtain

$$\delta = \frac{H_0 (H_{ref} - H_{sple})}{H_{sple} H_{ref}} \times 10^6$$

But H_{sple} is very nearly the same as H_0, so that the ratio H_0/H_{sple} is very nearly unity. Thus,

$$\delta \cong \frac{H_{ref} - H_{sple}}{H_{ref}} \times 10^6 \qquad (7\text{-}15)$$

The quantity δ is dimensionless and expresses the relative shift in parts per million; for a given peak δ will be the same regardless of whether a 40-, 60-, or 100-MHz instrument is employed. Most proton peaks lie in the δ range of 1 to 12.

Another chemical shift parameter τ is defined as

$$\tau = 10 - \delta \qquad (7\text{-}16)$$

Recently, the effort has been made to adopt a standard way of presenting nmr data; the two spectra for ethanol shown in Figure 7-11 illustrate the proposed method. There is general agreement that nmr plots should have linear scales in δ and τ and that the data should be plotted with the field increasing from left to right. Thus, if TMS is employed as the reference, its peak will appear on the far right-hand side of the plot since σ for TMS is large. As shown, the zero value for the δ scale corresponds to the TMS peak and the value of δ increases from right to left. The τ scale, of course, changes in the opposite way. Note that the various peaks appear at the same values of δ and τ in spite of the fact that the two spectra were obtained with instruments having different fixed fields.

For the purpose of reporting spin-spin splitting, it is desirable to utilize scalar units of milligauss or hertz. The latter scale is now coming into more general use and is the one shown in Figure 7-11. The position of the reference TMS peak is arbitrarily taken as zero, with frequencies increasing from right to left. The effect of this choice is to make the frequency of a given peak the increase in oscillator frequency that would be needed to bring that proton into resonance if the field was maintained constant at the level required to produce the TMS peak.

It can be seen in Figure 7-11 that the spin-spin splitting in frequency units (J) is the same for the 60-MHz and the 100-MHz instruments. Note, however, that the chemical shift *in frequency units* is enhanced with the higher frequency instrument.

The Chemical Shift

As noted earlier, chemical shifts arise from the secondary magnetic fields produced by the circulation of electrons in the molecule. These elec-

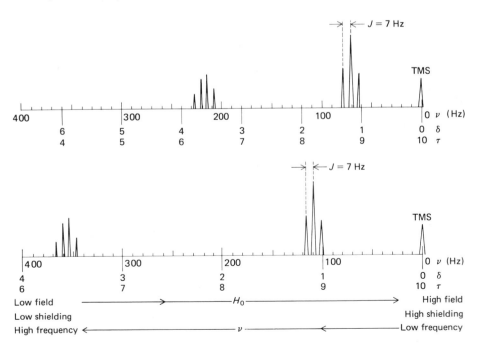

Figure 7-11. Abscissa scales for nmr spectra.

tronic currents (local *diamagnetic currents*[3]) are induced by the fixed magnetic field and result in secondary fields that may either reduce or enhance the field to which a given proton responds. The effects are complex, and we consider only the major aspects of the phenomenon here. More complete treatments can be found in several reference works.[4]

Under the influence of the magnetic field, electrons bonding the proton tend to precess around the nucleus in a plane perpendicular to the magnetic field (see Figure 7-12). As a consequence of this motion a secondary field is produced which opposes the primary field; the behavior here is analogous to the passage of electrons through a wire loop. The nucleus then experiences a resultant field, which is smaller (the nucleus is said to be *shielded* from the full effect of the primary field); as a consequence, the external field must be increased to cause nuclear resonance. The frequency of the

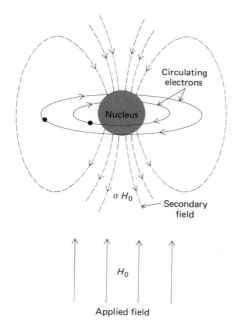

Figure 7-12. Diamagnetic shielding of a nucleus.

[3]The intensity of magnetization induced in a *diamagnetic* substance is smaller than that produced in a vacuum with the same field. Diamagnetism is the result of motion induced in bonding electrons by the applied field; this motion (a *diamagnetic current*) creates a secondary field that opposes the applied field. *Paramagnetism* (and the resulting *paramagnetic currents*) operate in just the opposite sense.

[4]L. M. Jackman and S. Sternhall, *Nuclear Magnetic Resonance Spectroscopy*, 2d ed. (New York: Pergamon Press, Inc., 1969); J. R. Dyer, *Applications of Absorption Spectroscopy of Organic Compounds* (Englewood Cliffs, N.J.: Prentice-Hall, Inc., 1965), Chap. 4; J. A. Pople, W. G. Schneider and H. J. Bernstein, *High-resolution Nuclear Magnetic Resonance* (New York: McGraw-Hill Book Company, Inc., 1959), Chap. 7.

precession, and thus the magnitude of the secondary field, is a direct function of the external field.

The shielding experienced by a given nucleus is directly related to the electron density surrounding it. Thus, in the absence of the other influences, shielding would be expected to decrease with increasing electronegativity of adjacent groups. This effect is illustrated by the δ values for the protons in the methyl halides, CH_3X, which lie in the order I (2.16), Br (2.68), Cl (3.05), and F (4.26). Here, iodine (the least electronegative) is the least effective of the halogens withdrawing electrons from the protons; thus, the electrons of iodine provide the largest shielding effect. Similarly, electron density around the methyl protons of methanol is greater than around the proton associated with oxygen because oxygen is more electronegative than carbon. Thus, the methyl peaks are upfield from the hydroxyl peak. The position of the proton peaks in TMS is also explained by this model, since silicon is relatively electropositive. Finally, acidic protons have very low electron densities, and the peak for the proton in RSO_3H or $RCOOH$ lies far downfield ($\delta > 10$).

Effect of magnetic anisotropy. It is apparent from an examination of the spectra of compounds containing double or triple bonds that local diamagnetic effects do not suffice to explain the position of certain of the proton peaks. Consider, for example, the irregular change in δ values for protons in the following hydrocarbons, arranged in order of increasing acidity (or increased electronegativity of the groups to which the protons are bonded): CH_3-CH_3 ($\delta = 0.9$), $CH_2=CH_2$ ($\delta = 5.8$), and $HC\equiv CH$ ($\delta = 2.9$). Furthermore, the aldehydic proton $RC(=O)H$ ($\delta \sim 10$) and the protons on benzene ($\delta \sim 7.3$) appear considerably farther downfield than might be expected on the basis of the electronegativity of the groups to which they are attached.

The effects of multiple bonds upon the chemical shift can be explained by taking into account the anisotropic magnetic properties of these compounds. For example, the magnetic susceptibilities[5] of crystalline aromatic compounds have been found to differ appreciably depending upon the orientation of the ring with respect to the applied field. This anisotropy is readily understood from the model shown in Figure 7-13. Here the plane of the ring is perpendicular to the magnetic field; in this position the field can induce a flow of the π electrons around the ring (a ring current). The consequence is similar to that of a current flow in a wire loop; namely, a secondary field is produced that acts in opposition to the applied field. This secondary field, however, exerts a magnetic effect on the proton attached to the ring, and as shown in Figure 7-13, this effect is in the direction of the field. Thus, the aromatic protons require a lower external field to bring them into resonance. This effect is either absent or self-canceling in other orientations of the ring.

[5]The magnetic susceptibility of a substance can be thought of as the extent to which it is susceptible to induced magnetization by an external field.

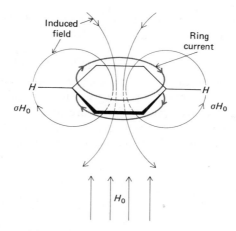

Figure 7-13. Deshielding of aromatic protons brought about by ring current.

A somewhat analogous model can be envisioned for the ethylenic or carbonyl double bonds. Here one can imagine circulation of the π electrons in a plane along the axis of the bond when the molecule is oriented to the field as shown in Figure 7-14. Again, the secondary field produced acts upon the proton to reinforce the applied field. Thus, deshielding shifts the peak to larger values of δ. With an aldehyde, this effect combines with the deshielding brought about by the electronegative nature of the carbonyl group; a very large value of δ results.

In an acetylenic bond the symmetrical distribution of π electrons about the bond axis permits electron circulation around the bond (in contrast, such circulation is prohibited by the nodal plane in the electron distribution of a double bond). From Figure 7-14 it can be seen that in this orientation the protons are shielded. This effect is apparently large enough to offset the

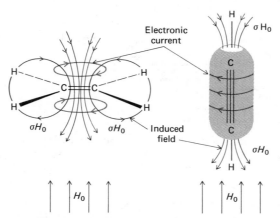

Figure 7-14. Deshielding of ethylene and shielding of acetylene brought about by electronic currents.

deshielding resulting from the acidity of the protons and from the electronic currents at perpendicular orientations of the bond.

Correlation of chemical shift with structure. The chemical shift is employed for the identification of functional groups and as an aid in determining structural arrangements of groups. These applications are based upon empirical correlations between structure and shift. A number of correlation charts[6] and tables[7] have been published. Two of these are shown in Figure 7-15 and Table 7-3. It should be noted that the exact values

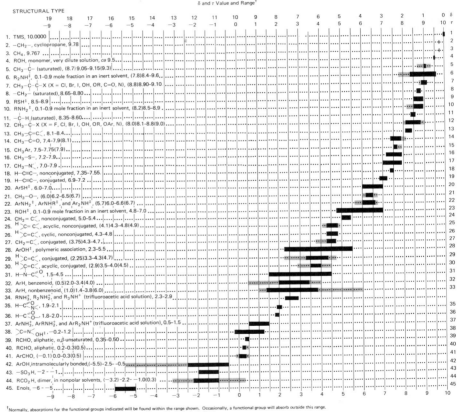

Figure 7-15. Absorption positions of protons in various structural environments. Table taken from J. R. Dyer, *Applications of Absorption Spectroscopy of Organic Compounds* (Englewood Cliffs, N. J.: Prentice-Hall, Inc., © 1965), p. 85. (With permission.)

[6]N. F. Chamberlain, *Anal. Chem.*, **31**, 56 (1959).

[7]R. M. Silverstein and G. C. Bassler, *Spectrometric Identification of Organic Compounds*, 2d ed. (New York: John Wiley & Sons, Inc., 1967), Chap. 4; L. M. Jackman and S. Sternhall, *Nuclear Magnetic Resonance Spectroscopy*, 2d ed. (New York: Pergamon Press, Inc., 1969).

for δ may depend upon the nature of the solvent as well as upon the concentration of the solute. These effects are particularly pronounced for protons involved in hydrogen bonding; an example is the hydrogen atom in the alcoholic functional group.

Table 7-3 Approximate Chemical Shifts for Certain Methyl, Methylene and Methine Protons

	δ, ppm		
Structure	$M = CH_3$	$M = CH_2$	$M = CH$
Aliphatic β substituents			
M—C—Cl	1.5	1.8	2.0
M—C—Br	1.8	1.8	1.9
M—C—NO₂	1.6	2.1	2.5
M—C—OH (or OR)	1.2	1.5	1.8
M—C—OC(=O)R	1.3	1.6	1.8
M—C—C(=O)H	1.1	1.7	—
M—C—C(=O)R	1.1	1.6	2.0
M—C—C(=O)OR	1.1	1.7	1.9
M—C—φ	1.1	1.6	1.8
Aliphatic α substituents			
M—Cl	3.0	3.5	4.0
M—Br	2.7	3.4	4.1
M—NO₂	4.3	4.4	4.6
M—OH (or OR)	3.2	3.4	3.6
M—O—φ	3.8	4.0	4.6
M—OC(=O)R	3.6	4.1	5.0
M—C=C	1.6	1.9	—
M—C≡C	1.7	2.2	2.8
M—C(=O)H	2.2	2.4	—
M—C(=O)R	2.1	2.4	2.6
M—C(=O)φ	2.4	2.7	3.4
M—C(=O)OR	2.2	2.2	2.5
M—φ	2.2	2.6	2.8

Spin-spin Splitting

As may be seen in Figure 7-10, the absorption bands for the methyl and methylene protons in ethanol consist of several narrow peaks that can be separated only with a high-resolution instrument. Careful examination of these peaks shows that the spacing for the three components of the methyl band is identical to that for the four peaks comprising the methylene band. Moreover, the areas of the peaks in a multiplet approximate an integral ratio to one another. Thus, for methyl triplet the ratio of areas is 1:2:1; for the quartet of methylene peaks it is 1:3:3:1.

Origin. It seems plausible to attribute these observations to the effect that the spins of one set of nuclei exert upon the resonance behavior of another. That is to say, there is a small interaction or coupling between the two groups of protons. This explanation presupposes that such coupling takes place via interactions between the nuclei and the bonding electrons rather than through free space. For our purpose, however, the details of this mechanism are not important.

Let us first consider the effect of the methylene protons in ethanol on the resonance of the methyl protons. We must first remember that the ratio of protons in the two possible spin states is very nearly one, even in a strong magnetic field. We can imagine, then, that the two methylene protons in a molecule can have four possible combinations of spin states and that in an entire sample each of these combinations will be approximately equally represented. If we represent the spin orientation of each nucleus with a small arrow, the four states are

$$\longrightarrow H_0$$

Field direction

Possible spin orientations
of methylene protons

In the first combination the spins of the two methylene protons are paired and aligned against the field while in the second combination the paired spins are reversed; there are also two combinations in which the spins are opposed to one another. The magnetic effect that is transmitted to the methyl protons on the adjacent carbon atoms is determined by the spin combinations that exist in the methylene group at any instant. If the spins are paired and opposed to the external field, the effective applied field on the methyl protons is slightly lessened; thus, a somewhat higher field is needed to bring them into resonance, and an upfield shift results. Spins paired and aligned with the field result in a downfield shift. Neither of the combinations of opposed spin have an effect on the resonance of the methyl protons. Thus, splitting into three peaks results. The area under the middle peak is twice that of either of the other two, since two spin combinations are involved.

Let us now consider the effect of the three methyl protons upon the methylene peak. Possible spin combinations for the methyl protons are

$$H_0 \longrightarrow$$

Here we have eight possible spin combinations; however, among these are two groups containing three combinations that have equivalent magnetic effects. The methylene peak is thus split into four peaks having areas in the ratio of 1:3:3:1.

The coupling constant J. The separation, in hertz, between adjacent peaks of a multiplet is called the *coupling constant J*. While the magnitude of *J* appears to be a measure of the effectiveness of the spin-spin interaction between groups, no entirely adequate theory has been developed to relate the size of the coupling constant to structural relationships. On the other hand, a large amount of empirical data relating *J* to structure is available, and these data are of great value in establishing the relationship among groups in a molecule.

It is found empirically that the analysis of an nmr spectrum is greatly simplified if the chemical shift between two interacting groups is large with respect to their coupling constant; that is, when $\Delta\nu/J > 7$, where $\Delta\nu$ is the difference in chemical shift peaks expressed in hertz. The ethanol spectrum (Figure 7-10) falls in this category; here, *J* for the methyl and methylene peaks is approximately 7 Hz, while the separation between the centers of the two multiplets is about 140 Hz. The discussion that follows is concerned primarily with this type of spectrum.

Rules governing multiplet spectra when $\Delta\nu/J > 7$. The following rules govern the appearance of spin-spin spectra.

1. Equivalent nuclei do not interact with one another to give multiple absorption peaks. The three protons in the methyl group in ethanol give rise to splitting of the adjacent methylene protons only and not to splitting among themselves.

2. Coupling constants decrease with separation of groups, and coupling is seldom observed at distances greater than three bond lengths.

3. The multiplicity of a band is determined by the number *n* of magnetically equivalent protons on the neighboring atoms and is given by $(n + 1)$. Thus, the multiplicity for the methylene band in ethanol is determined by the number of protons in the adjacent methyl groups and is equal to $(3 + 1)$.

4. If the protons on atom B are affected by protons on atoms A and C that are nonequivalent, the multiplicity of B is equal to $(n_A + 1)(n_C + 1)$, where n_A and n_C are the number of equivalent protons on A and C, respectively.

5. The approximate relative areas of a multiplet are symmetric around the midpoint of the band and are proportional to the coefficients of the terms in the expansion $(x + 1)^n$. The application of this rule is demonstrated in the examples that follow.

6. The coupling constant is independent of the applied field; thus, multiplets are readily distinguished from closely spaced chemical shift peaks.

   Example. For each of the following compounds, calculate the number of multiplets for each band and their relative areas.

   (a) $ClCH_2CH_2CH_2Cl$. The multiplicity of the band associated with the four equivalent protons on the two ends of the molecule would be determined by the number of protons on the central carbon; thus, the multiplicity is $(2 + 1) = 3$ and the areas would be 1:2:1. The multiplicity of two central methylene protons would be determined by the four equivalent protons at the ends and would thus be $(4 + 1) = 5$. Expansion of $(x + 1)^5$ gives the following coefficients, which are proportional to the areas of the peaks 1:4:6:4:1.

(b) $CH_3CHBrCH_3$. The band for the six methyl protons will be made up of $(1 + 1) = 2$ peaks having relative areas of 1:1; the proton on the central carbon atom has a multiplicity of $(6 + 1) = 7$. These peaks will have areas in the ratio of 1:6:15:20:15:6:1.

(c) $CH_3CH_2OCH_3$. The methyl protons on the right are separated from the other protons by more than three bonds so that only a single peak will be observed for them. The protons of the central methylene group will have a multiplicity of $(3 + 1) = 4$ and a ratio of 1:3:3:1. The methyl protons on the left have a multiplicity of $(2 + 1) = 3$ and an area ratio of 1:2:1.

The foregoing examples are relatively simple because all of the protons influencing the multiplicity of any one of the peaks are magnetically equivalent. A more complex splitting pattern results when a set of protons is affected by two or more nonequivalent protons. As an example, consider the spectrum of 1-iodopropane, $CH_3CH_2CH_2I$. If we label the three carbon atoms (a), (b), and (c) from left to right, the chemical shift bands are found at $\delta_{(a)} = 1.02$, $\delta_{(b)} = 1.86$, and $\delta_{(c)} = 3.17$. The band at $\delta_{(a)} = 1.02$ will be split by the two methylene protons on (b) into $(2 + 1) = 3$ peaks having relative areas of 1:2:1. A similar splitting of the band at $\delta_{(c)} = 3.17$ will also be observed. The experimental coupling constants for the two shifts are $J_{(ab)} = 7.3$ and $J_{(bc)} = 6.8$. The band for the methylene protons (b) is affected by two groups of protons which are not magnetically equivalent, as is evident from the difference between $J_{(ab)}$ and $J_{(bc)}$. Thus, invoking rule 4, the number of peaks will be $(3 + 1)(2 + 1) = 12$. In cases such as this, derivation of a splitting pattern, as shown in Figure 7-16, is helpful. Here, the effect of the (a) proton is first shown and leads to four peaks of relative areas 1:3:3:1 spaced at 7.3 Hz. Each of these is then split into three new peaks spaced at 6.8 Hz, having relative areas of 1:2:1. (The same final pattern is produced if the original band is first split into a triplet.) At very high resolution the spectrum for 1-iodopropane exhibits a series of peaks that approximates the series shown at the bottom of Figure 7-16. At lower resolution (so low that the instrument does not detect the difference between $J_{(ab)}$ and $J_{(bc)}$) only six peaks are observed with relative areas of 1:5:10:10:5:1.

Spectra when $\Delta\nu/J < 7$. Coupling constants are usually smaller than 20, whereas chemical shifts may be as high as 1000. Therefore, the splitting behavior described by the rules in the previous section is common. When, however, $\Delta\nu/J$ becomes less than 7, these rules no longer apply. Generally, as $\Delta\nu$ approaches J the peaks on the inner side of two multiplets tend to be enhanced at the expense of the peaks on the outer side, and the symmetry of each multiplet is thus destroyed. Analysis of a spectrum under these circumstances is considerably more difficult.

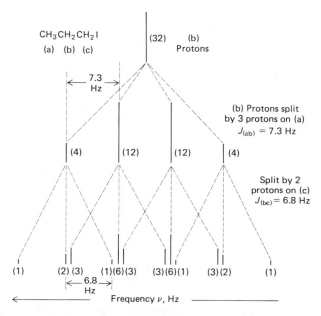

Figure 7-16. Splitting pattern for methylene (b) protons in $CH_3CH_2CH_2I$. Figures in parentheses are relative areas under peaks.

Effect of chemical exchange on spectra. Turning again to the spectrum of ethanol (Figure 7-10) it is interesting to consider why the OH proton appears as a singlet rather than a triplet. The methylene protons and the OH proton are separated by only three bonds; coupling should occur to increase the multiplicity of both OH and the methylene peaks. Actually, as shown in Figure 7-17, the expected multiplicity can be observed by employing a highly purified sample of the alcohol. Note the triplet OH peaks and the eight methylene peaks in this spectrum. If, now, a trace of acid or base is added to the pure sample, the spectrum reverts to the form shown in Figure 7-10.

The exchange of OH protons among alcohol molecules is known to be catalyzed by both acids and bases, as well as by the impurities that commonly occur in alcohol. It is thus plausible to associate the decoupling observed in the presence of these catalysts to an exchange process. If exchange is rapid, each OH group will have several protons associated with it during any brief period; within this interval all of the OH protons will experience the effects of the three spin arrangements of the methylene protons. Thus, the magnetic effects on the alcoholic proton are averaged and a single sharp peak is observed. Spin decoupling always occurs when the exchange frequency is greater than the separation (in frequency units) between the interacting components.

Chemical exchange can affect not only spin-spin spectra but also chemical-shift spectra. Purified alcohol-water mixtures have two well-

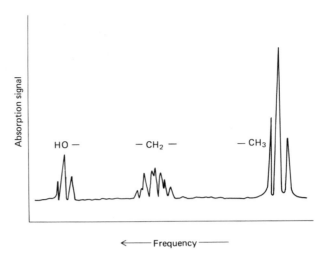

Figure 7-17. Spectrum of highly purified ethanol showing additional splitting of OH and CH_2 peaks (compare with Figure 7-10).

defined and easily separated OH proton peaks. Upon addition of an acidic or basic catalyst, however, the two peaks coalesce to form a single sharp line. Here the catalyst enhances the rate of proton exchange between the alcohol and the water, and thus averages the shielding effect. A single sharp line is obtained when the exchange rate is significantly greater than the separation frequency of the individual lines of alcohol and water. On the other hand, if the exchange frequency is about the same as this frequency difference, shielding is only partially averaged and a broad line results. The correlation of line breadth with exchange rates has provided a direct means for investigating the kinetics of such processes and represents an important application of the nmr experiment.

Identification of Compounds

An nmr spectrum, like an infrared spectrum, seldom by itself suffices for the identification of an organic compound. Used in conjunction with other observations such as elemental analysis as well as ultraviolet, infrared, and mass spectra, however, it provides an important tool for the characterization of a pure compound. The simple examples that follow give some idea of the kind of information that can be extracted from nmr studies.

Example. The nmr spectrum shown in Figure 7-18 is for an organic compound having the empirical formula $C_5H_{10}O_2$. Identify the compound. From the integral plot we obtain relative areas from left to right of about 6.1, 4.2, 4.2, and 6.2. These figures suggest a distribution of the 10 protons of 3, 2, 2, and 3. The single peak at $\delta = 3.6$ must be due to an isolated methyl group, and upon inspection of Figure 7-15 and Table 7-3, the functional group $CH_3OC(=O)-$ is suggested. The empirical formula

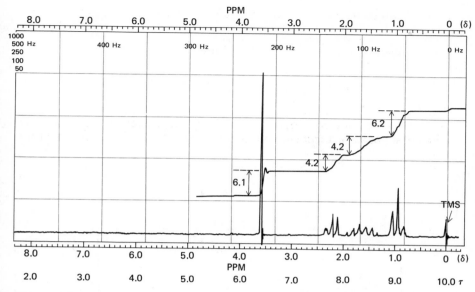

Figure 7-18. nmr spectrum and peak integral curve for the organic compound $C_5H_{10}O_2$ in CCl_4. From R. M. Silverstein and G. C. Bassler, *Spectrometric Identification of Organic Compounds*, 2d ed. (New York: John Wiley & Sons, Inc., 1967), p. 201. (With permission.)

and the 2:2:3 distribution of the remaining protons indicate the presence of an *n*-propyl group as well. The structure $CH_3OC(=O)CH_2CH_2CH_3$ is consistent with all of these observations. In addition, the positions and the splitting patterns of the three remaining peaks are entirely compatible with this hypothesis. The triplet at $\delta = 0.9$ is typical of a methyl group adjacent to a methylene. From Table 7-3 the two protons of the methylene adjacent to the carboxylate peak should yield the observed triplet peak at about $\delta = 2.2$. The other methylene group would be expected to produce a pattern of 12 peaks (3 × 4) at about $\delta = 1.7$. Only six are observed, presumably because the resolution of the instrument is insufficient.

Example. The spectra shown in Figure 7-19 are for colorless, isomeric, liquids containing only carbon and hydrogen. Identify the two compounds.

The single peak at about $\delta = 7.2$ in the upper figure suggests an aromatic structure; the relative area of this peak corresponds to 5 protons; from this we conclude that we may have a monosubstituted derivative of benzene. The seven peaks for the single proton appearing at $\delta = 2.9$ and the six-proton doublet at $\delta = 1.2$ can only be explained by the structure

$$\begin{array}{c} CH_3 \\ | \\ -C-CH_3 \\ | \\ H \end{array}$$

Thus, we conclude that this compound is cumene.

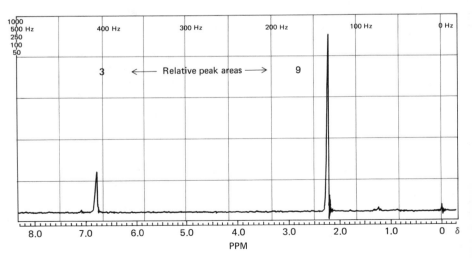

The isomeric compound has an aromatic peak at $\delta = 6.8$; its relative area suggests a trisubstituted benzene, which can only mean that the

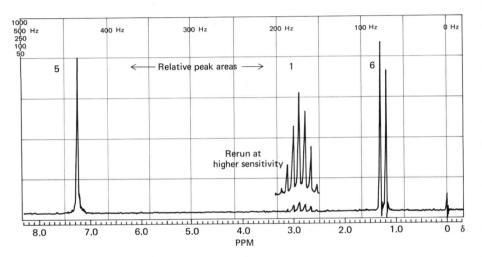

Figure 7-19. nmr spectra for two organic isomers in $CDCl_3$ solution. (Courtesy of Varian Associates, Palo Alto, Calif.)

compound is $C_6H_3(CH_3)_3$. The relative peak areas confirm this diagnosis. We cannot, however, decide which of the three trimethyl benzene derivatives we have from the nmr data.

Example. The spectrum shown in Figure 7-20 is for an organic compound having a molecular weight of 72 and containing carbon, hydrogen, and oxygen only. Identify the compound.

The triplet peak at $\delta = 9.8$ appears, from Figure 7-15, to be that of an aliphatic aldehyde, RCHO. If this is the case R has a molecular weight of 43, which corresponds to a C_3H_7 fragment. The triplet nature of the peak at $\delta = 9.8$ requires that there be a methylene group adjacent to the carbonyl. Thus, the compound would appear to be *n*-butyraldehyde, $CH_3CH_2CH_2CHO$. The triplet peak at $\delta = 0.97$ appears to be that of the terminal methyl. The protons on the adjacent methylene would be expected to show a complicated splitting pattern of 12 peaks (4×3); the grouping of peaks around $\delta = 1.7$ is compatible with this prediction. Finally, the peak for the protons on the methylene group adjacent to the carbonyl should appear as a sextet downfield from the other methylene proton peaks. The group at $\delta = 2.4$ is consistent with this conclusion.

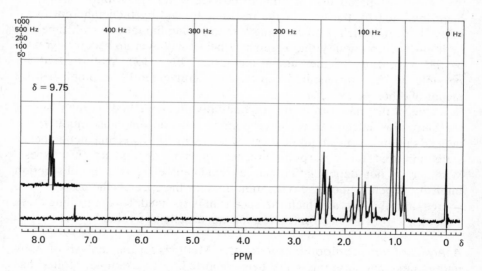

Figure 7-20. nmr spectrum of a pure organic compound containing C, H, and O only. (Courtesy of Varian Associates, Palo Alto, Calif.)

OTHER APPLICATIONS OF nmr SPECTROSCOPY

As we have pointed out, the primary application of nmr spectroscopy has been the characterization of organic molecules by means of their proton resonance. In this section we consider briefly other applications and potential applications of the method.

Quantitative Analysis

One of the unique aspects of nmr spectra is the direct proportionality between peak areas and the number of nuclei responsible for the peak. As a consequence, the quantitative determination of a specific compound need not require pure samples of the compound for calibration. Thus, if an identifiable peak for one of the constituents of a sample does not overlap the peaks of the other constituents, the area of this peak can be employed to establish the concentration of the species directly, provided only that the signal area per proton is known. This latter parameter can be obtained conveniently from a known concentration of an internal standard. For example, if the solvent, present in a known amount, were benzene, cyclohexane, or water, the areas of the single proton peak for these compounds could be used to give the desired information; of course, the peak of the internal standard should not overlap with any of the sample peaks. Organic silicon derivatives are uniquely attractive for calibration purposes, owing to the high upfield location of their proton peaks.

The widespread use of nmr spectroscopy for quantitative work has been inhibited by the cost of the instruments and by their inherent instability (both of which are becoming less imposing limitations with time). In addition, the probability that resonance peaks will overlap becomes greater as the complexity of the sample increases. Often, too, analyses that are possible by the nmr method can be more conveniently accomplished by means of other techniques.

One of the main problems in quantitative nmr methods is the result of the saturation effect. As we have pointed out, the nmr absorption signal depends upon a very minute excess of nuclei in the lower magnetic energy state and that the absorption process tends to depopulate this excess. Whether or not depopulation has a significant effect on the absorption intensity depends upon the relaxation time for the species, the power of the source, and the rate at which the spectrum is scanned. By controlling these variables, errors arising from saturation can usually be avoided.

Analysis of multicomponent mixtures.   Methods for the analysis of many multicomponent mixtures have been reported.[8] For example, Hollis[9] has described a method for the determination of aspirin, phenacetin, and caffeine in commercial analgesic preparations. The procedure requires about twenty minutes and the relative errors are in the range of 1 to 3 percent. Chamberlain[10] describes a procedure for the rapid analysis of benzene,

---

[8]References to many of these can be found in the following series of review articles: *Anal. Chem.*, **42**, 418R (1970); **40**, 560R (1968); **38**, 331R (1966); **36**, 266R (1964); **34**, 255R (1962); **32**, 221R (1960); **30**, 839 (1958).

[9]D. P. Hollis, *Anal. Chem.*, **35**, 1682 (1963).

[10]N. F. Chamberlain in I. M. Kolthoff and P. J. Elving, Eds., *Treatise on Analytical Chemistry*, Part I, Vol. 4 (New York: Interscience Publishers, 1963), p. 1932.

heptane, ethylene glycol, and water in mixtures. A wide range of mixtures of this type was analyzed with a precision of 0.5 percent.

One of the important applications of the nmr technique has been to the quantitative determination of water in food products, pulp and paper, and agricultural material. The water in these substances is sufficiently mobile to give a narrow peak that is suitable for quantitative measurement.[11]

Elemental analysis. nmr spectroscopy can be employed for the determination of the total concentration of a given kind of magnetic nucleus in a sample. For example, Jungnickel and Forbes[12] have investigated the integrated nmr intensities of the proton peaks for numerous organic compounds and have concluded that accurate quantitative determination of total hydrogen in organic mixtures is possible. Paulsen and Cooke[13] have shown that the resonance of fluorine-19 can be used for the quantitative analysis of that element in an organic compound—an analysis that is very difficult to carry out by classical methods. For quantitative work a low-resolution or *wide-line* spectrometer can be employed.

Study of Isotopes Other Than the Proton

Table 7-1 lists several nuclei, in addition to the proton, which have magnetic moments and can thus be studied by the magnetic resonance technique. More than one hundred other isotopes also possess magnetic moments;[14] the resonance behavior of only a few of these has been investigated to date.

Fluorine. Fluorine, with an atomic number of 19, has a spin quantum number of $\frac{1}{2}$, and a magnetic moment of 2.6285 nuclear magnetons. Thus, the resonance frequency of fluorine in similar fields is only slightly lower than the proton (56.5 MHz, as compared with 60.0 MHz at 14,000 gauss). Therefore, with relatively minor changes, a proton nmr spectrometer can be adapted to the study of fluorine resonance. With the exception of the proton, more magnetic resonance studies have been made of ^{19}F than of any other isotope.

It is found experimentally that the fluorine absorption is also sensitive to environment; the resulting chemical shifts, however, extend over a range of about 300 ppm compared with a maximum of 20 ppm for the proton. In

[11]T. M. Shaw and R. H. Elsken, *J. Chem. Phys.,* **18**, 1113 (1950); *J. Appl. Physics,* **26**, 313 (1955); T. M. Shaw, R. H. Elsken, and C. H. Kunsman, *J. Assoc. Offic. Agr. Chemists,* **36**, 1070 (1953).

[12]J. L. Jungnickel and J. W. Forbes, *Anal. Chem.,* **35**, 938 (1963).

[13]P. J. Paulsen and W. D. Cooke, *Anal. Chem.,* **36**, 1721 (1964).

[14]See, for example, J. A. Pople, W. G. Schneider, and H. J. Bernstein, *High-resolution Nuclear Magnetic Resonance* (New York: McGraw-Hill Book Company, Inc., 1959), pp. 480-485.

addition, the solvent plays a much more important role in determining fluorine peak positions than with the proton.

Empirical correlations of the fluorine shift with structure are relatively sparse when compared with information concerning proton behavior. It seems probable, however, that the future will see further developments in this field, particularly for structural investigation of organic fluorine compounds.

Phosphorus. Phosphorus-31, with spin number $\frac{1}{2}$, also exhibits sharp nmr peaks with chemical shifts extending over a range of 700 ppm. The resonance frequency of ^{31}P at 14,000 gauss is 24.3 MHz. Several investigations correlating the chemical shift of the phosphorus nucleus with structure have been reported.

Other nuclei. Other nuclei that appear to offer considerable potential for nmr studies include carbon-13, oxygen-17, hydrogen-2, boron-11, and silicon-29; an increasing amount of work is being reported for each of these.

ELECTRON SPIN RESONANCE

Electron spin resonance spectroscopy[15] (esr) is based upon the splitting of magnetic energy levels caused by the action of a magnetic field on an unpaired electron contained in an ion, a molecule, or an atom. The principles of electron spin resonance are closely related to those of nmr spectroscopy. The free electron behaves as a spinning, charged particle with a resulting magnetic moment (opposite in sign to the proton and most other nuclei). The spin number of the electron is $\frac{1}{2}$; this particle, like the proton, thus has two magnetic energy levels. In contrast to the proton, the lower energy level corresponds to $m = -\frac{1}{2}$ and the higher to $m = +\frac{1}{2}$.

The magnetic moment of the electron is nearly 1000 times greater than that of the proton, and thus the absorption frequency in a given magnetic field is much larger. For example, when the appropriate value from Table 7-2 is substituted into equation (7-5)[16] we find that

$$\nu = 2.803 \, H_0 \tag{7-17}$$

Instruments for esr measurement consist of an electromagnet having a field of about 3500 gauss and sweep coils that permit variation of the field over a small range. The sample is contained in a microwave cavity that is held in the field of the magnet. The source is a Klystron tube that produces radiation at a constant frequency of about 9500 MHz. As with nmr spectrometers, resonance peaks are located by varying the field.

Most molecules fail to exhibit an esr spectrum because they contain

---

[15]Also called electron paramagnetic resonance and electron magnetic resonance.

[16]Note that the sign for the right-hand side becomes negative for the electron because the lower-energy state corresponds to $m = -\frac{1}{2}$.

an even number of electrons, and the number in the two spin states is identical (that is, the spins are paired); as a consequence, the magnetic effects of electron spin are canceled. Such substances are diamagnetic because of the small fields induced from the *orbital precession* of the electrons around the nuclei; these fields act in opposition to the applied field (see p. 201). In contrast, the spin of an unpaired electron induces a field that reinforces the applied field. This paramagnetic effect is much larger than the diamagnetic effect arising from the orbital motion of the electron; as a result, species containing an unpaired electron exhibit a net paramagnetic behavior. In a magnetic field, splitting of the energy levels of the electrons occurs, and this splitting can be observed, as noted, by microwave absorption. Furthermore, the electron spin can couple with the spin of nuclei in the species to give splitting patterns analogous to those observed for nuclear spin-spin coupling. When an electron interacts with n equivalent nuclei, its resonance peak is split into $(2nI + 1)$ peaks, where I is the spin quantum number of the nuclei. A study of this so-called hyperfine splitting provides useful structural information.

Common substances that contain unpaired electrons include free radicals, molecules with triplet-state electrons, and transition-metal ions that have unpaired electrons in their d and f orbitals. ESR spectrometry has been most widely employed in the study of chemical, photochemical, and electrochemical reactions which proceed via free radical mechanisms; the technique has permitted the detection and identification of these species in many instances. Electron-spin resonance spectroscopy has also proved useful in the study of transition-metal complexes.

Problems

1. Predict the appearance of the high-resolution nmr spectrum of propanoic acid.
2. Predict the appearance of the high-resolution nmr spectrum of
    (a) acetaldehyde,
    (b) acetic acid,
    (c) ethyl bromide.
3. Predict the appearance of the high-resolution nmr spectrum of
    (a) acetone,
    (b) methyl ethyl ketone,
    (c) methyl *i*-propyl ketone.
4. Predict the appearance of the high-resolution nmr spectrum of
    (a) α-chloropropanoic acid,
    (b) β-chloropropanoic acid,
    (c) 1-chloropropane,
    (d) 2-chloropropane.
5. Predict the appearance of the high-resolution nmr spectrum of
    (a) toluene,
    (b) ethyl benzene,
    (c) *i*-butane.

6. The spectrum in Figure 7-21 is for an organic compound containing a single atom of bromine. Identify the compound.
7. The spectrum in Figure 7-22 is for a compound having an empirical formula $C_4H_7BrO_2$. Identify the compound.

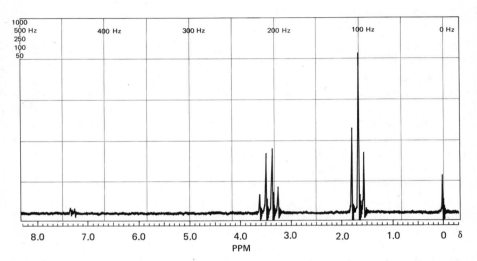

Figure 7-21. (Courtesy of Varian Associates, Palo Alto, Calif.) See Problem 6.

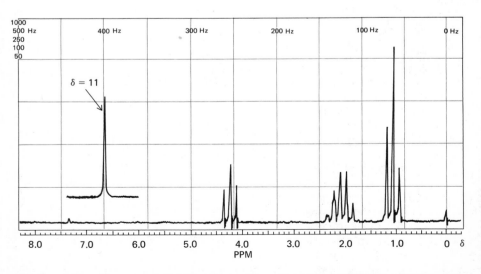

Figure 7-22. (Courtesy of Varian Associates, Palo Alto, Calif.) See Problem 7.

8. The spectrum in Figure 7-23 is for a compound of empirical formula C_4H_8O. Identify the compound.
9. The spectrum in Figure 7-24 is for a compound having an empirical formula $C_4H_8O_2$. Identify the compound.

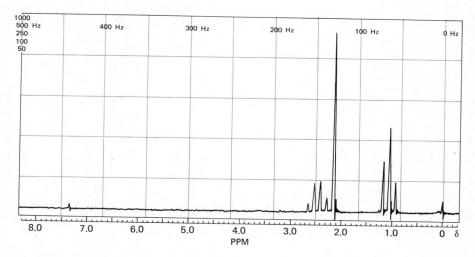

Figure 7-23. (Courtesy of Varian Associates, Palo Alto, Calif.) See Problem 8.

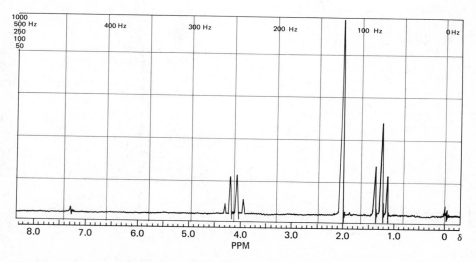

Figure 7-24. (Courtesy of Varian Associates, Palo Alto, Calif.) See Problem 9.

10. The spectra in Figures 7-25(a) and 7-25(b) are for compounds with empirical formulas C_8H_{10}. Identify the compounds.

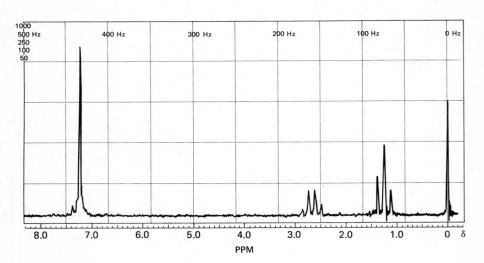

Figure 7-25a. (Courtesy of Varian Associates, Palo Alto, Calif.) See Problem 10.

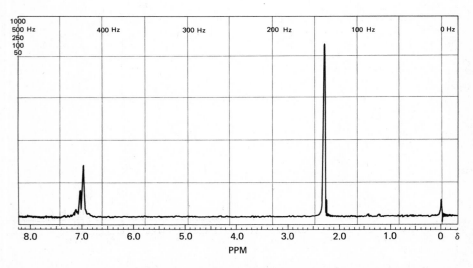

Figure 7-25b. (Courtesy of Varian Associates, Palo Alto, Calif.) See Problem 10.

11. The spectrum in Figure 7-26 is for a compound having the empirical formula C_9H_{12}. Identify the compound.
12. The spectrum in Figure 7-27 is for a compound having the empirical formula C_4H_8S. Identify the compound.

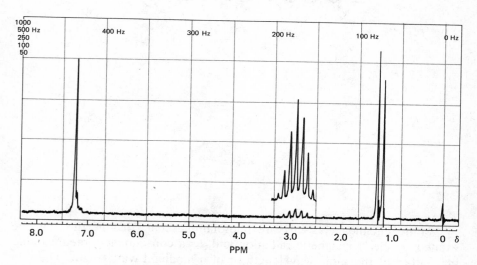

Figure 7-26. (Courtesy of Varian Associates, Palo Alto, Calif.) See Problem 11.

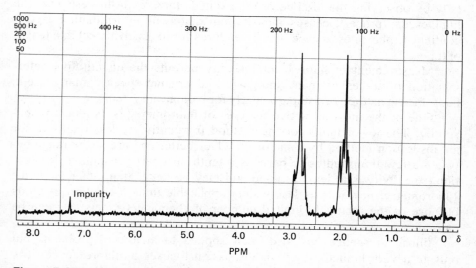

Figure 7-27. (Courtesy of Varian Associates, Palo Alto, Calif.) See Problem 12.

8 Fluorescence Spectrometry

Many chemical systems are photoluminescent; that is, they can be excited by electromagnetic radiation and, as a consequence, reemit radiation either of the same wavelength or of a modified wavelength. The two most common manifestations of photoluminescence are *fluorescence* and *phosphorescence* which, as we shall see, are produced by mechanistically different processes. The two phenomena can be distinguished experimentally by observing the lifetime of the excited state. With fluorescence, the luminescent process ceases almost immediately after irradiation is discontinued; phosphorescence usually endures for an easily detectable length of time.

Measurement of fluorescent intensity permits the quantitative determination of traces of many inorganic and organic species; many useful fluorometric methods exist, particularly for biological systems.

One of the most attractive features of fluorometry is its inherent sensitivity. The lower limits for the method frequently are less than that for an absorption method by a factor of 0.1 or better and are in the range between a few thousandths to perhaps a tenth of a part per million. In addition, selectivity is at least as good and may be better than other methods. Fluorometry, however, is less widely applicable than absorption methods because of the relatively limited number of chemical systems that can be caused to fluoresce.

Phosphorescence has also been applied to analytical problems but only in a very limited way. Our discussion focuses on fluorescence; it is necessary, however, to consider phosphorescence in presenting the theory of the fluorescence phenomenon.

THEORY OF FLUORESCENCE[1]

Examples of fluorescent behavior can be found in simple as well as in complex chemical systems in the gaseous, liquid, and solid states. The simplest kind of fluorescence is that exhibited by dilute atomic vapors. For example, the 3s electrons of vaporized sodium atoms can be excited to the 3p state by absorption of radiation of 5895 and 5890 Å. After an average time lapse of about 10^{-8} sec, the electrons return to the ground state, and in so doing emit radiation of the same two wavelengths in all directions. This type of fluorescence, in which the adsorbed radiation is reemitted without alteration, is known as *resonance radiation* or *resonance fluorescence*.

With polyatomic molecules or ions, resonance radiation also occurs; in addition, characteristic radiation of longer wavelengths is emitted. This phenomenon is called the *Stokes shift*.

Nearly all fluorescent systems that are useful for analysis are complex organic compounds containing one or more aromatic functional groups. Furthermore, the absorption process that leads to the most intense fluorescence in these compounds generally involves a $\pi \rightarrow \pi^*$ transition (see p. 71), although other transitions such as $n \rightarrow \pi^*$ and $n \rightarrow \sigma^*$ are occasionally of interest. Only $\pi \rightarrow \pi^*$ transitions are considered here.

Excited States

In order to understand the characteristics of the fluorescence and phosphorescence phenomena we shall again need to employ qualitatively some of the ideas of molecular orbital theory. To recapitulate these briefly, the pair of electrons forming a bond between two atoms can be considered to occupy a molecular orbital that is formed from the overlap of two of the atomic orbitals of the atoms making up the bond. Combination of two atomic orbitals gives rise to both a bonding and an antibonding orbital; the former has the lower energy and is thus occupied by the electrons in the ground state. Superimposed upon the electronic energy level of each molecular orbital is a series of closely spaced vibrational energy levels. As a consequence, each electronic absorption band contains a series of closely spaced vibrational peaks corresponding to transitions from the ground state to the several vibrational levels of the excited electronic state.

Most molecules contain an even number of electrons; in the ground state these electrons exist as pairs in the various atomic or molecular orbitals. The Pauli exclusion principle demands, however, that the spins of

[1] For further discussion of fluorescence and phosphorescence theory see D. M. Hercules, Ed., *Fluorescence and Phosphorescence Analysis* (New York: Interscience Publishers, 1966); W. West, *Chemical Applications of Spectroscopy* (New York: Interscience Publishers, Inc., 1956), Chap. 6.

the two electrons in a given orbital be opposite to one another (the spins are then said to be paired). As a consequence of spin pairing, most molecules have no net electron spin and are, therefore, diamagnetic. A molecular electronic state in which all of the electron spins are paired is called a *singlet* state, and no splitting of the energy level occurs when the molecule is exposed to a magnetic field (we are here neglecting the effects of nuclear spin). The ground state for a free radical, on the other hand, is a *doublet* state; here, the odd electron can assume two orientations in a magnetic field and thus give rise to a splitting of the energy level.

When one of the electrons of a molecule is excited to a higher energy level, a singlet or a *triplet* state can result. In the excited singlet state the spin of the electron is still paired with the electron in the ground-state orbital; in the triplet state, however, the spins of the two electrons have become unpaired and are thus parallel. These states can be represented as follows:

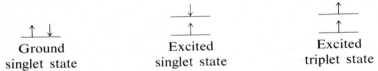

Ground	Excited	Excited
singlet state	singlet state	triplet state

The nomenclature singlet, doublet, and triplet comes from spectroscopic multiplicity considerations with which we need not be concerned.

The properties of a molecule in the excited triplet state differ significantly from those of the corresponding singlet state. For example, a molecule is paramagnetic in the triplet state and diamagnetic in the singlet state. More important, however, is the fact that a singlet–triplet transition (or the reverse) that also involves a change in electronic state is a significantly less probable event than the corresponding singlet–singlet transition. As a consequence, the average lifetime of an excited triplet state may be as long as a second or more, as compared with a lifetime of about 10^{-8} sec for an excited singlet state. Furthermore, radiation-induced excitation of a ground-state molecule to an excited triplet state does not occur readily, and absorption peaks due to this process are several orders of magnitude less intense than the analogous singlet–singlet transition. We shall see, however, that an excited triplet state can be populated from an *excited* singlet state of certain molecules; the consequence of this process is phosphorescent behavior.

Figure 8-1 is a partial energy-level diagram for a typical complex molecule. The energies of the ground state and the two electronic excited states are shown by heavy horizontal lines. Note that the energies of the triplet states are somewhat lower than the corresponding singlet states. Also shown by lighter horizontal lines is a series of closely spaced, excited vibrational states for each electronic state.

The energies required for excitation of the molecule to the two singlet excited states are indicated by the two groups of vertical lines labeled

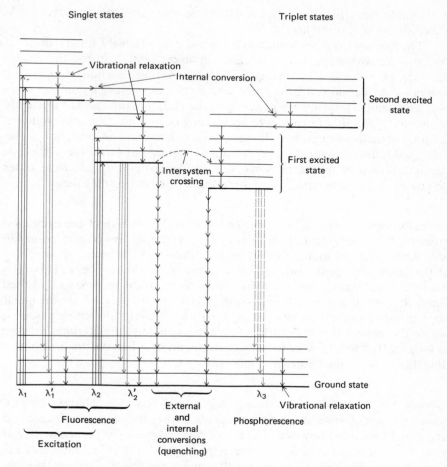

Singlet states

Triplet states

Vibrational relaxation

Internal conversion

Second excited state

First excited state

Intersystem crossing

λ_1 λ_1' λ_2 λ_2' λ_3

Ground state

Fluorescence

Excitation

External and internal conversions (quenching)

Phosphorescence

Vibrational relaxation

Figure 8-1. Partial energy level diagram for a photoluminescent system.

λ_1 and λ_2 in Figure 8-1. Note that the molecule can be converted to any one of the several vibrational levels; thus radiation of a number of wavelengths will cause excitation to each of the two electronic levels. Direct excitation of the molecule to the triplet levels by absorption has not been indicated in Figure 8-1 because this process involves a change in multiplicity, an event that does not occur to any great extent.

Deactivation Processes

An excited molecule can return to its ground state by any of several mechanistic paths; the favored route is the one that minimizes the lifetime of the excited state. Thus, if deactivation by fluorescence is rapid with respect to other processes, such emission is observed. On the other hand, if a radiationless path has a more favorable rate constant, fluorescence is either absent or less intense.

The fluorescence phenomenon is limited to a relatively small number of systems incorporating structural and environmental features that cause the rate of radiationless relaxation or deactivation reactions (*quenching reactions*) to be slowed to a point where the emission reaction can compete kinetically. Information concerning the emission process is sufficiently complete to permit a quantitative accounting of its rate. Our understanding of other deactivation routes, however, is rudimentary at best; for these, only qualitative statements or speculations can be put forth. Nevertheless, the interpretation of fluorescence requires consideration of these other processes, even though much remains to be learned concerning them.

Emission rate. Since fluorescent emission is the reverse of the excitation process, it is perhaps not surprising that a simple, inverse relationship exists between the lifetime of an excited state and the molar absorptivity of the absorption peak corresponding to the excitation process. This relationship (and experiment as well) shows that typical lifetimes of excited states that are deactivated by emission are 10^{-7} to 10^{-9} sec, where the molar absorptivities range between 10^3 and 10^5; for weakly absorbing systems, where the probability of the transition process is low, the lifetimes may be as long as 10^{-6} to 10^{-5} sec. Any deactivation process that occurs in a shorter time thus reduces the fluorescent intensity.

Vibrational relaxation. As shown in Figure 8-1, a molecule may be promoted to any of several vibrational levels during the electronic excitation process. In solution, however, excess vibrational energy is immediately lost as a consequence of collisions between the molecules of the vibrationally excited species and the solvent; the result is an energy transfer and a minuscule increase in temperature of the solvent. This relaxation process is so efficient that the average lifetime of a vibrationally excited molecule is only 10^{-13} to 10^{-10} sec, a period that is significantly shorter than the average lifetime of an electronically excited state. As a consequence, fluorescence from solution, when it occurs, always involves a transition *from the lowest vibrational level of the excited state*. Several closely spaced peaks are produced, however, since the electron can return *to any one of the vibrational levels of the ground state* (Figure 8-1), whereupon it will rapidly fall to the lowest ground state by further vibrational relaxation.

A consequence of the efficiency of vibrational relaxation is that the fluorescent band for a given electronic transition is displaced toward longer wavelengths from the absorption band; overlap occurs only for the resonance peak corresponding to the transition from the lowest ground state to the lowest vibrational level of the excited state.

Internal conversion. The term *internal conversion* is employed to describe intermolecular processes by which a molecule passes to a lower energy *electronic* state without emission of radiation. These processes are neither well-defined nor well-understood, but it is apparent that they are often highly efficient since relatively few compounds exhibit fluorescence.

Internal conversion appears to be particularly efficient when two electronic energy levels are sufficiently close for an overlap of vibrational levels to exist. This situation is depicted for the two singlet excited states and the two triplet excited states in Figure 8-1. At the overlaps shown, the potential energies of the two excited states are identical; this equality apparently results in an efficient transition. Internal conversion through overlapping vibrational levels is usually more probable than the loss of energy by fluorescence from a higher excited state. Thus, referring again to Figure 8-1, excitation by radiation of λ_1 frequently produces fluorescence of wavelength λ_2' to the exclusion of the band at λ_1'. Here the excited molecule proceeds from the higher electronic state to the lowest vibrational state of the lower electronic excited state via a series of vibrational relaxations, an internal conversion and then further relaxations. Under these circumstances the fluorescence would be of λ_2' *only*, regardless of whether radiation of wavelength λ_1 or λ_2 was responsible for the excitation.

Internal conversion may also result in the phenomenon of *predissociation*. Here the electron moves from a higher electronic state to an upper vibrational state of a lower electronic state in which the vibrational energy is great enough to cause rupture of a bond. In a large molecule with many bonds the probability for the existence of bonds with strengths less than the electronic excitation energy of the chromophores is appreciable. Rupture of these bonds can then occur as a consequence of absorption by the chromophore followed by internal conversion of the electronic energy to vibrational energy associated with the weak bond.

A predissociation should be differentiated from a *dissociation*, in which the absorbed radiation excites the electron of a chromophore directly to a sufficiently high vibrational level to cause the chromophoric bond to break; no internal conversion is involved. Dissociation processes also compete with the fluorescence process.

External conversion. Deactivation of an excited electronic state frequently involves interaction and energy transfer between the excited molecule and the solvent or other solutes. These processes are called *external conversions*. Evidence for external conversion includes the marked effect upon fluorescent intensity exerted by the solvent; furthermore, those conditions that tend to reduce the number of collisions between particles (low temperature and high viscosity) generally lead to enhanced fluorescence. The details of external conversion processes are not well-understood.

Radiationless transitions to the ground state from the lower singlet state as well as the lower triplet state (Figure 8-1) probably involve external conversions, and perhaps internal conversions.

Intersystem crossing. *Intersystem crossing* is a process in which the spin of an excited electron is reversed and a change in multiplicity of the molecule results. As with internal conversion, the probability of this transition is enhanced if the vibrational levels of the two states overlap. The singlet-triplet transition shown in Figure 8-1 is an example; here the lowest singlet vibrational state overlaps one of the upper triplet vibrational states and a change in spin state is thus more probable.

Intersystem crossings are most common in molecules that contain heavy atoms, such as iodine or bromine. Apparently in atoms of higher atomic number, interaction between the spin and orbital motions becomes large and a change in spin is thus more favorable. The presence in solution of paramagnetic species such as molecular oxygen also enhances intersystem crossing and a consequent decrease in fluorescence.

Phosphorescence. Deactivation may also involve phosphorescence. After intersystem crossing to a triplet excited state, further deactivation can occur either by internal or external conversion or by phosphorescence. A triplet-singlet transition is much less probable than a singlet-singlet conversion, and the average lifetime of the excited triplet state with respect to emission ranges from 10^{-4} to several seconds. Thus, emission from such a transition may persist for some time after irradiation has been discontinued.

External and internal conversions compete so successfully with phosphorescence that this kind of emission is ordinarily observed only at very low temperatures or in viscous media.

Variables That Affect Fluorescence and Phosphorescence

Both molecular structure and chemical environment are influential in determining whether a substance will or will not fluoresce (or phosphoresce), as well as in determining the intensity of any emission that does occur. The effects of some of these variables are considered briefly in this section.

Quantum yield. The *quantum yield*, or *quantum efficiency*, for a fluorescent process is simply the ratio of the number of molecules that fluoresce to the total number of excited molecules (the quantum yield for phosphorescence can be defined in an analogous way). For a highly fluorescent molecule such as fluorescein, the quantum efficiency under some conditions approaches unity. Chemical species that do not fluoresce appreciably have an efficiency that approaches zero.

From a consideration of Figure 8-1 and our discussion of deactivation processes, it is apparent that the fluorescent quantum yield ϕ for a compound must be determined by the relative rates for the processes by which the lowest excited singlet state is deactivated—namely fluorescence, intersystem crossing, external and internal conversion, predissociation, and dissociation. We may express these relationships by the equation

$$\phi = \frac{k_f}{k_f + k_i + k_{ec} + k_{ic} + k_{pd} + k_d} \tag{8-1}$$

where the k's are the respective rate constants for the several processes enumerated above.

Equation (8-1) permits a qualitative interpretation of many of the structural environmental factors that influence fluorescent intensity. Clearly, those variables that lead to high values for k_f and low values for the other k's enhance fluorescence. The magnitude of k_f, k_{pd}, and k_d are mainly dependent upon chemical structure; the remaining constants are strongly influenced by environment and to a somewhat lesser extent by structure.

Transition types in fluorescence. It is important to note that fluorescence seldom results from absorption of ultraviolet radiation of wavelengths lower than 250 nm because such radiation is sufficiently energetic to cause deactivation of the excited states by predissociation or dissociation. For example, 200-nm radiation corresponds to about 140 kcal per mole, and most molecules have at least some bonds that can be ruptured by energies of this magnitude. As a consequence, fluorescence due to $\sigma^* \rightarrow \sigma$ transitions is seldom observed; instead, the emission is confined to the less energetic $\pi^* \rightarrow \pi$ and $\pi^* \rightarrow n$ processes.

As we have noted, an electronically excited molecule ordinarily returns to its *lowest excited state* by a series of rapid vibrational relaxations and internal conversions that produce no emission of radiation. Thus, any fluorescence observed most commonly arises from a transition from the first excited state to the ground state. For the majority of fluorescent compounds then, radiation is produced by deactivation of either the n,π^* or the π,π^* excited state, depending upon which of these is the less energetic.

Quantum efficiency and transition type. It is observed empirically that fluorescent behavior is more commonly found in compounds in which the lowest energy excited state is of a π,π^* type than in those with a lowest energy n,π^* state; that is, the quantum efficiency is greater for $\pi^* \rightarrow \pi$ transitions.

The greater quantum efficiency associated with the π,π^* state can be rationalized in two ways. First, the molar absorptivity of a $\pi \rightarrow \pi^*$ transition is ordinarily 100- to 1000-fold greater than for an $n \rightarrow \pi^*$ process, and this quantity represents a measure of transition probability in either direction; thus, the inherent lifetime associated with a $\pi \rightarrow \pi^*$ transition is shorter (10^{-7} to 10^{-9} sec compared with 10^{-5} to 10^{-7} sec for an n,π^* state) and k_f in equation (8-1) is larger.

It is also believed that the rate constant for intersystem crossing k_i is smaller for π,π^* excited states because the energy difference between the singlet-triplet states is larger; that is, more energy is required to unpair the electrons of the π,π^* excited state. As a consequence, overlap of triplet vibrational levels with those of the singlet state is less, and the probability of an intersystem crossing is smaller.

In summary, then, fluorescence is more commonly associated with π,π^* states than with n,π^* states because the former possess shorter average lifetimes and because the deactivation processes that compete with fluorescence are less likely to occur.

Fluorescence and structure. The most intense and most useful fluorescent behavior is found in compounds containing aromatic functional groups with low-energy $\pi \rightarrow \pi^*$ transition levels. Compounds containing aliphatic and alicyclic carbonyl structures or highly conjugated double-bond structures may also exhibit fluorescence, but the number of these is small compared with the number in the aromatic systems.

Most unsubstituted aromatic hydrocarbons fluoresce in solution, the quantum efficiency usually increasing with the number of rings and their degree of condensation. The simplest heterocyclics, such as pyridine, furan, thiophene, and pyrrole, do not exhibit fluorescent behavior; on the other hand, heterocyclics with fused-ring structures ordinarily do. With nitrogen heterocyclics the lowest-energy electronic transition is believed to involve an $n \rightarrow \pi^*$ system which rapidly converts to the triplet state and prevents fluorescence. Fusion of benzene rings to a heterocyclic nucleus, however, results in an increase in the molar absorptivity of the absorption peak. The lifetime of an excited state is shorter in such structures; fluorescence is thus observed for compounds such as quinoline, isoquinoline, and indole.

Substitution on the benzene ring causes shifts in the wavelength of absorption maxima and corresponding changes in the fluorescence peaks. In addition, substitution frequently affects the fluorescent efficiency; some of these effects are illustrated by the data for benzene derivatives in Table 8-1.

The effects of halogen substitution are striking; the decrease in fluorescence with increasing atomic number of the halogen is thought to be due in part to the heavy atom effect (p. 228), which increases the rate of intersystem crossing to the triplet state. Predissociation is thought to play an important role in iodobenzene and in nitro derivatives as well; these compounds have easily ruptured bonds that can absorb the excitation energy following internal conversion.

Substitution of a carboxylic acid or carbonyl group on an aromatic ring generally leads to an inhibition of fluorescence. In these compounds the energy of the n,π^* system is less than in the π,π^* system; as we have pointed out earlier, the fluorescent yield from the former type of system is ordinarily low.

Effect of structural rigidity. It is found experimentally that fluorescence is particularly favored in molecules that possess rigid structures. For example, the quantum efficiencies for fluorene and biphenyl are nearly 1.0 and 0.2, respectively, under similar conditions of measurement.

fluorene   biphenyl

The difference in behavior appears to be largely a result of the increased rigidity furnished by the methylene group to fluorene. Many similar examples can be cited. In addition, when fluorescing dyes are adsorbed on a solid surface, enhanced emission frequently results; here again, the added rigidity induced by the solid surface may account for the observed effect.

The influence of rigidity has also been invoked to account for the increase in fluorescence of certain organic chelating agents when they are complexed with a metal ion. For example, the fluorescent intensity of 8-hydroxyquinoline is much less than that of the zinc complex:

Table 8-1 Effect of Substitution on the Fluorescence of Benzene[a,b]

Compound	Formula	Wavelength of Fluorescence (nm)	Relative Intensity of Fluorescence
Benzene	C_6H_6	270–310	10
Toluene	$C_6H_5CH_3$	270–320	17
Propylbenzene	$C_6H_5C_3H_7$	270–320	17
Fluorobenzene	C_6H_5F	270–320	10
Chlorobenzene	C_6H_5Cl	275–345	7
Bromobenzene	C_6H_5Br	290–380	5
Iodobenzene	C_6H_5I	–	0
Phenol	C_6H_5OH	285–365	18
Phenolate ion	$C_6H_5O^-$	310–400	10
Anisole	$C_6H_5OCH_3$	285–345	20
Aniline	$C_6H_5NH_2$	310–405	20
Anilinium ion	$C_6H_5NH_3^+$	–	0
Benzoic acid	C_6H_5COOH	310–390	3
Benzonitrile	C_6H_5CN	280–360	20
Nitrobenzene	$C_6H_5NO_2$	–	0

[a] In ethanol solution.

[b] Taken from W. West, *Chemical Applications of Spectroscopy* (*Techniques of Organic Chemistry,* Vol. IX) (New York: Interscience Publishers, Inc., 1956), p. 730. With permission.

Lack of rigidity in a molecule probably causes an enhanced internal conversion rate (k_{ic} in equation 8-1) and a consequent increase in the likelihood for radiationless deactivation. One part of a nonrigid molecule can undergo low frequency vibrations with respect to its other parts; such motions undoubtedly account for some energy loss.

Temperature and solvent effects. The quantum efficiency of most molecules decreases with increasing temperature because the increased frequency of collisions at elevated temperature improves the probability of deactivation by external conversion. A decrease in solvent viscosity also increases the likelihood of external conversion for the same reason, and leads to the same result.

The polarity of the solvent may also have an important influence. In Chapter 4 we pointed out that the energy for $n \rightarrow \pi^*$ transitions is often increased in polar solvents while that for a $\pi \rightarrow \pi^*$ transition suffers the opposite effect. In some instances such shifts may be great enough to lower the energy of the $\pi \rightarrow \pi^*$ process below that of the $n \rightarrow \pi^*$ transition; enhanced fluorescence results.

The fluorescence of a compound is decreased by solvents containing heavy atoms or other solutes with such atoms in their structures; carbon tetrabromide and ethyl iodide are examples. The effect is similar to that found when these species are substituted into fluorescent compounds; orbital spin interactions result in an increase in the rate of triplet formation and a corresponding decrease in fluorescence. Compounds containing heavy atoms are frequently incorporated into solutions when enhanced phosphorescence is desired.

The effect of pH on fluorescence. The fluorescence of an aromatic compound with acidic or basic ring substituents is usually *p*H-dependent. Both the wavelength and the emission intensity are likely to be different for the ionized and nonionized forms of the compound. The data for phenol and aniline shown in Table 8-1 illustrate this effect. The changes in emission exhibited by these compounds is analogous to the absorption changes that occur with acid-base indicators; indeed, fluorescent indicators have been proposed for acid-base titrations in highly colored solutions. For example, the fluorescence of the phenolic form of 1-naphthol-4-sulfonic acid is not detectable by the eye since it occurs in the ultraviolet region. When the compound is converted to the phenolate ion by the addition of base, the emission peak shifts to visible wavelengths where it can readily be detected. The change in spectral behavior occurs in a small *p*H range. It is of interest that this change occurs at a significantly lower *p*H than would be predicted from the acid dissociation constant for the phenol. The explanation of this discrepancy is that the acid dissociation constant for the *excited* molecule is lower than that for the same species in its ground state. Changes in acid or base dissociation constants with excitation are common and are occasionally as large as four or five orders of magnitude.

It is clear from these observations that analytical procedures based on fluorescence frequently require close control of $p\mathrm{H}$.

Effect of dissolved oxygen. The presence of dissolved oxygen often reduces the emission intensity of a fluorescent solution. This effect may be the result of a photochemically-induced oxidation of the fluorescent species. More commonly, however, the quenching takes place as a consequence of the paramagnetic properties of molecular oxygen that can be expected to promote intersystem crossing and conversion of excited molecules to the triplet state. Other paramagnetic species also tend to quench fluorescence.

Effect of concentration on fluorescent intensity. The power of fluorescent radiation F is proportional to the radiant power of the excitation beam that is absorbed by the system. That is,

$$F = K' (P_0 - P) \tag{8-2}$$

where P_0 is the power of the beam incident upon the solution and P is its power after traversing a length b of the medium. The constant K' depends upon the quantum efficiency of the fluorescence process. In order to relate F to the concentration c of the fluorescing particle we write Beer's law in the form

$$\frac{P}{P_0} = 10^{-\epsilon bc} \tag{8-3}$$

where ϵ is the molar absorptivity of the fluorescent molecules and ϵbc is its absorbance A. By substitution of (8-3) into (8-2) we obtain

$$F = K'P_0 (1 - 10^{-\epsilon bc}) \tag{8-4}$$

The exponential term in equation (8-4) can be expanded to yield

$$F = K'P_0 \left[2.3 \ \epsilon bc - \frac{(-2.3 \ \epsilon bc)^2}{2!} - \frac{(-2.3 \ \epsilon bc)^3}{3!} - \cdots \right] \tag{8-5}$$

Provided $\epsilon bc = A < 0.05$, all of the subsequent terms in the brackets become small with respect to the first; under these conditions we may write

$$F = 2.3 \ K' \ \epsilon bcP_0 \tag{8-6}$$

or at constant P_0,

$$F = Kc \tag{8-7}$$

Thus, a plot of the fluorescent power of a solution versus concentration of the emitting species should be linear at low concentrations c. When c becomes great enough so that the absorbance is larger than about 0.05, however, linearity is lost and F lies below an extrapolation of the straight-line plot.

Two other factors responsible for further negative departures from linearity at high concentration are *self-quenching* and *self-absorption*. The former is the result of collisions between fluorescing molecules. Radiation-less transfer of energy occurs, perhaps in a fashion analogous to the transfer to solvent molecules that occurs in an external conversion. Self-quenching can be expected to increase with concentration.

Self-absorption occurs when the wavelength of emission overlaps the absorption peak of the compound; the fluorescence is then decreased as the emitted beam traverses the solution.

The effects of these phenomena are such that a plot of fluorescent power versus concentration often exhibits a maximum at high concentrations.

MEASUREMENT OF FLUORESCENCE

The various components of instruments for the measurement of fluorescence are similar to those in photometers and spectrophotometers (see Chapter 3). Figure 8-2 shows a typical arrangement of these components. Radiation from a suitable source is passed through a monochromator or filter which serves to transmit that part of the beam which will excite fluorescence but exclude wavelengths that are subsequently produced by the irradiated sample. Fluorescent radiation is emitted by the sample in all directions but is most conveniently observed at right angles to the excitation beam; at other angles increased scattering from the solution and the cell walls is likely to result in large errors in the measurement of fluorescent intensity. The emitted radiation reaches a photoelectric detector after

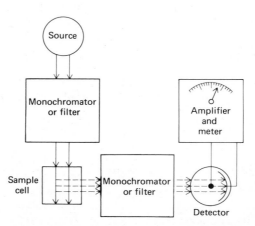

Figure 8-2. Components of a fluorometer or spectrofluorometer.

passing through a second filter or monochromator system that isolates the fluorescent peak. The output of the detector is amplified and displayed on a meter, a recorder, or an oscilloscope.

A variety of instruments for fluorescent measurements have been described, and a dozen or more designs are produced commercially. Their sophistication, performance characteristics, and costs differ as widely as do the corresponding instruments for absorption measurements (Chapter 3). *Fluorometers* are analogous to photometers in that filters are employed to restrict the wavelengths of the excitation and emission beams. *Spectrofluorometers* are of two types. The first employs a suitable filter to limit the excitation radiation, and a grating or prism monochromator to isolate a peak of the fluorescent emission spectrum. Several commercial spectrophotometers can be purchased with adapters that permit their use in this way. The adapters consist of a separate light source, a filter holder, a cell compartment, and a means of focusing the fluorescent radiation upon the entrance slit of the monochromator. True spectrofluorometers are specialized instruments equipped with two monochromators. One of these restricts the excitation radiation to a narrow band; the other permits the isolation of a particular fluorescent wavelength. The selectivity provided by these instruments is of prime importance to investigations concerned with the electronic and structural characteristics of molecules, and is of value in analytical work as well. For most analytical purposes, however, the information provided by simpler instruments is entirely satisfactory. Indeed, relatively inexpensive fluorometers that have been designed specifically to meet the measurement problems peculiar to fluorescent analysis are frequently as specific and selective as modified spectrophotometers.

The discussion that follows is largely focused on the simpler instruments for fluorescence analysis.

Components of Fluorometers and Spectrofluorometers

The components of fluorometers and spectrofluorometers differ only in detail from those of photometers and spectrophotometers; we need to consider only these differences.

Sources. In most applications a more intense source is needed than the tungsten or hydrogen lamp employed for the measurement of absorption. A mercury or xenon arc lamp is commonly employed.

The xenon arc lamp produces intense radiation by the passage of current through an atmosphere of xenon. The spectrum is continuous over the range between about 250 and 600 nm, with the peak intensity occurring at about 470 nm (see Figure 2-9, p. 23).

Mercury arc lamps produce an intense line spectrum. High-pressure lamps ($\sim$ 8 atmospheres) give lines at 365, 398, 436, 546, 579, 690, and 734 nm. Low-pressure lamps, equipped with a silica envelope, additionally provide intense radiation at 254 nm. Inasmuch as fluorescent behavior can be induced in most fluorescing compounds by a variety of wavelengths, at least one of the mercury lines ordinarily proves suitable.

Filters and monochromators. Both interference and absorption filters have been employed in fluorometers. Most spectrofluorometers are equipped with grating monochromators.

Detectors. The typical fluorescent signal is of low intensity; large amplification factors are thus required for its measurement. Photomultiplier tubes have come into widespread use as detectors in sensitive fluorescence instruments.

Cells and cell compartments. Both cylindrical and rectangular cells constructed of glass or silica are employed in fluorescence measurements. Care must be taken in the design of the cell compartment to reduce the amount of scattered radiation reaching the detector. Baffles are often introduced into the compartment for this purpose.

Fluorometer Designs

Fluorometers, like photometers, may be of single-beam or double-beam design. The double-beam instruments have many advantages, and most commercial equipment incorporates this feature.

Figure 8-3 is a schematic diagram of a simple double-beam fluorometer. A collimated beam from a mercury lamp is filtered to remove visible but not ultraviolet radiation. A part of the beam then passes through a reduction plate, which reduces the intensity of all wavelengths by as much as 99 percent; this attenuation of the reference beam is necessary in order that its power be of the same order of magnitude as the weak fluorescent beam. The attenuated beam is reflected onto the surface of a reference photovoltaic cell which is mounted on a table that can be rotated. By adjustment of the angle of the photocell, the fraction of the beam absorbed by the photoelectric surface, and thus the current output, can be adjusted to a convenient level.

The undeflected beam passes into a cuvette, which has a pair of photocells mounted at right angles to the excitation beam. Fluorescent radiation passes through filters that remove scattered ultraviolet radiation but pass visible light. The two photocells, connected in parallel, are employed in order to increase the sensitivity of the instrument for the weak fluorescent radiation.

A bridge circuit, with a sensitive galvanometer or a null-detector, is employed to compare the outputs of the pair of photocells with the output of the reference cell. Note that the circuit is arranged so that the current from the reference photocell opposes that generated by the photocells that respond to the fluorescent radiation. Thus, the current indicated by the galvanometer is

$$i = i_{g_1} - i_{g_2}$$

and when contact B is adjusted until $i = 0$,

$$i_{g_1} = i_{g_2}$$

The current i_1 resulting from the fluorescence is divided into two components by the circuit, with i_{g_1} passing through the galvanometer and i_{r_1} through the resistance AC; the galvanometer component is given by

$$i_{g_1} = i_1 \frac{R_{AC}}{R_{AC} + R_g}$$

where R_g is the resistance of the galvanometer.

The current from the reference photocell i_2 is likewise divided, with i_{r_2} passing through the resistance AB, and i_{g_2} through resistance BC and then the galvanometer. The galvanometer component is given by

$$i_{g_2} = i_2 \frac{R_{AB}}{R_{AB} + R_{BC} + R_g} = i_2 \frac{R_{AB}}{R_{AC} + R_g}$$

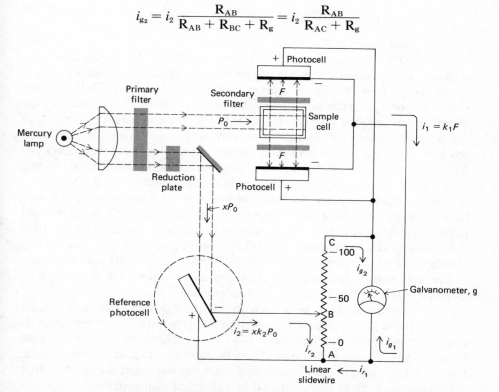

Figure 8-3. A double-beam fluorometer. The Lumetron Model 402-EF Fluorometer. (Courtesy of Photovolt Corporation, New York, N.Y.)

When the instrument is brought to the null condition, $i_{g_1} = i_{g_2}$ and

$$i_1 = i_2 \frac{R_{AB}}{R_{AC}} \qquad (8\text{-}8)$$

If the photocells respond linearly to radiant power we may write

$$i_1 = k_1 F$$

where F is the power of the fluorescent radiation. Substitution of equation (8-6) into this equation gives

$$i_1 = 2.3 \, k_1 K' \epsilon b c P_0 \qquad (8\text{-}9)$$

The radiant power of the beams reaching the reference photocell is some fraction or multiple x of P_0. The resulting current then is

$$i_2 = x k_2 P_0 \qquad (8\text{-}10)$$

where k_2 is a proportionality constant. Substitution of equations (8-9) and (8-10) into (8-8) yields, upon rearrangement,

$$c = \frac{x k_2}{2.3 \, k_1 K' \epsilon b} \times \frac{R_{AB}}{R_{AC}} = k \frac{R_{AB}}{R_{AC}} \qquad (8\text{-}11)$$

Note that the constant k is independent of P_0.

Ordinarily, an instrument of this type is calibrated before each analysis with a standard solution that has a fluorescence somewhat stronger than any of the samples; a solution of quinine sulfate or a standard solution of the substance being determined serves this purpose. With the standard in the cell, R_{AB} can be set at full scale (often 100) and x can be varied by rotation of the reference photocell until the null point is reached. With x fixed, one or more standards are then measured to determine the relationship between R_{AB} and c; finally, the fluorescence of the unknown is determined. Providing k_1, k_2, and x are invariant, c is independent of fluctuations in P_0. Normally, k_1 and k_2 remain constant for short periods but most certainly not for long times; nor do they necessarily change in the same way. Thus, calibration against the fluorescent standard must be performed regularly; in addition, it is good practice to check the calibration constant against one or two standard solutions from time to time. Although c is proportional to F only at low concentrations, preparation of a suitable calibration curve permits analysis in the nonlinear regions of fluorescence.

An instrument such as shown in Figure 8-3 suffers from a limited sensitivity due to the use of photovoltaic cells; in addition, the photocells cannot be perfectly matched; thus, k_1/k_2 can be expected to fluctuate in an irregular way. Frequent calibration is thus required for reliable performance. Figure 8-4 is a schematic diagram of a more sophisticated fluorometer that avoids many of these limitations. It is a double-beam instrument that employs a single photomultiplier tube as a detector. A rotating light interrupter causes the reference beam and the fluorescent beam to strike the photosensitive surface alternately. The output of the detector is thus an alternating current that can be readily amplified. Depending upon whether the fluorescent beam or the reference beam is stronger, this signal will either be positive or negative. A phase-sensitive device is employed to translate the difference and its sign into a deflection of a meter needle. The power of the reference beam can be varied by rotation of the light cam, which mechanically increases or decreases the fraction of the reference beam that reaches the detector. The cam is equipped with a linear dial, each increment of which corresponds to an equal fraction of light.

In order for an optical-null device of this type to be accurately adjusted, it must be possible to approach the null point from both directions. For a totally nonfluorescent sample, however, no light would reach the detector in one of the phases and the null point could then be approached from one direction only. To avoid the resultant error, a third beam of constant intensity (the forward light path) is directed to the detector *in phase* with the fluorescent beam so that, under all conditions, some radiation strikes the photomultiplier. Correction for the effect of the third beam on the measured fluorescence is accomplished by setting the fluorescence dial on zero with a solvent blank or a nonfluorescing dummy cuvette in the cell compartment; the intensity of the reference beam can then be varied by means of the blank shutter until an optical null is indicated; this operation should be carried out regularly during a set of analyses.

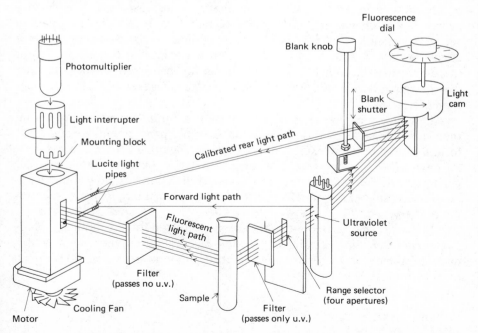

Figure 8-4. Optical design of the Turner Model 110 Fluorometer. (Courtesy of G. K. Turner Associates, Palo Alto, California.)

The single detector aspect of this double-beam instrument imparts high reproducibility even with long-term changes in the sensitivity of the detector and the output of the source. Thus, only occasional checks of calibration curves with a standard are required.

The two instruments just described are representative of the dozen or more fluorometers available commercially. These cost from a few hundred dollars to perhaps $1500. True double-monochromator spectrofluorometers are also available; the added feature may increase the cost by a factor of ten.

APPLICATIONS OF FLUOROMETRY

Fluorometric methods are inherently applicable to lower concentration ranges than are spectrophotometric determinations, and are thus among the most sensitive analytical techniques available to the scientist. The basic difference in the sensitivity between the two methods arises from the fact that the concentration-related parameter for fluorometry F can be measured independently of the power of the source P_0. In spectrophotometry, on the other hand, not only must P be evaluated, but P_0 must be measured as well, for the concentration-dependent parameter A is dependent upon the ratio between these quantities. The sensitivity of a fluorometric method can be improved by increasing P_0 or by further amplifying the signal produced by the fluorescence. In spectrophotometry an increase in P_0 results in a proportionate change in P and therefore fails to affect A; thus, no gain in sensitivity results. Similarly, amplification of the detector signal has the same effect on both P and P_0 and results in no net gain with respect to A. For these reasons fluorometric methods generally have sensitivities that are two to four orders of magnitude better than the corresponding spectrophotometric procedures.

Few inorganic species possess sufficient natural fluorescence to permit their direct analysis by this method. On the other hand, a number of cations form fluorescent complexes with appropriate chelating agents and can thus be determined. As we shall see, the cations that behave in this way are generally colorless and tend to form complexes that fail to show characteristic absorption behavior. Thus, fluorometry often complements spectrophotometry.

Many organic compounds that are of interest in food technology, pharmacology, and clinical medicine exhibit strong fluorescence; as a consequence, numerous important fluorometric methods have been developed for use in these fields.

Inorganic Analysis

Inorganic fluorometric methods are of two types. Direct methods involve the formation of a fluorescent chelate and the measurement of its emission. A second group is based upon the diminution of fluorescence resulting from the quenching action of the substance being determined. The latter technique has been most widely used for anion analysis.

Cations that form fluorescent chelates. Two factors greatly limit the number of transition-metal ions that form fluorescent chelates. First, many of these ions are paramagnetic; this property increases the rate of intersystem crossing to the triplet state. Deactivation by fluorescence is thus unlikely, although phosphorescent behavior may be observed. A second reason is that transition-metal complexes are characterized by many closely spaced energy levels, which enhance the likelihood of deactivation by intersystem conversion. Nontransition-metal ions are less susceptible to the foregoing deactivation processes; it is for these elements that the principal applications of fluorometry are to be found.

Fluorometric reagents.[2] The most successful fluorometric reagents for cation analyses have aromatic structures with two or more donor functional groups that permit chelate formation with the metal ion. The structures of four common reagents follow:

8-Hydroxyquinoline
(Reagent for Al, Be, and
other metal ions)

Alizarin garnet R
(Reagent for Al, F⁻)

Flavanol
(Reagent for Zr and Sn)

Benzoin
(Reagent for B, Zn, Ge, and Si)

Selected fluorometric reagents and their applications are presented in Table 8-2. For a more complete summary see Meites[3] or the review articles by White and Weissler.[4]

[2]For a more detailed discussion of fluorometric reagents see T. S. West in W. W. Meinke and B. E. Scribner, Eds., *Trace Characterization, Chemical and Physical* (Washington, D.C.: National Bureau of Standards Monograph 100, 1967), pp. 237–266; C. E. White in J. H. Yoe and H. J. Koch, Eds., *Trace Analysis* (New York: John Wiley & Sons, Inc., 1957), Chap. 7.

[3]L. Meites, *Handbook of Analytical Chemistry* (New York: McGraw-Hill Book Company, Inc., 1963), pp. 6-178 to 6-181.

[4]C. E. White and A. Weissler, *Anal. Chem.*, **42**, 57R (1970); **40**, 116R (1968); **38**, 155R (1966); **36**, 116R (1964); **34**, 81R (1962).

Table 8-2 Selected Fluorometric Methods for Inorganic Species[a]

| Ion | Reagent | Wavelength, nm | | Sensitivity $\mu g/ml$ | Interference |
		Absorption	Fluorescence		
Al^{3+}	Alizarin garnet R	470	500	0.007	Be, Co, Cr, Cu, F⁻, NO_3^-, Ni, PO_4^{3-}, Th, Zr
F^-	Al complex of Alizarin garnet R (quenching)	470	500	0.001	Be, Co, Cr, Cu, Fe, Ni, PO_4^{3-}, Th, Zr
$B_4O_7^{2-}$	Benzoin	370	450	0.04	Be, Sb
Cd^{2+}	2-(o-Hydroxyphenyl)-benzoxazole	365	Blue	2	NH_3
Li^+	8-Hydroxyquinoline	370	580	0.2	Mg
Sn^{4+}	Flavanol	400	470	0.1	F⁻, PO_4^{3-}, Zr
Zn^{2+}	Benzoin	—	Green	10	B, Be, Sb, Colored ions

[a]L. Meites, *Handbook of Analytical Chemistry* (New York: McGraw-Hill Book Company, Inc., 1963), pp. 6-178 to 6-181. With permission.

Organic Species

The number of applications of fluorometric analysis to organic problems is impressive. Weissler and White have summarized the most important of these in several tables.[5] Under a heading of *Organic and General Biochemical Substances* are found over 100 entries that include such diverse substances as adenine, anthranilic acid, aromatic polycylic hydrocarbons, cysteine, guanidine, indole, naphthols, certain nerve gases, proteins, salicylic acid, skatole, tryptophan, uric acid, and warfarin. Some 50 medicinal agents that can be determined fluorometrically are listed. Included among these are adrenaline, alkylmorphine, chloroquin, digitalis principles, lysergic acid diethylamide, penicillin, phenobarbital, procaine, and reserpine.

Methods for the analysis of 10 steroids and an equal number of enzymes and coenzymes are also found in these tables. Some of the plant products listed include chlorophyll, ergot alkaloids, rawolfia serpentina alkaloids, flavonoids, and rotenone. Some 18 listings for vitamins and vitamin products are also included; among these are ascorbic acid, folic acid, nicotinamide, pyridoxal, riboflavin, thiamin, vitamin A, and vitamin B_{12}.

Without question, the most important applications of fluorometry are to be found in the analyses of food products, pharmaceuticals, clinical samples, and natural products. The sensitivity and selectivity of the method make it particularly valuable in these fields.

[5]A. Weissler and C. E. White in L. Meites, Ed., *Handbook of Analytical Chemistry* (New York: McGraw-Hill Book Company, Inc., 1963), pp. 6-182 to 6-196.

9 Methods Based upon Light Scattering

Three analytical methods that are based upon the scattering of radiation are considered in this chapter. Two of these, *nephelometry* (from the Greek *nephelē*, meaning cloud) and *turbidimetry* are so closely related that they are conveniently discussed together. The third method, *Raman spectroscopy*, is sufficiently unique to warrant separate treatment.

NEPHELOMETRY AND TURBIDIMETRY[1]

Light is scattered in all directions when it passes through a transparent medium containing a particulate second phase. As a consequence, the mixture has a cloudy or turbid appearance; in addition, a loss in power is suffered by the beam along its axis of travel. If other variables are held constant, the extent of this loss can be related to the weight concentration of the particles responsible for the scattering. *Turbidimetric analysis* is based upon measurement of the diminution in power of a collimated beam as a result of scattering.

The concentration of particulate matter can also be assessed from a

[1]For a more complete discussion see F. P. Hochgesang, in I. M. Kolthoff and P. J. Elving, Eds., *Treatise on Analytical Chemistry*, Part I, Vol. 5 (New York: Interscience Publishers, Inc., 1964), Chap. 63.

measurement of the radiant power at 90 deg (or some other angle) to the incident beam. Such a procedure is inherently more sensitive than the turbidimetric method for the same reasons that the measurement of fluorescence is more sensitive than absorption. *Nephelometry* is based upon the measurement of the power of scattered radiation at right angles to a collimated beam.

The choice between a nephelometric and a turbidimetric measurement depends upon the fraction of light scattered. When scattering is extensive, owing to the presence of many particles, a turbidimetric measurement is the more satisfactory. If the suspension is less dense and the diminution in power of the incident beam is small, nephelometric measurements provide more satisfactory results.

Theory of Nephelometry and Turbidimetry

It is important to appreciate that the scattering associated with nephelometry and turbidimetry (in contrast to Raman spectroscopy) involves no net loss in radiant power; only the direction of propagation is affected. The intensity of radiation appearing at any angle is dependent upon the number of particles, their size and shape, the relative refractive indexes of the particles and the medium, and the wavelengths of the radiation. The relationship among these variables is complex. A theoretical treatment is feasible, but because of its complexity it is seldom applied to specific analytical problems. In fact, most nephelometric and turbidimetric procedures tend to be highly empirical.

Effect of concentration on scattering. In a dilute suspension the attenuation of a parallel beam of radiation by scattering is given by the relationship

$$P = P_0 \, e^{-\tau b} \qquad (9\text{-}1)$$

where P_0 and P are the power of the beam before and after passing through the length b of the turbid medium. The quantity τ is called the *turbidity coefficient*, or the turbidity; its value is often found to be linearly related to the concentration c of the scattering particles. Under these circumstances, a relationship analogous to Beer's law results; that is,

$$\log_{10} \frac{P_0}{P} = kbc \qquad (9\text{-}2)$$

where $k = 2.3 \, \tau/c$.

Equation (9-2) is employed in turbidimetric analysis in exactly the same way as Beer's law is used in photometric analysis. The relationship (hopefully linear) between $\log_{10} P_0/P$ and c is established with standard samples, the solvent being used as a reference to determine P_0. The resulting calibration curve is then used to determine the concentration of samples from turbidimetric measurements.

For nephelometric measurements the power of the beam scattered at right angles to the incident beam is normally plotted against concentration; a linear relationship is frequently obtained. The procedure here is entirely analogous to a fluorometric method.

Effect of particle size on scattering. The fraction of radiation scattered at any angle depends upon the size and shape of the particles responsible for the scattering; the effect is large. Since most analytical applications of scattering involve the generation of a colloidal dispersed phase in a solution, those variables that influence particle size during precipitation influence both turbidimetric and nephelometric measurements. Thus, such factors as concentration of reagents, rate and order of mixing, length of standing, temperature, pH, and ionic strength are important experimental variables. Care must be exercised to reproduce all conditions likely to affect particle size during calibration and analysis.

Effect of wavelength on scattering. It has been shown experimentally that the turbidity coefficient varies with wavelength as given by the equation

$$\tau = s\lambda^{-t}$$

where s is a constant for a given system. The quantity t is dependent on particle size and has a value of 4 when the scattering particles are significantly smaller than the wavelength of radiation (Rayleigh scattering); for particles with dimensions similar to the wavelength (the usual situation in a turbidimetric analysis) t is found to be about 2.

For purposes of analysis, ordinary white light is employed. If the solution is colored, it is necessary to select a portion of the spectrum in which absorption by the medium is minimized.

Instruments

Nephelometric and turbidimetric measurements can be readily made with the various fluorometers and photometers discussed in earlier chapters. Rectangular cells are ordinarily employed. With the exception of those areas through which the radiation is transmitted, the cell walls have a dull black coating; this eliminates reflection of unwanted radiation to the detector.

Figure 9-1 shows a simple visual turbidimeter. The viewing tube is adjusted in the suspension until the special S-shaped lamp filament just disappears. The length of solution is then related to concentration by calibration. This very elementary device yields remarkably accurate data for the analysis of sulfate in low concentrations. Here a $BaSO_4$ suspension is formed by the addition of $BaCl_2$.

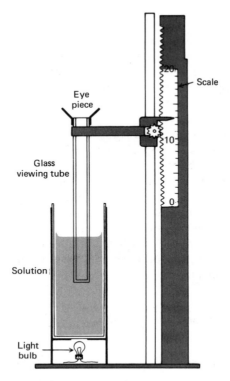

Figure 9-1.   A simple turbidimeter.

Applications of Scattering Methods

Turbidimetric or nephelometric methods are widely used in the analysis of water for the determination of clarity and for the control of treatment processes. In addition, the concentration of a variety of ions can be determined by the use of suitable precipitating reagents. Conditions must be chosen so that the solid phase forms as a stable colloidal suspension. Surface active agents (such as gelatin) are frequently added to the sample to prevent coagulation of the colloid. As noted earlier, reliable analytical data are obtained only when care is taken to control scrupulously all of the variables that affect particle size.

Table 9-1 lists some of the species that have been determined by turbidimetric or nephelometric methods. Perhaps the most widely employed is the method for sulfate ion. Nephelometric methods permit the determination of concentrations as low as a few parts per million with a precision of 1 to 5 percent. Turbidimetric methods are reported to give the same degree of reproducibility with more concentrated solutions.

Turbidimetric measurements have also been employed for the determination of the end point in precipitation titrations. The apparatus can be

Table 9-1 Some Turbidimetric and Nephelometric Methods[a]

Element	Method[b]	Suspensions	Reagent	Interferences
Ag	T, N	AgCl	NaCl	—
As	T	As	KH_2PO_2	Se, Te
Au	T	Au	$SnCl_2$	Ag, Hg, Pd, Pt, Ru, Se, Te
Ca	T	CaC_2O_4	$H_2C_2O_4$	Mg, Na, SO_4^{2-} (in high concentration)
Cl^-	T, N	AgCl	$AgNO_3$	Br^-, I^-
K	T	$K_2NaCo(NO_2)_6$	$Na_3Co(NO_2)_6$	SO_4^{2-}
Na	T, N	$NaZn(UO_2)_3(OAc)_9$	$Zn(OAc)_2$ and $UO_2(OAc)_2$	Li
SO_4^{2-}	T, N	$BaSO_4$	$BaCl_2$	Pb
Se	T	Se	$SnCl_2$	Te
Te	T	Te	NaH_2PO_2	Se

[a] Data taken from L. Meites, *Handbook of Analytical Chemistry* (New York: McGraw-Hill Book Company, Inc., 1963), p. 6-175. With permission.
[b] T = turbidimetric; N = nephelometric.

very simple, consisting of a light source and a photocell located on opposite sides of the titration vessel. The photocurrent is then plotted as a function of volume of reagent. Ideally, the turbidity increases linearly with volume of reagent until the end point is reached, whereupon it remains constant.[2]

RAMAN SCATTERING

In Chapter 2 it was noted that particles of molecular dimensions scatter radiation (Rayleigh scattering); the intensity of this scattering, however, is much weaker than for particles whose sizes approach the wavelength of the radiation. In 1928 the Indian physicist C. V. Raman discovered that, under some circumstances, the wavelength of part of the radiation scattered by molecules differs from that of the incident beam, and furthermore that the shift in wavelength was dependent upon the chemical structure of the molecule.[3]

Since its discovery, the Raman effect has been the subject of much study, and the theory of the wavelength shifts is now well-understood. Raman spectroscopy has provided a useful tool for certain limited types of quantitative analysis; it has also been applied to structural studies of both organic and inorganic systems.

[2] For an analysis of the factors affecting turbidity titration curves see E. J. Meehan and G. Chiu, *Anal. Chem.*, **36**, 536 (1964).
[3] Raman was awarded the 1931 Nobel prize in physics for his investigation of this phenomenon.

Theory

Nature of Raman spectra. Raman spectra are obtained by irradiating a sample with very intense monochromatic radiation and examining the wavelength characteristics of the radiation scattered at right angles to the source. At the very most, the intensities of Raman lines are 0.01 percent of the source; as a consequence, their detection and measurement is experimentally difficult.

Figure 9-2(b) shows the Raman spectrum for carbon tetrachloride excited by an intense mercury arc. For reference, the spectrum of the source is shown in Figure 9-2(a). Note that the two mercury lines appear unchanged in wavelength on the lower spectrum as a consequence of normal Rayleigh scattering. In addition, however, two sets of five lines appear on the longer wavelength side of the two intense mercury lines at 404.7 and 435.8 nm. If it is assumed that each set of Raman lines is excited by the nearer of the two intense mercury lines, it is found that the pattern of displacement in wave numbers (cm^{-1}) is identical for the two sets. Thus, for the 404.7-nm mercury line (24,710 cm^{-1}) a Raman line appears at a wave number that is 218 cm^{-1} smaller (24,492 cm^{-1}), which corresponds to a wavelength of 408.3 nm; similarly, the mercury line at 435.8 nm (22,946 cm^{-1}) has a Raman line at 22,728 cm^{-1}, or at a wavelength of 440.0 nm. Weaker mercury lines (not shown) are also capable of exciting sets of Raman lines; generally, however, these are not observed because of their low intensity.

Superficially, the appearance of Raman spectral lines at lower energies (longer wavelengths) is analogous to what is observed in a fluorescence

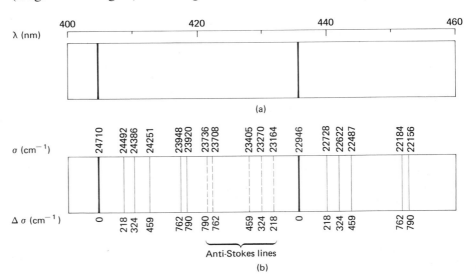

Figure 9-2. (a) Intense lines of a mercury arc spectrum and (b) Raman spectrum for CCl_4.

experiment; thus Raman shifts in this direction are sometimes called *Stokes shifts* (see p. 223). As we shall see, however, the two phenomena arise from fundamentally different processes; for this reason the application of this nomenclature to both fluorescent and Raman spectra is unfortunate.

In Figure 9-2 a further set of Raman lines occurs at energies that are higher than that of the excitation line at 435.8 nm. The displacements among these lines follow the same pattern as the displacements in the other direction. A similar set (not shown) appears at wavelengths shorter than the 404.7 mercury line. Shifts toward higher energies are termed *anti-Stokes*; quite generally, anti-Stokes lines have appreciably weaker intensities than the corresponding Stokes lines.

Raman spectra are characteristic of the substances responsible for the scattering and are thus useful for analytical purposes. Most commonly the spectra are plotted in terms of the energy *shift* with respect to source line. Since Stokes peaks are the more intense, $\Delta\sigma$ (the shift in units of cm^{-1}) is evaluated on the basis of the shift to lower wave numbers; that is,

$$\Delta\sigma = \sigma_{Hg} - \sigma \qquad (9-3)$$

where σ_{Hg} is the wave number (in cm^{-1}) of the particular source line and σ is the wave number for the Raman peak. Thus, the spectrum for carbon tetrachloride in Figure 9-2 reveals Raman peaks at 218, 314, 459, 762, and 790 cm^{-1}.

Transitions responsible for Raman peaks. The discrete energy shifts that characterize Raman spectra indicate the presence of quantized energy transitions. These shifts can be rationalized by assuming that the electrical field of the transmitted radiation interacts with the electrons of the sample and causes periodic polarization and depolarization. As a consequence, the energy of the radiation is momentarily retained in a distorted polarized species. The energetics of this process is shown by the vertical arrow on the left in Figure 9-3; note that the interaction, in contrast to absorption, does not involve transition to a higher quantized energy level. After a retention time of perhaps 10^{-15} to 10^{-14} sec, the species ordinarily returns to its ground state, as shown, with the release, in all directions, of radiation with exactly the same energy as the source. The small fraction of this radiation that is transmitted at right angles to the beam is Rayleigh scattering. Under some circumstances, however, the molecule may return from the distorted state to the *first excited vibrational level of the ground state*. The frequency emitted in this transition will be less by a quantized amount corresponding to the difference in energies between the ground state and the first vibrational level, ΔE. That is,

$$h\nu = h\nu_{Hg} - \Delta E \qquad \text{or} \qquad \Delta\nu = \Delta E/h$$

where ν and ν_{Hg} are the frequencies of the Raman peak and the source peak, respectively, and h is Planck's constant.

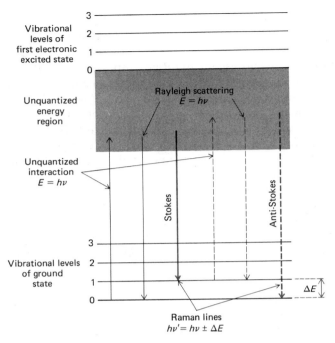

Figure 9-3. Origin of Raman lines.

As shown to the right in Figure 9-3, anti-Stokes scattering arises from the interaction of the radiation with a molecule that is initially in its first excited vibrational level. If the resulting distorted species returns to the ground state, the resulting emission is *increased* in frequency by the amount ΔE. Since the number of molecules in the first excited state is small at room temperature, anti-Stokes radiation is weak.

Vibrational modes associated with the Raman effect. It is of interest to note that the transitions responsible for the Raman effect are of the same type as those described in Chapter 6 for infrared absorption; that is, transitions between the ground state of a molecule and its first vibrational states are involved in both phenomena. Therefore, for a given molecule the *energy shifts* observed in a Raman experiment should be identical to the *energies* of its infrared absorption bands provided that the vibrational modes involved are active toward both infrared absorption and Raman scattering. Figure 9-4 illustrates the similarity of the two types of spectra. Here it is seen that several peaks for the two compounds appear at values for σ and $\Delta\sigma$ that are identical. It is also noteworthy, however, that the relative sizes of the corresponding peaks are frequently quite different; furthermore, certain peaks that occur in one spectrum are absent in the other.

The differences between a Raman and an infrared spectrum are not surprising when it is considered that the basic mechanisms, although dependent upon the same vibrational modes, arise from processes that are mechanistically different. Infrared absorption requires that a vibrational mode of the molecule have associated with it a change in dipole or charge distribution. Only then can radiation of the same frequency interact with the molecule and thus promote it to an excited vibrational state. In contrast, scattering involves a momentary, elastic distortion of the electrons distributed around a bond in a molecule, followed by reemission of the radiation in all directions as the bond returns to its normal state. In its distorted form the molecule is temporarily polarized; that is, it develops, momentarily, an induced dipole which disappears upon relaxation and reemission. The effectiveness of a bond toward scattering is directly dependent upon the ease with which the electrons can be distorted from their normal positions (its polarizability); polarizability decreases as the electron density increases or as the length of the bond decreases. The Raman shift in scattered radiation, then, requires that there be a *change in polarizability* associated with the vibrational mode of the molecule rather than a change in dipole; as a consequence, the Raman activity of a given mode may differ markedly from its infrared activity. For example, a homonuclear molecule such as nitrogen, chlorine, or hydrogen has no dipole moment

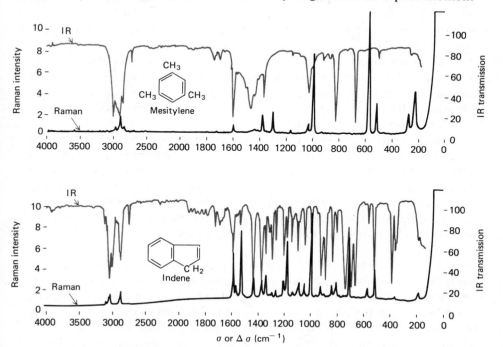

Figure 9-4. Comparison of Raman and infrared spectra. (Courtesy Perkin-Elmer Corp., Norwalk, Conn.)

either in its equilibrium position or when a stretching vibration causes the two nuclei to be separated at a greater or lesser distance. Thus, absorption of radiation of the vibration frequency cannot occur. On the other hand, the polarizability of the bond between the two atoms of such a molecule varies periodically in phase with the stretching vibrations, reaching a maximum at the greatest separation and a minimum at the closest approach. A Raman shift corresponding in frequency to that of the vibrational mode thus results.

It is of interest to compare the infrared and the Raman activities of coupled vibrational modes such as those described earlier (p. 142) for carbon dioxide. In the symmetric mode, shown in Figure 9-5(a), it is seen that no change in dipole occurs as the two oxygen atoms move away from or toward the central carbon atom; thus, this vibrational mode is infrared inactive. The polarizability, however, fluctuates in phase with the vibration since distortion of bonds becomes easier as they lengthen and more difficult as they shorten; Raman activity is associated with this mode.

As shown in Figure 9-5(b), the dipole moment of carbon dioxide fluctuates in phase with the asymmetric vibrational mode; thus, an infrared absorption peak arises from this mode. The polarizability also changes during the vibration, reaching a maximum at the equilibrium position and a minimum at either vibrational extreme. Note, however, that the fluctuations in polarizability are not in phase with the vibrational changes but occur at twice the frequency. As a consequence, Raman scattering is not associated with these asymmetric vibrations.

As shown earlier (p. 143), a dipole change, and thus infrared absorption, is also associated with the two degenerate scissoring vibrational modes of carbon dioxide. Here again, change in polarizability does not occur at the vibrational frequency but at twice the frequency; these modes do not, therefore, produce Raman scattering.

Often, as in the foregoing examples, parts of Raman and infrared spectra are complementary, each being associated with a different set of vibrational modes within a molecule. In other instances a vibrational mode may be both Raman and infrared active. Here the two spectra resemble one another with peaks involving the same energies. The relative intensities of corresponding peaks may differ, however, because the probability of the transition occurring is different for the two mechanisms.

Intensity of Raman peaks. The intensity or power of a Raman peak depends in a complex way upon the polarizability of the molecule, the intensity of the source, and the concentration of the active group, as well as other factors. In the absence of absorption, the power of Raman emission increases with the fourth power of the frequency of the source; however, advantage can seldom be taken of this relationship because of the likelihood that ultraviolet irradiation will cause photodecomposition.

Raman intensities are directly proportional to the concentration of the active species. In this regard Raman spectroscopy more closely resembles fluorescence than absorption, where the concentration intensity relationship is logarithmic.

Instrumentation

The apparatus for Raman spectroscopy consists of three components, namely an intense source, a special cell, and a suitable spectrometer or spectrograph.

Light source and cell. Until recently a high-current, low-pressure mercury arc has been used for most Raman studies. To increase the intensity of the radiation reaching it, the sample is located along the axis of a lamp formed in the configuration of a three- or four-turn helix (see Figure 9-6).

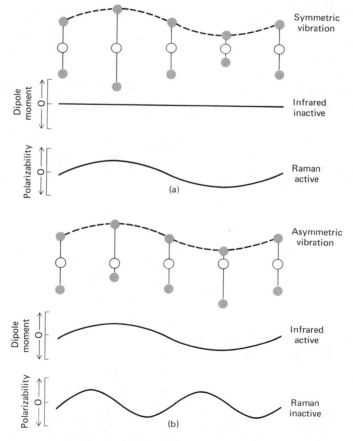

Figure 9-5. Raman and infrared activity of two vibrational modes of carbon dioxide: (a) symmetric, (b) asymmetric.

The assembly may be surrounded by a reflecting surface to maximize exposure of the sample to the output of the source; provision for cooling the lamp compartment is often necessary. The curved end of the sample cell is frequently blackened to reduce reflection from the glass surfaces. The usual volume of a sample tube is from 5 to 15 ml.

For most Raman work, the intense line at 453.8 nm is employed for excitation; removal of the other mercury lines is required only where an overlap of Raman peaks complicates interpretation of the spectra. If necessary, lines at wavelengths shorter than 453.8 nm can be filtered out by surrounding the cell with an annular tube containing a concentrated solution of sodium nitrate.

Recently, there has been a shift away from the mercury arc to a helium-neon laser. This source provides a very narrow beam with a wavelength of 632.8 nm. It has several advantages, among which is the very small divergence of the beam. As a consequence, the beam is sufficiently intense in a small area to excite Raman spectra; furthermore, much smaller samples and sample tubes (0.05 to 0.1 ml) can be examined. For example, the cell shown in the lower part of Figure 9-6 is a 1-mm o.d. tube that is about 5 cm long. The Raman radiation reflects off the walls of the tube and is collected by the lens mounted flush against one end of the tube. Despite the inverse fourth power dependence of Raman intensity with wavelength, adequate peak

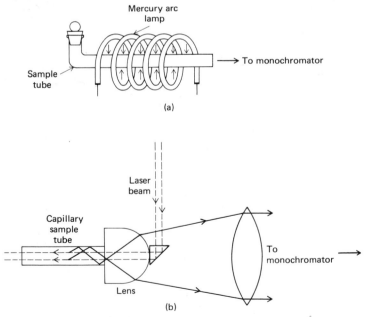

Figure 9-6. Excitation of Raman spectra. (a) Mercury arc lamp; (b) laser excitation. From R. C. Hawes, K. P. George, D. C. Nelson and R. Beckwith, *Anal. Chem.*, **38**, 1842 (1966). (With permission of the American Chemical Society.)

intensities are observed with the helium-neon laser. The longer wavelength has the advantage of minimizing fluorescence of the sample and also permits Raman studies of solutions that absorb at shorter wavelengths. Other laser sources are now available commercially.

Raman spectrophotometers. All early Raman studies were performed with prism spectrographs in which the spectra were recorded on photographic films or plates (see Chapter 10). Intensities were determined by measuring the blackness of the lines.

Several Raman spectrophotometers which employ photomultipliers to record peak intensities are now available. Their designs are not significantly different from the recording spectrophotometers discussed in Chapter 4. Most employ double monochromators to minimize spurious radiation that reaches the detector. In addition, a split-beam design is employed to compensate for the effects of fluctuations in the intensity of the source.

Applications of Raman Spectroscopy

Raman spectra are employed in much the same way as infrared data. For large organic molecules, however, a Raman spectrum is likely to have a less unique character than its infrared counterpart and thus be less useful for identification purposes. On the other hand, and in contrast to infrared spectra, Raman spectra are readily obtained from aqueous solutions. Here the most suitable concentration range is usually from about 0.1 to 1 molar.

Raman spectra of inorganic species.[4] The Raman technique is superior to the infrared for studying inorganic systems because the polarizability of water changes little during its vibrations. Its Raman scattering is thus weak and does not seriously affect studies in aqueous solutions. In addition, the vibrational energies of metal-ligand bonds are in the range of 100 to 700 cm^{-1}, a region of the infrared that is experimentally difficult to study. These vibrations are frequently Raman-active, however, and peaks with $\Delta\sigma$ values in this range are readily observed.

The covalent bonds between metal ions and ligands are frequently Raman active; as a result, Raman studies are potentially useful sources of information concerning the composition, the structure, and the stability of coordination compounds. For example, numerous halogen and halogenoid complexes produce Raman spectra and are thus susceptible to investigation by this means. Metal-oxygen bonds are also active; spectra for such species as VO_4^{3-}, $Al(OH)_4^-$, $Si(OH)_6^{2-}$, and $Sn(OH)_6^{2-}$ have been obtained. In several instances Raman studies have led to conclusions regarding the probable nature of such species. For example, in perchloric acid solu-

---

[4]For a review of inorganic applications see: R. S. Tobias, *J. Chem. Ed.*, **44**, 2 (1967); *ibid.*, **44**, 70 (1967); R. E. Hester, *Anal. Chem.*, **42**, 231R (1970), **40**, 320R (1968); R. N. Jones and M. K. Jones, *ibid.*, **38**, 393R (1966); A. C. Jones, *ibid.*, **36**, 296R (1964).

tions vanadium(IV) appears to be present as VO^{2+}(aq) rather than as $V(OH)_2^{2+}$(aq); studies of boric acid solutions show that the anion formed by acid dissociation is the tetrahedral $B(OH)_4^-$ rather than $H_2BO_3^-$. In addition, dissociation constants for strong acids such as sulfuric, nitric, selenic, and periodic acid have been obtained by Raman measurement.

It seems probable that the future will see even wider use of Raman spectroscopy for theoretical studies of inorganic systems.

Raman spectra of organic compounds.[5]   From Figure 9-4 it is apparent that Raman and infrared spectra are often quite similar. For identification purposes, infrared data are usually preferred because of the ease with which they can be obtained and also because they provide a greater degree of uniqueness for large molecules. Nevertheless, Raman spectra yield more information for some compounds than do infrared spectra. For example, the double-bond stretching vibration for olefins results in weak and sometimes undetected infrared absorption. On the other hand, the Raman band (which, like the infrared band, occurs at about 1600 cm^{-1}) is intense, and its position is sensitive to the nature of substituents as well as their geometry. Thus, Raman studies are likely to yield useful information about the olefinic functional group that may not be revealed by infrared spectra. This statement applies to cycloparaffin derivatives as well; these compounds have a characteristic Raman peak in the region of 700 to 1200 cm^{-1}. This peak has been attributed to a "breathing" vibration, in which the nuclei move in and out symmetrically with respect to the center of the ring. The position of the peak decreases continuously from 1190 cm^{-1} for cyclopropane to 700 cm^{-1} for cyclooctane; Raman spectroscopy thus appears to be an excellent diagnostic tool for the estimation of ring size of paraffins. The infrared peak associated with this vibration is weak or nonexistent.

Quantitative applications.   Raman spectra tend to be less cluttered with peaks than do infrared spectra. Peak overlap in mixtures is less likely and quantitative measurements are simpler as a consequence. In addition, Raman instrumentation is not subject to attack by moisture, and small amounts of water in a sample do not interfere. Despite these advantages, Raman spectroscopy has not been widely exploited for quantitative analysis.

An example of the potentialities of the method for the analysis of mixtures is provided in a paper by Nicholson[6] in which a procedure for the determination of the constituents in an eight-component mixture is described. The components included benzene, isopropyl benzene, three di-

[5]For a comparison of Raman and infrared activities of organic compounds see R. N. Jones and C. Sandorfy in W. West, Ed., *Chemical Applications of Spectroscopy* (New York: Interscience Publishers, Inc., 1956), Chap. 4; R. E. Hester, *Anal. Chem.,* **42**, 231R (1970), **40**, 320R (1968); R. N. Jones and M. K. Jones, *ibid.,* **38**, 393R (1966); A. C. Jones, *ibid.,* **36**, 296R (1964).
[6]D. E. Nicholson, *Anal. Chem.,* **32**, 1634 (1960).

isopropyl benzenes, two triisopropyl derivatives, and 1, 2, 4, 5-tetraisopropyl benzene. The power of the characteristic peaks, compared with the power of a reference peak (CCl_4), was assumed to vary linearly with volume percent of each component. Analysis of synthetic mixtures of all the components by the procedure produced results that agreed with the preparatory data to about one percent (absolute).

10 Emission Spectroscopy

Gaseous ions or molecules, when thermally or electrically excited, emit characteristic radiations in the ultraviolet and visible regions. Emission spectroscopy is concerned with the characterization of the wavelengths and the intensities of radiation produced in this manner. Such studies have yielded much information of theoretical interest to the chemist and the physicist. In addition, data derived from emission spectroscopy can be employed for qualitative and quantitative elemental analysis; such applications are the concern of this chapter and the one that follows.

While all elements can be induced to emit ultraviolet or visible radiations, practical difficulties in the excitation of the nonmetals have limited the application of emission spectroscopy largely to the analysis of some 70 metal and metalloid elements. Two characteristics of the method account for its importance in analysis. The first is its specificity, which arises from the unique character of the wavelength pattern produced by each element. The second is high sensitivity, a consequence of the efficiency not only in the excitation process but also in the detection devices that are available.

For qualitative analysis, emission spectroscopy offers advantages that are virtually without parallel. The sample ordinarily requires little or no preliminary treatment. Detection limits lie between the parts-per-million and parts-per-billion ranges.

The quantitative applications of emission spectroscopy are more limited, primarily because of difficulties encountered in reproducing radiation intensities. Only with the greatest care can relative errors be reduced to 1 to 2 percent; uncertainties on the order of 10 to 20 percent or greater are not uncommon. Where traces are being determined, errors of such

258

magnitudes are often tolerable and not significantly greater than those associated with other methods. In the determination of a major constituent, however, emission spectroscopy suffers by comparison. Nevertheless, the method has the virtue of providing information rapidly. Frequently, several elements can be determined spectroscopically within a few minutes of receipt of the sample; situations exist (in the control of industrial processes, for example) where this speed is more important than a high level of accuracy.

THE EMISSION PROCESS

In this chapter we are concerned with radiant emission stimulated by the thermal or electrical energy of an arc or a spark.[1] The radiation produced in a hot flame is considered in Chapter 11.

The central problem in analytical emission spectroscopy is the large influence of the source upon both the pattern and the intensity of lines produced by a species. It is apparent that the source must serve two functions. First, it must provide sufficient energy to volatilize the sample and convert the individual components to gaseous atoms or ions; in this process it is essential that the distribution of the elements in the vapor be reproducibly related to their concentration or distribution in the sample. The second function is to supply sufficient energy to cause electronic excitation of the elementary particles in the gas, again in a reproducible manner. Each of these processes undoubtedly involves a complex set of mechanisms, the nature of which depends upon the characteristics of both source and sample. Unfortunately, our knowledge of these mechanisms and the part they play in determining the intensity of a given line is rudimentary; a study of variables that affect the characteristics of a source must be largely empirical.

The Spectra Produced by Electrical Excitation

An examination of the emission produced by an electrical arc or spark reveals three types of superimposed spectra. First, there is a *continuous background radiation* emitted by the heated electrodes and perhaps also by hot particulate matter detached from the electrode surface. The frequency distribution of this radiation depends upon the temperature and approximates that of a black body (see p. 23).

[1]For a more complete discussion of emission spectroscopy, see B. F. Scribner and M. Margoshes in I. M. Kolthoff and P. J. Elving, Eds., *Treatise on Analytical Chemistry,* Part I, Vol. 6 (New York: Interscience Publishers, Inc., 1965), Chap. 64; R. A. Sawyer, *Experimental Spectroscopy,* 3d ed. (New York: Dover Publications, Inc., 1963); C. E. Harvey, *Spectrochemical Procedures* (Glendale, Calif.: Applied Research Laboratories, 1950); N. H. Nachtrieb, *Principles and Practice of Spectrochemical Analysis* (New York: McGraw-Hill Book Company, Inc., 1950).

Band spectra, made up of a series of closely spaced lines, are also frequently observed in certain wavelength regions. This type of emission is due to molecular species in the vapor state which produce bands as a result of the superposition of vibrational energy levels upon electronic levels. The cyanogen band, caused by the presence of CN radicals, is always observed when carbon electrodes are employed in an atmosphere containing nitrogen. Samples containing a high silicon content may yield an additional molecular band spectrum due to SiO. If these bands obscure line spectra of interest, precautions must then be taken to eliminate them.

Line spectra. In general, emission methods are based upon the line spectrum produced when the outer electrons of an excited atom or a simple ion return to the ground state in one or more steps. The wavelength pattern of lines is unique for each element, but it is important to appreciate that the pattern is also dependent upon the energy of the excitation source.

An energy-level diagram for atomic sodium is shown in Figure 10-1, with those transitions responsible for the more intense lines indicated. Note that the sodium spectrum is a system containing closely spaced doublet lines which arise from a splitting of the *p* orbitals into two levels that

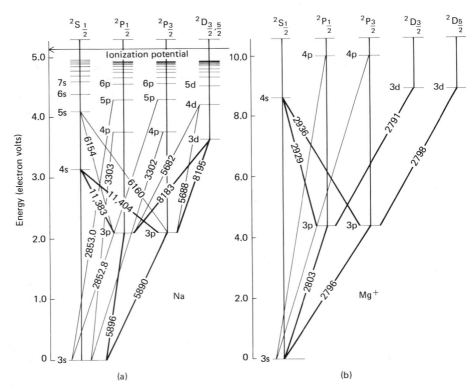

Figure 10-1. Energy level diagrams showing similarity of (a) Na and (b) Mg⁺.

differ slightly in energy. These energy differences can be traced to the interaction between the magnetic moment due to the spin of the outer p electron and the net magnetic field resulting from the orbital motions of all of the electrons. The spin magnetic moment then acts either to oppose or to reinforce the orbital field, and leads to two energy states. Similar splitting occurs in the d and f levels as well, but the energy differences are too small to detect.

The appearance of doublet lines is characteristic of all species containing a single external electron. Thus, the energy-level diagram for the singly-charged magnesium ion, shown in Figure 10-1(b), has much the same general appearance as that of the sodium atom. So also does the spectrum for the dipositive aluminum ion and the remainder of the alkali metal atoms. It is important to note, however, that the wavelengths for corresponding sets of transitions are much shorter for the magnesium ion than for the atomic sodium ion because the energy differences between the principal quantum levels are much greater for the magnesium species with its larger nuclear charge.

The spectrum for atomic magnesium, which has two outer electrons, is markedly different from the singly-charged cation, as may be seen in Figure 10-2. Here, a singlet and a triplet series of lines can be discerned. In the first of these, transitions between principal quantum levels result in single lines. Three lines are produced in the triplet system, however, because of a splitting of levels into three states that differ slightly in energy.

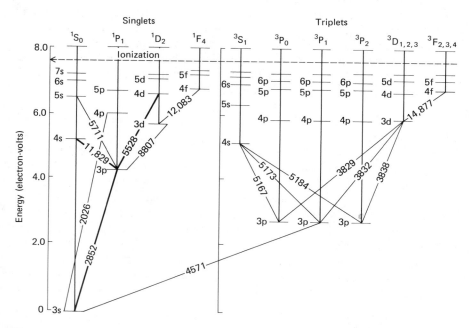

Figure 10-2. Energy level diagram for Mg.

As before, these phenomena can be rationalized by taking into account the interaction between the fields associated with the spins of the two outer electrons and the net field arising from the orbital motions of all of the electrons. In the singlet state the two spins are paired, and their respective magnetic effects cancel; thus, no energy-splitting is observed. In the triplet state, however, the two spins are unpaired (that is, their spin moments lie in the same direction); the effect of the orbital magnetic moment on the magnetic field of the combined spins produces a splitting of the *p* level into a triplet. This behavior is characteristic of all of the alkaline earth atoms, singly-charged aluminum and beryllium ions, and so forth.

Similar splitting should be associated with two electrons in *d* or *f* levels; such effects are difficult to observe, however, owing to the small energy differences that are involved.

As the number of electrons outside the closed shell increases, the corresponding spectra become more and more complex. Thus, with three outer electrons a system of doublet lines as well as a series with four lines is produced. With four outer electrons, singlets, triplets, and quintets are observed.

Although the ultraviolet and visible emission spectra for elements such as sodium or magnesium are relatively simple and amenable to theoretical interpretation, the same cannot be said for the heavier elements, and particularly the transition metals. These species have large numbers of closely-spaced energy levels; as a consequence, the number of lines produced upon excitation can be enormous. For example, Harvey[2] has listed the number of lines observed in the arc and spark spectra of neutral and singly-ionized atoms for a variety of elements: for the alkali metals this number ranges from 39 for lithium to 645 for cesium; for the alkaline earths, magnesium has 173, calcium 662, and barium 472. Typical of the transition series, on the other hand, are chromium, iron, and cerium, with 2277, 4757, and 5755 lines, respectively.

Effect of source on spectra. A comparison of Figures 10-1 and 10-2 reveals that the character of the emission spectrum for magnesium is quite different if the excited species is ionic instead of the neutral atom. The energy of the source determines which spectrum predominates. It is seen in Figure 10-2 that the first ionization potential for magnesium is 7.62 eV. Thus, if the source supplies energy in excess of this amount to most of the atoms, the spectrum of the singly-charged ion predominates. If the energy exceeds about 15 eV, doubly-ionized magnesium is produced, and yet another spectrum results.

Arc sources generally provide lower energies than do spark sources; as a result, lines of neutral atoms tend to predominate in an arc spectrum. On the other hand, spectra excited by a spark typically contain lines asso-

[2]C. E. Harvey, *Spectrochemical Procedures* (Glendale, California: Applied Research Laboratories, 1950), Chap. 4.

ciated with excited ions. Because the energy of each type of source may vary considerably, most spectra contain a mixture of lines from both atoms and ions. The relative intensities of the two types depends upon the energy provided by the source.

Line widths. An emission line is not truly monochromatic but includes a finite group of wavelengths surrounding the theoretical value. Two effects, *pressure broadening* and *Doppler broadening*, affect the width of a line. Experimentally it is found that the breadth of a line increases with concentration (or as the pressure of the atomic or ionic gas increases). This effect results from interaction among the more closely packed particles, which changes slightly or *perturbs* the energy levels of the individual atoms or ions. Small variations in wavelengths are the result.

Doppler broadening results from the rapid and random motion of the excited particles. Some move toward the detector and some away from it as emission takes place; depending upon the direction of this motion, the emitted frequency is slightly greater or slightly smaller. As mentioned in Chapter 5, Doppler broadening is observed not only with emission peaks but with atomic absorption peaks as well.

Self-absorption. Radiation emitted by an excited particle is readily absorbed by a similar species in the unexcited state. Thus, *self-absorption* may prevent an emitted photon from reaching the detector. Since only a small fraction of the gaseous atoms are in the excited state at any instant, those lines that involve transitions to the ground state (resonance lines) are most subject to self-absorption.

The center of most sources is hotter than the exterior; thus, the emitting species at the center tend to be surrounded by a cooler layer of unexcited atoms that are available for absorption. The Doppler broadening of an emission line is greater than the broadening of the corresponding absorption line, however, because the particles are moving at a greater rate in the emission zone. Thus, self-absorption tends to affect the center of a line more than the edges. In the extreme, the center may become less intense than the edges or it may even disappear, and result in the phenomenon of *self-reversal.*

Self-absorption often becomes troublesome when the element being determined is in high concentration. Under these circumstances measurements involving resonance lines are avoided.

Arc Sources and Arc Spectra

Formation of the arc. The usual arc source for a spectrochemical analysis is formed with a pair of graphite or metal electrodes that are separated by a space of about 1 to 20 mm. The arc is initially ignited by a low-current spark that causes formation of ions for current conduction in the gap.

Once the arc is started, thermal ionization maintains the current flow. Alternatively, the arc can be started by bringing the electrodes together to provide the heat for ionization; they are then separated to the desired distance.

In the typical arc, currents are in the range of 1 to 30 amp dc or ac. A dc source usually has an open circuit voltage of about 200 V; ac source voltages in the range of 2200 to 4400 are commonly employed.

Arc excitation. Although thermal ionization does occur in an arc, the energies are such that atom excitation predominates; consequently, neutral particles rather than ions are responsible for most of the radiation. Thermal excitation is governed by the Boltzmann equation (equation 5-1); it is clear from this relationship that the number of atoms excited to a given quantum level increases exponentially with temperature. Since concentration measurements are based upon the number of excited atoms, it is also apparent that small temperature fluctuations in the arc plasma will profoundly decrease the reproducibility of such measurements. Unfortunately, not enough is known about the mechanisms of current passage in an arc to allow for a prediction of the variables that affect its temperature. The assumption is usually made that close control of the current strength is required; while it is probable that constant current is a necessary condition, it is not obvious that control of this variable alone will suffice.

Current is carried in an arc by the motion of the electrons and ions formed by thermal ionization; the high temperature that develops is a result of the resistance to this motion by the atoms in the arc gap. Thus, the arc temperature depends upon the composition of the plasma, which in turn depends upon the rate of formation of atomic particles from the sample and the electrodes. Little is known of the mechanisms by which a sample is dissociated into atoms and then volatilized in an arc. It can be shown experimentally, however, that the rates at which various species are volatilized into the arc differ widely. The spectra of some appear early and then disappear; other species reach their maximum intensities at a later time. Thus, the composition of the plasma, and therefore the temperature, may undergo variation with time.

Another peculiarity of a dc arc that lessens reproducibility is the failure of the arc column to cover the entire surface of the electrode; instead, contact is made only at spots. These contact spots wander erratically and slowly with time; thus, only a part of the electrode surface (and an irregular one at that) is sampled. In addition, the fraction of radiation reaching the detector, and thus the observed line intensities, may fluctuate as a consequence of the movement.

Employment of alternating current or intermittent dc current improves the reproducibility of arc sources. Sufficiently high potentials are used to assure the spontaneous reformation of an ac arc with each alteration in the direction of the current. A low-power spark is needed to restart a dc arc after each interruption. The interruption of an arc source, regardless of type, is presumed to allow the escape of vapors from the gap. With reestablishment of the arc, a fresh plasma is formed. Hopefully, this process produces a more reproducible source because it minimizes many of the time-dependent changes that occur in the normal dc arc. In addition, with each reignition, contact is made at a different spot on the surface; and since the frequency of interruption can be made high, a statistically more uniform sampling and illumination of the detector results.

The precision obtainable with an arc is generally poorer than that obtainable with a spark. On the other hand, an arc source is more sensitive to traces of an element in a sample. For this reason it is often preferred for qualitative analysis as well as for the quantitative determination of trace constituents.

Spark Sources

Formation of a spark. Figure 10-3 shows a typical electrical circuit for driving an alternating current across a spark gap. The transformer converts line power to 15,000 to 40,000 V which then charges the capacitor. When the potential becomes large enough to break down the two air gaps shown on the right, a series of oscillating discharges follows (see Figure 10-4). The voltage then drops until the capacitor is incapable of forcing further current across the gaps, following which the cycle is repeated. The frequency and the current of the discharge are determined by the magnitude of the capacitance, inductance, and resistance of the circuit in Figure 10-3; these parameters also affect the relative intensities of the neutral atom and ion lines as well as the background intensity.

Most spark sources employ a pair of spark gaps arranged in series (Figure 10-3). The analytical gap is formed by a pair of carbon or metal electrodes that contain the sample. The control gap may be formed from smoothly rounded and carefully spaced electrodes over which is blown a stream of air. Alternatively, a so-called rotary gap is employed; this consists of a rotating disk upon which are mounted equally spaced electrodes. The disk is rotated past one or more fixed electrodes. Discharge occurs only as the moving electrodes pass the fixed electrode.

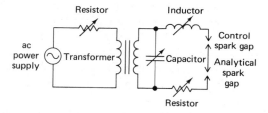

Figure 10-3. Power supply for a high-voltage spark source.

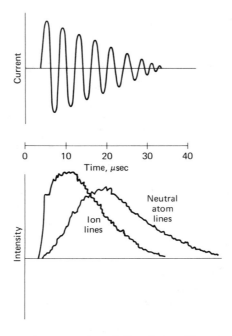

Figure 10-4. Variation in current and line intensities during a single-spark discharge. Taken from B. F. Scribner and M. Margoshes in I. M. Kolthoff and P. J. Elving, Eds., *Treatise on Analytical Chemistry*, Part I, Vol. 6 (New York: Interscience Publishers, Inc., 1965), p. 3367. (With permission.)

Conditions at the analytical gap are subject to continual change during operation. Accordingly, electrode spacing is arranged so that the voltage breakdown is determined by the more reproducible control gap. The employment of air-cooling (or rotation) tends to maintain constant operating conditions for the control gap; a higher degree of reproducibility is thus imparted to the analytical results than would otherwise be attained.

Spark excitation. The *average* current with a high-voltage spark is usually significantly less than that of the typical arc, being on the order of a few tenths of an ampere. On the other hand, during the initial pulse of the discharge the *instantaneous* current may exceed 1000 amp; this current is carried by a narrow streamer that involves but a minuscule fraction of the total electrode area. The temperature within this streamer is estimated to be as great as 40,000°K. Thus, while the average electrode temperature of a spark source is much lower than that of an arc, the energy in the small volume of the streamer may be several times greater. As a consequence, ionic spectra are more pronounced in a high-voltage spark than in an arc. Furthermore, as shown in Figure 10-4, the relative intensities of the ion lines and the neutral atom lines vary continuously with the current fluctuation during a discharge. Note that the intensity of an ion line reaches a maximum during the early pulses when the current (and thus the temperature) is high. Near the end of the discharge, neutral atom excitation reaches its peak.

The mechanisms by which neutral atoms and ions are formed when an individual spark streamer contacts the surface of an electrode are as obscure as those for the arc. Because of the high frequency of discharge and because each pulse presumably strikes a different spot on the electrode, random sampling of the surface and averaging of the radiant intensity reaching the detector results. As a consequence, spark excitation generally provides better precision than the arc (as good as one percent relative in favorable circumstances). On the other hand, the total amount of sample volatilized during a pulse is smaller; a spark spectrum is thus less sensitive than one produced by an arc.

EMISSION SPECTROSCOPIC EQUIPMENT AND PROCEDURES

The equipment for an emission spectroscopic analysis includes a means of sample excitation (the source), a suitable monochromator for dispersing the radiation, and a detector for the determination of line intensities.

Sample Excitation

Samples for a spectrographic analysis may be solid or liquid; they must be distributed more or less regularly upon the surface of at least one of the electrodes that serves as the source.

Metal samples. If the sample is a metal or an alloy, one or both electrodes can be formed from the sample by milling or turning, or by casting the molten metal in a mold. Ideally, the electrode will be shaped as a cylindrical rod that is $\frac{1}{8}$ inch to $\frac{1}{4}$ inch in diameter and tapered at one end. For some samples it is more convenient to employ a polished, flat surface of a large piece of the metal as one electrode and a graphite or metal rod as the other. Regardless of the ultimate shape of the sample, care must be taken to avoid contamination of the surface while it is being formed into an electrode.

Electrodes for nonconducting samples. For the analysis of nonmetallic materials, the sample is supported on an electrode whose emission spectrum will not interfere with the analysis. Carbon is an ideal electrode material for many applications. It can be obtained in a highly pure form, is a good conductor, has good heat resistance, and is readily shaped. Manufacturers offer carbon electrodes in many sizes, shapes, and forms (see Figure 10-5). Frequently one of the electrodes is shaped as a cylinder with a small crater drilled in one end; the sample is then packed into this cavity. The other

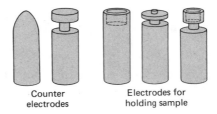

Counter
electrodes

Electrodes for
holding sample

Figure 10-5. Some typical graphite electrode shapes. Narrow necks are to reduce thermal conductivity.

electrode is commonly a tapered carbon rod with a slightly rounded tip. This configuration appears to produce the most stable and reproducible arc or spark.

Silver or copper rods are also employed to hold samples when these elements are not of analytical interest. The surfaces of these electrodes can be cleaned and reshaped after each analysis.

Excitation of the constituents of solutions. Several techniques are employed for excitation of the components of solutions or liquid samples. One common method is to evaporate a measured quantity of the solution in a small cup cut into the surface of a graphite or a metal electrode. Alternatively, a porous graphite electrode may be saturated by immersion in the solution; it is then dried before excitation. The rotating-disk technique can be employed with spark excitation. One of the electrodes is a disk (perhaps $\frac{1}{2}$ inch in diameter and $\frac{1}{8}$ inch thick) which is rotated in the vertical plane by a motor; the counter electrode is a fixed graphite rod. The lower portion of the disk is immersed in the sample; with rotation, a film of the sample is carried continuously into the spark gap.

Methods have also been developed for spraying liquid samples into a spark by means of an atomizer.

Dispersing Elements of Spectrographs

A *spectrograph* is an instrument that disperses radiation and then photographically records the resulting spectrum. Photoelectric recording with a *spectrometer* is also employed for emission work, but the photographic method is much more common. It is important to differentiate the spectrograph and the spectrometer from the spectrophotometer, which was considered in Chapter 3. The spectrograph or the emission spectrometer records the intensities of all or part of the lines of an entire spectral region *simultaneously*, whereas a spectrophotometer requires a wavelength-by-wavelength measurement to obtain the information. The spectrograph is particularly well-suited for emission spectral analysis because it permits the detection and the determination of several elements by excitation of a single small sample.

To facilitate simultaneous intensity recording by photographic means, the exit slit of a monochromator (see Figure 3-8) is replaced by a photographic plate (or film) located along the focal curve of the instrument.[3] After exposure and development of the emulsion, the various spectral lines of the source appear as a series of black images of the entrance slit distributed along the length of the plate (see Figure 10-6). The location of the lines provides qualitative information concerning the sample; the darkness of the images can be related to line intensities, and hence concentrations.

The radiation that appears at several chosen locations along the focal curve of a photoelectric spectrometer is focused upon a series of photomultiplier tubes to permit the direct and simultaneous measurement of the intensities for a number of spectral lines. Figure 10-7 is a schematic diagram of a commercial photoelectric spectrometer.

Optical requirements. The spectral region of greatest importance for emission analysis lies between 2500 and 4000 Å, although both longer and shorter wavelengths are occasionally useful. Thus, the optical components through which radiation is transmitted must be of quartz or fused silica, and the detector must be sensitive to ultraviolet radiation.

The dispersion requirements for a spectrographic instrument depend upon the nature of the sample to be analyzed. For elements with numerous, closely spaced lines, such as iron, nickel, and other transition metals, these requirements are stringent; reciprocal dispersions in the range of 3 to 4 Å per millimeter (along the focal curve) may be needed. For elements with simpler spectra, an instrument with a reciprocal dispersion of 8 to 10 Å per millimeter usually suffices. As shown in Chapter 11, simple filter instruments can be employed under some conditions for the determination of the alkali metals.

Spectroscopic laboratories that must contend with widely differing samples are frequently equipped with both a high- and a low-dispersion spectrograph. When it can be employed, the low-dispersion instrument has the advantage of requiring shorter exposure times. In addition, a high-dispersion instrument spreads the spectrum to such an extent that only a portion (perhaps 1000 Å) can be focused on the photographic plate at one time. Acquisition of an entire spectrum, then, requires repeated exposures after adjustment of the monochromator element.

Types of monochromators. Both prisms and gratings are employed in spectrographs. High-dispersion prism instruments employ a Littrow-mount monochromator; a Cornu-type (p. 43) is often used in lower-dispersion spectrographs. Prism spectrographs suffer from the disadvantage of nonlinear dispersion, which complicates the identification of lines and also crowds the longer wavelength lines together.

---

[3]The focal points of a prism spectrometer lie on a very slightly curved surface. In order to compensate for this curvature, spectrographic plate holders are designed to force a slight bend in the photographic plate.

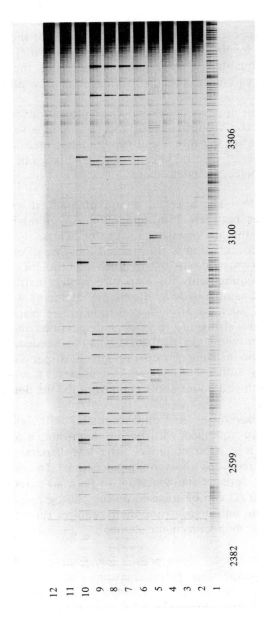

Figure 10-6. Typical spectra obtained with a 3.4-meter grating spectrograph. Exposures made with a step sector disk (Figure 10-11). Numbers on horizontal axis are wavelengths in Å. Spectra: (1) iron standard; (2)-(5) casein samples; (6)-(8) Cd-Ge arsenide samples; (9)-(11) pure Cd, Ge, and As respectively; (12) pure graphite electrode.

Reflection gratings are widely used in the manufacture of spectrographs. One of the common arrangements is shown in Figure 10-8.[4]

Spectrographs are generally provided with a masking device called a *Hartmann diaphragm,* by means of which only a fraction of the vertical length of the slit is illuminated (Figure 10-9). If the diaphragm is adjusted, several spectra can be recorded successively along the vertical axis of the plate or film without the need for moving the film.

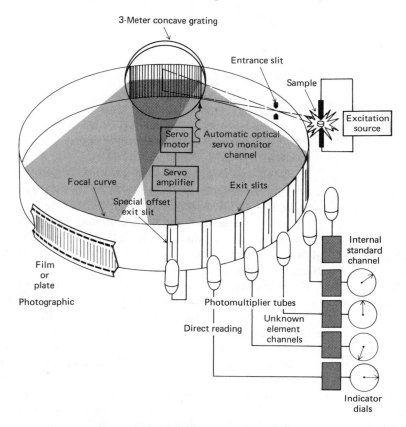

Figure 10-7. Optical system of a direct-reading grating spectrometer. (Courtesy of Baird-Atomic, Inc., Cambridge, Mass.)

DETERMINATION OF LINE INTENSITIES

Provided self-absorption does not occur, the intensity of a spectrographic line at any instant is directly proportional to the number of atoms

---

[4]Here also, the spectral lines are focused on a curved surface; with smaller instruments, the curvature may be so great as to preclude the use of glass photographic plates and the more flexible film must be employed.

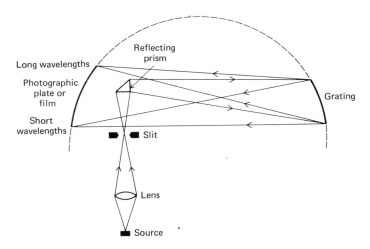

Figure 10-8. The Eagle mounting for a grating spectrograph.

or ions of the emitting species that exist in the gaseous plasma of the arc or spark gap. Under ideal conditions this number is in turn proportional to the weight percent of the element in the original sample; thus, ideally, the line intensity is linearly related to elemental concentration. We have noted, however, that continuous and random fluctuations in line intensities are characteristic of both arc and spark sources, owing to the large number of variables that affect the volatilization and excitation of the constituents of a sample. Thus, the intensity of a line at any given instant is virtually useless for the assessment of concentration. To overcome this difficulty it is necessary to integrate the line intensity over a period that is long with respect to the frequency of the random fluctuations; in this way such fluctuations are averaged out to a meaningful and reproducible quantity that can be related to concentration. In practice, this averaging procedure involves the determination of the intensity-time integral for a period ranging from 10 sec to a minute or more. Integration can be accomplished by a photographic emulsion or by means of a suitable electrical integrating circuit activated by the signal from a photomultiplier tube.

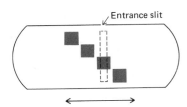

Figure 10-9. The Hartmann diaphragm.

Photographic Method

A photographic emulsion can serve as a detector, an amplifier, and an integrating device for evaluation of the intensity-time integral for a spectral line. When a photon of radiation is absorbed by a silver halide particle of the emulsion, a latent image is formed. Treatment with a reducing agent results in the formation of a large number of silver atoms. This process can be considered an example of chemical amplification; the number of black silver particles, and thus the darkness of the exposed area, is a function of the *exposure E*, which is defined by

$$E = I_\lambda t \tag{10-1}$$

where I_λ is the intensity of radiation and t is the exposure time.

Optical density.[5]   Optical density describes the degree of blackness of an area in a photographic emulsion. Before we define the term it is convenient to describe how an instrument called a *microphotometer* is used to measure this quantity.

A microphotometer is closely analogous to the simple single-beam photometer described in Chapter 3. It consists of a tungsten filament source, a narrow slit, and a holder for the plate or film that permits the blackened area of interest to be moved across the beam that emerges from the slit. After the beam has passed through the emulsion it is focused on a photoelectric detector, the output of which is indicated on a meter or chart. If we let I_0 be the intensity of the beam after it has passed through a clear part of the emulsion and I be its intensity after passage through the blackened portion, we may define the *transparency* (or transmittance) T of the darkened area as

$$T = I/I_0 \tag{10-2}$$

or
$$\text{percent } T = 100 \; I/I_0 \tag{10-3}$$

The *optical density D* of the area under study is then defined by

$$D = \log_{10} I_0/I \tag{10-4}$$

[5]We believe it worthwhile to employ the nomenclature and symbolism currently used by emission spectroscopists despite the fact that some of the symbols and names employed differ from those defined for analogous quantities in the chapters on absorption spectroscopy. For example, intensity I in this chapter is identical to the power P defined in Chapter 2; similarly, optical density as used here is quite analogous to absorbance as defined earlier. It would be desirable to have a standardized set of symbols and names for all of science; in the absence of this ideal, the student must learn to adapt to the language employed by the specialist in a given field.

If the response of the detector of the microphotometer is linear, then

$$T = \frac{I}{I_0} = \frac{\cancel{k}\,G}{\cancel{k}\,G_0}$$

where G and G_0 represent the magnitude of the respective electrical signals and k is a proportionality constant related to the radiation intensity and the electrical response. [It is important here to make a clear distinction between the intensities of the microphotometer beam (I and I_0) and the intensity of the spectral line (I_λ) which is ultimately to be determined.] The close analogy between D and the absorbance A of a solution should be obvious.

Optical density-exposure relationship. The relationship between optical density and exposure is dependent upon the wavelength of radiation that caused the blackening, the chemical characteristics of the emulsion, and details of the procedure used in development of the plate or film. Unfortunately, the relationship varies between different batches of the same kind of photographic film or plate; it also undergoes changes with the age of the solutions employed in development. Thus, an empirical determination of the exposure-density relationship must be performed periodically at each wavelength region of interest.

Plate calibration curves consist of a plot of optical density as a function of the logarithm of exposure by radiation of a given wavelength. Typical curves are shown in Figure 10-10. Emulsion calibration curves usually have a region that is nearly linear; in emission spectroscopy the attempt

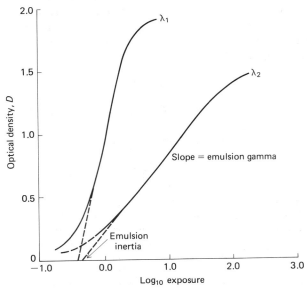

Figure 10-10. Typical plate calibration curves at two wavelengths.

is made to adjust exposures so that this portion of the curve will be involved. The slope of the linear section of the calibration curve is called the *gamma* of the emulsion. A high-contrast film or plate has a high gamma value. The intercept of the straight section of the plot with the abscissa gives the *inertia* of the emulsion; the most sensitive plates or films have low inertias.

The gamma of an emulsion increases with development time and temperature. It depends as well upon the composition and the strength of the developer solution. As shown in Figure 10-10, the gamma for a given emulsion also varies with the wavelength of radiation that caused the darkening.

Emulsion calibration for quantitative analysis. Under ideal conditions the intensity of an emission line is directly proportional to the concentration of the element responsible for the line. From equation (10-1) it is apparent that the exposure E should also be proportional to concentration, provided that the excitation time t is held constant. An emulsion calibration curve such as that shown in Figure 10-10 thus permits translation of experimental optical densities to exposures, and these in turn can be related empirically to concentrations. It is neither practical nor necessary for analytical work, however, to employ absolute values of exposure in plate calibration because absolute radiation intensities are difficult to determine. Instead, *relative* exposures serve equally well.

Emulsion calibration is accomplished by measurement of optical densities after exposure under conditions in which either I_λ or t in equation (10-1) is varied in a known way while the other variable is held constant. Figure 10-11(a) shows a *step-sector disk* which permits variation in t while I_λ is held constant. The disk is rotated at high speed in front of the slit of the spectrograph during illumination by a line source such as an iron arc or spark. The appearance of each spectral line after development is shown in Figure 10-11(b). (Note that the line width has been somewhat exaggerated to emphasize its stepwise nature.) The relative exposure time t for each step is governed by the length of the corresponding arc of the disk, a quantity that can be determined exactly. Since the disk is rotated rapidly (600 to 1800 rpm), only short-term fluctuations in the source intensity cause a variation in the average value of I_λ for each step. Another reason for rapid rotation of the disk is that of overcoming the *intermittency effect* that is characteristic of all emulsions. It is found that the blackening of an emulsion caused by, say, ten one-second exposures is not identical with the blackening produced by a single ten-second exposure even though the intensity is constant. It is observed, however, that above about 1000 interruptions per minute, blackening becomes essentially independent of frequency; a rapidly rotating step-sector disk can then be used without interference from the intermittency effect.

In employing the step-sector method it is common practice to measure the optical densities of the steps for a series of lines in the spectral region of interest. Plots of optical density or percent transmission versus the loga-

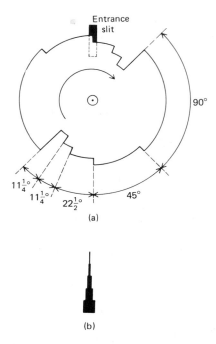

(a)

(b)

Figure 10-11. (a) Step-sector disk for exposure ratios of 1:2:4:8:16. (b) Appearance of a spectral line on the plate after exposure with the step-sector (enlarged).

rithm of the relative exposure time then yield a family of parallel curves displaced from one another along the abscissa (provided, of course, the gamma of the emulsion is constant over the wavelength region in question). These are combined into a single curve by applying arbitrary shifts along the exposure time scales (see Figure 10-12). The resulting single-calibration curve has the appearance of one of the curves in Figure 10-10, but the abscissa is a relative rather than an absolute exposure scale. It can also be considered to be a relative average intensity scale.

A second calibration technique employs a step-filter, which consists of a quartz plate upon which has been deposited parallel bands of aluminum or platinum. The transmission of the bands is dependent upon the thickness of the metal layer and is largely *independent* of wavelength. A typical filter might consist of three bands with transmittances of 1, 0.5, and 0.25. With the filter in place a spectrum similar to that obtained with step-sector results, and calibration curves are readily constructed from measured optical densities and the known transmission characteristics of the filter.

In order to cover a wide wavelength region several calibration curves may be necessary; the number is determined empirically. Careful calibration must be established for each new batch of films or plates; an occasional

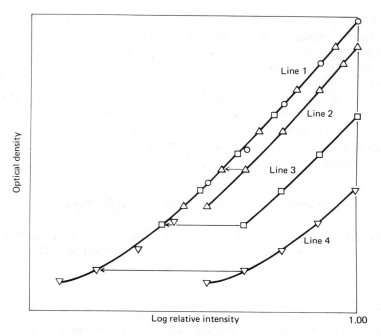

Figure 10-12. Emulsion calibration with a sector disk. Lines 2, 3, and 4 are merged with line 1 through application of arbitrary constants, as indicated by arrows.

calibration within a batch is useful to determine whether or not variations in the gamma have resulted from uncontrolled changes in the developer composition.

Photoelectric Method

For routine emission analyses, photoelectric detection of integrated line intensities is employed. Instruments for this purpose are equipped with as many as 24 photomultiplier tubes which are located behind fixed slits along the focal curve of the spectrometer (see Figure 10-7). The position of each slit must correspond exactly to the position of a line of interest.

The output of each photomultiplier is fed into a suitable capacitor-resistor circuit which integrates the output over a period of 10 to 40 sec. The capacitor voltage at the end of the exposure is a function of its total charge which, in turn, is the product of the detector current multiplied by time; thus, this voltage provides a measure of the time integral of the line intensity.

Photoelectric instruments are complex, expensive, and permit measurement of only a limited number of lines for any given positioning of the exit slits. For rapid routine analysis, however, such spectrometers are often ideal. For example, in the production of alloys, quantitative analyses for 5 to 20 elements can be accomplished within five minutes of receipt of sample; close control over the composition of the final product thus becomes possible.

In addition to speed, photoelectric spectrometers offer the advantage of better analytical precision. Under ideal conditions reproducibilities of the order of one percent relative of the amount present have been demonstrated. Such precision is as good as or better than that obtainable by routine chemical methods.

APPLICATIONS OF EMISSION SPECTROSCOPY

Spectroscopic instruments that rely upon photographic detection are most commonly encountered in the laboratory; our consideration of applications is confined to methods based upon such instrumentation.

Qualitative Analysis

Emission spectroscopy permits the detection of some 70 elements by brief excitation of a few milligrams of sample. Furthermore, from a subjective estimate of line blackening, the chemist with a little experience can rapidly estimate the concentration of an element with an accuracy of perhaps one order of magnitude. In this application, emission spectroscopy is unsurpassed.

The sensitivity of spectrographic methods is dependent upon the nature and the amount of sample, the type of excitation, and the instrument employed. Table 10-1 lists some representative data for lower limits of detection that can be achieved with dc arc excitation.

Excitation technique. For qualitative work a dc arc is most often employed. The current and exposure times are adjusted so that complete volatilization of the sample occurs; currents of 5 to 30 amp for 20 to 100 sec are common. Generally, 2 to 50 mg of sample are packed into the cavity of a graphite electrode or evaporated upon its surface; a counter electrode, also of graphite, then completes the arc. The sample electrode is ordinarily made the anode because of its higher temperature. On the other hand, an enhancement of spectral intensities may be observed in the vicinity of a sample electrode that is operated as a cathode, as a result of retarded diffusion by positive ions in the arc plasma. Advantage is sometimes taken of this phenomenon by using only the radiation adjacent to the cathode.

Identification of elements. For a qualitative analysis a standard iron spectrum is usually recorded as a reference on each plate or film. Identification of unknown lines is greatly facilitated with a projector that permits the simultaneous comparison of this plate with a master iron spectrum upon which the wavelengths of the more intense lines have been identified. Also indicated on the master spectrum are the positions of several of the most sensitive lines for each of the elements. The analytical plate and the comparison plate are then aligned so that the two iron spectra coincide exactly. Identification of three or four of the prominent lines of a particular element then constitutes strong evidence for its presence in the sample.

If comparison equipment is not available, line identification becomes much more tedious. Under these circumstances the wavelengths of a series of the iron lines is first determined by reference to a published compendium.[6] The wavelength of a line of interest is then established by measurement of its distance from one or more of the known iron lines located nearby. If a grating spectrograph has been employed, linear interpolation suffices to calculate the wavelength. If, on the other hand, the dispersion is prismatic, the *Hartmann formula* must be employed. That is,

$$\lambda = \lambda_0 + \frac{c}{d - d_0} \qquad (10\text{-}5)$$

where λ is the wavelength of the line and d is its distance from some fixed point on the plate or film. The constants λ_0, c, and d_0 are calculated from

Table 10-1 Detection Limits With a DC Arc[a]

Element	Wavelength, Å	Lower Limit of Detection[b]	
		Percent	*Micrograms*
Ag	3280.68	0.0001	0.01
As	3288.12	0.002	0.2
B	2497.73	0.0004	0.04
Ca	3933.66	0.0001	0.01
Cd	2288.01	0.001	0.1
Cu	3247.54	0.00008	0.008
K	3446.72	0.3	30
Mg	2852.12	0.00004	0.004
Na	5895.92	0.0001	0.01
P	2535.65	0.002	0.2
Pb	4057.82	0.0003	0.03
Si	2516.12	0.0002	0.02
Sr	3464.45	0.02	2
Ti	3372.80	0.001	0.1
Zn	3345.02	0.003	0.3

[a]Data taken from N. W. H. Addink, *Spectrochim. Acta*, **11**, 168 (1957). With permission.
[b]Values are based upon complete volatilization of a 10-mg sample in a dc arc.

[6]For example, H. Jaffe, *Atlas of Analysis Lines* (London: Hilger and Watts, Ltd., 1962); G. R. Harrison, *Massachusetts Institute of Technology Wave Length Tables* (New York: John Wiley & Sons, Inc., 1939).

measurements of d for three known wavelengths in the reference iron spectrum. Identification of lines by this means requires some means of magnification of the plate or film, and can be very tedious.

Quantitative Applications

Quantitative spectrographic analysis requires precise control of the many variables involved in sample preparation, excitation, and film processing. In addition, it requires a set of carefully prepared standards for calibration; these standards should approximate as closely as possible the composition and physical properties of the samples to be analyzed.

Internal standards and data treatment. As we have pointed out, the central problem of quantitative spectrographic analysis is the very large number of variables that affect the blackness of the image of a spectral line on a photographic plate. Most of these variables are associated with the excitation and photographic processes and are difficult or impossible to control completely. In order to compensate for the effects of these uncontrolled variables, an *internal standard* is generally employed in quantitative work.

An internal standard is an element incorporated in a fixed concentration into each sample and each calibration standard. The relative exposure of one of the lines of the internal standard is then determined for each sample or standard at the same time the relative exposures of the elements of interest are measured. A ratio of exposures is then employed for the concentration determination. If we let the subscripts x and s refer to the element to be determined and the internal standard, respectively, we may derive the following relationship from equation (10-1):

$$\frac{E_x}{E_s} = \frac{I_x \mathcal{Y}}{I_s \mathcal{Y}} = \frac{I_x}{I_s}$$

If the two intensities are proportional to the concentrations of the two elements and if, further, the concentration of the internal standard is fixed, then

$$\frac{E_x}{E_s} = \frac{I_x}{I_s} = kc_x \tag{10-6}$$

where k is a proportionality constant that is determined by calibration with a set of standards and c_x is the concentration of the element of interest.

Experimentally, a direct proportionality often exists between the ratio of relative exposures or intensities and concentration. In some instances, however, the relationship is nonlinear; the resulting curve is then employed for concentration determinations.

Criteria in the choice of an internal standard. The ideal internal standard has the following properties.

1. Its concentration in samples and standards is always the same.
2. Its chemical and physical properties are as similar as possible to those of the element being determined; only under these circumstances will the internal standard provide adequate compensation for those variables associated with volatilization.
3. It should have an emission line that has about the same excitation energy as one for the element being determined so that the two lines are similarly affected by temperature fluctuations in the source.
4. The ionization energies of the internal standard and the element of interest should be similar to assure that both have the same distribution ratio of atoms to ions in the source.
5. The lines of the standard and the element of interest should be in the same spectral region in order to provide adequate compensation for emulsion variables (this requirement does not apply to photoelectric spectrometers).
6. The intensity of the two lines that are selected should not be too widely divergent in order for compensation for photographic variables to be most effective.

It is seldom possible to find an internal standard that will meet all of these criteria, and compromises must be made, particularly where the same internal standard is used for the determination of several elements.

If the samples to be analyzed are in solution, considerable leeway is available in the choice of internal standards since a fixed amount can be introduced volumetrically. Here, an element must be chosen whose concentration in the sample is small relative to the amount to be added as the internal standard.

The introduction of a measured amount of an internal standard is seldom possible with metallic samples. Here, instead, the element in largest concentration is chosen, and the assumption is made that its concentration is essentially invariant. For example, in the spectrographic analysis of the minor constituents in a brass, either zinc or copper might be employed as the internal standard.

For powdered samples the internal standard is sometimes introduced as a solid. Weighed quantities of the finely ground sample and the internal standard are thoroughly mixed prior to excitation.

The foregoing criteria provide theoretical guidelines for the selection of an internal standard; nevertheless, experimental verification of the effectiveness of a particular element and the line chosen is necessary. These experiments involve determining the effects of excitation times, source temperatures, and development procedures on the relative intensities of the lines of the internal standard and the elements being determined.

Spectrographic standards. In addition to the choice of an internal standard, a most critical phase in the development of a spectrographic method involves the preparation or acquisition of a set of standard samples from which a calibration curve is prepared. For the ultimate in accuracy the standards must closely approximate the samples both in chemical composition and in physical form; their preparation often requires a large expenditure of time and effort, an expenditure that can be justified economically only if a large number of analyses is anticipated.

In some instances standards can be synthesized from pure chemicals; solution samples are most readily prepared by this method. Standards for an alloy analysis might be prepared by melting together weighed amounts of the pure elements.

Another common method involves chemical analysis of a series of typical samples encompassing the expected concentration range of the elements of interest. A set of spectrographic standards is then chosen on the basis of these results.

The National Bureau of Standards has available a large number of carefully analyzed metals, alloys, and mineral materials; occasionally, suitable spectrographic standards can be found among these. Standard samples are also available from commercial sources.[7]

Outline of the development and application of a quantitative spectrographic procedure. The following outline describes in a general way the steps involved in the development of a spectrographic procedure.

1. *General investigation of the nature of the samples.* This step may involve a qualitative analysis (by spectrographic and other means), a study of the physical properties of the sample, and a careful consideration of the source of the samples and their likely composition.

2. *An experimental study to determine the best method of excitation, the most suitable internal standard, and the best spectral lines upon which to base the analysis.*

3. *Preparation or acquisition of a set of standards.*

4. *Preparation of a working curve.* Employing the conditions chosen in (2), the spectra of the standard samples are obtained on a plate or film which also contains spectra that permit plate calibration. Plots of relative exposure versus optical density are then obtained from the latter for various wavelength regions.

The optical density of each line chosen in the second step is then measured. In addition, it is necessary to measure the background darkening of the emulsion by obtaining optical densities in areas free from lines and as close as possible to the spectral lines of interest. All of the optical density data are then converted to relative exposures or intensities, and the background exposures are subtracted from the corresponding line exposures. Working curves of I_x/I_s versus c_x are then prepared. (If the concentration range is large, logarithmic plots of these variables may be more convenient.)

5. *Sample analysis.* The samples are treated in an identical way to that

[7]For a listing of sources see R. E. Michaelis, *Report on Available Standard Samples and Related Materials for Spectrochemical Analysis, Am. Soc. Testing Materials Spec. Tech. Publ. No. 58-E* (Philadelphia: American Society for Testing and Materials, 1963).

of the standards, and the concentrations are determined from the working curves. Plate calibration should be performed periodically; occasional checks of the working curves with one or more of the standard samples is also desirable.

The method of additions. When the number of samples is too small to justify the extensive preliminary work outlined in the previous section, the *method of additions* may be employed. To apply this method it must be possible to add varying known amounts of the element of interest to the sample. The addition method is particularly well-suited to solution samples; it may also be applied to solids if the element can be introduced in the same form in which it occurs in the sample.

To identical aliquots of the sample are added known amounts of the element to be determined; these, plus the untreated sample, are excited. A plot of I_x/I_s versus the concentration of added element is then constructed. Hopefully, a straight line or a smooth curve results, with I_x/I_s for the untreated sample at zero on the added concentration axis. From the smooth curve and the intensity ratio for the untreated sample the concentration of the element can be readily calculated. A reasonable degree of confidence can be placed upon the results of this method provided the original sample falls upon the smooth working curve and provided further that identical concentrations are yielded from plots of the intensity ratios for two or three different lines of the element of interest.

Semiquantitative methods. Numerous semiquantitative methods have been described which provide concentration data reliable to within 30 to 100 percent of the true amount of an element present in the sample.[8] Such methods are useful where the preparation of good standards is not economic. Several such procedures are based upon the total vaporization of a measured quantity (1 to 10 mg) of sample in the arc. The concentration estimate may then be based on a knowledge of the minimum amount of an element required to cause the appearance of each of a series of lines. In other methods optical densities of elemental lines are measured and compared with the line of an added matrix material, or with the background. Concentration calculations are then based on the assumption that the line intensity is independent of the state in which the element occurs. The effects of other elements on the line intensity can be minimized by mixing a large amount of a suitable matrix material (a *spectroscopic* buffer) with the sample.

It is sometimes possible to estimate the concentration of an element that occurs in small amount by comparing the blackness of several of its lines with a number of lines of a major constituent. Matching densities are then used to indicate the concentration of the minor constituent.

---

[8]For references to a more complete description of some semiquantitative procedures see C. E. Harvey, *Spectrochemical Procedures* (Glendale, California: Applied Research Laboratories, 1950), Chap. 7.

11 Flame Emission Spectroscopy

Flame emission spectroscopy (also called flame photometry) is a spectral method in which excitation is brought about by spraying a solution of the sample into a hot flame. In general, the techniques and the equipment required for a flame emission analysis are simpler and quite different from those for a spectrographic analysis; discussion of the two in separate chapters thus is warranted.

The most important applications of flame photometry have been to the analysis of sodium and potassium, particularly in biological fluids and tissue. For reasons of convenience, speed, and relative freedom from interference, flame emission spectroscopy has become the method of choice for those otherwise difficult-to-determine elements. The method has also been applied, with varying degrees of success, to the determination of perhaps half the elements in the periodic table. Thus, flame emission spectroscopy must be considered to be one of the important tools for analysis.[1]

FLAMES AND FLAME SPECTRA

Much of the information in Chapter 5 (pp. 115 to 130) regarding the properties of flames, burners, and the fate of the sample in a flame applies

---

[1]For a more complete discussion of the theory and applications of flame emission spectroscopy see J. A. Dean, *Flame Photometry* (New York: McGraw-Hill Book Company, Inc., 1960); B. L. Vallee and R. E. Thiers in I. M. Kolthoff and P. J. Elving, Eds., *Treatise on Analytical Chemistry*, Part I, Vol. 6 (New York: Interscience Publishers, Inc., 1965), Chap. 65.

equally well to flame emission spectroscopy as to atomic absorption spectroscopy. Atomic absorption, however, is based upon the behavior of particles that exist in the ground state in the flame. Flame emission spectroscopy, on the other hand, is dependent upon those particles that are electronically excited in the medium. It is important to recall that unexcited species predominate by a wide margin at the temperatures that exist in most flames.

FLAME CHARACTERISTICS

In flame emission spectroscopy the flame serves (1) to convert the constituents of the liquid sample into the vapor state; (2) to decompose these constituents into atoms or simple molecules; and (3) to electronically excite a fraction of the resulting atomic or molecular species. As with other types of emission spectroscopy, control of the source is the most critical aspect of the method.

Flame Temperature

Table 11-1 lists the common fuels employed in flame emission spectroscopy and the maximum temperatures resulting from their combustion with oxygen as the oxidant. In addition to these, acetylene-nitrous oxide flames produce temperatures of about 2950°C.

The temperatures produced by the burning of natural or manufactured gas in air are so low that only the alkali and alkaline-earth metals, with very low excitation energies, produce useful spectra. Acetylene-air mixtures give a somewhat higher temperature. In order to excite the spectra of many metals, oxygen or nitrous oxide must be employed as the oxidant; with the common fuels temperatures of 2700 to 3000°C are obtained. The hottest practical flame results from the combustion of cyanogen in oxygen:

$$C_2N_2 + O_2 \rightarrow 2CO + N_2$$

Table 11-1 Maximum Flame Temperatures for Various Fuels[a]

	Temperature, °C	
Fuel	*In Air*	*In Oxygen*
Illuminating gas	1700	2700
Propane	1925	2800
Butane	1900	2900
Hydrogen	2100	2780
Acetylene	2200	3050
Cyanogen	–	4550

[a]Data taken from J. A. Dean, *Flame Photometry* (New York: McGraw-Hill Book Company, Inc., 1960), p. 17. With permission.

This reaction is highly exothermic and the products of the combustion are not appreciably dissociated at temperatures less than 4000°C. The cyanogen flame produces spectra that are nearly arc-like in quality, and permits the determination of elements with high excitation energies.

Temperature and excitation profiles in flames. In the typical flame, the maximum temperature occurs in the region somewhat above the inner cone (see Figure 11-1). The temperature distribution, however, varies considerably with the fuel and the oxidant used, as well as their concentration ratio. As noted earlier (p. 117), the fraction of excited atoms increases exponentially with temperature; thus, large differences in emission intensities are observed in different parts of a flame. Figure 11-2, for example, shows the intensity profile for a calcium line in a cyanogen flame. Note that the emission maximum is found just above the inner cone.

Figure 11-2 also demonstrates that the intensity of emission is critically dependent upon the rate at which the sample is introduced into the flame. Initially the line-intensity rises rapidly with increasing flow rate as a consequence of the increasing number of calcium particles. A rather sharp maximum in the intensity is observed, however, beyond which the water that is aspirated with the sample lowers the flame temperature.

Band, rather than line, spectra occasionally are employed in flame

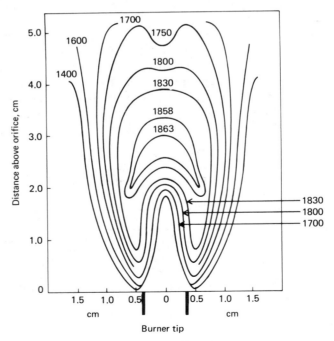

Figure 11-1. Temperature profiles (in °C) for a natural gas-air flame. From B. Lewis and G. von Elbe, *J. Chem. Phys.*, **11**, 75 (1943). (With permission.)

photometry. For example, calcium produces a useful band in the region of 540 to 567 nm, probably due to the presence of the vaporized oxide in the flame. The maximum in the intensity of band emission often occurs in the inner cone and decreases rapidly in the mantle as the molecules responsible for the emission dissociate at the higher temperatures of the latter region.

The more sophisticated instruments for flame emission spectroscopy are equipped with monochromators that sample the radiation from a relatively small part of the flame; adjustment of the position of the flame with respect to the entrance slit is thus critical. Filter photometers, on the other hand, scan a much larger portion of the flame; here control of flame position is less important.

Optimum flame temperature. The best flame temperature for an analysis must be determined empirically and depends upon the excitation energy of the element, how it is combined in the sample, the sensitivity required, and what other elements are present. The high temperatures achieved with oxygen and nitrous oxide as oxidants are needed to excite many elements and may also be required for more easily excited elements when these occur in the sample as refractory compounds. Although high temperatures usually provide enhanced emission intensities, and thus higher sensitivity, there are some notable instances where the use of a low-temperature flame is an advantage. For example, the intensity of the atomic doublet line for potassium at about 767 nm is limited by the ease of ionization of this element. As a consequence, little enhancement in sensitivity accompanies temperature increases; in fact, the ionization of potassium leads to a nonlinear calibration curve. In addition, the likelihood of interference by other elements becomes greater at higher temperatures. Thus, the analysis of

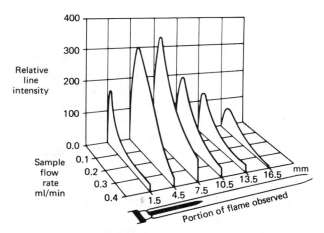

Figure 11-2. Flame profile for calcium line in a cyanogen-oxygen flame for different sample flow rates. Taken from K. Fuwa, R. E. Thiers, B. L. Vallee, and M. R. Baker, *Anal. Chem.,* **31,** 2039 (1959). (With permission of the American Chemical Society.)

potassium (and the other alkali metals as well) is advantageously performed with a low-temperature flame.

Emission by Fuels

All of the fuels employed in flame emission spectroscopy produce both continuous background radiation as well as certain band spectra. The nature and intensity of the spectrum for a given fuel is strongly dependent upon the fuel-to-oxidant ratio and the flame temperature. In analysis it is usually necessary to correct line intensities for the emission by the fuel.

Hydrogen flames are advantageous because they tend to produce a high metal-to-background signal. A flame lacks the energy to cause significant excitation of the continuous hydrogen spectrum that is observed in a discharge tube; thus, the continuum is very weak. Band spectra occur, however, in the region between 300 and 350 nm due to OH radicals formed in the combustion process.

The continuous spectra associated with hydrocarbon flames are more intense than those of hydrogen. Undoubtedly, several processes contribute to this type of emission, but little is known of their nature. At high fuel-to-oxidant ratios black-body radiation from carbon particles probably contributes. The combination of carbon monoxide with atomic oxygen,

$$CO + O \rightarrow CO_2 + h\nu$$

is known to produce a continuum that may, under some circumstances, appear in a hydrocarbon flame. A number of band spectra due to such species as OH radicals, CH radicals, and C_2 molecules also occur. Again, the intensities are strongly dependent upon the combustion conditions.

Chemical Reactions in Flames

Both theoretical and experimental evidence suggest that many of the processes occurring in the mantle of a flame are in approximate thermodynamic equilibrium. As a consequence, it becomes possible to regard the burned gases of the flame as a solvent medium to which thermodynamic calculations can be applied. The processes of interest include translational, vibrational, and rotational motion as well as excitation, ionization, and dissociation. The last two types of processes are of principal concern in this section.

Ionization in flames.   Ionization of atoms and molecules is small in combustion mixtures that involve air as the oxidant, and generally can be neglected. At the higher temperatures of oxygen or nitrous oxide flames, however, ionization becomes important, and a significant concentration of free electrons is present as a consequence of the equilibrium

$$M \rightleftharpoons M^+ + e^- \qquad (11\text{-}1)$$

where M represents a neutral atom or molecule and M^+ is its ion. We will focus upon equilibria in which M is a metal atom.

The equilibrium constant K for this reaction may take the form

$$K = \frac{[M^+][e^-]}{[M]} = \frac{x^2}{1-x}p \qquad (11\text{-}2)$$

where the bracketed terms are activities, x is the fraction of M that is ionized, and p is the partial pressure of the metal before ionization in the gaseous solvent.

The effect of temperature on K is given by the *Saha equation*; namely

$$\log K = \frac{-5040\,E_i}{T} + \frac{5}{2}\log T - 6.50 + \log \frac{g_{M^+}\,g_{e^-}}{g_M} \qquad (11\text{-}3)$$

where E_i is the ionization potential for the metal in electron volts, T is the absolute temperature of the medium, and g is the statistical weight for each of the three species indicated by the subscripts. For the alkali metals the final term has a value of zero while for the alkaline earths it is equal to 0.6.

Table 11-2 shows the calculated degree of ionization for several of the common metals under conditions that approximate those used in flame emission spectroscopy. The three temperatures correspond roughly to conditions that exist in an air-fuel, an oxygen-acetylene, and an oxygen-cyanogen flame.

It is important to appreciate that treatment of the ionization process as an equilibrium with free electrons as one of the products immediately implies that the degree of ionization of a metal will be strongly influenced by the presence of other ionizable metals in the flame. Thus, if the medium contains

Table 11-2 Degree of Ionization of Metals at Flame Temperatures[a]

Element	Ionization Potential, eV	Temperature °K at					
		$p = 10^{-4}$ atm			$p = 10^{-6}$ atm		
		2000	3500	5000	2000	3500	5000
Cs	3.893	0.01	0.86	> 0.99	0.11	> 0.99	> 0.99
Rb	4.176	0.004	0.74	> 0.99	0.04	> 0.99	> 0.99
K	4.339	0.003	0.66	0.99	0.03	0.99	> 0.99
Na	5.138	0.0003	0.26	0.98	0.003	0.90	> 0.99
Li	5.390	0.0001	0.18	0.95	0.001	0.82	> 0.99
Ba	5.210	6×10^{-4}	0.41	0.99	6×10^{-3}	0.95	> 0.99
Sr	5.692	1×10^{-4}	0.21	0.97	1×10^{-3}	0.87	> 0.99
Ca	6.111	3×10^{-5}	0.11	0.94	3×10^{-4}	0.67	0.99
Mg	7.644	4×10^{-7}	0.01	0.83	4×10^{-6}	0.09	0.75

[a]Data calculated from equation (11-3); from B. L. Vallee and R. E. Thiers, in I. M. Kolthoff and P. J. Elving, Eds., *Treatise on Analytical Chemistry*, Part I, Vol. 6 (New York: Interscience Publishers, Inc., 1965), p. 3500. With permission.

not only species M, but species B as well, and if B ionizes according to the equation

$$B \rightleftharpoons B^+ + e^-$$

then the degree of ionization of M will be decreased by the mass-action effect of the electrons. Determination of the degree of ionization under these conditions requires a calculation involving the dissociation constant for B and the mass-balance expression

$$[e^-] = [B^+] + [M^+]$$

Practical consequence of ionization. The presence of atom-ion equilibria in flames has a number of important consequences in flame emission spectroscopy. For example, the intensity of the *atomic* emission lines for the alkali metals, particularly potassium, rubidium, and cesium, are affected in a complex way by temperature. Increased temperature causes an increase in the population of excited atoms, according to the Boltzmann relationship (p. 117); counteracting this effect, however, is a decrease in concentration of atoms as a result of ionization. Thus, under some circumstances a decrease in emission intensity may be observed in hotter flames. For this reason, lower excitation temperatures are usually specified for analysis of the alkali metals. With calcium or magnesium, on the other hand, increased sensitivity can be gained by the use of oxygen-fuel mixtures since ionization is still low at these flame temperatures.

Ionization may also cause marked departures from linearity for intensity-concentration curves. For example, the calibration curve for potassium is often nonlinear at low concentrations and linear at high concentrations. The curvature is directly attributable to the significant increase in the degree of ionization (and thus lowered intensities) at low concentrations of the element. It is of interest that the curvature can be caused to disappear by the addition of an easily ionized element (an *ionization suppressor*) such as cesium. Here the mass-action effect of electrons produced from the ionization of cesium represses the formation of potassium ions.

From the foregoing remarks it is clear that the presence of one easily ionized element may interfere with the quantitative determination of a second unless precautions are taken to repress completely the ionization of the element being determined. Such interference can often be avoided by the addition of an excess of an ionization suppressor to both the calibration standards and the sample.

Dissociation equilibria. In the hot, gaseous environment of a flame a variety of dissociation and association reactions lead to conversion of the metallic constituents to the elemental form. It seems probable that at least some of these reactions are reversible and can be treated by the laws of thermodynamics. Thus, in theory, it should be possible to formulate reactions such as

$$MO \rightleftharpoons M + O$$
$$M(OH)_2 \rightleftharpoons M + 2OH$$

or, more generally,

$$MA \rightleftharpoons M + A$$

In practice, not enough is known about the nature of the chemical reactions in a flame medium to permit a quantitative treatment such as that described for the ionization process. Instead, reliance must be placed on empirical observations.

Dissociation reactions involving metal oxides and hydroxides clearly play an important part in determining not only the intensity but also the nature of emission by an element. For example, the alkaline-earth oxides are relatively stable, with dissociation energies of the order of 5 eV. Molecular bands arising from the presence of metal oxides or hydroxides in the flame thus constitute a prominent feature of their spectra. Except at very high temperatures, these bands are more intense than the lines for the atoms or ions. In contrast, the oxides and hydroxides of the alkali metals are much more readily dissociated so that line intensities for these elements are high, even at relatively low temperatures.

It seems probable that dissociation equilibria involving anions other than oxygen may also be of importance in flame emission. For example, the line intensity for sodium is markedly decreased by adding HCl to the flame. A likely explanation is the mass-action effect on the equilibrium

$$NaCl \rightleftharpoons Na + Cl$$

Chlorine atoms formed from the added HCl decrease the sodium concentration and thereby lower the line intensity.

Metallic Spectra in Flames

In flame emission spectroscopy both line and band spectra are useful for analysis. Moreover, the presence of metals in a flame often results in a marked enhancement of the continuous background radiation. The origins of line and band spectra have been considered in an earlier section; their principal characteristics, along with those of continuous spectra, are considered in the paragraphs that follow.

Line spectra. Flame excitation of many elements produces line spectra that ordinarily are much less complex than the corresponding arc or spark spectra. In general, only the lowest-energy resonance lines of atoms rather than ions have sufficient intensity to be useful. These characteristics of flame spectra arise directly from relatively low excitation energies involved.

The simplicity of flame spectra is a significant advantage for quantitative work since it permits the use of lower dispersion monochromators (or perhaps filters) and reduces the likelihood of interference from overlapping spectra.

Inasmuch as resonance lines are employed, self-absorption (see p. 263) is common in flame emission spectroscopy. This phenomenon manifests itself in nonlinear calibration curves, with negative deviations from a straight line appearing at high concentrations.

Self-absorption and ionization sometimes result in S-shaped calibration curves with three distinct segments. Thus, at intermediate concentrations of potassium for example, a linear relationship between intensity and concentration is observed (Figure 11-3). At low concentration, however, curvature arises because of the increased degree of ionization in the flame. Self-absorption, on the other hand, causes negative departures from a straight line at higher concentrations.

Band spectra. In contrast to the distinct line character of arc and spark spectra, flame spectra are often characterized by the appearance of bands originating from excitation of metal oxides and hydroxides. Presumably such compounds are largely dissociated in the higher temperature of the arc or spark. For perhaps one-third of the elements that can be determined by flame emission spectroscopy, a band spectrum is employed. In this group the rare-earth and alkaline-earth metals predominate.

Continuous radiation. An increase in the background continuum is often observed in the presence of high concentrations of some metallic species.

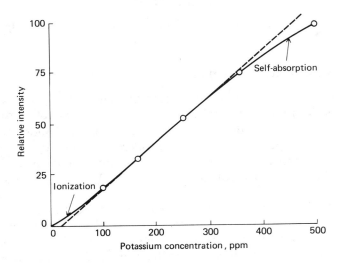

Figure 11-3. Effects of ionization and self-absorption on a calibration curve for potassium.

Continuous radiation is the result of transitions in which one of the energy-level states is unquantized and possesses free kinetic energy. For example, the emission can arise from a reaction of the type

$$M^+ + e^- \rightarrow M + h\nu$$

Here, the free electron in the flame can have an infinite number of kinetic energies; thus, the frequency of radiation varies continuously.

From the analytical standpoint, it is important to appreciate that high concentrations of a metal increase the background radiation; line intensities must be corrected for this effect.

Effect of Organic Solvents on Flame Spectra

Lower molecular weight alcohols, ketones, and esters, alone or mixed with water, have been used to convey the sample into the flame. Their employment tends to enhance line intensities by as much as a factor of 10 or more.

The intensification of lines by organic solvents is the result of several effects. Their low viscosity and surface tension result in higher flow rates of sample into the flame. In addition, cooling by the solvent is lessened; thus, the net effect is to increase the number of particles available for excitation without a corresponding adverse effect on the flame temperature (p. 286). Combustion of the organic substance contributes to the temperature; increases as great as 200°C have been reported for certain solvents.

Although organic solvents usually increase the sensitivity of a flame photometric analysis, they also increase the likelihood of interference.

FLAME SPECTROPHOTOMETERS

The optical and electronic systems employed in flame photometers and spectrophotometers do not significantly differ from the photometers and spectrophotometers discussed in Chapter 3. In fact, most of these instruments can be adapted readily to flame photometric measurements by replacing the source and cell compartment with a suitable burner and a means for bringing the emitted radiation to the entrance slit. Flame emission spectrophotometers are even more closely related to atomic absorption spectrophotometers; many commercial instruments are designed for either use.

Flame Sources

The source system in flame photometry consists of gas flow regulators, an atomizer, and a burner; in many designs the latter two components are integral. As in other forms of emission spectroscopy, rigid control of the source is crucial to a successful analysis. Important variables include the flow rates of the fuel and the oxidant, the rate of introduction of the sample, and the droplet size of the atomized solution.

Regulation of gas flow rates. In order to achieve a stable and reproducible flame the flow rates of oxidant and fuel must be both constant and reproducible. Most instruments incorporate pressure regulators to control the flow rate of the gases; it is assumed that the resistance-to-flow of the burner system remains constant. Ordinarily two-stage diaphragm regulators are employed to reduce the tank pressures of the gases. Additional needle-valve regulators and narrow-range gauges (10 to 25 lb) are needed to provide for control of the gas flows. For the most precise work, flow meters should be employed in order to compensate for changes in the resistance-to-flow caused by deposits in the burner system.

Atomizers and burners. The atomizer-burners for flame emission spectroscopy are similar to those employed in atomic absorption spectroscopy (see p. 125). The total consumption burner shown in Figure 5-5(a) has been widely employed. Figure 11-4 shows a typical premix atomizer suitable for flame emission spectroscopy.

Optical and Electronic Systems

Isolation of flame emission lines or bands can be accomplished with filters or with monochromators. Instruments of the former type are properly called flame photometers, while those of the latter type would be more aptly termed flame spectrophotometers.

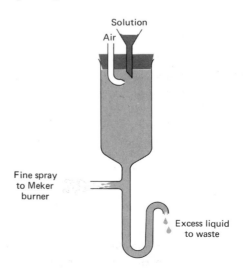

Figure 11-4. Spray chamber atomizer.

Photometers.[2]   When a low-temperature flame is employed as an excitation source, only the most prominent lines of the alkali and alkaline-earth elements appear. As a consequence, a filter photometer is frequently adequate to isolate a low-energy line for one of these elements. Attractive features of these instruments include their simplicity and relatively low cost. Glass, gelatin, and interference filters have all been employed in flame photometers, the last being the most satisfactory because of their narrow band widths.

Vacuum phototubes are ordinarily used as detectors for the radiant energy passing through the filter. Some photometers are equipped with two filters and two detectors to compensate for fluctuations in the output of the source; these instruments are analogous in design to the two-cell photometers discussed in Chapter 3. With such instruments an *internal standard,* consisting of an element whose spectral behavior is similar to the element being determined, is incorporated into both sample and standards. A measured quantity of a lithium salt, for example, might serve as an internal standard for a sodium analysis. The radiant energy emitted by the flame is split, with one-half of the beam passing through a filter that transmits only a sodium line and the other half passing through a filter that transmits only a lithium line. The power of the two beams is then compared by means of a simple bridge circuit such as that shown in Figure 3-22. In principle, the internal-standard method offers substantial advantages. The general requirements for an internal standard were discussed in connection with emission spectroscopy (Chapter 10); many of these same considerations and limitations are applicable to flame emission spectroscopy.

Flame photometers are generally satisfactory for the determination of alkali or alkaline-earth elements when these ions represent major constituents of a sample. If several elements are present in high concentration, the inadequacy of the filter system becomes manifest, and substantial errors can be expected. The principal application of flame photometers has been to the determination of sodium and potassium in biological fluids.

Spectrophotometers.   The specificity of methods based upon flame excitation can be greatly enhanced by employment of a prism or grating monochromator to isolate the spectral lines; several such instruments are manufactured commercially. A flame attachment is available for the Beckman DU-2 Spectrophotometer described on page 60. A unit consisting of an atomizer and a burner replaces the light-source housing shown in Figure 3-23.

Flame spectrophotometers are more complex and expensive than the corresponding filter instruments. They do, however, offer the advantages of wider application and greater specificity.

[2]For a discussion of instrumental requirements for flame photometers and spectrophotometers see R. Herrman and C. T. J. Alkemade, *Chemical Analysis by Flame Photometry,* 2d ed., trans. by P. T. Gilbert (New York: Interscience Publishers, Inc., 1963).

QUANTITATIVE ANALYSIS BY FLAME EMISSION SPECTROSCOPY

While flame emission spectroscopy has, in special situations, some applications in qualitative analysis, its primary importance lies in quantitative elemental analysis. Here it has both advantages and disadvantages when compared with arc and spark emission methods. In terms of precision, flame photometry is usually superior; standard deviations of about 2 percent relative can be expected in a typical analysis. In addition, the equipment is ordinarily much simpler and less expensive. On the other hand, a limitation of flame emission spectroscopy is the need to dissolve the sample in a suitable solvent. The sensitivity of the procedure varies widely, depending upon the element and the flame temperature. Often it lies in a range from a few tenths of a part per billion to less than a part per million in the solution employed for excitation.[3] Finally, fewer elements can be excited in a flame than in an arc or a spark.

Errors in Flame Photometry

Errors in flame photometry arise from two sources. Instrumental errors are traceable to fluctuations in the behavior of the source or the detector. Errors also arise from differences in composition between the sample and the standards against which they are compared. Errors of the latter type are the more troublesome.

Instrumental errors. A stable flame is needed for the attainment of reproducible results. Stability requires that the flow rate of the fuel and the oxidant be reproducible to approximately one percent. With properly designed flow gauges or pressure regulators, flame stability need not be a serious source of uncertainty.

The atomizer must also perform in a perfectly reproducible manner so that the sample or standard is introduced at a constant rate and in droplets of constant size. Since clogging of the capillary can cause serious errors, every effort must be made to guard against irregularities from this source.

Clearly, any drift or fluctuation in the performance of the detector and the amplifier leads to analytical errors.

Errors from radiation of foreign elements. Analytical errors arise when other components of the sample emit radiation of a wavelength that is not completely removed by the monochromator system. The magnitude of the effect is dependent upon the quality of the monochromator, the temperature of the source, and the concentration ratio between the contaminant and the element sought.

---

[3]For a comparison of the sensitivity of flame emission and atomic absorption spectroscopy, see E. E. Pickett and S. R. Koirtyohann, *Anal. Chem.,* **41**(14), 28A (1969).

Figure 11-5 shows flame emission spectra of the alkali and alkaline-earth metals. Note that the emission lines for lithium, sodium, and potassium are widely spaced; as a consequence, a simple filter system eliminates interference by any two of these elements during an analysis for the third. On the other hand, calcium and strontium emit bands in the vicinity of both the sodium and the lithium lines; these bands cause serious errors in an alkali analysis unless a very good monochromator system is available. Where sodium is to be determined and the calcium-to-sodium ratio is high, the problem of interference becomes particularly acute. Clearly, sodium and lithium are likely to induce errors in a calcium analysis.

As the temperature of the source is increased, the need for a high-quality monochromator becomes more stringent, not only because additional lines for the alkali and alkaline-earth elements appear, but also because excitation of other elements in the sample gives rise to more lines. When hydrogen or acetylene is employed in conjunction with oxygen, a prism or grating system is needed to isolate the desired radiation.

Cation enhancement. In higher-temperature flames it is not uncommon to find that the intensity of a line for one cation is increased by the presence of a second. For example, in an acetylene flame the intensity of the rubidium line may be as much as doubled by the presence of a potassium ion. Smaller interactions between other alkali metal pairs have also been reported. Cation enhancement arises from a decrease in the ionization of the cation of interest as a result of the presence of the second cation (p. 290).

Errors from cation enhancement are minimized by adding a *radiation-buffer* to both standards and samples. The buffer contains a high concentration of the potentially interfering ion, and thus minimizes the effect of small concentrations of the ion in the sample.

Anion interference. Certain anions have a strong depressant effect on the intensities of a number of cation lines. For example, oxalate, phosphate,

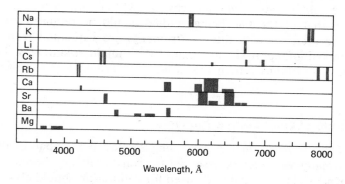

Figure 11-5. Flame spectra of the alkali and alkaline-earth elements. From R. B. Barnes, D. Richardson, J. W. Berry, and R. L. Hood, *Ind. Eng. Chem., Anal. Ed.,* **17**, 605 (1945). (With permission of the American Chemical Society.)

sulfate, and aluminate may decrease the emission intensities of the alkaline earths by 50 percent or more. The best evidence indicates that these anions repress emission by forming compounds with the cation which volatilize only slowly at the flame temperatures. Thus, a smaller concentration of the cation is available for excitation in the flame. A slow rate of dissociation of the compound to the elemental state may also contribute. The repressive effect of chloride ion on atomic sodium lines (p. 291) is an example of this phenomenon.

Where the anion effect is small, compensation can be achieved with a radiation buffer. When it is severe, however, more drastic measures may be needed. A simple and straightforward method is to replace the interfering anion with another by passing the solution through an anion-exchange column. Precipitating agents can also be used. Alternatively, so-called *releasing agents* can be employed. Releasing agents function either by combining firmly with the interfering anion or displacing it by forming a stable complex with the cation. For example, the phosphate interference in a calcium determination can be eliminated by the addition of zirconium or lanthanum ions; these releasing agents strongly bond with phosphate and thereby render the calcium available for excitation. Ethylenediaminetetraacetic acid would also act as a releasing agent by forming a stable and readily volatilized calcium complex.

Background radiation. The line spectra observed by flame excitation are superimposed upon a continuous background radiation arising in part from the flame itself. In addition, the continuous spectra of the various compounds in the sample may contribute to the background. The magnitude of the latter effect depends upon the kind and the amount of salts present, and can cause serious analytical errors if uncorrected.

With a flame spectrophotometer, correction for the background is readily accomplished by measurement of the intensity on either side of the emission peak being used. In principle this same technique could be employed with a filter instrument; a narrow-band filter with a transmission peak adjacent to that of the line being studied would be required. Ordinarily, however, corrections with filter instruments are estimated from a blank whose composition approximates that of the sample or standard. This technique is less satisfactory than the direct measurement of the background intensity.

Methods of Quantitative Analysis

From the remarks in the preceding section it is apparent that close control of many variables is essential for reliable flame photometric data. Whenever possible, the standards used for calibration should closely match the overall composition of the unknown solution. Ordinarily, it is best to perform a calibration concurrently with the analysis. Even with these precautions, measurements with a filter photometer can be expected to yield good results only where the sample solution has a relatively simple composition and the element being determined is a major constituent.

Several techniques have been suggested for the performance of a flame photometric analysis[4]; three of these are described briefly.

Analyses based upon calibration curves. A series of standards containing various known concentrations of the element sought is prepared; their overall composition should approximate that of the unknown. The instrument is first adjusted to give a zero response when pure water is aspirated into the flame. Next, the optical or electric system is adjusted to give the maximum response with the most concentrated standard sample. Without altering this adjustment, emission intensities for the remaining standards and the unknown are measured. A background correction to the calibration data can be made with a spectrophotometer by measuring the radiation intensity on either side of the peak at points sufficiently removed so that the line spectrum contributes nothing to the radiation. Otherwise, the background is estimated from a blank containing all components of the sample except the one being determined. A plot of the corrected instrument response versus concentration of the element sought is ideally a straight line; however, a straight line is often not obtained. The concentration of the unknown solution is evaluated with the aid of this curve.

With most instruments, fresh calibration curves must be prepared each day. Even within this time interval, checks of the curve with one or two of the standards should be made periodically to insure that instrument drift has not occurred.

Standard addition method. In the standard addition technique a calibration curve is prepared as before. Emission readings are then obtained for an aliquot of the unknown and for an identical aliquot to which a known quantity of the element sought has been added. The concentration of the element in each solution is then calculated from the calibration curve. The difference in concentration between the two aliquots should be equal to the known quantity added unless there has been an enhancement or a depression of the line intensity. If a discrepancy is found, the apparent concentration of the unknown is multiplied by the ratio of the true quantity added to the apparent amount added. This correction assumes that the calibration curve is essentially linear in the region of measurement and that the relative enhancement or depression of the line intensity is the same for the two aliquots.

A somewhat more complicated variant of this procedure is reported to give better results. Equal volumes of the unknown are added to a series of solutions containing known and varying quantities of the element of interest. Emission readings for these solutions are then corrected for background and

[4]See J. A. Dean, *Flame Photometry* (New York: McGraw-Hill Book Company, Inc., 1960), pp. 110-122.

plotted against the concentration of the standard present in the solution. The resulting line is extrapolated to zero concentration; the concentration of the unknown can then be obtained by dividing the extrapolated reading by the slope of the line.

Internal standard method. The internal standard method described on page 284 may be applied to flame emission spectroscopy as well. The requirements for the ion chosen as a standard are the same as those listed for an arc or a spark determination. A calibration curve of intensity ratios versus concentration is then prepared and employed for the analysis.

Applications

Flame photometric methods have been applied to the analysis of a wide variety of materials, including biological fluids, soils, plant materials, cements, glasses, and natural waters.[5] The most important applications are for the determination of the alkali metals.

[5]For a summary of applications, see J. A. Dean, *Flame Photometry* (New York: McGraw-Hill Book Company, Inc., 1960).

12 Refractometry

When radiation passes through a transparent medium, interaction occurs between the electric field of the radiation and the bound electrons of the matter; as a consequence, the rate of propagation of the beam is less than in a vacuum (see pp. 14 to 16). The refractive index of a substance n_i is given by the relationship

$$n_i = \frac{c}{v_i} \qquad (12\text{-}1)$$

where v_i is the velocity of propagation in the medium and c is the velocity in vacuum (a constant under all conditions). The refractive index of most liquids is in the range between 1.3 and 1.8; it is 1.3 to 2.5 or higher for solids.

In common with density, melting point, and boiling point, the refractive index is one of the classical physical constants that can be used to describe a chemical species. While it is a nonspecific property, few substances have identical refractive indexes at a given temperature and wavelength. Thus, this constant is useful for confirming the identity of a compound and measuring its purity. In addition, refractive index determinations are conveniently employed for the quantitative analysis of binary mixtures. Finally, combined with other measurements, the refractive index provides structural and molecular weight information about a substance.[1]

[1] For a more complete discussion of refractometry see N. Bauer, K. Fajans and S. Z. Lewin in A. Weissberger, Ed., *Physical Methods of Organic Chemistry,* Vol. I, Part II (New York: Interscience Publishers, Inc., 1960). Chap. 28: L. W. Tilton and J. K. Tayler, in W. G. Berl, Ed., *Physical Methods in Chemical Analysis,* Vol. 1 (New York: Academic Press, Inc., 1950), pp. 486 ff.; S. Z. Lewin and N. Bauer, in I. M. Kolthoff and P. J. Elving, Eds., *Treatise on Analytical Chemistry,* Part I, Vol. 6 (New York: Interscience Publishers, Inc., 1965), Chap. 70.

GENERAL PRINCIPLES

The Measurement of Refractive Index

The refractive index of a substance is ordinarily determined by measuring the change in direction (refraction) of collimated radiation as it passes from one medium to another. As shown in Figure 2-6 (p. 16),

$$\frac{n_2}{n_1} = \frac{v_1}{v_2} = \frac{\sin\theta_1}{\sin\theta_2} \qquad (12\text{-}2)$$

where v_1 is the velocity of propagation in the less dense medium M_1 and v_2 is the velocity in medium M_2; n_1 and n_2 are the corresponding refractive indexes and θ_1 and θ_2 are the angles of incidence and refraction, respectively. When M_1 is a vacuum, n_1 is unity (equation 12-1) and

$$n_2 = n_{\text{vac}} = \frac{c}{v_2} = \frac{\sin\theta_1}{\sin\theta_2} \qquad (12\text{-}3)$$

where n_{vac} is the *absolute refractive index* of M_2. Thus, n_{vac} can be obtained by measuring the two angles θ_1 and θ_2.

It is much more convenient to measure the refractive index with respect to some medium other than vacuum, and air is often employed as a standard for this purpose. Most compilations of n for liquids and solids in the literature are with reference to air at laboratory temperatures and pressures. Fortunately, the change in the refractive index of air with respect to temperature and pressure is small enough so that a correction from ambient laboratory conditions to standard conditions is needed for only the most precise work. A refractive index n_D measured with respect to air under the usual laboratory conditions with the D line of sodium can be converted to n_{vac} with the equation

$$n_{\text{vac}} = 1.00027\, n_D \qquad (12\text{-}4)$$

This conversion is seldom required.

It is usually necessary to measure the refractive index with an accuracy of at least 2×10^{-4}. Accuracies on the order of 6 to 7×10^{-5} may be required for the routine analysis of solutions. For the detection of impurities, a difference in refractive index between the sample and a pure standard is measured; here, the ability to detect a difference on the order of 1×10^{-6} or better is required.

Specific and Molar Refraction

The rate at which a beam of radiation is propagated depends upon the density of the electrons in the medium as well as upon their environment. In a gas, for example, the electron density is low and minimal interaction occurs; as a result, the refractive index is only slightly greater than unity for a gas under ordinary conditions. An increase in gas pressure, and hence gas density, increases the concentration of electrons encountered by the radiation and correspondingly increases the refractive index. Thus, n for air changes from about 1.00027 to about 1.03 when the pressure is raised from 1 to 100 atmospheres. In liquids and solids, where densities are much greater than in gases, the radiation is influenced by a much larger number of electrons, and higher refractive indexes are the consequence.

The existence of a relationship between refractive index and density d has been recognized since the time of Newton, who observed that the relationship $(n^2 - 1)/d$ was approximately constant for a number of substances. It has further been recognized that the refractive index, although dependent upon density, must also be affected by the arrangement of the electrons in the medium. Thus, if refractive indexes could be corrected for density differences, the resulting parameter, the *specific refraction*, should be a measure of the electronic environment in the medium. Several empirical and theoretical formulas for specific refraction have been developed over the last century. The Lorentz and Lorenz relationship has a theoretical basis for certain classes of liquids and has been widely employed in structural studies. It defines the specific refraction r as

$$r = \frac{(n^2 - 1)}{(n^2 + 2)} \cdot \frac{1}{d} \qquad (12\text{-}5)$$

The *molar refraction R* is equal to rM, where M is the molecular weight of the substance.

Specific and molar refractions have proved useful for analytical purposes since they are found to vary in a systematic way within homologous series of compounds. In addition, molar refraction has been of value in structural studies and in providing insights into the nature of chemical bonds. Some of these applications are considered in a later section.

Variables That Affect Refractive Index Measurements

Temperature, wavelength, and pressure are the most common experimentally controllable variables that affect a refractive index measurement.

Temperature. Temperature influences the refractive index of a medium primarily because of the accompanying change in density. For many liquids the temperature coefficient of refractive index lies in the range of -4 to -6×10^{-4} degree^{-1}. Water is an important exception, with a coefficient of about -1×10^{-4}; aqueous solutions behave similarly. Solids have temperature coefficients that are roughly an order of magnitude smaller than the typical liquid.

It is apparent from the foregoing that the temperature must be controlled closely for accurate refractive index measurements. For the average liquid, temperature fluctuations should be less than $\pm 0.2°C$ if fourth place accuracy is required, and $\pm 0.02°C$ for measurements to the fifth place.

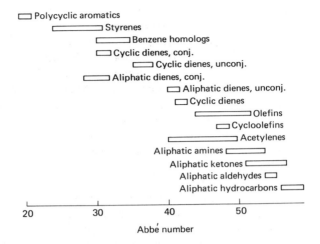

Figure 12-1. Correlation chart for Abbé numbers. From S. Z. Lewin and N. Bauer, in I. M. Kolthoff and P. J. Elving, Eds. *Treatise on Analytical Chemistry,* Part I, Vol. 6 (New York: Interscience Publishers, Inc., 1965), p. 3927. (With permission.)

Wavelength of radiation. As noted on p. 15, the refractive index of a transparent medium gradually decreases with increasing wavelength; this effect is referred to as *normal disperson*. In the vicinity of absorption bands, however, rapid changes in refractive index occur; here the dispersion is referred to as *anomalous* (see Figure 2-5).

Dispersion phenomena make it essential that the wavelength employed be specified in quoting a refractive index. The D line from a sodium vapor lamp ($\lambda = 5893$ Å) is most commonly used as a source in refractometry, and the corresponding refractive index is designated as n_D (often the temperature in °C is also indicated by a superscript; for example n_D^{20}). Other lines commonly employed for refractive index measurements include the C an F lines from a hydrogen source ($\lambda = 6563$ Å and 4861 Å, respectively) and the G line of mercury ($\lambda = 4358$ Å).

Measurements of dispersion are sometimes useful in the characterization of organic functional groups. For this purpose the so-called *Abbé number ν* is often employed:

$$\nu = \frac{n_D - 1}{n_F - n_C} \tag{12-6}$$

Figure 12-1 shows the range of Abbé numbers for several types of organic compounds.

Pressure. The refractive index of a substance increases with pressure because of the accompanying rise in density. The effect is most pronounced in gases, where the change in *n* amounts to about 3×10^{-4} per atmosphere;

for liquids, the figure is less by a factor of 10 and for solids it is yet smaller. Thus, only for precise work with gases and for the most exacting work with liquids and solids is the variation in atmospheric pressure important.

INSTRUMENTS FOR MEASURING REFRACTIVE INDEX

Two types of instruments for measuring refractive index are available from commercial sources. *Refractometers* are based upon measurement of the so-called *critical angle* or upon the determination of the displacement of an image. *Interferometers* utilize the interference phenomenon to obtain differential refractive indexes with very high precision.

Critical Angle Refractometers

The most widely used instruments for the measurement of refractive index are of the critical-angle type. The *critical angle* is defined as the angle of refraction in a medium when the angle of the incident radiation is 90 deg (the *grazing angle*); that is, when θ_1 in equation (12-2) is 90 deg, θ_2 becomes the critical angle θ_c. Thus,

$$\frac{n_2}{n_1} = \frac{\sin 90}{\sin \theta_c} = \frac{1}{\sin \theta_c} \qquad (12\text{-}7)$$

Figure 12-2(a) illustrates the critical angle that is formed when the critical ray approaches the surface of the medium M_2 at 90 deg to the normal and is then refracted at some point O on the surface. Note that if the medium could be viewed end-on, as in Figure 12-2(b), the critical ray would appear as the boundary between a dark and a light field. It should be noted, however, that the illustration is unrealistic in that the rays are shown as entering the medium at but one point O; in fact, they would be expected to enter at all points along the surface and thus create an entire family of

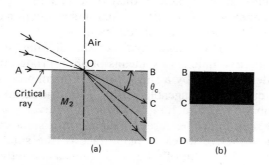

Figure 12-2. (a) Illustration of the critical angle, θ_c, and the critical ray AOC; (b) end-on view showing sharp boundary between the dark and light fields formed at the critical angle.

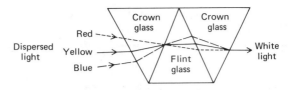

Figure 12-3. Amici prism for compensation of dispersion by sample. Note that the path of the yellow radiation (sodium D line) is not deviated by the prism.

critical rays with the same angle θ_c. A condensing or focusing lens is needed to produce a single dark-light boundary such as shown in Figure 12-2(b).

It is important to realize that the critical angle is dependent upon wavelength. Thus, if polychromatic radiation is employed, no single sharp boundary such as that in Figure 12-2(b) is observed. Instead, there is a diffuse chromatic region between the light and dark areas; the precise establishment of the critical angle is thus impossible. This difficulty is often overcome by the use of monochromatic radiation.

As a convenient alternative, many critical-angle refractometers are equipped with a compensator that permits the use of radiation from a tungsten source, but compensates for the resulting dispersion in such a way as to give a refractive index in terms of the sodium D line. The compensator consists of one or two *Amici prisms*, as shown in Figure 12-3. The properties of this complex prism are such that the dispersed radiation is converged to give a beam of white light that travels in the path of the yellow sodium D line.

The dipping refractometer.   Figure 12-4 is a schematic diagram of a dipping refractometer, perhaps the simplest of the commercial critical-angle instruments. The prism is immersed in the thermostatted sample, and white light is reflected from a mirror onto the prism face, as shown. After compensation by a single Amici prism, the refracted radiation is focused on a linear scale which is then observed through the eyepiece. The scale is equipped with a vernier that permits estimation of the boundary position to a small fraction of a scale division.

The readings from the linear scale of the dipping refractometer are converted to refractive indexes by means of appropriate tables. The range of refractive index covered by a single prism is small and amounts to only about 0.04. Several interchangeable prisms permit measurements over the range between about 1.32 and 1.54, however. Whenever a prism is changed, the zero of the scale must be reset by means of a standard solution or a solid of known refractive index (a plane face of the solid is held on the surface of the prism by a film of a liquid that has a refractive index greater than that of the solid).

The dipping refractometer is capable of high reproducibility ($\pm 1 \times 10^{-5}$); the accuracy, however, is likely to be about $\pm 4 \times 10^{-5}$ because of the

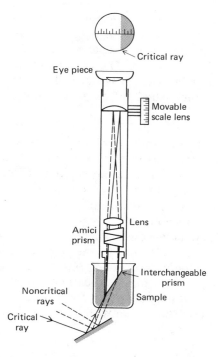

Figure 12-4. Dipping refractometer.

uncertainty introduced in setting the zero of the scale and in calibrating the scale divisions in terms of refractive index. The dipping refractometer has been used extensively for quantitative analysis of aqueous solutions where the range of refractive indexes is small. The instrument suffers from the disadvantages of limited range, the requirement of large samples, and the resulting greater problems of temperature control.

Abbé refractometer. The Abbé instrument is undoubtedly the most convenient and widely used refractometer. Figure 12-5 shows a schematic diagram of its optical system. The sample is contained as a thin layer (~ 0.1 mm) between two prisms. The upper prism is firmly mounted on a bearing that permits its rotation by means of the side arm shown in dotted lines. The lower prism is hinged to the upper to allow separation for cleaning and for introduction of sample. The lower prism face is rough-ground; when light is reflected into the prism, this surface effectively becomes the source for an infinite number of rays that pass through the sample at all angles. The radiation is refracted at the interface of the sample and the smooth-ground face of the upper prism, whereupon it passes into the fixed telescope. Two Amici prisms, which can be rotated with respect to one another, serve to collect the divergent critical angle rays of different colors into a single white beam which corresponds in path to that of the sodium D ray. The eyepiece of the telescope is equipped with crosshairs; in making a measurement, the prism angle is changed until the light-dark interface just coincides with the crosshairs. The position of the prism is then established from the fixed scale (which is normally graduated in units of n_D). Thermostatting is accomplished by circulation of water through jackets surrounding the prisms.

The Abbé refractometer permits a rough estimate of the dispersion of the sample during the determination of n_D. It is apparent that the greater the dispersion, the greater must be the compensation provided by the Amici prisms. A calibrated scale indicates the position of the prisms required to provide a colorless critical boundary; a value for $(n_F - n_C)$ can be calculated from this information. Uncertainties of 1 to 30 percent in dispersions determined in this way are common.

The Abbé refractometer owes its popularity to its convenience, its wide range ($n_D = 1.3$ to 1.7), and to the minimal sample required. The accuracy of the instrument is about $\pm$ 0.0002; its precision is half this figure. The most serious error in the Abbé instrument is caused by the fact that the nearly grazing rays are cut off by the arrangement of the two prisms; the boundary is thus less sharp than with a dipping instrument.

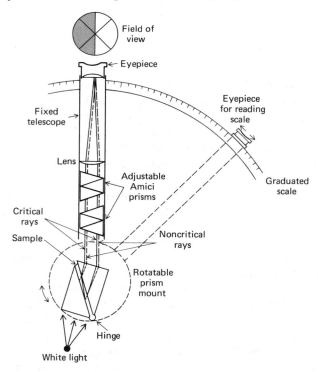

Figure 12-5. The Abbé refractometer.

A *precision* Abbé refractometer, which reduces the uncertainties of the ordinary instrument by a factor of about three, is also available. The improvement in accuracy is obtained by replacing the compensator with a monochromatic source and by using larger and more precise prism mounts. The former provides a much sharper critical boundary, and the latter permits a more accurate determination of the prism position.

The Pulfrich refractometer. In the Pulfrich refractometer the sample is held on the horizontal surface of a prism by means of a closely fitting glass cylinder (see Figure 12-6). The critical boundary is particularly sharp with this arrangement, since the full area of the refractory prism can be illuminated by radiation that just grazes the surface. In this respect the Pulfrich instrument is superior to the Abbé.

The critical boundary is observed with a telescope mounted on a calibrated turntable; the refractive index is determined from the position of the telescope with respect to the prism.

The Pulfrich refractometer is somewhat more accurate than the Abbé. On the other hand, it is less convenient to use, particularly in the matter of thermostatting of the sample.

Other Instruments for Refractive Index Measurements

Image displacement refractometer. The most straightforward method for determining refractive index involves measurement of the angles of incidence and refraction by means of a spectrometer arrangement somewhat like that shown in Figure 3-8. Liquid samples are contained in a prism-shaped container mounted at the center of a large circular metal table; solid samples are cut into the shape of a prism and are similarly mounted. A light source, a slit, and a collimator are employed to direct a parallel beam of radiation onto one surface of the prism. The refracted image of the slit is then viewed with a telescope mounted on the circle. Since the slit image can be made very sharp, the accuracy of the determination is dependent only upon the accuracy of the angular measurements and the

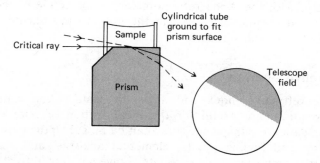

Figure 12-6. Prism arrangement in the Pulfrich refractometer.

control of temperature; uncertainties of 1×10^{-6} or smaller can be attained. On the other hand, this type of refractometer is not amenable to routine measurements and is thus employed only for special purposes.

Interferometers. Interferometers permit the comparison of refractive indexes with a very high precision (in some instances to the seventh decimal place with liquids and to the eighth decimal place with gases). In an interferometer the radiation from a slit is collimated and is then passed through a pair of secondary slits. The two parallel beams are then passed through identical cells containing the two solutions to be compared. The beams are subsequently merged by a second lens and a pattern of interference bands results. The position of the bands depends upon the difference in propagation rates or refractive indexes between the two media. For reference, the pattern of bands yielded by identical path lengths is displayed below that for the two solutions. A variable glass compensator is then imposed in the path containing the sample with lower refractive index until the band pattern is identical to the reference pattern. The difference in refractive index is then related to the position of the compensator device.

APPLICATIONS OF REFRACTOMETRY

Refractometric measurements are employed for qualitative and quantitative analysis as well as for structural investigations. Regardless of use, periodic calibrations of the refractometer are essential.

Refractometers can be conveniently calibrated with purified liquids such as water ($n_D^{20} = 1.3330$), toluene ($n_D^{20} = 1.4969$), and methylcyclohexane ($n_D^{20} = 1.4231$). The latter two compounds can be obtained from the National Bureau of Standards as certified samples with five-decimal indexes at 20, 25, and 30°C and for each of seven wavelengths. A glass test piece, supplied with most refractometers, can also be employed as a reference. The difference between the refractive index of the standard and the instrument scale reading is applied as an arithmetic correction to subsequent determinations. Alternatively, with the Abbé refractometer the objective of the telescope can be adjusted mechanically so that the instrument indicates the proper refractive index for the standard.

Qualitative Analysis

Precise refractive index and dispersion data, when combined with other information, are useful for the identification of pure compounds. While these data are nonspecific, they can be most helpful when correlated with melting and boiling points, elemental analysis, and spectroscopic evidence. It is important to appreciate, however, that refractive indexes and dispersions may be highly sensitive to the purity of the sample; unfortunately, some published data are not as reliable as might be desired. Thus, in qualitative work, considerable care must be taken in sample purification and a critical evaluation of the data in the literature must be made.

As we have pointed out earlier, measurement of the dispersion of a system is also useful in qualitative work. Thus, as shown in Figure 12-1, the Abbé number can provide valuable information regarding the probable presence or absence of certain functional groups in a compound.

Quantitative Analysis

A refractive index measurement is often the simplest, most convenient, and most rapid procedure for evaluating the composition of a binary liquid or a gaseous mixture. A linear relationship between refractive index and some concentration parameter frequently exists over at least a limited concentration range. A straight-line calibration curve is often obtained if the concentrations of aqueous solutions are expressed in grams of solute per 100 ml of solution. For mixtures of organic liquids, on the other hand, linearity is more often observed when volume percent is employed. Linearity, of course, is not a requirement for quantitative work since a suitable calibration curve can always be prepared.

Many quantitative applications of refractometry can be cited. For example, the method is widely used for the determination of the concentration of aqueous sugar solutions. Most of the common sugars have about the same effect on the refractive index of an aqueous solution; the procedure thus gives a measure of total carbohydrate concentration. Refractometric procedures have also proved useful in determining the concentration of sulfur in unvulcanized rubber and the bound styrene content of certain synthetic rubbers.

An important application of refractometry has been to the evaluation of apparatus and methods for separations such as distillation, extraction, adsorption chromatography, and diffusion. For example, the number of theoretical plates for a distillation column can be evaluated from refractive index measurements of fractions collected during the separation of a binary mixture such as heptane (n.b.p. 98.4°C) and methylcyclohexane (n.b.p. 100.9°C). The number of theoretical plates for the column can be calculated from the resulting analytical data.

Specific refraction and molar refraction data are also employed in quantitative analysis. For example, the specific refraction of certain silicate glasses is found to vary nearly linearly with the mole percent of SiO_2 over a range of 50 to 90 percent. Thus, accurate determination of the silicon content for such glasses is feasible from refractive index and density measurements. Similarly, specific refraction has been employed for the estimation of unsaturation in vegetable oils and for the degree of fluorination in paraffin oils. Molar refraction has been widely used in the petroleum industry for determining the percent carbon incorporated in aromatic structures in hydrocarbon mixtures.

Other Uses of Refractometry Data

For many years it has been recognized that the molar refraction increases in regular increments with the number of carbon atoms within a homologous series. This finding has led to the notion that the molar refraction of a compound can be considered to be the sum of atomic increments and that, within certain limits, the contribution of each atom is the same in every molecule. These atomic increments can be calculated from the molar refraction of a series of pure compounds and then can be employed to estimate the molar refractions of other compounds without recourse to experiment. For example, atomic increments for a saturated carbon, a nonpolar hydrogen, and an ether oxygen are 2.42, 1.10, and 1.64, respectively. Thus, for diethyl ether, the calculated molar refraction is

$$R = 4 \times 2.42 + 10 \times 1.10 + 1 \times 1.64$$
$$= 22.32$$

The molar refraction obtained from experimental values for density and refractive index is 22.58.

As an alternative to the foregoing, refraction increments have been assigned to bonding electrons. Thus, the molar refraction of C_3H_8 is the sum of increments for two C—C bonds and eight C—H bonds. Both of these methods allow reasonably good approximations of molar refractivities for many compounds and have been useful to organic chemists in verifying structural formulas. In some instances, however, marked differences between experimental and predicted values have been noted. On the basis of extensive studies of such nonadditive systems, Fajans[2] has shown how refractometric studies are valuable for the investigation of polarizability and intramolecular binding forces.

[2]K. Fajans, in A. Weissberger, Ed., *Physical Methods of Organic Chemistry*, Vol. I, Part II (New York: Interscience Publishers, Inc., 1960), pp. 1169 ff.

13 Polarimetry, Optical Rotatory Dispersion, and Circular Dichroism

Optical activity is a measure of the ability of certain substances to rotate plane-polarized light. The phenomenon, first reported for quartz in 1811, has been intensely studied since that time. By the mid-nineteenth century, many of the laws relating to optical activity had been formulated; these in turn played a direct part in the development of the ideas of organic stereochemistry and structure later in the same century. Some of the early concepts of optical activity have stood the test of time, and remain essentially unaltered today. It is of interest, however, that despite this long history, the interactions of radiation with matter that cause the rotation of polarized light are less clearly understood than the processes responsible for absorption, emission, or nuclear magnetic resonance.

The term *polarimetry* as it is used by most chemists can be defined as the study of the rotation of polarized light by transparent substances. The direction and the extent of rotation (the *optical rotatory power*) is useful for both qualitative and quantitative analysis, and for the elucidation of chemical structures as well.[1]

[1]For a more complete discussion of the various aspects of polarimetry, see W. Heller and D. D. Fitts in A. Weissberger, Ed., *Physical Methods of Organic Chemistry*, Vol. I, Part III (New York: Interscience Publishers, Inc., 1960), Chap. 33; W. A. Struck and E. C. Olson, in I. M. Kolthoff and P. J. Elving, Eds., *Treatise on Analytical Chemistry*, Part I, Vol. 6, (New York: Interscience Publishers, Inc., 1965), Chap. 71.

It has long been recognized that the optical rotatory power of a substance depends in a characteristic way upon the wavelength of light employed. Only recently, however, have instruments been developed that permit the convenient study of this effect. Beginning in the mid-1950s, photoelectric spectropolarimeters for visible and ultraviolet radiation became available; the advent of this equipment has clearly demonstrated that the study of the wavelength dependence of optical rotation (called *optical rotatory dispersion*) is a powerful tool for elucidating the structural details of complex molecules.[2]

A third instrumental method that is based upon the interaction of polarized radiation with matter is called *circular dichroism*; its development is even more recent. Circular dichroism involves the study of the differential absorption of light that is circularly polarized in opposite directions. Circular dichroism also provides useful structural information about optically active species.[3]

PROPERTIES AND INTERACTIONS OF POLARIZED RADIATION WITH MATTER

Some of the properties of plane-polarized radiation and its interaction with matter were considered briefly in Chapter 2. It is now necessary to treat these topics in greater depth.

Transmission and Refraction of Plane-polarized Radiation

Homogeneous liquids and gases transmit radiation at equal velocities in all directions and are termed optically *isotropic* as a consequence. Solids that crystallize in the cubic form, noncrystalline solids such as glasses, and many polymers are also isotropic toward radiation. Noncubic crystals are, however, optically *anisotropic* and transmit radiation at different velocities depending upon the orientation of the crystal with respect to the beam. This anisotropy arises from the variations in spatial distribution of the atoms in the crystal with orientation; as a result, the force fields encountered by a beam of radiation passing through the crystal vary with direction.

[2]For a discussion of optical rotatory dispersion see C. Djerassi, *Optical Rotatory Dispersion: Applications to Organic Chemistry* (New York: McGraw-Hill Book Company, Inc., 1960); W. Klyne and A. C. Parker in A. Weissberger, Ed., *Physical Methods of Organic Chemistry*, Vol. I, Part III (New York: Interscience Publishers, Inc., 1960), Chap. 34.

[3]For a discussion of circular dichroism see P. Crabbé, *Optical Rotatory Dispersion and Circular Dichroism* (San Francisco: Holden-Day, 1965); A. Abu-Shumays and J. J. Duffield, *Anal. Chem.*, **38** (7), 29A (1966); L. Velluz, M. Legrand and M. Grosjean, *Optical Circular Dichroism* (New York: Academic Press, Inc., 1965).

Transmission of radiation by optically anisotropic crystals. When a beam of nonpolarized, monochromatic radiation strikes the surface of a transparent, optically isotropic crystal (such as sodium chloride) at a normal angle, the radiation can be pictured as spreading outward from each point of interaction in a spherical wave as indicated in Figure 13-1(a). The comparable process in an anisotropic crystal oriented in two different directions with respect to the beam is illustrated in Figure 13-1(b) and 13-1(c). Here, a spherical wave O spreads out from the intersection B, as in the isotropic crystal; in addition, however, a second wave is also propagated in the ellipsoidal path E from the point of origin. Furthermore, the orientation of the axis for the ellipse with respect to the original beam path is dependent upon the orientation of the crystal. Thus, the beam of radiation is divided into two rays upon entry into an anisotropic crystal. The *ordinary ray* O travels with the same velocity in all directions, while the *extraordinary ray* E moves more rapidly in some directions than in others. As shown in Figure 13-1(d) and Figure 13-1(e) the extraordinary ray is transmitted in some anisotropic crystals at a slower rate rather than a faster one.

In every anisotropic crystal there is at least one direction, called the *optic axis*, along which both the ordinary and the extraordinary rays move at the same rate. As shown in Figure 13-1(b) through 13-1(e), the wave

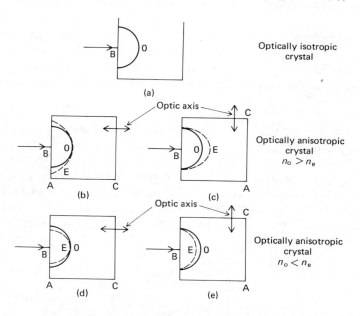

Figure 13-1. Propagation of an unpolarized monochromatic beam in isotropic and anisotropic crystals. In (b) and (c), ordinary ray O travels slower than extraordinary ray E. In (d) and (e) the reverse is the case.

fronts of the two rays coincide along the optic axis; in all other directions, however, separation of the ordinary and the extraordinary components occurs. It should be noted in passing that some anisotropic crystals have two optic axes.

The index of refraction of the extraordinary ray in an anisotropic crystal is clearly dependent upon direction, being identical with that for the ordinary ray along the optic axis and changing continuously to a maximum or a minimum along a perpendicular axis. Refractive indexes for the extraordinary ray are normally reported in terms of this perpendicular axis. Table 13-1 compares the refractive index for the ordinary ray n_o with that for the extraordinary ray n_e in some common anisotropic crystals. In calcite the extraordinary ray is clearly propagated at a greater rate than the ordinary ray; in quartz the reverse is the case. Because anisotropic crystals have two characteristic refractive indexes, they are said to be *double refracting*.

Table 13-1   Refractive Index Data for Selected Anisotropic Crystals

Crystal	n_o	n_e
Calcite	1.6583	1.4864
Quartz	1.544	1.553
Ice	1.306	1.307

Double refraction and polarization.   The ordinary ray and the extraordinary ray also differ in their planes of polarization, *the two planes being mutually perpendicular to one another.* This important difference makes possible the production of polarized radiation with anisotropic crystals.

The Nicol prism.   Figure 13-2 shows a *Nicol prism*, a device that is widely used for producing polarized light. The prism is manufactured from a crystal of calcite ($CaCO_3$). The end faces of a natural crystal are trimmed slightly to give the angle of 68 deg shown, and the crystal is then cut across its short diagonal. A layer of Canada balsam, a transparent substance with a refractive index between the two refractive indexes of calcite, is placed between the two crystal halves. This layer is totally reflecting

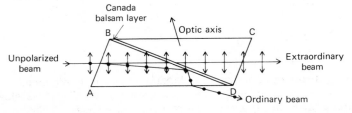

Figure 13-2. The Nicol prism showing polarization of radiation.

for the ordinary ray with its greater refractive index but, as shown in the figure, transmits the extraordinary ray almost unchanged.

Pairs of Nicol prisms are employed in measurements involving the rotation of the plane of polarized light. One prism serves to produce a polarized beam that is passed into the medium under study. A second analyzer prism then determines the extent of rotation caused by the medium. If the two Nicols are identically oriented with respect to the beam, and if the medium has no effect, the extraordinary ray, comprising nearly 50 percent of the original intensity, is emitted from the analyzer [see Figure 13-3(a)]. If the polarizer is rotated [Figure 13-3(b)], only the vertical component *m'n'* of the beam *mn* emitted from the polarizer is transmitted through the analyzer, the horizontal component being now reflected off of the Canada balsam layer. Thus, less than 50 percent of the incident beam appears at the face of the analyzer. Rotation of the polarizer by 90 deg, results in a beam from the analyzer that has no vertical component. Thus, no radiation is observed at the face of the analyzer. If a medium affecting the rotation of light is interposed between the two Nicols, the relative orientations needed to achieve a maximum or a minimum in transmission are changed by an amount corresponding to the rotatory power of the medium.

Interference Effects with Polarized Radiation

To account for many of the experimental observations regarding the interactions of polarized radiation, it is necessary to assume that inter-

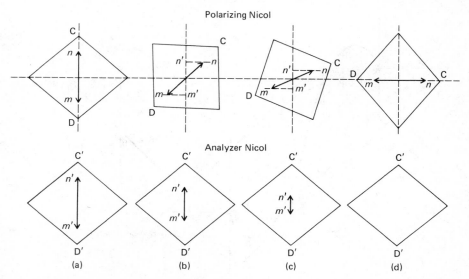

Figure 13-3. End view of a polarizer and analyzer Nicol. *mn* = electrical vector of beam transmitted by polarizer. *m'n'* = vertical component of beam transmitted by polarizer and analyzer.

ference between polarized beams can occur provided the beams are *coherent* (p. 10). The effect of interference can then be visualized by vector addition of the electromagnetic components of the individual beams.

Figure 13-4(a) illustrates the interference between two plane-polarized beams that are in phase but oriented 90 deg to one another. Addition of the electrical vectors of the two beams is shown schematically in the end-

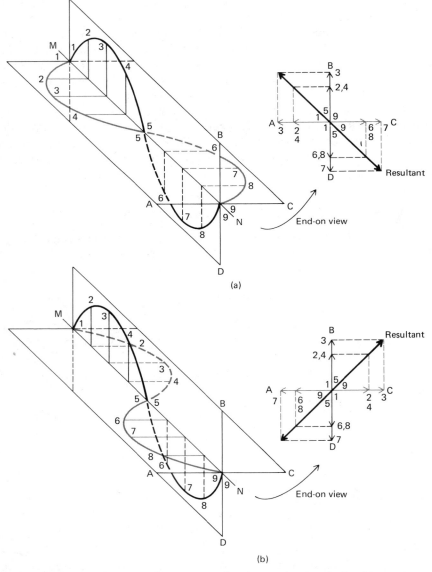

Figure 13-4. Interference between two in-phase plane-polarized beams.

on view to the right. Note that the numbered points along the axis MN have been projected onto the plane ABCD which is perpendicular to MN, thus giving a two-dimensional representation of the resultant and its components. It will be seen that the resultant is a vector that oscillates in a plane oriented at 45 deg to the planes of the two component beams; *interference thus produces a single plane-polarized beam.*

Figure 13-4(b) demonstrates interference when the phase-relationship between the two plane-polarized beams differs by one-half wavelength (180 deg). Here, the waves are again in phase, and the resultant is a plane-polarized beam which, however, is perpendicular with respect to the resultant of Figure 13-4(a).

Circularly and elliptically polarized radiation. It is of interest now to examine the behavior of a plane-polarized beam of monochromatic radiation as it passes through an anisotropic crystal. In Figure 13-5(a), the

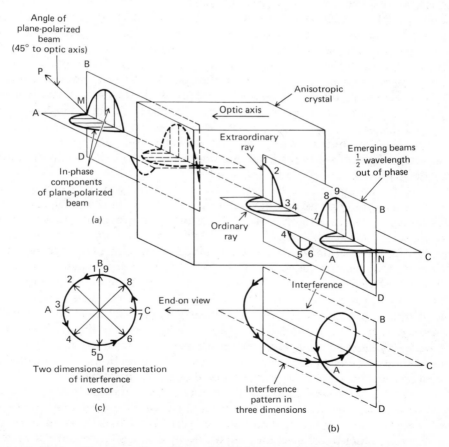

Figure 13-5. Circular polarization of light by an anisotropic crystal.

path of the incident beam is normal to the optic axis of the crystal with the plane of the polarized radiation oriented at 45 deg to that axis (the angle of the plane of polarization is given by the arrow MP). As shown by Figure 13-4, however, the plane-polarized beam can be considered to consist of two *coherent* components lying in perpendicular planes oriented along MB and MA. These components are also indicated in Figure 13-5. Upon entering the crystal, the component lying in the MA direction travels at the rate of an ordinary ray since this direction lies along the optic axis of the crystal; the orientation of component MB, on the other hand, corresponds to that for an extraordinary ray, and its rate of propagation is thus different. As a result of the velocity difference, *the two components are no longer coherent* and can thus not interfere. That is, within the crystal the beam can be considered to be two entities having different velocities.

When the two rays leave the crystal, their velocities again become equal in the isotropic air medium; thus, they can again interfere since they are once more coherent. The nature of the resultant will, however, depend upon the phase relationship between the two that exists at the instant they emerge from the crystal surface. This phase relationship is determined by the relative velocities of the two rays in the medium as well as by the length of traverse. If, for example, the path in the crystal is such that the two rays are completely in phase upon exiting, then constructive interference similar to that shown in Figure 13-4(a) occurs. That is, the resultant beam will be polarized at the same angle as the entering beam. If, on the other hand, the crystal thickness is such that the phase relationship between the two rays at the face is shifted exactly one-half wavelength, interference such as shown in Figure 13-4(b) results. Here, the plane of the exit beam is oriented 90 deg with respect to the entering beam.

In Figure 13-5(a) we have shown the emerging waves as one-quarter wavelength out of phase and have indicated their relationship to one another assuming that no interference takes place. In fact, however, interaction does occur as the two rays enter the air medium, and the path of the electrical vector for the resulting wave can be obtained by addition of the two vectors. The resulting vector quantity is seen to travel in a helical pathway around the direction of travel [Figure 13-5(b)]. If the vector sum is plotted in two-dimensional form [Figure 13-5(c)], a circle is obtained. This condition is in distinct contrast to the original linearly polarized radiation in which the electrical vector lay in a single plane. A helical beam of this type is called *circularly polarized light*. Note that if the two waves had been out of phase one-quarter wavelength in the other sense, the direction of travel of the vector would have been clockwise rather than counterclockwise.

Thus far we have considered the nature of the exit beam from the aniso-tropic crystal when a phase difference created was 0, $\frac{1}{4}$, $\frac{1}{2}$, or some multiple of these fractions. If the light path in the crystal is such as to produce phase differences other than these, the path traced by the resultant electrical vector

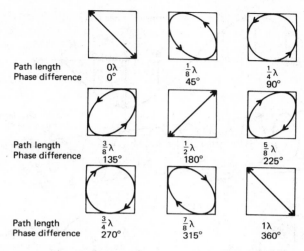

Path length	0λ	$\frac{1}{8}λ$	$\frac{1}{4}λ$
Phase difference	0°	45°	90°
Path length	$\frac{3}{8}λ$	$\frac{1}{2}λ$	$\frac{5}{8}λ$
Phase difference	135°	180°	225°
Path length	$\frac{3}{4}λ$	$\frac{7}{8}λ$	1λ
Phase difference	270°	315°	360°

Figure 13-6. Effect of an anistropic crystal on plane-polarized radiation. Each diagram corresponds to a different path length in the crystal.

is an ellipse, and the radiation is called *elliptically* polarized light. Figure 13-6 summarizes the states that result when the components of a plane-polarized beam are emitted from an anisotropic crystal with various phase differences.

Anisotropic crystals of suitable length are employed experimentally for producing circularly-polarized radiation. Such crystals are called *quarter-wave plates* and find use in circular dichroism studies.

Relationship between plane- and circularly polarized radiation. In the last section we have seen that the behavior of plane-polarized radiation upon passage through an anisotropic crystal can be rationalized by considering the beam to be the resultant of two plane-polarized rays that are in phase and oriented at 90 deg to one another. It is of equal importance to understand that plane-polarized radiation can also be treated as the interference product of *two coherent circular rays of equal amplitude that rotate in opposite directions.* Figure 13-7 shows how the vectors for the *d* and *l* circular components are added to produce the equivalent vectors of a plane-polarized beam. From the middle figure it can be seen that each of the two rotating vectors describes a helical path around the axis of travel of the beam.

The rationalization of many phenomena to be considered in this chapter is based upon the idea that plane-polarized radiation consists of a *d* and an *l* circular component. Here, *d* (*dextrorotatory*) refers to the clockwise rotation as the beam approaches the observer; *l* (*levorotatory*) is the counterclockwise component.

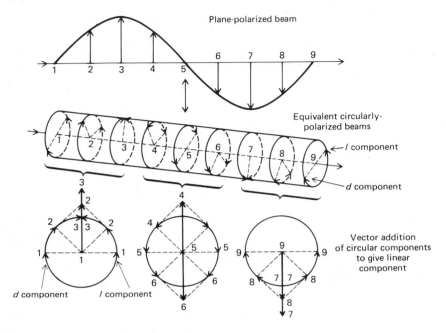

Figure 13-7. Equivalency of plane-polarized beam to two (*d,l*) circularly polarized beams.

Circular Double Refraction

The rotation of plane-polarized light by an optically active species can be explained if one assumes that the rates of propagation of the *d* and the *l* circular components of a plane-polarized beam are different in the presence of such a species; that is, the refractive index of the substance with respect to *d* radiation (n_d) is different from that for *l* (n_l). Thus, optically active substances are anisotropic with respect to circularly polarized light and show *circular double refraction*. Note that the two circular components are no longer coherent in the anisotropic medium and cannot interfere until they again reach an isotropic medium.

The rotation of a beam of light by an optically active medium is shown schematically in Figure 13-8. Initially in Figure 13-8(a) the beam is polarized in a vertical plane with the circular *d* and *l* components rotating at equal velocities. Upon entering the anisotropic medium, the rate of propagation of the *d* component is slowed more than that of the *l* because $n_d > n_l$. Thus, at some location [Figure 13-8(b)] in the medium, the *d* vector will lag behind the *l*; if at this point the two rays could interfere, the resultant would still be a plane, but one that was rotated from the vertical. At a farther point [Figure 13-8(c)], the *d* component would be still further retarded and greater rotation would result. Figure 13-8 has been drawn to show that, upon emerging from the medium, retardation has been such that the resultant

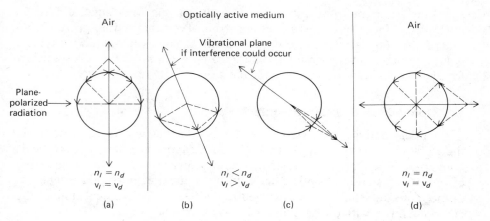

Figure 13-8. Rotation of plane-polarized light in a medium in which $n_l < n_d$.

plane is now horizontal. As shown in Figure 13-8(d), the d and l components are again propagated at identical rates in the isotropic air medium, and interference of the coherent rays occurs. Passage through the anisotropic medium has had the effect, however, of shifting the phase relationship such that the observed plane of polarization is now 90 deg with respect to the original.

Quantitative relationships. It is readily shown that the rotation (ψ) in radians caused by an optically active substance is given by the relationship

$$\psi = \frac{\pi \, \mathbf{l}}{\lambda} (n_l - n_d) \tag{13-1}$$

The rotation in degrees is then $180 \, \psi/\pi$. In equation (13-1), **l** is the path length of the medium in centimeters and λ is the wavelength of the plane-polarized radiation (in vacuum), also in centimeters. The quantity $(n_l - n_d)$ is termed the *circular birefringence*. The following calculation demonstrates the magnitude of the circular birefringence required to bring about a typical rotation.

A solution contained in a 10-cm cell is found to rotate the plane-polarized radiation of the sodium D line by 100 deg. What is the difference in refractive index of the medium for the l and d circularly polarized components?

$$\lambda = 589 \text{ nm} = 5.89 \times 10^{-5} \text{ cm}$$

Substituting into equation (13-1),

$$100 \text{ deg} = \frac{180 \text{ deg} \times 10 \text{ cm}}{5.89 \times 10^{-5} \text{ cm}} (n_l - n_d)$$

$$(n_l - n_d) = 3.1 \times 10^{-6}$$

It is apparent from this calculation that a relatively small difference in refractive index has a large effect in terms of optical rotatory power. Note that if the refractive index for the sodium D line is about 1.5 (a typical value), then a 2 part per million difference between n_d and n_l is responsible for the 100-deg rotation shown in the example.

Optically Active Compounds

Optical activity may be of two kinds: (1) that which occurs only in the crystalline form of a compound and disappears when the crystal is converted to a liquid, a gas, or a solution (quartz is the classic example of this manifestation of optical activity); (2) optical activity that is inherent in the molecule itself and is observed regardless of the physical state of the compound. It is with this second type that we are concerned.

The structural requirements for an optically active molecule are well understood and treated in detail in organic textbooks. It is sufficient to note that the optically active forms of a molecule are mirror images which cannot be superimposed, regardless of the orientation of one with respect to the other. Optical isomers thus bear the same relationship to one another as the left to the right hand. The two isomers then rotate polarized light equally but in opposite directions; if one is in excess in a mixture or is isolated from the other, rotation is observed. The most common form of optical activity is the result of asymmetric substitution on a tetrahedral carbon atom in an organic compound. Other types of asymmetric centers can also occur both in organic and inorganic structures.[4]

It is important to appreciate that the interactions which lead to rotation of polarized light are not peculiar to molecules with asymmetric centers, but instead are characteristic of most molecules. However, rotation is not *observed* in noncrystalline samples lacking asymmetric character because the molecules are randomly oriented; rotation caused by a molecule in one orientation is canceled by the equal and opposite rotation of another oriented as its mirror image. For the same reason, samples containing *d* and *l* isomers in equal concentration exhibit no net rotation because their individual effects cancel. It is only when one form of an asymmetric molecule is present in excess that a compensating mirror image will be absent, and a net rotation can be observed.

Variables that Affect Optical Rotation

The rotation of plane-polarized radiation by optically active compounds can range from several hundred to a few hundredths of a degree. Experimental variables that influence the observed rotation include the

---

[4]For a discussion of optical activity in inorganic systems, see J. C. Bailar and D. H. Busch, *The Chemistry of the Coordination Compounds* (New York: Reinhold Publishing Corporation, 1956).

wavelength of radiation, the optical path length, the temperature, the density of the substance if undiluted, and the concentration if in solution. For solutions, the rotation may also vary with the solvent.

The *specific rotation* or the *specific rotatory power* $[\alpha]_\lambda^t$ is widely employed to describe the rotatory characteristics of a liquid. It is defined as

$$[\alpha]_\lambda^t = \frac{\alpha}{l\,c} \tag{13-2}$$

where α is the observed rotation in degrees, l is the path length in decimeters, and c is the weight of solute (g) in one cubic centimeter of solution. The wavelength and temperature are usually specified with a subscript and a superscript, as shown. Most specific rotations are measured at 20°C with the sodium D line and are thus reported as $[\alpha]_D^{20}$. For a pure liquid, c is replaced by the density. By convention counterclockwise, or l rotation as the observer faces the beam, is given the negative sign. Clockwise (d) rotation is positive.

The term *molecular rotation* $[M]$ is also encountered; it is defined as

$$[M] = M[\alpha]/100 \tag{13-3}$$

where M is the molecular weight.

It is frequently necessary to measure the optical rotatory power of a substance as a solute. Unfortunately, the specific rotation of a compound is nearly always found to vary with the nature of the solvent. Because of solubility considerations, no single standard solvent can be designated. In addition, the specific rotation in a given solvent may not be entirely independent of concentration, although the variation in dilute solution is usually small. Because of these effects it is common practice to designate both the kind of solvent and the solute concentration in specifying a specific rotation.

The variation in specific rotation with temperature is approximately linear, but the temperature coefficient differs widely from substance to substance. For example, the specific rotation of tartaric acid solutions may vary as much as 10 percent per degree; for sucrose, on the other hand, the variation is less than 0.1 percent per degree.

The effect of wavelength on rotation is considered in a later section on optical rotatory dispersion. The D line of sodium is the standard for single wavelength measurements; the green line of mercury is occasionally employed.

Mechanism of Optical Rotation

While the structural requirements for optical activity can be precisely defined, the mechanism by which a beam of circularly polarized radiation interacts with matter and is thus retarded is much less obvious. That is, we can readily predict that a compound such as 2-iodobutane should have optical isomers because it possesses an asymmetric center. On the other hand, it is difficult to account for specific rotations of $\pm$ 32 deg for these isomers, while those for isomers of 2-butanol have values of $\pm$ 13.5 deg.

Several theories concerning the mechanism of optical rotation have been developed. Generally these theories are couched in terms of quantum mechanics and are of sufficient complexity to evade all but the most mathematically oriented chemist. Furthermore, while in principle it may be possible to predict optical rotatory power for a specific compound with the aid of these theories, such calculations are not practical in terms of effort; nor has it been demonstrated that the values obtained from such calculations can be sufficiently precise to be useful.

POLARIMETRY

Discussion in this section is confined to the instrumentation for and the applications of optical rotation measurements at single wavelengths.

Polarimeters

The basic components of a polarimeter include a monochromatic light source, a polarizing prism for producing polarized radiation, a sample tube, an analyzer prism with a circular scale, and a detector (see Figure 13-9). The eye serves as the detector for most polarimeters. The sensitivity of such an instrument is increased with a so-called half-shadow device, often another small Nicol prism, introduced into part of the beam as shown in the Figure 13-9.

Sources. Because optical rotation varies with wavelength, monochromatic radiation is employed. Historically, the sodium D line was obtained by introducing a sodium salt into a gas flame. Suitable filters then removed other lines and background radiation. Sodium vapor lamps with a filter to remove all but the D line are now employed. Mercury vapor lamps are also useful, the line at 546 nm being isolated by a suitable filter system.

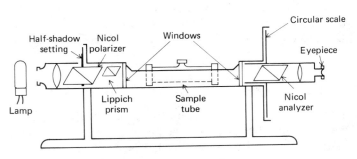

Figure 13-9. Typical visual polarimeter.

Polarizer and analyzer. Nicol prisms (p. 316) are most commonly employed to produce plane-polarized light and to determine the angle through which the light has been rotated by the sample. In principle, the measurement could be made by first adjusting the two prisms to a crossed position that yields a minimum in light intensity in the absence of sample. With a sample in place, rotation of the beam would cause an increase in light intensity which could then be offset by rotation of the analyzer prism. The angular change required to minimize the intensity would correspond to the rotatory power of the sample. Unfortunately, however, the position of minimum intensity cannot be determined accurately with the eye (nor with a photoelectric detector, for that matter) because the rate of change in intensity per degree of rotation is at a minimum in this region. Therefore, polarimeters are equipped with *half-shadow* devices which permit the determination to be made by matching two halves of a field at a radiation intensity greater than the minimum.

Half-shadow devices. Figure 13-9 shows a typical half-shadow device consisting of a small Nicol prism (called a *Lippich prism*) that intercepts about half the beam emerging from the polarizer. The position of the Lippich prism is adjusted to alter the plane of polarization by a few degrees; thus, in the absence of sample and with the analyzer Nicol at 90 deg with respect to the polarizer, a split, light-dark field is observed. The light portion, of course, corresponds to that half of the beam that has been rotated by the auxiliary prism and the dark part of the field corresponds to the unobstructed beam. The intensity of the two halves is then balanced by rotation of the analyzer. The analyzer scale is adjusted to read zero at this point. With the sample in place the analyzer is rotated until the same balance is obtained. The rotation of the sample can then be read directly from the circular analyzer scale.

Other end-point devices, which operate upon the same principle as the Lippich prism, permit the determination of optical rotatory power with a precision of 0.005 to 0.01 deg under ideal conditions.

Sample tubes. The sample for polarimetry is contained in cylindrical tubes, usually 10 to 20 cm in length. The ends are plane-parallel glass disks that are either fused to the tube walls or held in place with screw-cap holders. For precise measurements the tubes are surrounded by a jacket for temperature control. Tubes can be calibrated for length by measuring the rotation of a liquid of known rotatory power; nicotine-alcohol mixtures are often used for this purpose.

Applications of Polarimetry

Qualitative analysis. The optical rotation of a pure compound under a specified set of conditions provides a basic physical constant that is useful for identification purposes in the same way as its melting point, boiling point, or refractive index. Optical activity is characteristic of many naturally occurring substances such as amino acids, steroids, alkaloids, and carbohydrates; polarimetry represents a valuable tool for the identification of such compounds.

Structural determination. In this application the change in optical rotation resulting from a chemical transformation is measured. Empirical correlations obtained from the study of known structures are then employed to deduce information about an unknown compound. Details of steroid structures, in particular, have been acquired from polarimetric measurements; similar information has been obtained for carbohydrates, amino acids, and other organic compounds.

Quantitative analysis. Polarimetric measurements are readily adapted to the quantitative analysis of optically active compounds. Empirical calibration curves that relate optical rotation to concentration are employed. These plots may be linear, parabolic, or hyperbolic.

The most extensive use of optical rotation for quantitative analysis is in the sugar industry. For example, if sucrose is the only optically active constituent, its concentration can be determined from a simple polarimetric analysis of an aqueous solution of the sample. The concentration is directly proportional to the measured rotation. If other optically active materials are present, a more complex procedure is required; here the change in rotation resulting from the hydrolysis of the sucrose is determined. The basis for this analysis is shown by the equation

$$C_{12}H_{22}O_{11} + H_2O \xrightarrow{\text{acid}} C_6H_{12}O_6 + C_6H_{12}O_6$$

$$\underset{[\alpha]_D^{20} = +66.5°}{\text{sucrose}} \qquad \underset{+52.7°}{\text{glucose}} \quad \underset{-92.4°}{\text{fructose}}$$

This reaction is termed an *inversion* because of the change in sign of the rotation that occurs. The concentration of sucrose is directly proportional to the difference in rotation before and after inversion.

OPTICAL ROTATORY DISPERSION AND CIRCULAR DICHROISM

Optical rotatory dispersion and circular dichroism are two closely related physical methods that are based upon the interaction of circularly polarized radiation with an optically active species. The former method measures the wavelength dependence of the molecular rotation of a compound. As we have already indicated, optical rotation at any wavelength depends upon the difference in refractive index of a substance toward d and l circularly polarized radiation—that is, upon the *circular birefringence* $(n_l - n_d)$; this quantity is found to vary in a characteristic way as a function of wavelength. In contrast, circular dichroism is dependent upon the fact that the *molar absorptivity* of an optically active compound is different for the two types of circularly polarized radiation. Here the wavelength dependence of $(\epsilon_l - \epsilon_d)$ is studied, where ϵ_l and ϵ_d are the respective molar absorptivities.

The inequality of molar absorptivities was first reported by A. Cotton in 1895, and the whole complex relationship between absorptivity and refractive index differences is now termed the *Cotton effect*. As noted at the outset of this chapter, the widespread application of the Cotton effect to chemical investigations was delayed until convenient instruments for its study became available.

General Principles

Employing solutions of potassium chromium tartrate which absorb in the visible region, Cotton demonstrated that right circularly polarized radiation was not only refracted but also absorbed to a different extent than the left circular beam; that is, $\epsilon_d \neq \epsilon_l$. At the same time, Cotton observed that dramatic changes occurred in the optical rotation $(n_l - n_d)$ as well as the difference in absorptivities $(\epsilon_l - \epsilon_d)$ in the region of an absorption maximum. These effects are shown in Figure 13-10. It is important to appreciate that the $(n_l - n_d)$ curve for a substance is similar in shape to curves showing the change in its refractive index as a function of the wavelength of unpolarized light [see Figure (2-5)]. Here, too, marked changes in the refractive index (anomalous dispersion) occur in the region of absorption.

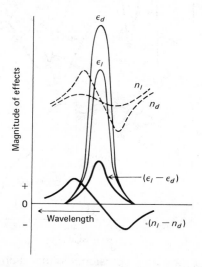

Figure 13-10. The Cotton effect. Adapted from W. Heller in A. Weissberger, Ed., *Physical Methods of Organic Chemistry*, Vol. I, Part 3 (New York: Interscience Publishers, Inc., 1960), p. 2164. (With permission.)

Optical Rotatory Dispersion Curves

Optical rotatory dispersion curves consist of a plot of optical rotation as a function of wavelength. Two types of curvature can be discerned. One, the normal dispersion range, is a region (or regions) wherein [α] changes only gradually with wavelength. The second, the region of anomalous dispersion, occurs near an absorption peak. If one peak is isolated from others, the anomalous part of the dispersion curve will have the appearance of the curve labeled ($n_l - n_d$) in Figure 13-10. That is, the rotation undergoes rapid change to some maximum (or minimum) value, alters direction to a minimum (or maximum), and then finally reverts to values corresponding to normal dispersion. As indicated in Figure 13-10, a change in the sign of the rotation may accompany these changes.

If molecules have overlapping absorption peaks, as is usually the case, overlapping regions of anomalous dispersion lead to optical rotatory dispersion curves that are complex, as shown in Figure 13-11. Note that the ultraviolet absorption spectrum for the compound is also included for reference.

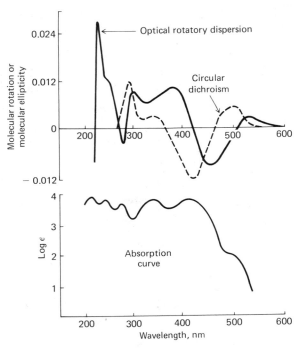

Figure 13-11. Optical rotatory dispersion, circular dichroism, and absorption curves for (+)-camphor trithione. Adapted from H. Wolf, E. Bunnenberg, C. Djerassi, A. Lüttringhaus, and A. Stockhausen, *Ann. Chem.*, **674**, 62 (1964). (With permission.)

Circular Dichroism Curves

In circular dichroism one of the circular components of a plane-polarized beam is more strongly absorbed than the other. The effect of this differential absorption is to convert the plane-polarized radiation to an elliptically polarized beam. Figure 13-12 illustrates how two circular components of unequal amplitude, which results from the differential absorption by a medium, are combined to give a resultant that travels in an elliptical path. The l component of the original beam is shown as retarded more than the d component because $n_l > n_d$; on the other hand, the amplitude of the d component is less than that of the l component because we have assumed that its molar absorptivity is greater; that is, $\epsilon_d > \epsilon_l$.

The angle of rotation α is taken as the angle between the major axis of the emergent elliptical beam and the plane of polarization of the incident beam. The *ellipticity* is given by the angle θ; the tangent of θ is clearly equal to the ratio of the minor axis of the elliptical path to the major (that is, OB/OA).

It can be shown that the ellipticity is approximated by the relation

$$\theta = \frac{1}{4}(k_l - k_d) \tag{13-4}$$

where k_l and k_d are the absorption coefficients[5] of the l and d circularly

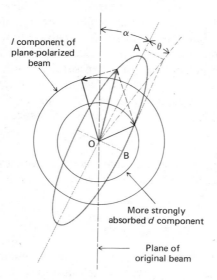

l component of plane-polarized beam

More strongly absorbed *d* component

Plane of original beam

Figure 13-12. Elliptically polarized radiation after leaving a medium in which $\epsilon_d > \epsilon_l$ and $n_l > n_d$.

[5]The absorption coefficient k is related to the more common molar absorptivity ϵ by the equation $k = 2.303\ \epsilon c$, where c is the concentration in moles per liter.

polarized radiation and θ is expressed in radians. The quantity $(k_l - k_d)$ is termed the *circular dichroism*. The molecular ellipticity $[\theta]$ can be shown to be equal to

$$[\theta] = 3305 \, (\epsilon_l - \epsilon_d) \tag{13-5}$$

where $[\theta]$ has the units of degree-cm² per decimole and the ϵ's are the molar absorptivities of the respective circular components.

Circular dichroism curves consist of a plot of $[\theta]$ versus wavelength. Note that $[\theta]$ can be either negative or positive depending on the relative magnitudes of ϵ_l and ϵ_d. The dotted line in Figure 13-11 is a typical curve.

Instrumentation

Optical rotatory dispersion. A number of recording spectropolarimeters are now manufactured that directly provide optical rotatory dispersion curves in the ultraviolet and visible regions. In these instruments radiation from a conventional monochromator is passed through a polarizer, the sample, an analyzer, and then to a photomultiplier tube. The signal from the detector is amplified; it is then employed to adjust the analyzer position to compensate for rotation caused by the sample and to position a recorder pen as well. As with a visual polarimeter, it is most efficient to use some modification of the half-shade method to determine the null position of the analyzer. In one instrument the polarizer is mechanically rocked through a small angle at low frequency. The amplifier system of the detector responds to the resulting ac signal and adjusts the analyzer until the signal is symmetric around the null point. Another spectropolarimeter employs an ordinary double-beam spectrometer with two sets of polarizer-analyzer prisms. The two analyzers are offset from one another by a few degrees, and both beams are passed through the sample. The ratio of the power of the two beams is compared electronically and gives a measure of the optical rotation of the sample.

Circular dichroism. A conventional spectrometer can be adapted to measure molecular ellipticity. From Beer's law, equation (13-5) can be written in the form

$$[\theta] = \frac{3305}{bc} \left(\log \frac{P_{l0}}{P_l} - \log \frac{P_{d0}}{P_d} \right)$$

where P_{l0} and P_l represent the power of the circularly polarized l beam before and after it has passed through a solution of length b and containing a molar concentration c of the sample. The terms P_d and P_{d0} have equivalent meanings for the d radiation. If now $P_{d0} = P_{l0}$, then

$$[\theta] = \frac{3305}{bc} \log \frac{P_d}{P_l} \tag{13-6}$$

Thus, the molecular ellipticity can be obtained directly by comparing the power of the two transmitted beams, provided the intensities of the incident *l* and *d* circularly polarized beams are identical.

In order to employ equation (13-6) with an ordinary spectrophotometer a device for producing *d* and *l* circularly polarized radiation must be provided. We have noted (p. 319) that circularly polarized radiation can be obtained by passing plane-polarized radiation through an anisotropic crystal which has a thickness such that the extraordinary and ordinary rays are one-quarter wavelength out of phase. Rotation of the optic axis of the quarter-wave plate by 90 deg yields either *d* or *l* circularly polarized radiation. In employing a single-beam spectrophotometer for circular dichroism a polarizer followed by a quarter-wave plate is inserted in the cell compartment of the instrument, provision being made to allow rotation of the plate by ± 45 deg. The sample cell is then placed between the plate and the detector; with the plate set to produce *d* circular radiation, the instrument is set on 100 percent transmission, or zero absorbance. The plate is then rotated 90 deg; the new absorbance reading then corresponds to log (P_d/P_l) for the sample. In order to cover a very wide spectral range several plates of different thickness must be employed.

Other methods exist for producing circularly polarized radiation. One of these, incorporated as an adapter for a double-beam spectrophotometer, is shown in Figure 13-13. Here a *Fresnel rhomb* is employed. When a polarized beam undergoes internal reflection in this device, one of the perpendicular components is retarded with respect to the other. The retardation depends upon the refractive index of the medium, the angle of incidence of the reflected beam, and the number of reflections. Quarter-wave retardation, and hence circular polarization, can be attained through proper adjustment of these variables.

The unit shown in Figure 13-13 is fitted into the sample compartment of a double-beam spectrophotometer, with a similar unit in the reference beam. The units are so adjusted that *d* radiation is emitted from one and *l* radiation from the other. The two beams then pass through identical cells containing the sample and their relative powers are compared photometrically.

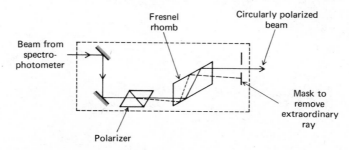

Figure 13-13. Spectrophotometer adapter for production of circularly polarized radiation.

APPLICATIONS OF OPTICAL ROTATORY DISPERSION AND CIRCULAR DICHROISM

Optical rotatory dispersion and circular dichroism studies often provide spectral details for optically active compounds that are absent in their ultraviolet spectra. Thus, in the lower plot in Figure 13-11, the absorption spectrum is seen to consist of a group of overlapping peaks that would be difficult to interpret. On the other hand, the molecular rotation and ellipticity curves for the optically active groups are much more clearly defined and lend themselves to detailed analysis.

Optical rotatory dispersion curves have been mainly applied to structural determination in two major areas: (1) amino acids, polypeptides, and proteins; and (2) complex natural products such as steroids, terpenes, antibiotics, and so forth. Most of the structural conclusions from this work are empirical, being based upon spectral observations of known structures. The curves can provide information concerning the configuration of angular substituents at ring junctures, the location of ketone groups, conformational analysis of substituents exerting a vicinal action on an optically active chromophore, the degree of coiling of protein helices, and the type of substitution in amino acids.

The applications of circular dichroism are less developed than optical rotatory dispersion; it appears, however, that the technique will also provide much useful structural information regarding organic and biological systems as well as metal-ligand complexes.

14 X-ray Methods

The analytical uses of electromagnetic radiation in the x-ray region are similar in many respects to the applications of ultraviolet and visible radiation, which were discussed in earlier chapters. Thus, methods based upon emission (including fluorescent emission), absorption, and diffraction of x-rays now find wide use in science and industry.[1]

FUNDAMENTAL PRINCIPLES

Before describing the applications of x-rays to analytical problems it is desirable to consider certain theoretical aspects that relate to the emission, the absorption, and the diffraction of radiation in the wavelength range between about 0.1 and 25 Å.

Emission of X-rays

For analytical purposes x-rays are obtained in three ways, namely: (1) by bombardment of a metal target with a beam of high-energy electrons, (2) by exposure of a substance to a primary beam of x-rays in order to generate a secondary beam of fluorescent x-rays, and (3) by employment of a radioactive source whose decay process results in x-ray emission.

[1] For a more extensive discussion of the theory and analytical applications of x-rays see H. A. Liebhafsky, H. G. Pfeiffer, E. H. Winslow and P. D. Zemany, *X-ray Absorption and Emission in Analytical Chemistry* (New York: John Wiley & Sons, Inc., 1960); also by the same authors in I. M. Kolthoff and P. J. Elving, Eds., *Treatise on Analytical Chemistry,* Part I, Vol. 5 (New York: Interscience Publishers, Inc., 1964), Chap. 60; and W. T. Sproull, *X-rays in Practice* (New York: McGraw-Hill Book Company, Inc., 1946).

X-ray sources, like ultraviolet and visible emitters, often produce both a continuous and a discontinuous (line) spectrum; both types are of importance in analysis.

Continuous spectra from electron beam sources. The most common source of x-rays is the *Coolidge tube,* which consists of a heated cathode and a massive, water-cooled anode called the target; the electrodes are contained in an evacuated glass or metal housing (see Figure 14-7). Electrons produced at the cathode are accelerated toward the anode by potentials as great as 100 kV, and upon striking the anode, produce x-rays. Under some conditions only a continuous spectrum such as that shown in Figure 14-1 results; under others, a line spectrum is superimposed upon the continuum (see Figure 14-2).

As may be seen in Figure 14-1, the continuous x-ray spectrum is characterized by a well-defined, short wavelength limit (λ_0) which is dependent upon the accelerating voltage V but independent of the target material. Thus, λ_0 for the spectrum produced with a molybdenum target at 35 kV (Figure 14-2) is identical to λ_0 for the tungsten target at the same voltage (Figure 14-1).

The sharp wavelength limit can be explained only by assuming that the photon corresponding to radiation of this wavelength results from conversion of the entire kinetic energy of an electron (Ve) to radiant energy. That is, the electron is instantly deaccelerated to zero kinetic energy. Thus,

$$hv_0 = \frac{hc}{\lambda_0} = Ve \qquad (14\text{-}1)$$

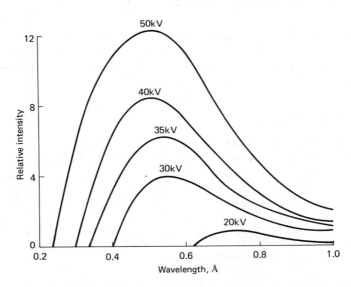

Figure 14-1. Continuous x-radiation at various voltages from a Coolidge tube with a tungsten target.

where V is the accelerating voltage, e is the charge on the electron, λ_0 is the frequency of the limiting radiation, and h and c have their usual meanings. This relationship is known as the *Duane-Hunt law*. Upon substituting numerical values for the constants and rearranging, equation (14-1) becomes

$$\lambda_0 = 12{,}393/V$$

where λ_0 and V have units of Ångstroms and volts. It is of interest that the substitution of experimental values for λ_0 and V in equation (14-1) has provided a direct means for the highly accurate determination of Planck's constant h.

In order to account for the longer wavelength portion of the continuous spectrum it is necessary to assume that the bombarding electrons lose different amounts of their energy when they strike an atom of the target. Thus, even if all of the electrons are initially monoenergetic, some are instantly and totally decelerated as a result of collision with an atom; others undergo a series of collisions, losing energy in steps corresponding to $h\nu$, where $\nu < \nu_0$. A wave-mechanical treatment of the processes responsible for the continuous x-ray spectrum is difficult, but nevertheless has provided relationships that agree reasonably with experimental observations.

The production of x-rays by electron bombardment is a highly inefficient process. Less than one percent of the electrical power is converted to radiant power, the remaining being degraded to heat. As a consequence, special cooling of x-ray tubes is required.

Characteristic line spectra from electron beam sources. As shown in Figure 14-2, bombardment of a molybdenum target produces intense emis-

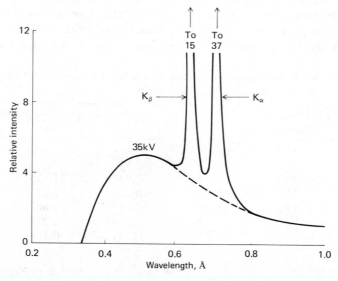

Figure 14-2. Line spectrum for tube with a molybdenum target.

sion lines at about 0.63 and 0.71 Å. If the spectrum for this element is examined at longer wavelengths, an additional simple series of lines is found in the range of 4 to 6 Å.

The emission behavior of molybdenum is typical of all elements having atomic numbers larger than 23; that is, the x-ray spectra are very simple when compared with ultraviolet emission and consist of two series of lines.[2] The lower-wavelength group is called the K series and the other the L series. Elements with atomic numbers smaller than 23 produce only a K series. Table 14-1 presents wavelength data for the emission spectra of a few elements.

A second characteristic of x-ray spectra is that the minimum acceleration voltage required for the excitation of the lines of each element increases with atomic number. Thus, the line spectrum for molybdenum (atomic number = 42) disappears if the excitation voltage drops below 20 kV. As shown in Figure 14-1, bombardment of tungsten (atomic number = 74) produces no lines in the region of 0.1 to 1.0 Å even at 50 kV; if the voltage is raised to 70 kV, however, characteristic K lines appear at 0.18 and 0.21 Å.

Figure 14-3 illustrates the linear relationship between the square root of the frequency for a given (K or L) line and the atomic number of the element responsible for the radiation. This property was first discovered by H. G. S. Moseley in 1914.

X-ray line spectra result from electronic transitions that involve the innermost atomic orbitals. The short wavelength K series is produced when the high-energy electrons from the cathode remove electrons from those orbitals nearest to the nucleus of the target atom (in x-ray terminology, the orbital of principal quantum number $n = 1$ is called the K shell; the orbital of quantum number $n = 2$ is called the L shell, and so forth). The collision results in the formation of an excited *ion* which then loses quanta of x-radiation as electrons from outer orbitals undergo transitions to the vacated orbital. As shown in Figure 14-4, the lines in the K series involve elec-

Table 14-1 Wavelengths in Ångstrom Units of the More Intense Emission Lines for Some Typical Elements

Element	Atomic Number	K Series		L Series	
		α_1	β_1	β_1	α_1
Na	11	11.909	11.617	—	—
K	19	3.742	3.454	—	—
Cr	24	2.290	2.085	21.323	21.714
Rb	37	0.926	0.829	7.075	7.318
Cs	55	0.401	0.355	2.683	2.892
W	74	0.209	0.184	1.282	1.476
U	92	0.126	0.111	0.720	0.911

[2] For the heavier elements, additional series of lines are found at larger wavelengths. The line intensities are low, however, and little use is made of them.

tronic transitions between higher-energy levels and the K shell. The L series of lines results when an electron is lost from the second principal quantum level either as a consequence of ejection by an electron from the cathode or from the transition of an L electron to the K level that accompanies the production of a quantum of K radiation. It is important to appreciate that the energy scale in Figure 14-4 is logarithmic. Thus, the energy difference between the L and K levels is significantly larger than that between the M and L levels. The K lines therefore appear at shorter wavelengths. It is also important to note that the energy differences between the transitions labeled α_1 and α_2 as well as those between β_1 and β_2 are so small that only a single line is observed in all but the highest resolution spectrometer. Thus in Figure 14-2 the K_α and K_β lines appear as singlets.

The energy level diagram in Figure 14-4 would be applicable to any element with sufficient electrons to permit the number of transitions shown. The differences in energies between the levels increase regularly with atomic number because of the increasing charge on the nucleus; thus, the radiation for the K series appears at shorter wavelengths for the heavier elements (see Table 14-1). The effect of nuclear charge is also reflected in the increase in the minimum voltage required to excite the spectra of these elements.

It is important to note that, because x-ray line spectra result from electronic transitions of the innermost electrons only, the wavelengths of the characteristic lines are essentially independent of chemical combination in all elements other than those with the lowest atomic weights. Thus, the position of the K_α lines for molybdenum is the same regardless of whether the target is the pure metal, its sulfide, or its oxide.

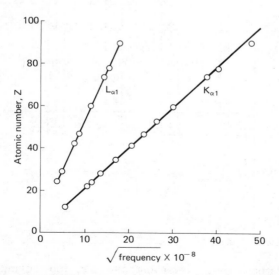

Figure 14-3. Relationship between x-ray emission frequency and atomic number (K_{α_1} and L_{α_1} lines).

Fluorescent line spectra. Another convenient way of producing a line spectrum is to irradiate the element or one of its compounds with the continuous radiation from an x-ray tube. This process is considered in detail in a later section.

Radioactive sources. X-radiation occurs in two radioactive decay processes. *Gamma rays,* which are indistinguishable from x-rays, owe their production to intranuclear reactions. *Electron capture* or *K capture* also produces x-radiation. The process involves capture of a K electron (less commonly an L or an M electron) by the nucleus, and formation of an element of the next lower atomic number. As a result of K capture, elec-

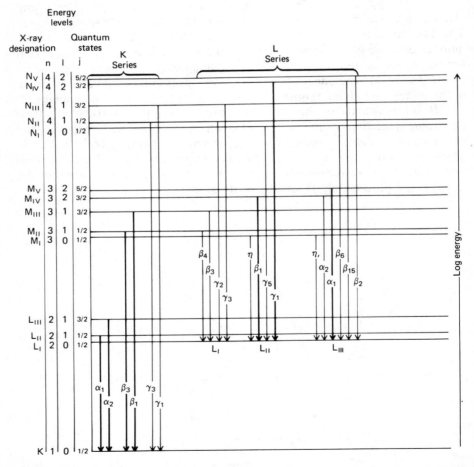

Figure 14-4. Partial energy level diagram showing common transitions leading to x-radiation. The most intense lines are indicated by the widest arrows.

tronic transitions to the vacated orbital occur, and the x-ray line spectrum of the newly formed element is observed. The half-lives (p. 386) of K-capture processes range from a few minutes to several thousands of years.

Artificially produced radioactive isotopes provide a very simple source of monoenergetic radiation for certain analytical applications. The best-known example is iron-55, which undergoes a K-capture reaction with a half-life of 2.6 years:

$$^{55}\text{Fe} \rightarrow {}^{54}\text{Mn} + h\nu$$

The resulting manganese K_α line at about 2.1 Å has proved useful for absorption analyses.

Absorption of X-rays

When a narrow beam of x-rays is passed through a thin layer of matter, its intensity or power is generally diminished as a consequence of absorption and of scattering. Ordinarily, the effect of scattering is small and can be neglected in those wavelength regions where appreciable absorption occurs. As shown in Figure 14-5, the absorption spectrum of an element, like its emission spectrum, is simple and consists of a few broad, but well-defined, absorption peaks. Here again, the wavelengths of the peaks are characteristic of the element, and are largely independent of its chemical state.

A peculiarity of x-ray absorption spectra is the appearance of sharp discontinuities, called *absorption edges,* at wavelengths immediately beyond absorption maxima.

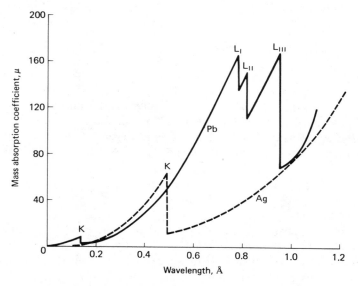

Figure 14-5. X-ray absorption spectrum for lead and silver.

The absorption process. Absorption of an x-ray quantum causes ejection of one of the innermost electrons from an atom and the consequent production of an excited ion. In this process the entire energy $h\nu$ of the radiation is partitioned between the kinetic energy of the electron (the *photoelectron*) and the potential energy of the excited ion. The highest probability of absorption arises when the energy of the quantum is exactly equal to the energy required to remove the electron just to the periphery of the atom (that is, the kinetic energy of the ejected electron approaches zero).

The absorption spectrum for lead, shown in Figure 14-5, exhibits four peaks, the first occurring at 0.14 Å. The energy of the quantum corresponding to this wavelength exactly matches the energy required to cause ejection of the highest-energy K electron of lead; immediately beyond this wavelength the radiant-energy quantum is insufficient to bring about removal of a K electron and an abrupt decrease in absorption occurs. At wavelengths lower than 0.14 Å the probability of interaction between the electron and the radiation diminishes, and results in a smooth decrease in absorption. In this region the kinetic energy of the photoelectron increases continuously with the decrease in wavelength.

The additional peaks at longer wavelengths correspond to the removal of an electron from the L energy levels of lead. Since three sets of L levels, differing slightly in energy, exist (see Figure 14-4), three peaks are observed. Another set of peaks, arising from interactions involving M electrons, will be located at still longer wavelengths.

The mass absorption coefficient. Beer's law is as applicable to the absorption of x-radiation as to other types of electromagnetic radiation; thus,

$$\ln \frac{P_0}{P} = \mu_1 \, x$$

where x is the sample thickness in centimeters and P and P_0 are the powers of the transmitted and incident beams. The constant μ_1 is called the *linear absorption coefficient* and is characteristic of the element as well as the number of its atoms in the path of the beam. A more convenient form of Beer's law is

$$\ln \frac{P_0}{P} = \mu \, \rho \, x \qquad (14\text{-}2)$$

where ρ is the density of the sample and μ is the *mass absorption coefficient,* a quantity that is *independent* of the physical and chemical states of the element. Thus, the mass absorption coefficient for bromine has the same value in gaseous HBr as in solid sodium bromate.

It has been found empirically that the mass absorption coefficient for an element can be approximated from the relationship

$$\mu = \frac{C \, N \, Z^4 \, \lambda^3}{A} \qquad (14\text{-}3)$$

where N is Avogadro's number, A is the atomic weight of the element, Z is its atomic number, and λ is the wavelength of radiation in Å. The quantity C is a constant for all elements within a limited wavelength region.

Mass absorption coefficients are additive functions of the weight fractions of elements contained in a sample. Thus,

$$\mu_M = W_A \, \mu_A + W_B \, \mu_B + W_C \, \mu_C + \cdots \qquad (14\text{-}4)$$

where μ_M is the mass absorption coefficient of a sample containing the weight fractions W_A, W_B, and W_C of elements A, B, and C. The terms μ_A, μ_B, and μ_C are the respective mass absorption coefficients for each of the elements.

X-ray Fluorescence

The absorption of x-rays produces electronically excited ions that return to their ground state by transitions involving electrons from higher-energy levels. Thus, an excited ion with a vacant K shell is produced when lead absorbs radiation of wavelengths shorter than 0.14 Å; after a brief period the ion returns to its ground state via a series of electronic transitions characterized by the emission of x-radiation (fluorescence) of wavelengths identical to those that result from excitation produced by electron bombardment. The wavelengths of the fluorescent lines are always somewhat greater than the wavelength of the corresponding absorption edge, however, because absorption requires a complete removal of the electron (that is, ionization), whereas emission involves transitions of an electron from a higher-energy level within the atom.

Diffraction of X-rays

In common with other types of electromagnetic radiation, interaction between the electric vector of x-radiation and the electrons of the matter through which it passes results in scattering. When x-rays are scattered by the ordered environment in a crystal, interference (both constructive and destructive) takes place among the scattered rays because the distances between the scattering centers is of the same order of magnitude as the wavelength of the radiation. Diffraction is the result.

Bragg's law. When an x-ray beam strikes a crystal surface at some angle θ, a portion is scattered by the layer of atoms at the surface. The unscattered portion of the beam penetrates to the second layer of atoms where again a fraction is scattered and the remainder passes on to the third layer. The cumulative effect of this scattering from the regularly spaced centers of the crystal is a diffraction of the beam in much the same way as visible radiation is diffracted by a reflection grating (p. 44). The requirements for diffraction are: (1) the spacing between layers of atoms must be roughly the same as the wavelength of the radiation and (2) the scattering centers must be spatially distributed in a highly regular way.

W. L. Bragg treated the diffraction of x-rays by crystals as shown in Figure 14-6. Here a narrow beam strikes the crystal surface at angle θ; scattering occurs as a consequence of interaction of the radiation with atoms located at O, P, and R. If the distance

$$AP + PC = n\lambda$$

where $\mathbf{n}$ is an integer, the scattered radiation will be in phase at OCD, and the crystal will appear to reflect the x-radiation. But it is readily seen that

$$AP = PC = d \sin \theta$$

where d is the interplanar distance of the crystal. Thus, we may write that the conditions for constructive interference on the beam at angle θ are

$$\mathbf{n}\lambda = 2d \sin \theta \qquad \textit{Bragg eq} \qquad (14\text{-}5)$$

Equation (14-5) is called the *Bragg equation* and is of fundamental importance. Note that x-rays appear to be reflected from the crystal only if the angle of incidence satisfies the condition that

$$\sin \theta = \mathbf{n}\lambda/2d$$

At all other angles destructive interference occurs.

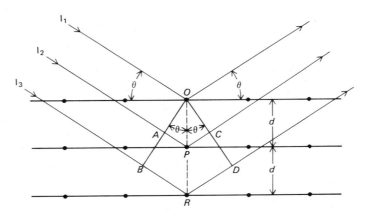

Figure 14-6. Diffraction of x-rays by a crystal.

X-RAY INSTRUMENTATION

The absorption, emission, and diffraction of x-rays all find applications in analytical chemistry. Instruments for each of these applications require a source and a detector of x-radiation; often, in addition, a device for restricting the radiation to a narrow and known wavelength region is needed.

X-ray Tubes

The most common source of x-rays for analytical work is the Coolidge tube, which can take a variety of shapes and forms. Basically, however, it is a highly evacuated tube in which is mounted a tungsten filament cathode and a massive anode constructed of tungsten, copper, molybdenum, chromium, silver, nickel, cobalt, or iron (Figure 14-7). Separate circuits are used to heat the filament and to accelerate the electrons to the target. The former provides the means for controlling the intensity and the latter, the energy of the x-rays that are emitted. The Coolidge tube is normally self-rectifying, and a high-voltage ac source is connected directly to the cathode to provide the accelerating potential.

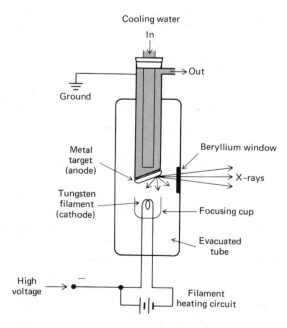

Figure 14-7. Schematic diagram of the Coolidge tube.

X-ray Detection and Measurement

The detection and determination of the relative intensity of x-radiation is accomplished in three different ways: (1) by the darkening of a photographic film or plate, (2) by measurement of ion currents produced when the beam is absorbed by a gas, and (3) by measurement of the visible radiation produced when the beam strikes a suitable phosphor.

Photographic detectors.   X-rays affect a photographic emulsion in much the same way as ordinary light. Thus, the photographic techniques discussed in Chapter 10 are applicable to x-ray spectroscopy as well.

Gas-ionization detectors.   Gas-ionization detectors are widely used for the detection and measurement of both x-radiation and the radiation produced by the decay of radioactive isotopes. These detectors are of three types: (1) *ionization chambers*; (2) *proportional detectors*; and (3) *Geiger-Müeller*, or *Geiger, tubes*; each is described in Chapter 16.

Scintillation counters.   A number of crystalline substances fluoresce in the ultraviolet and visible regions when exposed to x-radiation. This phenomenon has long been used in radiology to observe visually the presence of absorbing bodies in media that are transparent to x-rays (portions of the body, for example). A *scintillation counter* employs a photomultiplier tube to detect and measure the fluorescence produced by x-rays (or, for that matter, by the radiation emanating from a radioactive source). Scintillation counters are considered in detail in Chapter 16.

Filters for X-ray Beams

In many applications it is desirable to employ an x-ray beam that is restricted in its wavelength range. As in the visible region, both filters and monochromators are employed to provide radiation of limited band width.

Figure 14-8 illustrates a common technique for producing a relatively pure monochromatic beam by use of a filter. Here the K_β line and most of the continuous radiation from the emission of a molybdenum target is removed by a zirconium filter having a thickness of about 0.01 cm. The pure K_α line is then available for analytical purposes. Several other target-filter combinations of this type have been developed, each of which serves to isolate one of the intense lines of a target element. Monochromatic radiation produced in this way is widely used in x-ray diffraction studies. The choice of wavelengths available by this technique is limited by the relatively small number of target-filter combinations that exist.

Filtration of the continuous radiation from a Coolidge tube is also

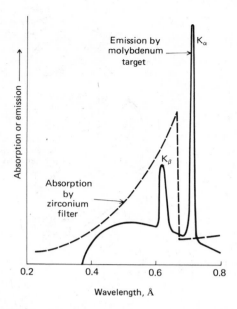

Figure 14-8.   Use of a filter to produce monochromatic radiation.

feasible with thin strips of metal. As with glass filters for visible radiation, relatively broad bands are produced with a significant loss in intensity of the desired wavelengths.

X-ray Monochromators

Figure 14-9 shows the essential components of an x-ray monochromator. The dispersing element is a single crystal mounted on a device (a *goniometer*) that permits the accurate determination of the angle of its face with respect to the incident beam and the detector. Radiation emerges from the monochromator at an angle that is twice the angle of incidence θ. Thus, as the spectrum is scanned it is necessary to rotate the detector at twice the angular velocity of the crystal. Generally it is the angle of the detector with respect to the incident beam (2θ) that is recorded and subsequently related to wavelength by means of the Bragg equation; clearly, the interplanar spacing d for the crystal must be known.

Beam collimation.   In order to collimate the diverging beam from the x-ray source a series of closely spaced metal tubes or plates that absorb all but the parallel beams is imposed in the beam path.

The loss of intensity is high in a monochromator employing a flat crystal because as much as 99 percent of the radiation is sufficiently divergent to be absorbed in the collimators. Increased intensities, by as much as a factor of ten, have been realized by employing a curved crystal surface

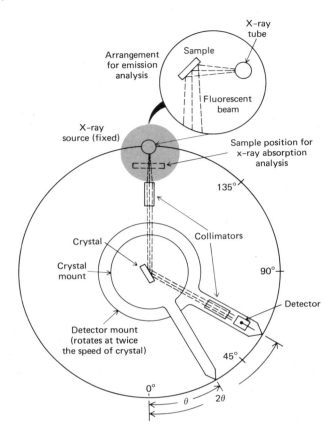

Figure 14-9. An x-ray monochromator and detector. Note that the angle of the detector with respect to the beam is always twice that of the crystal face. For absorption analysis, the source is an x-ray tube and the sample is located in the beam as shown. For emission work, the source and sample are arranged as shown in the insert.

which acts not only to diffract but also to focus the divergent beam from the source upon an exit slit.

Crystals for diffraction. As illustrated in Table 14-1, most analytically important x-ray lines lie in the region between about 0.1 and 10 Å. A consideration of the data in Table 14-2, however, leads to the conclusion that no one single crystal satisfactorily disperses radiation over this entire range. As a consequence, an x-ray monochromator must be provided with at least two (and preferably more) interchangeable crystals.

The useful wavelength range for a crystal is determined by its lattice spacing d and the problems associated with the detection of the radiation at values of 2θ that approach zero or 180 deg. When a monochromator is

Table 14-2 Properties of Typical Diffracting Crystals

Crystal	Lattice Spacing d, Å	Wavelength Range[a], Å		Dispersion $d\theta/d\lambda$, deg/Å	
		λ_{max}	λ_{min}	at λ_{max}	at λ_{min}
Topaz	1.356	2.67	0.24	2.12	0.37
LiF	2.014	3.97	0.35	1.43	0.25
NaCl	2.820	5.55	0.49	1.02	0.18
EDDT[b]	4.404	8.67	0.77	0.65	0.11
ADP[c]	5.325	10.50	0.93	0.54	0.09

[a]Based on assumption that the measurable range of 2θ is from 160 deg for λ_{max} to 10 deg for λ_{min}.
[b]Ethylenediamine d-tartrate.
[c]Ammonium dihydrogen phosphate.

set at angles of 2θ that are much less than 10 deg, the amount of polychromatic radiation scattered from the surface becomes prohibitively high; this consideration then was used to approximate the minimum wavelength for the crystals shown in Table 14-2. Generally, values of 2θ greater than about 160 deg cannot be measured because the location of the source unit prohibits positioning of the detector at such an angle (see Figure 14-9). The values for λ_{max} in Table 14-2 were determined from this consideration.

It will be seen from Table 14-2 that a crystal such as ammonium dihydrogen phosphate, with a large lattice spacing, has a much greater wavelength range than a crystal in which this parameter is small. The advantage of larger values of d is offset, however, by the consequent lower dispersion. This effect can be seen by differentiation of equation (14-5), which leads to

$$\frac{d\theta}{d\lambda} = \frac{n}{2d \cos \theta}$$

Here $d\theta/d\lambda$ is a measure of dispersion and is seen to be inversely proportional to d. Table 14-2 has dispersion data for the various crystals at their maximum and minimum wavelengths. The low dispersion of ammonium dihydrogen phosphate prohibits its use in the region of low wavelengths; here a crystal such as topaz or lithium fluoride must be substituted.

Nondispersive X-ray Spectrometers

It will be shown in Chapter 16 that gamma ray and x-ray detectors are ordinarily operated so that an incident photon produces a pulse of electrical current; a count of the number of pulses per unit time thus gives a measure of the power of the beam. With a certain class of these detectors, called *proportional counters*, the number of electrons per pulse (the *peak height*) is directly proportional to the energy of the photon—that is, to $h\nu$. Thus, the peak height of a pulse produced by a 1.00 Å photon (3×10^{18} Hz) is twice the peak height for a 2.00 Å photon (1.5×10^{18} Hz).

Proportional counters provide a means by which x-ray fluorescence lines from different elements can be separated and measured without the use of a crystal monochromator. An electronic circuit that can discriminate between pulses of different peak heights is employed. Such a device, called a *pulse-height analyzer* (p. 396), counts only those pulses lying within a predetermined height range. In a multichannel analyzer, pulses in several different ranges are counted simultaneously. A pulse-height analyzer then is a kind of spectrometer in which wavelength selection is accomplished electronically rather than with a crystal.

As shown in Figure 14-10 a nondispersive x-ray spectrometer consists of a polychromatic source, a sample holder, a proportional detector, and a pulse-height analyzer. The source is either a Coolidge tube or a small amount of a radioactive material that produces x-rays in its decay. The detector may be a proportional or a scintillation counter tube, or a semiconductor detector.

Nondispersive systems offer the advantages of simplicity and relatively low cost when compared with the conventional crystal monochromator. Furthermore, radiation losses are several orders of magnitude lower, not only because collimation is unnecessary but also because air, which absorbs longer wavelengths, is readily eliminated from the short beam path (see Figure 14-10). The main disadvantage of the nondispersive system is its poorer resolution, which results in interference by elements with atomic numbers that lie close to the number of the element being determined. Marked improvements in the resolving power of pulse height analyzers have occurred in recent times; it seems probable that the future will see an ever-increasing use of nondispersive equipment in lieu of crystal monochromators.

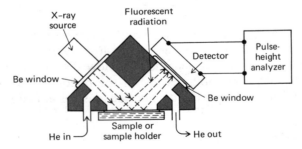

Figure 14-10. Nondispersive x-ray fluorescence apparatus.

X-RAY EMISSION METHODS

The last two decades have seen a growing use of x-ray emission for the qualitative, semiquantitative, and quantitative determination of the elements in a variety of naturally occurring and manufactured products.

In its applications this technique is perhaps most closely related to optical emission analysis, and it is of interest to compare the two. An x-ray emission spectrum is usually simpler and thus easier to interpret for qualitative identification purposes. On the other hand, the light elements (atomic number < 20) are either not detectable or are, at best, difficult to detect. Both optical and x-ray methods are rapid, and semiquantitative estimates of concentration can be made readily from a spectrum obtained for qualitative purposes; the x-ray method is somewhat more reliable in this application. For quantitative work the x-ray method appears capable of yielding greater accuracy under comparable conditions, with uncertainties on the order of 1 to 2 percent relative often being reported. A particular advantage of the x-ray method is that it is nondestructive of the sample. On the other hand, it is generally less sensitive than optical emission methods; although sensitivities in the parts-per-million range have been reported for x-ray emission under ideal conditions, the technique appears to be better applied at concentrations above 0.01 to 0.1 percent.

Equipment and Spectral Excitation

A spectrometer such as that shown in Figure 14-9 is employed for x-ray emission analysis. For convenience, the crystal and detector mounts are ordinarily driven mechanically and their motions are synchronized with the chart drive of a recorder; the detector signal is then fed to the pen drive, thus providing a spectrum automatically.

Sample excitation. Although it is feasible to excite the x-ray spectrum of a sample by incorporating the substance into the target area of an x-ray tube, the inconvenience of this technique discourages its general use. Instead, excitation is brought about by irradiation of the sample with a beam of x-rays from a Coolidge tube. Under these circumstances the elements in the sample that are excited by the radiation from the tube emit their own characteristic fluorescent x-rays. This method is often called *x-ray fluorescence* analysis because of the nature of the excitation process; *x-ray emission* is also a proper description.

In order to assure excitation of the heavier elements, a source that produces a continuous spectrum down to relatively short wavelengths is generally employed; a tube with a tungsten target operated at 50 to 100 kV ordinarily suffices. There is no need for the excitation beam to be monochromatic. The typical excitation arrangement is shown in the inset in the upper part of Figure 14-9.

Samples may be in the form of metals, powdered solids, evaporated films, pure liquids, or solutions. Where necessary, the materials are held in a cell with a plastic or a cellophane window. The amount of sample required can be quite small.

Qualitative and Semiquantitative Analysis

Figure 14-11 illustrates an interesting qualitative application of the x-ray method. Here, the untreated sample was excited in the x-ray beam by irradiation; it was subsequently recovered unchanged. Note that the abscissa is plotted in terms of the angle 2θ, which can be readily converted to wavelength with knowledge of the crystal spacing of the monochromator (equation 14-5). Identification of peaks is then accomplished by reference to tables for emission lines of the elements.

Qualitative information, such as that shown in Figure 14-11, can be converted to semiquantitative data by the careful measurement of peak heights. This measurement ordinarily involves setting the goniometer to the appropriate angle 2θ and then counting for a suitable length of time. To obtain a rough estimate of concentration the following relationship is used:

$$P_x = P_s W_x \qquad (14\text{-}6)$$

where P_x is the relative line intensity measured in terms of number of counts for a fixed period and W_x is the weight fraction of the element in the sample. The term P_s is the relative intensity of the line that would be observed under identical counting conditions if W_x were unity. The value of P_s is determined by employing a sample of the pure element or a standard sample of known composition with respect to the element of interest.

The use of equation (14-6) as outlined in the previous paragraph carries with it the assumption that the emission from the species of interest is unaffected by the presence of other elements in the sample. This assumption may not be justified; as a consequence, a concentration estimate based upon equation (14-6) may be in error by a factor of two or more. On the other hand, this uncertainty is significantly smaller than that associated

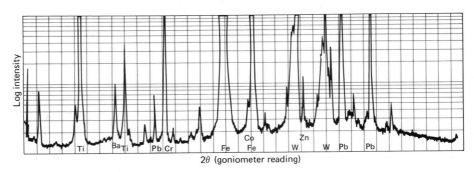

Figure 14-11. X-ray emission spectrum recorded for a genuine bank note. Taken from H. A. Liebhafsky, H. G. Pfeiffer, E. H. Winslow, and P. D. Zemany, *X-ray Absorption and Emission in Analytical Chemistry* (New York: John Wiley & Sons, Inc., 1960), p. 163. (With permission.)

with a semiquantitative analysis by optical emission where an order-of-magnitude error is not uncommon.

Quantitative Analysis

Nonlinearity of the concentration-intensity relationship. It is important to realize that the x-rays produced in the fluorescence process are generated not only from atoms at the surface of a sample but also from atoms well below the surface. Thus, a part of both the incident beam and the resulting fluorescent beam traverse a significant thickness of sample, within which absorption and scattering of either can occur. The extent to which either beam is attenuated depends upon the mass absorption coefficient of the medium, which in turn is determined not only by the mass absorption coefficient of the element excited but also by the coefficients of all of the other elements in the sample. Thus, while the net intensity of a line reaching the detector in an x-ray emission analysis depends upon the concentration of the element producing the line, it is dependent upon the concentration and mass absorption coefficients of the matrix elements as well.

Absorption effects by the matrix may cause results calculated by equation (14-6) to be either high or low. If, for example, the matrix contains a significant amount of an element that absorbs either the incident or the emitted beam more strongly than the element being determined, then W_x will be low since P_s was evaluated with a standard in which absorption was smaller. On the other hand, if the matrix elements of the sample absorb less than those in the standard, high values for W_x result.

A second matrix effect, called the *enhancement effect*, can also yield results that are greater than expected. This behavior is encountered when the sample contains an element whose characteristic emission spectrum is excited by the incident beam, and this spectrum in turn causes a secondary excitation of the analytical line.

Absorption and enhancement effects can clearly cause the intensity of an analytical line to depend not only upon the concentration of the element of interest but also, to a lesser degree, upon the concentrations of the various elements comprising the sample matrix. Several techniques have been developed to compensate for such effects.

Calibration against standards. Here the relationship between the analytical line intensity and the concentration is determined empirically with a set of standards that closely approximate the samples in overall composition. The assumption is then made that absorption and enhancement effects are identical for both samples and standards, and the empirical data are employed to convert emission data to concentrations. Clearly, the degree of compensation achieved in this way depends upon the closeness of the match between the samples and the standards.

Use of internal standards. In this procedure an element, absent in the samples, is introduced in known and fixed concentration into both the calibration standards and the samples. The ratio of the line intensity of the element being determined to the intensity of a line of the internal standard serves as the analytical parameter. The assumption here is that absorption and enhancement effects are the same for the two lines and that use of intensity ratios compensates for these effects.

Dilution of sample and standards. Here, both sample and standards are diluted with a substance that absorbs x-rays only weakly (that is, a substance comprised of elements with low atomic numbers). Examples of such diluents include water; organic solvents containing carbon, hydrogen, oxygen, and nitrogen only; starch; lithium carbonate; alumina; and boric acid or borate glass. By employing an excess of diluent, the matrix effect becomes essentially constant for both standards and samples, and adequate compensation is achieved. This procedure has proved particularly useful for mineral analyses, where the samples and standards are dissolved in molten borax; the fused mass is then excited in the usual way.

Analytical application of x-ray emission. Speed and relatively good precision make x-ray fluorescence spectroscopy a useful tool for quality control in the manufacture of metals and alloys. The method has been widely used for mineral analysis as well, and has proved valuable for the determination of the closely related constituents of rare-earth minerals.

X-ray emission methods are readily adapted to liquid samples. Thus, for example, methods for the direct quantitative determination of lead and bromine in aviation gasoline samples have been developed. Similarly, calcium, barium, and zinc have been determined in lubricating oils by excitation of fluorescence in liquid samples. The method is also convenient for the direct determination of the pigments in paint samples.

X-RAY ABSORPTION METHODS

The application of x-ray absorption to analysis is more limited than that of x-ray emission or of optical absorption primarily because of the probability that absorption spectra for various elements will overlap.

Direct Absorption Methods

An x-ray absorption analysis can be performed in much the same way as an optical absorption determination. It is ordinarily advantageous to employ a monochromatic beam, although absorption of polychromatic radiation has also proved useful in some applications.

Equipment. Monochromatic beams for absorption methods can be achieved with a monochromator such as that shown in Figure 14-9; alternatively, filtration of the output from a Coolidge tube has been employed (Figure 14-8). Finally, in some applications the x-rays produced by the radioactive decay of certain species has proved to be a convenient and inexpensive source of monochromatic x-radiation.

Applications. The direct absorption method is largely limited to the analysis of samples containing a single heavy element in a matrix of lighter elements. The reason for this limitation can be seen from equation (14-4), which shows that the absorption of a mixture at any given wavelength is sensitive to variations in concentrations of *all* of the elements present. Thus, the only condition under which the measurement of μ_M in equation (14-4) reflects the concentration of just one of the elements, say A, in a mixture is when $\mu_A W_A$ is much larger than $\mu_B W_B$, $\mu_C W_C$, and so forth; from the practical standpoint, then, it is necessary that $\mu_A \gg \mu_B, \mu_C, \cdots$. An examination of equation (14-3) indicates that this condition exists only when the atomic number of A is significantly larger than all of the rest.

An interesting example of the direct absorption method is the analysis of sulfur in hydrocarbons. Here the K_α line from a molybdenum target ($\lambda = 0.71$) can be used as a source of monochromatic radiation (see Figure 14-8). A more convenient and experimentally less complicated alternative is to employ the radiation resulting from the decay of iron-55 (p. 341).

Applying equation (14-4) to this system we find

$$\mu_M = W_C \mu_C + W_H \mu_H + W_S \mu_S \tag{14-7}$$

where the subscripts refer to the three elements. In the typical hydrocarbon sample, the ratio r of carbon to hydrogen concentrations is not affected by the sulfur content. Thus,

$$r = W_C/W_H \tag{14-8}$$

We can also write

$$W_S + W_C + W_H = 1 \tag{14-9}$$

By combining equations (14-7), (14-8), and (14-9), we obtain

$$\mu_M = \left(\mu_S - \frac{r \mu_C + \mu_H}{1 + r}\right) W_S + \frac{r \mu_C + \mu_H}{1 + r} \tag{14-10}$$

or

$$\mu_M = a W_S + b \tag{14-11}$$

where a and b are constants that depend upon the known mass absorption coefficients for the three elements and the carbon-hydrogen ratio r. Clearly, a linear relationship exists between W_S and the measured quantity μ_M. The slope a and intercept b of this line can be determined either by measurements of standard samples or from the literature values for μ_C, μ_S, and μ_H. In either case equation (14-11) provides a simple means for determining the concentration of sulfur from μ_M. (μ_M can be evaluated with equation 14-2.)

Additional applications of the direct absorption procedure include the determination of lead in gasoline. Here, the absorption is affected by the sulfur as well as by the lead concentration; thus, a separate determination for sulfur is required in order to provide a correction for the effects of this element.

The direct procedure has also proved useful for the analysis of chlorine in organic compounds.

Analysis of lighter elements. X-ray emission analysis for elements with atomic numbers lower than about 23 is more difficult than that for the heavier elements primarily because of several factors that tend to reduce the power of the fluorescent radiation reaching the detector. For one, the wavelengths of the K_α lines for these elements are all sufficiently long (> 2.7 Å) so that absorption of the emitted radiation by air occurs; as a consequence, the components of the monochromator and detector must be blanketed with a nonabsorbing gas such as hydrogen or helium. Attenuation of longer wavelengths by the windows of the source and detector units also becomes serious and requires the use of very thin and fragile sheets of Mylar or beryllium. As shown in Table 14-2, satisfactory dispersion of radiation in the 5 to 10 Å range requires a crystal having a large interplanar spacing; unfortunately, however, radiation losses with this type of crystal are great. Finally, the fluorescent yield for elements of low atomic number is generally small; thus, the power of the emitted radiation is low compared with that for a heavier element.

Nondispersive spectrometers are particularly useful for the x-ray determination of the lighter elements because, as we have mentioned earlier, radiant power losses are significantly less with this type of equipment.

The Absorption Edge Method

The absorption edge method largely avoids the matrix effects that influence the results of both the direct absorption method and the x-ray emission method.

Basis of the method.[3] In the absorption edge method the *change* in the mass absorption coefficient of a sample at an absorption edge is employed as a measure of the concentration of the element responsible for the edge. The difference in mass absorption coefficients on either side of the absorption edge is large (see Figure 14-5) when compared with the typical absorption change over the same wavelength region in the absence of an

[3]See H. W. Dunn, *Anal. Chem.,* **34**, 116 (1962).

edge. Thus, the difference parameter is not greatly affected by matrix variations except in the very unlikely situation where the matrix contains some other element that also has an absorption edge in the small wavelength region under study.

Method. The absorption edge method can be performed in a number of ways, one of which is illustrated by Figure 14-12. Here, the analysis of lead is based upon its L_{III} edge ($\lambda = 0.95$ Å). In order to bracket this edge, the absorption of the sample is measured with radiation of 0.93 and 1.04 Å provided by the fluorescence induced in a sample of sodium bromide. That is, the fluorescent bromine spectrum is excited with the source arrangement employed for emission analysis (see upper part of Figure 14-9). This radiation is then passed through the sample and into a monochromator arranged to measure first the intensity of the K_α line of bromine and then the intensity of the K_β line. In this way the difference in absorption coefficient for the sample can be determined on either side of the lead edge; these data are then readily corrected to give the difference in absorption at the wavelength of the edge. This difference is directly proportional to the concentration of the element and is essentially independent of the matrix. The high intensity and narrow band widths associated with bromine fluorescence provide distinct advantages over the direct use of the Coolidge tube as the source.

Suitable fluorescent sources for bracketing the absorption edges of

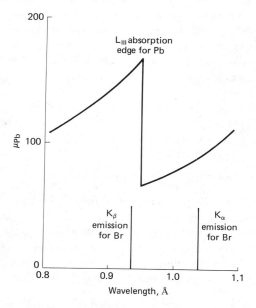

Figure 14-12. Data for the absorption edge analysis of lead. Absorption measurements are made with the K_α and K_β lines of bromine.

some 40 elements have been suggested. Application of the technique to the determination of 10 elements has shown that these analyses can be performed without recourse to standards of any kind; that is, proportionality constants for the concentration-absorption difference relationships can be calculated from literature values of mass absorption coefficients. Matrix variations appear to have little effect on the data and the results are nearly as accurate as more elaborate procedures employing standards. Typically, relative errors of 2 to 5 percent are observed.[4]

X-RAY DIFFRACTION METHODS

Since its discovery in 1912 by von Laue, x-ray diffraction has provided a wealth of important information to science and industry. For example, much that is known about the arrangement and the spacing of atoms in crystalline materials has been directly deduced from diffraction studies. In addition, such studies have led to a much clearer understanding of the physical properties of metals, polymeric materials, and other solids. Currently, x-ray diffraction work is of prime importance in the elucidation of the structures of such complex, natural products as steroids, vitamins, and antibiotics.

X-ray diffraction also provides a convenient and practical means for the qualitative identification of crystalline compounds. This application is based upon the fact that the x-ray diffraction pattern is unique for each crystalline substance. Thus, if an exact match can be found between the pattern of an unknown and an authentic sample, chemical identity can be assumed. In addition, diffraction data are sometimes employed for the quantitative measurement of a crystalline compound in a mixture. The method may provide data that are difficult or impossible to obtain by other means as, for example, the percentage of graphite in a graphite-charcoal mixture.

The field of x-ray diffraction and crystal structure is extensive; only a brief and empirical summary of this application of the method to the analysis of crystalline powders is described here.[5]

Measurement of X-ray Diffraction

Sample preparation. For analytical diffraction studies, the sample is ground to a fine, homogeneous powder. In such a form the enormous number of small crystallites are oriented in every possible direction; thus, when an x-ray beam traverses the material, a significant number of the particles can be expected to be oriented in such ways as to fulfill the Bragg condition for reflection from every possible interplanar spacing.

---

[4]E. P. Bertin, R. J. Longobucco and R. J. Carver, *Anal. Chem.,* **36**, 641 (1964).

[5]Recommended for further information: H. P. King and L. E. Alexander, *X-Ray Diffraction Procedures* (New York: John Wiley & Sons, Inc., 1954).

Samples may be held in the beam in thin-walled glass or cellophane capillary tubes. Alternatively, a specimen may be mixed with a suitable noncrystalline binder and molded into a suitable shape.

Photographic recording. The classical, and still widely used, method for recording diffraction patterns is photographic. Perhaps the most common instrument for this purpose is the *Debye–Scherrer* powder camera, which is shown schematically in Figure 14-13. Here, the beam from a Coolidge tube is filtered to produce a nearly monochromatic beam (often the copper or molybdenum K_α line), which is collimated by passage through a narrow tube. The undiffracted radiation then passes out of the camera via a narrow exit tube. The camera itself is cylindrical and equipped to hold a strip of film around its inside wall. The inside diameter of the cylinder usually is 11.46 cm so that each lineal millimeter of film is equal to $\frac{1}{2}$ deg in θ.

The sample is held in the center of the beam by an adjustable mount.

Figure 14-13(b) depicts the appearance of the exposed and developed film; each set of lines (D_1, D_2, and so forth) represents diffraction from one set of crystal planes. The Bragg angle θ for each line is easily evaluated from the geometry of the camera.

Electronic recording. Diffraction patterns can also be obtained with an instrument such as that shown in Figure 14-9. Here again, the fixed source

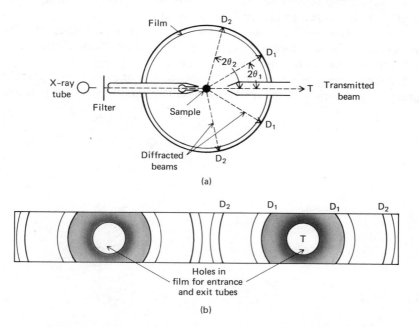

Figure 14-13. Schematic diagram of a powder camera (a); the film strip after development (b). D_2, D_1, and T indicate positions of the film in the camera.

is a filtered Coolidge tube. The sample, however, replaces the single crystal on its mount. The diffraction pattern is then obtained by automatic scanning in the same way as for an emission or absorption spectrum.

Interpretation of Diffraction Patterns

The identification of a species from its powder diffraction pattern is based upon the position of the lines (in terms of θ or 2θ) and their relative intensities. The diffraction angle 2θ is determined by the spacing between a particular set of planes; with the aid of the Bragg equation this distance d is readily calculated from the known wavelength of the source and the measured angle. Line intensities are dependent upon the number and the kind of atomic reflection centers that exist in each set of planes.

Identification of crystals is empirical. The American Society for Testing Materials (ASTM) publishes file cards that provide d spacings and relative line intensities for pure compounds; data for nearly 10,000 crystalline materials have been compiled.

For identification purposes the cards are arranged in order of d spacing for the most intense line; cards can then be chosen which identify compounds having d spacings for the most intense line that falls within a few hundredths of an Ångstrom of the experimental value. Further elimination of possible compounds is accomplished by consideration of the spacing for the second most intense line, then the third, and so forth. Ordinarily, three or four spacings serve to identify the compound.

If the sample contains two or more crystalline compounds, identification becomes more complex. Here, various combinations of the more intense lines are used until a match can be found.

By measuring the intensity of the diffraction lines and comparing with standards, a quantitative analysis of crystalline mixtures is also possible.

15 Mass Spectrometry

In mass spectrometry the sample is converted into rapidly moving positive ions which are then separated and characterized. Resolution is frequently based upon differences in ion paths in a magnetic field or an electrostatic field or both. Partition can also be based upon variations in ion velocities in a field-free space.

Mass spectrometry evolved from studies dating from the beginning of this century that were concerned with the behavior of positive ions in magnetic and electrostatic fields. During the following two decades the method was developed and refined, and provided important information concerning the isotopic abundance of the various elements. The first true analytical application of mass spectrometry was described in 1940 when reliable mass spectrometers became commercially available. The focus of this early analytical work was toward quantitative methods for the determination of the components in complex hydrocarbon mixtures. The mass spectrometer has since proved to be an invaluable tool for this work, and it is widely used in the petroleum industry.

Beginning in about 1960, interest in mass spectrometry shifted toward its use for the identification and structural analysis of complex compounds. Within three or four years it became recognized as a powerful and versatile tool for this purpose — as important or perhaps more important than infrared or nuclear magnetic resonance spectroscopy.[1]

[1]For a general review see R. W. Kiser, *Introduction to Mass Spectrometry and Its Applications* (Englewood Cliffs, N.J.: Prentice-Hall, Inc., 1965); C. A. McDowell, Ed., *Mass Spectrometry* (New York: McGraw-Hill Book Company, Inc., 1963); and F. W. Melpolder and R. A. Brown in I. M. Kolthoff and P. J. Elving, Eds., *Treatise on Analytical Chemistry*, Part I, Vol. 4 (New York: Interscience Publishers, Inc., 1963), Chap. 40.

THE MASS SPECTROMETER

The principles of mass spectral measurements are simple and easily understood; unfortunately, this simplicity does not extend to the instrumentation. Indeed, the mass spectrometer is one of the most complex electronic and mechanical devices employed by the chemist. As a consequence, costs are high, not only for the instrument but also for its operation and maintenance. Few chemists have the skills needed to repair and maintain a modern high resolution mass spectrometer.

Principles of the Measurement

Figure 15-1 shows schematically the essential parts of a typical analytical mass spectrometer. Its operation is based on the following sequence of events: (1) a micromole (or less) of sample is volatilized and allowed to leak slowly into the ionization chamber, which is maintained at a pressure of about 10^{-5} mm mercury; (2) the molecules of the sample are ionized by a stream of electrons flowing from the heated filament toward an anode (both positive and negative ions are formed by impact, but the former predominate, and all analytical methods are based upon positive particles); (3) the positive ions are separated from the negative by the small negative potential at slit A; they are then accelerated by a potential of a few hundred to a few thousand volts between A and B; a collimated beam of positive ions enter the separation area through the slit in B; (4) in the ion separator, which is maintained at a pressure of about 10^{-7} mm mercury, the fast-

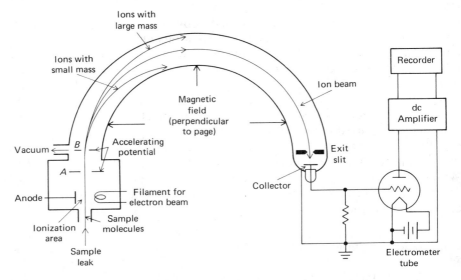

Figure 15-1. Schematic diagram of an analytical mass spectrometer.

moving particles are subjected to a strong magnetic field which causes them to travel in a curved path, the radius of which depends upon their velocity and mass[2] as well as the field strength; particles of different mass can be focused on the exit slit by varying their velocity (via the accelerating potential) or the field; (5) the ions passing through the exit slit fall upon a collector electrode; the ion current that results is amplified and recorded as a function of field strength or accelerating potential.

The utility of a mass spectrum for identification arises from the fact that the impact of an electron stream upon a given molecular species produces a family of positive particles whose mass distribution is characteristic of the parent species. As a consequence, a mass spectrum is somewhat comparable to an infrared or nuclear magnetic resonance spectrum. Mass spectral data are easier to interpret in some respects, since they provide information in terms of molecular mass of the structural components; in addition, an accurate measure of the molecular weight of the compound can usually be obtained.

Mass spectra can also be employed for the quantitative analysis of mixtures. Here the magnitude of ion currents at various mass settings are related to concentration.

Sample Handling System

Most commonly, the sample is introduced as a gas from a 1-to-5-liter reservoir. The pressure of the sample in the reservoir should be about one to two orders of magnitude greater than that within the ionization chamber to maintain a steady flow through a pin hole into the chamber; sample pressures of 0.01 mm of mercury are typical. For liquids boiling below 150°C a suitable quantity will evaporate into the evacuated reservoir at room temperature; for less volatile samples the sample and reservoir may be heated, provided the compound is thermally stable; otherwise the sample must be introduced directly into the ionization chamber, a procedure that requires special equipment. Gaseous samples are readily handled by expansion of a small volume into the reservoir.

Sample requirements with regard to volatility and quantity ($\sim$1 micromole) are similar for both gas chromatography (Chapter 24) and mass spectrometry; thus, the effluent from a chromatographic column can often be led directly into a mass spectrometer for analysis. In this way the excellent separatory capabilities of the former can be conveniently combined with the superior identification power of the latter. If it is desirable to remove the helium carrier before the mass analysis, the gases are passed through a thin-walled tube that is permeable to helium; most of the carrier diffuses away from the sample during this treatment.

---

[2]More correctly, the dependence is on the ratio of mass to charge (m/e). Generally, however, the ions of interest bear only a single charge; thus, the term mass is frequently used in lieu of the more cumbersome mass-to-charge ratio.

Ion Sources

Figure 15-2 is a schematic diagram of a typical ionization source sometimes called an *ion gun*. The positive ions produced on electron impact are forced through the slit of the first accelerating plate by a small potential difference between this plate and the repeller. The high potential between the first and second accelerators give the particles their final velocities; the third slit gives further collimation to the beam. In most spectrometers the potential between the accelerator slits provides the means whereby particles of a particular mass are focused on the collector.

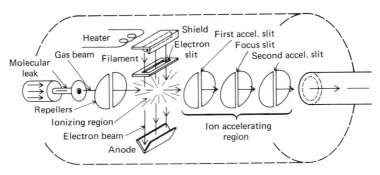

Figure 15-2. An ion source. From R. M. Silverstein and G. C. Bassler, *Spectrometric Identification of Organic Compounds,* 2d ed. (New York: John Wiley & Sons, Inc., 1967.) (With permission.)

Ion Separators

Several arrangements exist for the resolution of ions with different mass-to-charge ratios. Ideally, the separator should distinguish between very small mass differences; in addition, it should produce a high level of ion currents. As with radiation monochromators, to which the separator is analogous, these two properties are incompatible and a design compromise must always be made. Discrimination between integral mass numbers suffices for some applications. For others, however, much higher resolution is necessary; thus, for example, identification of the ions $C_2H_4^+$, CH_2N^+, N_2^+, and CO^+ (mass numbers 28.031, 28.019, 28.006, and 27.995, respectively) requires a resolving power of about 0.01 mass unit.

The main difference among the various mass spectrometers lies in their systems for separation of ions.

Single-focusing separators with magnetic deflection. Separators of this kind employ a circular beam path of 180 deg, 90 deg, or 60 deg. The first is illustrated in Figure 15-1. The path described by any given particle represents a balance between the forces that are acting upon it. The magnetic centripetal force F_M is given by

$$F_M = Hev \qquad (15\text{-}1)$$

where H is the magnetic field strength, v is the particle velocity, and e is the charge on the ion. The balancing centrifugal force F_c can be expressed as

$$F_c = \frac{mv^2}{r} \qquad (15\text{-}2)$$

where m is the particle mass and r is the radius of curvature. Finally, the kinetic energy of the particle E is given by

$$E = eV = \tfrac{1}{2}mv^2 \qquad (15\text{-}3)$$

where V is the accelerating voltage applied in the ionization chamber. Note that all particles of the same charge, regardless of mass, are assumed to have the same kinetic energy after acceleration in the electrical field. This assumption is only approximately valid, since the ions will possess a small distribution of energies before acceleration.

A particle must fulfill the condition that $F_M = F_c$ in order to traverse the circular path to the collector; thus

$$Hev = \frac{mv^2}{r} \qquad (15\text{-}4)$$

Substituting equation (15-4) into equation (15-3) and rearranging gives

$$\frac{m}{e} = \frac{H^2 r^2}{2V} \qquad (15\text{-}5)$$

In most spectrometers H and r are fixed; therefore, the mass-to-charge ratio of the particle reaching the slit is inversely proportional to the acceleration voltage. Most particles possess a single unit of positive charge; thus, any desired mass can be focused on the exit slit by suitable adjustment of V.

Double-focusing analyzers. The ability of the single-focusing instrument to discriminate between small mass differences (that is, its resolving power) is limited by the small variations in the kinetic energy of particles of a given species as they leave the ion source. These variations, which arise from the initial distribution of kinetic energies of the neutral molecules, cause a broadening of the ion beam reaching the collector and a loss in resolving power. In a double-focusing instrument, shown schematically in Figure 15-3, the beam is first passed through a radial electrostatic field. This field has the effect of focusing only particles of the same kinetic energy on slit 2, which then serves as the source for the magnetic separator. Resolution of particles differing by small fractions of a mass unit becomes possible. The ion-currents produced are extremely small, however, and require great amplification for detection and recording. Double-focusing

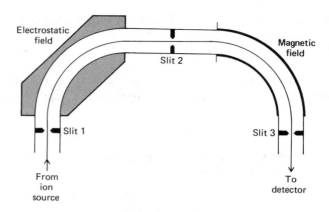

Figure 15-3. Design of a double-focusing separator.

mass spectrometers are available commercially; their cost is high and their maintenance difficult.

Another type of double-focusing instrument is the *cycloidal mass spectrometer* illustrated in Figure 15-4. Here the ion beam is simultaneously subjected to crossed electrostatic and magnetic fields, and acquires a cycloidal path as a result. This design permits the use of a relatively small magnet and a short ion path, thus providing an instrument that is considerably more compact than most other single- or double-focusing instruments; a small spectrometer of this type has a resolving power that approaches that of a much larger single-focusing instrument.

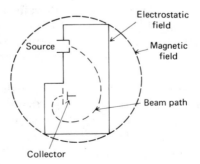

Figure 15-4. Schematic diagram of the cycloidal mass spectrometer. Note that the electrostatic and magnetic fields are crossed.

Time-of-flight separators. Ion separation in a time-of-flight instrument is achieved by nonmagnetic means. Here, the positive ions are produced intermittently by bombardment with a very short pulse of electrons. These pulses, which are controlled by a grid, typically have a frequency of 10,000

Hz and a lifetime of 0.25 μsec. The ions produced are then accelerated by an electrical field pulse that has the same frequency but lags behind the ionization pulse. The accelerated particles pass into a field-free *drift tube* of about a meter in length (Figure 15-5). As noted earlier, all particles entering the tube have the same kinetic energies; thus, their velocities in the drift tube must vary inversely with their masses (equation 15-3), with the lighter particles arriving at the collector earlier than the heavier ones.

The detector in a time-of-flight mass spectrometer is an electron multiplier tube similar in principle to a photomultiplier tube (p. 59), the output of which is fed across the vertical deflection plates of a cathode-ray oscilloscope; the horizontal sweep is synchronized with the accelerator pulses, and an essentially instantaneous display of the entire mass spectrum appears on the oscilloscope screen.

From the standpoint of resolution, reproducibility, and ease of mass identification, instruments employing time-of-flight mass separators are less satisfactory than those using magnetic focusing. On the other hand, several advantages partially offset these limitations. Included among these are the ruggedness and ease of accessibility to the ion source, which allows the ready placement of nonvolatile or heat-sensitive samples directly into the ion source. The instantaneous display feature is also useful for the study of short-lived species. In general, time-of-flight instruments are smaller, more mobile, and more convenient to use than their magnetic focusing counterparts.

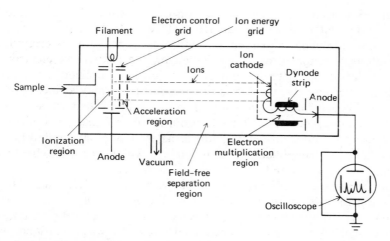

Figure 15-5. Schematic diagram of a time-of-flight mass spectrometer.

Other separators. Several other means of separating ions of different masses have been developed for use in mass spectrometers. The *quadrupole* spectrometer employs four short, parallel metal rods arranged symmetrically around the beam. The opposed rods are connected together, one pair being attached to the positive side of a dc source and the other pair to the negative terminal. In addition, a radio-frequency ac potential is applied to both pairs. Neither field acts to accelerate the positive particles ejected from an ion source into the space between the gun. The combined fields, however, cause the particles to oscillate about their central axis of travel, and only those with a certain mass-to-charge ratio can pass through the array without being removed by collision with one of the rods. Mass scanning is achieved by varying the frequency of the ac supply or by varying the potentials of the two sources while keeping their ratio constant. Quadrupole instruments are compact and less expensive than magnetic focusing instruments.

The *omegatron* spectrometer is based upon the cyclotron principle. Here, the ions are fed into the center of a box-shaped space and caused to travel in an expanding spiral under the influence of a constant magnetic field and a radio-frequency electrostatic field. A collector electrode is located at the periphery of the box; particles of various mass are focused on the collector by variation in the frequency or the magnetic field. Advantages of the omegatron include compactness and low cost.

Measurement of Ion Currents

The ions from the separator pass through a slit and are collected on an electrode that is well-shielded from stray ions. In many instruments the current produced is passed through a large resistor to ground, and the resulting potential drop is impressed on the grid of an electrometer tube. The resulting plate current is then further amplified before being recorded. Alternatively, the ions from the separator may strike a cathode surface and cause electron emission. The electrons formed are accelerated toward a dynode and, upon impact, produce several additional electrons. This process is repeated several times, as in a photomultiplier tube, to produce a highly amplified electron current which can then be further amplified electronically and recorded.

Ion currents are ordinarily small and vary over a wide range (10^{-9} to 10^{-15} amp). In order to record peaks of such diverse size, most mass spectrometers are equipped with several pens, with each recording a different current range. Thus, a peak of convenient size can always be found on the chart regardless of the current magnitude. Some instruments employ a recorder that consists of a set of mirrored galvanometers which record the spectrum on sensitized paper; each galvanometer has a different sensitivity to permit the recording of peaks of widely different sizes. A typical recorder tracing is shown in Figure 15-6.

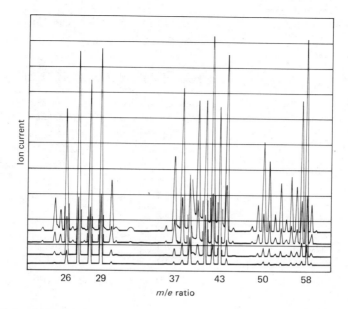

Figure 15-6. Mass spectra of *n*-butane recorded simultaneously by four galvanometers of different sensitivities. (From top to bottom, galvanometer sensitivities are in the ratio 30:10:3:1). Courtesy of E. I. du Pont de Nemours & Company, Monrovia, California. (With permission.)

Resolution

The capability of a mass spectrometer to differentiate between masses is usually stated in terms of its resolution $m/\Delta m$, where m and $m + \Delta m$ are the masses of two particles that give just separable peaks of equal size. Two peaks are considered to be separated if the valley between them is no more than 10 percent of their height.

The resolution needed in a mass spectrometer is greatly dependent upon its application. For example, discrimination between particles of the same nominal mass, such as N_2^+ (mass 28.006) and CO^+ (mass 27.995), requires an instrument with a resolution of several thousand. On the other hand, low-molecular-weight particles differing by a unit of mass or more (NH_3^+ and CH_4^+ for example) can be distinguished with an instrument having a resolution smaller than 50. An instrument with a resolution of perhaps five to ten times this figure is needed for unit separation where the sample yields large fragments.

Table 15-1 lists the resolution and mass ranges of various types of commercially available mass spectrometers. The cost of these instruments is related to their resolution and range. Unit resolution to mass number 250 requires an instrument with a purchase price of roughly $30,000. A high-resolution, double-focusing instrument costs over $100,000. Low-resolution instruments with limited range are priced in the range of $5000 to $10,000.

Table 15-1 Comparison of Some Typical Commercial Mass Spectrometers

Type	Approximate Mass Range	Approximate Resolution
Double focusing	2–5000	10,000–20,000
	1–240	1000–2500
Single focusing	1–1400	1500
	2–700	500
	2–150	100
Cycloidal	2–230	200
	2–150	100
Time-of-flight	1–700	150–250
	0–250	130
Quadrupole	2–100	100
	2–80	20–50
Omegatron	1–90	>50

MASS SPECTRA

Even for relatively simple compounds the mass spectrum generally contains an array of peaks of differing heights. The detailed nature of the spectrum depends upon the properties of the molecule as well as upon the ionization potential, the sample pressure, and the instrument design. Complete interpretation of a spectrum is seldom, if ever, possible.

The Ionization Process

Generation of a mass spectrum requires a minimum electron-beam energy (7 to 15 eV for most organic compounds). This represents the energy needed to initiate the ionization process

$$M + e \rightarrow M^+ + 2e$$

where M represents the molecule and M^+ is the *molecular ion,* or *parent ion.* Small additional increases in electron-beam energy produce a higher yield of molecular ions, owing to the greater probability for ion-producing collisions. Still further increases in electron-beam energies result in collisions that cause bond rupture, with the formation of fragments that have smaller masses (and occasionally larger masses as well) than the parent molecule. Typical of these processes are the following, illustrated with the hypothetical molecule ABCD:

$$ABCD + e \rightarrow ABCD^+ + 2e$$
$$ABCD^+ \rightarrow BCD^{\cdot} + A^+$$

Molecular ion formation (15-6)

$$
\left.
\begin{array}{l}
\qquad\qquad B^{\cdot} + A^+ \\
\xrightarrow{\hspace{1cm}} CD^{\cdot} + AB^+ \\
\qquad\qquad A^{\cdot} + B^+ \\
\qquad\qquad D^{\cdot} + C^+ \\
\xrightarrow{\hspace{1cm}} AB^{\cdot} + CD^+ \\
\qquad\qquad C^{\cdot} + D^+
\end{array}
\right\}
$$

Fragmentation (15-7)

$$
ABCD^+ \rightarrow ADBC^+
\left.
\begin{array}{l}
\nearrow BC^{\cdot} + AD^+ \\
\searrow AD^{\cdot} + BC^+
\end{array}
\right\}
$$

Rearrangement followed by fragmentation (15-8)

$$ABCD^+ + ABCD \rightarrow (ABCD)_2^+ \rightarrow BCD^{\cdot} + ABCDA^+$$

Collision followed by fragmentation (15-9)

For a molecule containing a large number of atoms, the number of different positive ions produced can be large. Their distribution depends upon the stability of the precursor ion and the energy imparted to the molecule by the electron beam. Fortunately, when the beam energy exceeds 50 to 70 eV, the pattern of products from a given molecule becomes more or less reproducible.

The neutral fragments in equations (15-7) to (15-9) are shown as radicals, but they may also occur as molecules. Neither will reach the detector.

The Molecular Ion

The molecular ion M^+ has a mass corresponding to the molecular weight of the compound from which it is generated. Thus, the mass of this ion is an important parameter in the identification of a compound. For perhaps 80 to 90 percent of organic substances, the molecular ion peak is readily recognizable; certain characteristics of this peak must be borne in mind, however.

Isotope peaks. Table 15-2 lists the natural isotopic abundance of elements that frequently occur in organic compounds. Because of the presence of these elements, there is often an observable peak one mass unit beyond the molecular ion peak (the M + 1 peak), and on occasion at M + 2 as well. Compounds containing chlorine or bromine will clearly have relatively large M + 2 peaks.

Peaks for collision products. Ion-molecule collision can produce peaks of higher mass number than the molecular ion peak (equation 15-9).

Table 15-2 Natural Abundance of Isotopes of Some Common Elements

Element[a]	Most Abundant Isotope	Abundance of Other Isotopes Relative to 100 Parts of the Most Abundant[b]	
Hydrogen	^1H	^2H	0.016
Carbon	^{12}C	^{13}C	1.08
Nitrogen	^{14}N	^{15}N	0.38
Oxygen	^{16}O	^{17}O	0.04
		^{18}O	0.20
Sulfur	^{32}S	^{33}S	0.78
		^{34}S	4.40
Chlorine	^{35}Cl	^{37}Cl	32.5
Bromine	^{79}Br	^{81}Br	98.0

[a]Fluorine (^{19}F), phosphorus (^{31}P), and iodine (^{127}I) have no additional naturally occurring isotopes.
[b]The numerical entries indicate the average number of isotopic atoms present for each 100 atoms of the most abundant isotope; thus, for every 100 ^{12}C atoms there will be an average of 1.08 ^{13}C atoms.

At ordinary sample pressures the only important reaction of this type is one in which the collision process transfers a hydrogen atom to the ion; an enhanced M + 1 peak results. The proton transfer process is a second-order reaction, and the amount of product is strongly affected by the reactant concentration. An M + 1 peak of this type increases at a more rapid rate with increased sample pressure than do other peaks; thus, detection of this reaction is usually possible.

Stability of the molecular ion. For a given set of conditions the intensity of a molecular ion peak depends upon the stability of the ionized particle; a minimum lifetime of about 10^{-5} sec is needed for a particle to reach the collector and be detected. The stability of the ion is strongly affected by structure; the size of molecular ion peaks will thus show great variability.

In general, the molecular ion is stabilized by the presence of π electron systems, which more easily accommodate a loss of one electron. Cyclic structures also give large parent peaks, since rupture of a bond does not necessarily produce two fragments. In general, the stability of the molecular ion decreases in the following order: aromatics, conjugated olefins, alicyclics, sulfides, unbranched hydrocarbons, mercaptans, ketones, amines, esters, ethers, carboxylic acids, branched hydrocarbons, and alcohols. These effects are illustrated in Table 15-3, which compares the height of the molecular ion peak for some C_{10} compounds relative to the total peak heights in the spectrum.

The sensitivity of a mass spectrometer is such that a parent ion peak is readily detectable provided the peak is at least one percent of the total; for some instruments this limit is lowered to 0.1 percent. In most cases, then, the mass of the molecular ion and thus the molecular weight of the compound can be determined.

Table 15-3[a]   Variation in Molecular Ion Peak with Structure

Compound	Formula	Relative Peak Height, (percent of total peak heights)
Naphthalene		44.3
n-Butylbenzene	C₄ H₉	8.3
trans-Decaline		8.2
Diamyl sulfide	$(C_5H_{11})_2S$	3.7
n-Decane	$C_{10}H_{22}$	1.41
Diamylamine	$(C_5H_{11})_2NH$	1.14
Methyl nonanoate	$C_9H_{17}COOCH_3$	1.10
Diamyl ether	$(C_5H_{11})_2O$	0.33
3,3,5-Trimethylheptane	$C_{10}H_{22}$	0.007
n-Decanol	$C_{10}H_{21}OH$	0.002

[a]Taken from K. Biemann, *Mass Spectrometry, Organic Chemical Applications* (New York: McGraw-Hill Book Company, Inc., 1962), p. 52. With permission.

The Base Peak

The largest peak in a mass spectrum is termed the *base peak*; it is common practice to report peak heights in terms of percentages of the base peak height. Alternatively, intensities are reported as percentages of the total peak heights, a more informative number, but one that is more laborious to calculate.

QUALITATIVE APPLICATIONS OF MASS SPECTROMETRY

The mass spectrum of a pure compound provides valuable information for qualitative identification purposes. In addition, mass spectrometry has proved useful for identifying the components of simple mixtures — in particular, gaseous mixtures. Some of these applications are outlined in the paragraphs that follow.

Molecular Weight Determination

For most compounds that can be volatilized, the mass spectrometer is unsurpassed for the determination of molecular weight. As we have noted earlier, the method requires the identification of the molecular ion peak, the mass of which gives the molecular weight to at least the nearest whole number — an accuracy that can not be realized by other molecular weight measurements. Caution must be used, however, for occasionally the molecular ion peak may either be absent or so small that it is confused with a peak caused by an impurity. In addition, collision processes may produce an $M + 1$ peak that is more intense than the parent ion peak.

There is no single method for establishing unambiguously that the peak of highest mass number (neglecting the small isotope peaks) is indeed produced by the molecular ion. On the other hand, from a series of observations the experienced mass spectroscopist can ordinarily make this judgment with reasonable assurance of being correct. As mentioned earlier, it is frequently possible to identify an $M + 1$ peak by observing its behavior as a function of sample size. Identification of the highest mass peak as a fragment rather than the molecular ion is more troublesome. Here, a knowledge of fragmentation patterns for various kinds of compounds is essential. For example, a peak at $M - 3$ immediately casts doubt upon the peak at M being caused by the molecular ion. This pattern could only occur as the result of abstraction of three hydrogen atoms from the molecular ion; such a fragmentation is most unlikely. On the other hand, a strong peak at $M - 18$ suggests that even a weak peak at M may be the parent ion since, for alcohols and aldehydes, the loss of water is a common occurrence.

It is of interest to point out that the molecular weight determined by mass spectroscopy will not be identical with that calculated from atomic weights on the chemical scale if the parent compound contains certain elements. For example, methyl bromide will have a peak at mass 96 that is nearly as strong as the one at mass 94 because of the high isotopic abundance of bromine-81 (see Table 15-2).

Determination of Molecular Formulas

Partial or exact molecular formulas can be determined from the mass spectrum of a compound, provided the molecular ion peak can be identified.

Molecular formulas from high-resolution instruments. A unique formula for a compound can often be derived from the exact mass of the parent ion peak. This application, however, requires a high-resolution instrument capable of detecting mass differences of a few thousandths of a mass unit. Consider, for example, the molecular weights of the following compounds: purine, $C_5H_4N_4$ (120.044); benzamidine, $C_7H_8N_2$ (120.069); ethyltoluene, C_9H_{12} (120.094); and acetophenone, C_8H_8O (120.157). If the measured mass of the parent ion peak is 120.069, then all but $C_7H_8N_2$ are excluded as possible formulas. Tables are available that list all reasonable combinations of C, H, N, and O by molecular weight to the third decimal place.[3]

[3]J. H. Beynon and A. E. Williams, *Mass and Abundance Tables for Use in Mass Spectrometry* (Amsterdam: Elsevier Publishing Company, 1963).

Formulas from isotopic ratios. The data from an instrument that can discriminate between whole mass numbers provides useful information about the formula of a compound provided only that the molecular ion peak is sufficiently intense so that its height and the heights of the (M + 1) and (M + 2) isotope peaks can be determined accurately. The following example illustrates this type of analysis.

> **Example.** Calculate the ratios of the (M + 1) to M peak heights for the following two compounds: dinitrobenzene, $C_6H_4N_2O_4$ (M = 168) and an olefin, $C_{12}H_{24}$ (M = 168).
>
> From Table 15-2 we see that for every 100 ^{12}C atoms there are 1.08 ^{13}C atoms. Since there are six carbon atoms in nitrobenzene, however, we would expect there to be 6.48 (6 × 1.08) molecules of nitrobenzene having one ^{13}C atom for every 100 molecules having none. Thus, from this effect alone the (M + 1) peak will be 6.48 percent of the M peak. The isotopes of the other elements also contribute to this peak; we may tabulate their effects as follows:

$C_6H_4N_2O_4$				$C_{12}H_{24}$		
^{13}C	6 × 1.08	= 6.48	percent	^{13}C	12 × 1.08	= 12.96 percent
^2H	4 × 0.016	= 0.064	percent	^2H	24 × 0.016	= 0.38 percent
^{15}N	2 × 0.38	= 0.76	percent	(M + 1)/M		= 13.34 percent
^{17}O	4 × 0.04	= 0.16	percent			
	(M + 1)/M	= 7.46	percent			

It is seen in this example that a measurement of the ratios of the (M + 1) to M peak heights would permit discrimination between two compounds having identical whole-number mass weights. If the ratios of the (M + 2) to M peaks are calculated for the same two compounds, the nitrobenzene has a value of 1.04 percent while the olefin ratio is 0.82 percent.

The use of relative isotope peak heights for the determination of molecular formulas is greatly expedited with the tables developed by Beynon[4]; a portion of a modified form of his tabulations is shown in Table 15-4. Here a listing for all reasonable combinations of C, H, O, and N is given for mass numbers 83 and 84 (the original tables extend to mass number 500); also tabulated are the heights of the corresponding (M + 1) and (M + 2) peaks reported as percentages of the M peak. If a reasonably accurate experimental determination of these percentages can be obtained from a spectrum, a likely formula can be ascertained. For example, a molecular ion peak at mass 84 and with (M + 1) and (M + 2) peaks of 5.6 and 0.3 percent would suggest that the formula of the compound is C_5H_8O (Table 15-4).

The isotopic ratio is particularly useful for the detection and estimation of the number of sulfur, chlorine, and bromine atoms in a compound because of their large contribution to the (M + 2) peak (see Table 15-2).

---

[4]J. H. Beynon and A. E. Williams, *Mass and Abundance Tables for Use in Mass Spectrometry* (Amsterdam: Elsevier Publishing Company, 1963.)

Table 15-4   Isotopic Abundance Ratios for Various Combinations of Carbon, Hydrogen, Oxygen, and Nitrogen[a,b]

M = 83			M = 84		
	M + 1	M + 2		M + 1	M + 2
C_2HN_3O	3.36	0.24	$C_2H_2N_3O$	3.38	0.24
$C_2H_3N_4$	3.74	0.06	$C_2H_4N_4$	3.75	0.06
C_3HNO_2	3.72	0.45	$C_3H_2NO_2$	3.73	0.45
$C_3H_3N_2O$	4.09	0.27	$C_3H_4N_2O$	4.11	0.27
$C_3H_5N_3$	4.47	0.08	$C_3H_6N_3$	4.48	0.81
$C_4H_3O_2$	4.45	0.48	$C_4H_4O_2$	4.47	0.48
C_4H_5NO	4.82	0.29	C_4H_6NO	4.84	0.29
$C_4H_7N_2$	5.20	0.11	$C_4H_8N_2$	5.21	0.11
C_5H_7O	5.55	0.33	C_5H_8O	5.57	0.33
C_5H_9N	5.93	0.15	$C_5H_{10}N$	5.95	0.15
C_6H_{11}	6.66	0.19	C_6H_{12}	6.68	0.19

[a]Taken from R. M. Silverstein and G. C. Bassler, *Spectrometric Identification of Organic Compounds*, 2d ed. (New York: John Wiley & Sons, Inc., 1967), p. 36. With permission.
[b]Data are given as percentages of the peak height of M.

Thus, for example, an (M + 2) peak that is about 65 percent of the M peak would be strong evidence for a molecule containing two chlorine atoms; an (M + 2) peak of about 4 percent, on the other hand, would suggest one atom of sulfur. By examination of the heights of the (M + 4) and (M + 6) peaks as well, it is sometimes feasible to identify combinations of chlorine and bromine atoms.

The nitrogen rule.   The *nitrogen rule* also provides information concerning possible formulas of compounds whose molecular weight has been determined. This rule states that all organic compounds having an even molecular weight must contain zero or an even number of nitrogen atoms; all compounds with odd molecular weights must have an odd number of nitrogen atoms. The fragments formed by cleavage of one bond, however, have an odd mass number if they contain zero or an even number of nitrogen atoms; conversely, such fragments have an even mass number if the nitrogens are odd in number. The rule is a direct consequence of the fact that with the exception of nitrogen, the valency and the mass number of the isotopes of most elements that occur in organic compounds are either both even or both odd. The nitrogen rule applies to all covalent compounds containing carbon, hydrogen, oxygen, sulfur, the halogens, phosphorus, and boron.

Identification of Compounds from Fragmentation Patterns

From Figure 15-7 it is evident that fragmentation of even simple molecules produces a large number of ions with different masses. A complex spectrum results which often permits identification of compounds or at least recognition of the presence of functional groups in compounds.

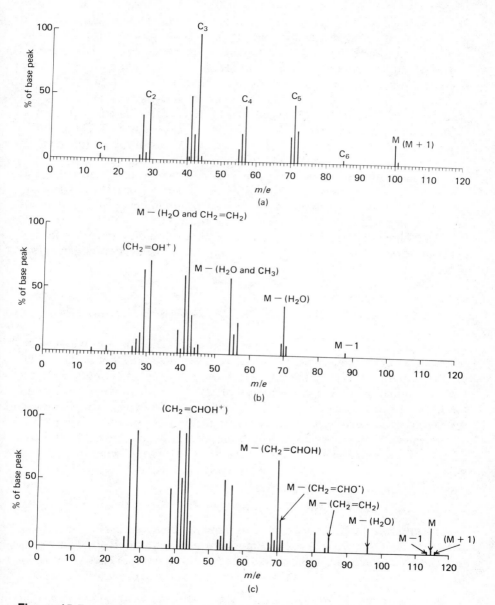

Figure 15-7. Mass spectra of some simple compounds. (a) *n*-Heptane; (b) 1-pentanol; (c) *n*-heptanal.

Systematic studies of fragmentation patterns for pure substances have led to rational fragmentation mechanisms and a series of general rules that are helpful in interpreting spectra.[5] It is seldom possible (or desirable) to account for all of the peaks in a spectrum. Instead, characteristic patterns of fragmentation are sought. For example, the top spectrum in Figure 15-7 is characterized by clusters of peaks differing in mass by 14. Such a pattern is typical of straight-chain paraffins in which the successive loss of a methylene group results in the observed mass decreases. This same pattern is evident in the left-hand parts of the two lower spectra as well. Quite generally, the most stable hydrocarbon fragments contain three or four carbon atoms and the corresponding peaks are thus the largest in magnitude.

Alcohols usually have a very weak or nonexistent parent ion peak but lose water to give a stronger peak at (M − 18) (see Figure 15-7). Cleavage of the C—C bond next to an oxygen is also common, and primary alcohols always have a strong peak at mass 31 due to the ion $CH_2\!\!=\!\!OH^+$.

The interested reader should refer to reference literature for further generalizations concerning the identification of organic compounds from mass-spectrometric data.[6]

QUANTITATIVE APPLICATIONS OF MASS SPECTROMETRY

The mass spectrometer is a powerful tool for the quantitative analysis of mixtures containing closely related components. As mentioned earlier, the first commercial mass spectrometers were developed in about 1940 for the determination of the constituents of volatile hydrocarbon samples of the kind encountered in the petroleum industry. Within a decade, however, the method had been applied successfully to the analysis of a large number of other volatile compound types. More recently, the method has been adapted to the analysis of both inorganic and organic mixtures of low volatility.

Principles of Quantitative Mass Spectrometry

Basic requirements.   The basic requirements for a successful mass spectrometric analysis are: (1) each component must exhibit at least one peak that differs markedly from the others, (2) the contribution of each component to a peak must be linearly additive, (3) the sensitivity (ion current per unit

[5]See, for example, R. M. Silverstein and G. C. Bassler, *Spectrometric Identification of Organic Compounds,* 2d ed. (New York: John Wiley & Sons, Inc., 1967), p. 13.

[6]R. M. Silverstein and G. C. Bassler, *Spectrometric Identification of Organic Compounds,* 2d ed. (New York: John Wiley & Sons, Inc., 1967), Chap. 1; K. Biemann, *Mass Spectrometry, Organic Chemical Applications* (New York: McGraw-Hill Book Company, Inc., 1962). H. Budzikiewicz, C. Djerassi and D. H. Williams, *Interpretation of Mass Spectra of Organic Compounds* (San Francisco: Holden-Day, Inc., 1964). R. I. Reed, *Applications of Mass Spectrometry to Organic Chemistry* (New York: Academic Press, Inc., 1966).

partial pressure) must be reproducible to perhaps one percent relative, and (4) suitable standards for calibration must be available.

Calibration. Quantitative mass spectrometry is based upon empirical calibration with standards. Under appropriate conditions, mass peak heights are directly proportional to the partial pressures of the components. For complex mixtures it is seldom possible to find a mass peak that is entirely unique for each component; a set of simultaneous equations must thus be solved to resolve the data from mixtures. That is,

$$i_{11}p_1 + i_{12}p_2 + \ldots + i_{1n}p_n = I_1$$
$$i_{21}p_1 + i_{22}p_2 + \ldots + i_{2n}p_n = I_2$$
$$\vdots \qquad \vdots \qquad \cdots \qquad \vdots \qquad \vdots$$
$$\vdots \qquad \vdots \qquad \cdots \qquad \vdots \qquad \vdots$$
$$i_{m1}p_1 + i_{m2}p_2 + \ldots + i_{mn}p_n = I_m$$

where I_m is the measured ion current at mass m in the spectrum of the mixture, and i_{mn} refers to the ion current at mass m for component n. The partial pressure of component n in the mixture is given by p_n. The value of i_{mn} is determined for each component by calibration with a standard at a known partial pressure p_n. Substitution of the measured ion currents for a mixture into the equations permits solution of the simultaneous equation to give the partial pressures of each of the components in the sample.

Component Analysis

The literature dealing with quantitative applications of mass spectrometry is so extensive as to make a summary difficult. The listing of typical applications assembled by Melpolder and Brown[7] demonstrates clearly the versatility of the method. For example, some of the mixtures that can be analyzed without sample heating include natural gas, C_3—C_5 hydrocarbons; C_6—C_8 saturated hydrocarbons; C_1—C_5 alcohols, aldehydes, and ketones; C_1—C_4 chlorides and iodides; fluorocarbons; thiophenes; atmospheric pollutants; exhaust gases; and many others. By employing higher temperatures, successful analytical methods have been reported for C_{16}—C_{27} alcohols; aromatic acids and esters; steroids; fluorinated polyphenyls; aliphatic amides; halogenated aromatic derivatives; and aromatic nitriles.

Mass spectrometry has also been used for the characterization and analysis of high-molecular-weight polymeric materials. Here, the sample is first pyrolyzed; the volatile products are then led into the spectrometer for examination. Some polymers yield essentially a single fragment; for example, isoprene from natural rubber, sytrene from polystyrene, ethylene

[7]F. W. Melpolder and R. A. Brown in I. M. Kolthoff and P. J. Elving, Eds., *Treatise on Analytical Chemistry*, Part I, Vol. 4 (New York: Interscience Publishers, Inc., 1963), p. 2047.

from polyethylene, and $CF_2=CFCl$ from Kel—F. Other polymers yield two or more products that depend in amount and kind upon the pyrolysis temperature. Studies of the temperature effect can provide information regarding the stabilities of the various bonds and also the approximate molecular weight distribution.

Component Type Determination

Because of the complex nature of petroleum products, quantitative data as to *types* of compounds are often more useful than analysis for individual components. Mass spectrometry can provide such information. For example, it has been found that paraffinic hydrocarbons generally give unusually strong peaks at masses 43, 57, 71, 85, and 99. Cycloparaffins and monoolefins, on the other hand, exhibit characteristically intense peaks at masses 41, 55, 69, 83, and 97. Another group of peaks is attributable to cycloolefins, diolefins, and acetylenes (67, 68, 81, 82, 95, and 96). Finally, alkylbenzenes are found to fragment to masses of 77, 78, 79, 91, 92, 105, 106, 119, 120, 133, and 134. A mathematical combination of the peak heights of a set provides an analytical parameter for assessing the concentration of each type of hydrocarbon. Type analyses have been used to characterize the properties and the behavior of gasolines, fuel oils, lubricating oils, asphalts, and mixtures of paraffins, olefins, alcohols, and ketones.

Inorganic Application

Much effort is currently being expended to apply mass spectrometry to inorganic analysis. Here, the central problem is that of converting the inorganic materials to the gaseous ionic form. Ion sources that have been investigated include the electric spark, high-temperature furnaces, a laser beam (for heating), a pulsed dc arc, and ion bombardment. At present, none of these is capable of the reproducibility of the electron beam employed with gases; the precision and accuracy of mass spectrometric analysis of solids is thus severely limited. On the other hand, the specificity and potential sensitivity of the method are so great that development of better sources is to be expected.

Isotope Abundance Measurement

The mass spectrometer was developed initially for the study of isotopic abundance, and the instrument continues to be the most important source for this kind of data. Information regarding the abundance of various isotopes is now employed for a variety of purposes. The determination of formulas of organic compounds cited earlier is one example. Other important applications include analysis by isotope dilution, tracer studies with isotopes, and dating of rocks and minerals by isotopic ratio measurements. The techniques employed are similar to those described in Chapter 16 for radioactive isotopes. Mass spectrometry, however, permits the extension of these procedures to nonradioactive species such as ^{13}C, ^{17}O, ^{18}O, ^{15}N, ^{34}S, and others.

16 Radiochemical Methods

The discovery and production of both natural and artificial radioactive isotopes have made possible the development of analytical methods (radiochemical methods) that are both sensitive and specific. These procedures are often characterized by good accuracy and widespread applicability; in addition, some permit a drastic reduction in the amount, or even the elimination, of chemical preparation that must precede the measurement step.[1]

Radiochemical methods are of three types. In *activation analysis*, activity is induced in one or more elements of the sample by irradiation with suitable particles (most commonly thermal neutrons from a nuclear reactor); the resulting radioactivity is then measured. In an *isotope dilution* procedure, a pure but radioactive form of the substance to be determined is mixed with the sample in known amount. After equilibration, a fraction of the component of interest is isolated by suitable means; the analysis is then based upon the activity of this isolated fraction. In a *radiometric analysis,* a radioactive reagent is employed to separate completely the component of interest from the bulk of the sample; the activity of the isolated portion is measured.

---

[1] For a detailed treatment of radiochemical methods see G. Friedlander, J. W. Kennedy and J. M. Miller, *Nuclear and Radiochemistry*, 2d ed. (New York: John Wiley & Sons, Inc., 1964); B. R. Tolbert and W. E. Siri, in A. Weissberger, Ed., *Physical Methods of Organic Chemistry*, Part IV, Vol. 1. (New York: Interscience Publishers, Inc., 1960), Chap. 50.

RADIOACTIVE DECAY PROCESSES

The products of a radioactive decay event include an altered nucleus, energy, and possibly an elementary particle as well. The number of particles associated with spontaneous decay is relatively limited.

Properties of Elementary Particles

Alpha particles. On a subatomic scale, alpha particles are massive, consisting of two protons and two neutrons. They occur as products in the four transuranium decay series; alpha decay is seldom observed in elements other than those of high atomic number.

Alpha particles from a particular decay process are either mono-energetic or are distributed among relatively few discrete energies. They progressively lose their energy as a result of collisions as they pass through matter and ultimately are converted into helium atoms through capture of two electrons from their surroundings. Their relatively large mass and charge render alpha particles highly effective in producing ion pairs within the matter through which they pass, but these same properties are responsible for their low penetrating power. The identity of an isotope that is an alpha emitter can be established by measuring the length (or range) over which the emitted alpha particles produce ion pairs within a particular medium.

Their low penetrating power renders alpha particles relatively ineffective for the production of artificial isotopes.

Beta particles. A common type of nuclear event involves the emission of beta particles (electrons). In contrast to alpha emission, beta decay is characterized by production of beta particles having a continuous spectrum of energies that range from nearly zero to some maximum that is characteristic of each decay process. Because of its small mass (about 1/8000 that of an alpha particle), the beta particle is not nearly as effective as the alpha particle in producing ion pairs in matter; at the same time, its penetrating power is substantially greater. Beta ranges in air are difficult to evaluate because of the high likelihood that scattering will occur. As a result, beta energies are based upon the thickness of an absorber, ordinarily aluminum, required to diminish the activity of the emission by a factor of two.

Beta particles that carry a unit positive charge are called *positrons*. Positrons have highly transient lifetimes; their ultimate fate is annihilation by interaction with ordinary electrons to give two photons.

Gamma ray emission. Many alpha and beta emission processes leave a nucleus in an excited state which then returns to the ground state in one or more quantized steps with the release of gamma rays (photons of very high energy). The gamma ray emission spectrum is characteristic for each nucleus and is thus useful for the identification of radioisotopes.

Not surprisingly, gamma radiation is highly penetrating. Upon interaction with matter, gamma rays lose energy by three mechanisms. The *photoelectric effect* involves ejection of a single electron from the target atom. This electron possesses kinetic energy that is equal to the energy of the bombarding photon less the binding energy between the ejected electron and the atom. The photoelectric effect is the principal mode of interaction involving relatively low-energy gamma photons and target atoms of high atomic weight. The *Compton effect* is a process in which a gamma photon and an electron participate in an elastic collision. The electron acquires only a portion of the photon energy and recoils at a corresponding angle with respect to the projected path of the photon. The photon, now with diminished energy, suffers further energy losses; finally, it causes the photoelectric ejection of an electron. The Compton effect is responsible for the absorption of relatively energetic gamma photons. If the photon possesses sufficiently high energy (at least 1.02 MeV), *pair production* can occur. Here the photon is converted into a positron and an electron in the field surrounding a nucleus.

X-ray emission. Two types of nuclear processes, *electron capture* and *internal conversion*, are followed by the emission of x-ray photons. In the former, one of the inner electrons is captured by the nucleus and thus decreases the atomic number of the atom by one. Ordinarily, the electron involved will be abstracted from the innermost, or K, orbital; for this reason the process is sometimes called *K capture*. To be sure, capture of electrons from the L and M levels is also possible. Following K capture, electrons from higher-energy levels fall into the vacated K shell; accompanying this process is the emission of x-rays that are characteristic of the element formed by the capture.

Internal conversion is a process in which an excited nucleus (resulting from a decay event) loses its excitation energy by ejection of an electron from one of the orbitals near the nucleus. This alternative to gamma ray emission results in a vacated K, L, or M shell, which is then filled by an electron from a higher-energy level. An x-ray photon characteristic of the element results from this transition.

Neutrons. The neutron (n) is a particle of unit mass and zero charge. As such, the neutron is a highly effective bombarding particle, being uninfluenced by the electrostatic charge barrier surrounding a target nucleus. In contrast to charged particles, which require high kinetic energies to surmount such barriers, slow (or thermal) neutrons are generally more effective than high-energy neutrons in entering into nuclear processes. It is for this reason that the neutrons emitted from the source (ordinarily a nuclear reactor) are caused to lose much of their kinetic energy through collisions with a moderating substance of low atomic weight; the result is a neutron flux containing low energies distributed about some average value.

Neutrons can interact with matter in any of several ways. The product (or products) depend in large measure on the energies of the bombarding neutrons. Irradiation of a stable isotope with thermal neutrons is most likely to give rise to a highly excited isotope with an atomic mass number one unit larger than that of the target. The product achieves greater stability through the prompt (within $\approx 10^{-12}$ sec) emission of a gamma ray photon. The process can be depicted by the equation

$$\underset{Z}{\overset{A}{X}} + \underset{0}{\overset{1}{n}} \rightarrow [\underset{Z}{\overset{(A\ +\ 1)}{X}}]^* \xrightarrow{\sim\ 10^{-12}\text{sec}} \underset{Z}{\overset{(A\ +\ 1)}{X}} + \gamma$$
$$\text{excited state} \qquad\qquad \text{ground state}$$

where the superscript refers to the mass of element X having an atomic number Z. Fast neutrons interact with matter by other mechanisms, but these reactions are not pertinent to our discussion.

Units of Radioactivity

The *curie* is the fundamental unit of radioactivity; it is defined as that quantity of nuclide in which there occur 3.7×10^{10} disintegrations per second. Note that the curie is an enumerative quantity only, and provides no information concerning the products of the decay process or their energies.

The *millicurie* and *microcurie*, corresponding respectively to 3.7×10^7 and 3.7×10^4 disintegrations per second, are frequently more convenient units.

The Decay Law

Radioactive decay is a completely random process. Thus, while no prediction can be made concerning the lifetime of an individual nucleus, we can describe the behavior of a large ensemble of like nuclei with the expression

$$-\frac{dN}{dt} = \lambda N \tag{16-1}$$

where N represents the number of radioactive nuclei in the sample at time t and λ is the characteristic *decay constant* for a particular radioisotope. Upon rearranging equation (16-1) and integrating over the interval between $t = 0$ and $t = t$ (during which the number of nuclei in the sample decreases from N_0 to N), we obtain

$$\ln\frac{N}{N_0} = -\lambda t \tag{16-2}$$

The *half-life* of a radioactive isotope is defined as the time required for the number of atoms to decrease to one-half of its original quantity; that is, for N to become equal to $N_0/2$. Thus, substitution of $N_0/2$ for N in equation (16-2) gives

$$t_{1/2} = \frac{0.693}{\lambda}$$

Counting Errors[2]

The randomness inherent in the decay process prevents an exact prediction of the number of disintegrations that will occur in any given time interval. Nevertheless, given a sufficiently long counting period, the extent of decay during the interval is found to be reproducible within a predetermined level of precision.[3]

Figure 16-1 shows the deviation from the true average count that would be expected if 1000 replicate observations were made on the same sample. Curve A gives the distribution for a substance for which the true average count r for a selected period is 5; curves B and C correspond to samples having true averages of 15 and 35. Note that the *absolute* deviations become greater as r becomes larger, but the *relative* deviations become smaller. Note also that for the smallest number of counts the distribution is distinctly not symmetric around the average; this lack of symmetry is a consequence of the fact that a negative count is impossible, while a finite likelihood always exists that a given count can exceed the average by a factor greater than two.

Standard deviation of counting data. When the total number of counts becomes large ($r > 100$), the distribution of deviations from the mean approaches that of a symmetrical Gaussian or normal error curve. Under these circumstances it can be shown that

$$\sigma_N = \sqrt{N} \qquad\qquad (16\text{-}3)$$

where N is the number of counts for any given period and σ_N is the standard deviation in N. The relative standard deviation $(\sigma_N)_r$ is given by

$$(\sigma_N)_r = \frac{\sigma_N}{N} = \frac{\sqrt{N}}{N} = \frac{1}{\sqrt{N}} \qquad\qquad (16\text{-}4)$$

---

[2]For a more complete discussion see G. Friedlander, J. W. Kennedy and J. M. Miller, *Nuclear and Radiochemistry*, 2d ed. (New York: John Wiley & Sons, Inc., 1964), Chap. 8.

[3]The counting period should be short, however, with respect to the half-life so that no significant change in the number of radioactive atoms occurs. Further restrictions include a detector that responds to the decay of a single isotope only, and an invariant counting geometry, so that the detector responds to a constant fraction of the decay events that occur.

Thus, although the absolute standard deviation increases with the number of counts, the relative standard deviation decreases.

In normal practice, activities of samples are more conveniently expressed in terms of counting rates R (counts per minute) than in counts for some arbitrary time. Clearly,

$$R = N/t \qquad (16\text{-}5)$$

where t is the time in minutes required to obtain N counts. The standard deviation in rate units σ_R can be obtained by dividing both sides of equation (16-3) by t; thus

$$\sigma_R = \frac{\sigma_N}{t} = \frac{\sqrt{N}}{t}$$

Substitution of equation (16-5) leads to

$$\sigma_R = \sqrt{R/t} \qquad (16\text{-}6)$$

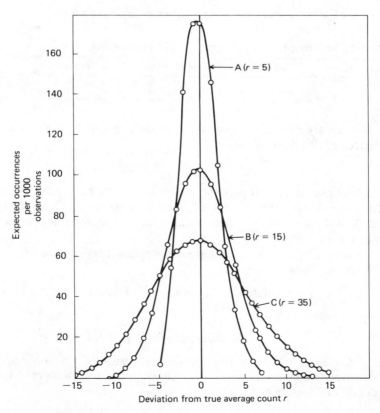

Figure 16-1. Distribution of counting data.

In relative terms

$$(\sigma_R)_r = \frac{\sqrt{R/t}}{R} = \sqrt{1/Rt} \qquad (16\text{-}7)$$

Uncertainty in a single measurement. The standard deviation can be used to define a range about the measured count within which the true average count (or the true average rate) can be expected to be with a given degree of confidence. For a Gaussian distribution it can be shown that

$$r = N \pm k\,\sigma_N = N \pm k\,\sqrt{N} \qquad (16\text{-}8)$$

where r is the true average count and k is a constant that depends upon the confidence level desired. The quantity $\pm k\,\sigma_N = \pm k\,\sqrt{N}$ is clearly the absolute uncertainty of the measurement; that is,

$$\text{absolute uncertainty} = \pm k\sqrt{N} \qquad (16\text{-}9)$$

Table 16-1 lists some values of k for various confidence levels.

Table 16-1 Values of k for Selected Confidence Levels

Confidence Level	50-Percent	90-Percent	95-Percent	99-Percent
k	0.68	1.65	1.96	2.58

Thus, the uncertainty associated with a single count N at the 50-percent confidence level would be

$$k\,\sigma_N = \pm 0.68\,\sigma_N = \pm 0.68\,\sqrt{N}$$

The uncertainty at the 50-percent confidence level ($\pm 0.68\,\sqrt{N}$) is often termed the *probable error* of a count. The probable error defines the interval around N within which the true mean r could be expected to be found 50 times out of 100.

The uncertainty in a counting measurement can also be expressed in relative terms. Thus,

$$\text{relative uncertainty} = \pm k\,(\sigma_N)_r \qquad (16\text{-}10)$$

and from equation (16-4),

$$\text{relative uncertainty} = \pm k/\sqrt{N} \qquad (16\text{-}11)$$

Example. Calculate the absolute and relative uncertainties at the 95-percent confidence level in the measurement of a sample that produced 675 counts in a given period.

$$\text{absolute uncertainty} = k\,\sigma_N = \pm 1.96\,\sqrt{675}$$
$$= \pm 51 \text{ counts}$$

Thus, for 95 out of 100 measurements the true average count *r* lies in the range 624 to 726. From equation (16-11),

$$\text{relative uncertainty} = \pm \frac{1.96}{\sqrt{N}} \times 100 = 7.5 \text{ percent}$$

Figure 16-2 illustrates the relationship between total counts and tolerable levels of uncertainty as calculated from equation (16-11). Note that the horizontal axis is logarithmic; it is clear that a tenfold decrease in the relative uncertainty requires an approximately hundredfold increase in the number of counts.

The foregoing treatment of uncertainty can be performed equally well in terms of counting rates rather than total counts; here, σ_R and $(\sigma_R)_r$, as defined by equations (16-6) and (16-7), are employed.

Background corrections. The count recorded in a radiochemical analysis includes a contribution from sources other than the sample. Background activity can be traced to the existence of minute quantities of radon isotopes in the atmosphere, to the materials used in construction of the laboratory, to accidental contamination within the laboratory, to cosmic radiation, and to the release of radioactive materials into the earth's atmosphere. In order to obtain a true assay, then, it is necessary to correct the total count for background. The counting period required to establish the background correction frequently differs from that for the sample; as a result, it is more convenient to employ counting rates. Then,

$$R_c = R_x - R_b \tag{16-12}$$

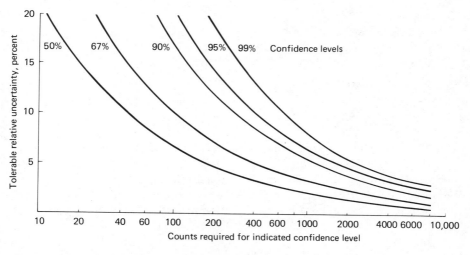

Figure 16-2. Relative uncertainty in counting.

where R_c is the corrected counting rate and R_x and R_b are the rates for the sample and the background, respectively.

The square of the standard deviation of a sum or difference is equal to the sum of the squares of standard deviations of the individual components making up the sum or the difference. Thus, we may write that

$$\sigma_c = \sqrt{\sigma_x^2 + \sigma_b^2} \qquad (16\text{-}13)$$

where σ_c is the standard deviation in R_c, and σ_x and σ_b are the corresponding standard deviations for the sample and the background. Substituting equation (16-6) yields

$$\sigma_c = \sqrt{\frac{R_x}{t_x} + \frac{R_b}{t_b}} \qquad (16\text{-}14)$$

The relative standard deviation $(\sigma_c)_r$ is given by

$$(\sigma_c)_r = \frac{\sqrt{\dfrac{R_x}{t_x} + \dfrac{R_b}{t_b}}}{R_x - R_b} \qquad (16\text{-}15)$$

Example. A sample yielded 1800 counts in a 10-min period. Background was found to be 80 counts in 4 min. Calculate the relative uncertainty in the corrected counting rate at the 95-percent confidence level.

$$R_x = 1800/10 = 180 \text{ cpm}$$
$$R_b = 80/4 = 20 \text{ cpm}$$

Thus,

$$(\sigma_c)_r = \frac{\sqrt{180/10 + 20/4}}{180 - 20} = 0.0300$$

At the 95-percent confidence level,

$$\text{relative uncertainty} = k(\sigma_c)_r = 1.96 \times 0.0300$$
$$= 0.059 \text{ or } 5.9 \text{ percent}$$

Thus, the chances are 95 in 100 that the error in the corrected count is smaller than 5.9 percent.

This example suggests that if the background counting rate is small with respect to the sample rate, the contribution of background to the standard deviation will likewise be small. Thus, the length of the background count can be shorter than that of the sample. It can be shown that the optimum division of time between background and sample counting is given by

$$\frac{t_b}{t_x} = \sqrt{\frac{R_b}{R_x}} \qquad (16\text{-}16)$$

RADIATION DETECTORS

Radiation can be detected and evaluated by measuring the energy released during its interaction with matter. Useful phenomena for this purpose include the darkening of photographic emulsions, the formation of ion pairs, and the induction of scintillation in phosphors.

Photographic Detection

Historically, the exposure of ordinary photographic emulsions formed the basis for radiation detection. Although largely supplanted by other methods in radioassay work, photographic detection remains useful for the monitoring of total dosage, in the detection of flaws in metals, and in the study of cosmic radiation.

Detection Based on Ionization

Alpha, beta, and gamma radiation (as well as ordinary x-rays) all interact with matter to give electrons and positive ions (ion pairs). As noted earlier, alpha particles are highly effective in causing ionization, and thus produce many more ion pairs per particle than do beta particles of the same energy. Gamma and x-ray photons are the least effective ionizing agents.

Several radiation detectors are based upon the electrical conductivity induced in a gas as a consequence of ion-pair formation; included among these are *ionization chambers, Geiger-Müeller tubes,* and *proportional counters.* All are gas-ionization chambers, a typical form of which is shown schematically in Figure 16-3. Radiation enters the chamber through a window of mica, aluminum, lead, or other transparent material. Each particle or photon of radiation interacts with the gas (usually argon) contained in the chamber to produce as many as several hundred primary ion pairs. Under the influence of an applied potential the mobile electrons migrate toward the central wire anode while the slower moving cations move toward the cylindrical metal cathode.

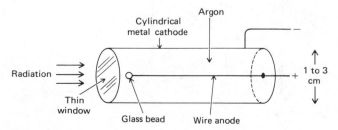

Figure 16-3. Typical gas-ionization chamber.

Figure 16-4 shows the effect of applied potential upon the number of electrons reaching the anode of an ionization chamber for each entering alpha or beta particle. Several characteristic voltage regions are indicated. At potentials less than V_1 the accelerating force on the ion pair is low and the rate of separation of the positive and negative species is insufficient to prevent partial recombination. As a consequence, the number of electrons reaching the anode is smaller than the number produced initially by the incoming radiation.

In the region between V_1 and V_2 the number of electrons reaching the anode is reasonably constant and represents the total number formed by one particle or one photon. Note that an alpha particle causes production of a significantly greater number of electrons than a beta particle.

In the region between V_2 and V_3 the number of electrons increases rapidly with applied potential. This increase is the result of secondary ion-pair production caused by collisions between the accelerated photoelectrons

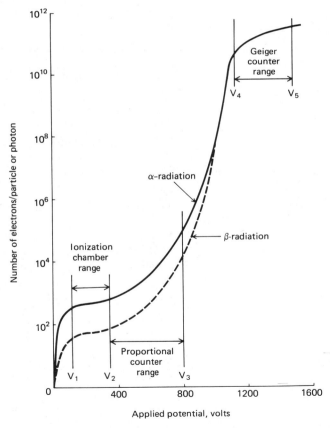

Figure 16-4. The effect of voltage and the type of particle on the number of electrons formed in a gas-ionization chamber.

and gas molecules. As a consequence, amplification of the ion current results.

In the range V_4 to V_5 amplification of the electrical pulse becomes enormous and is governed by the positive space charge created as the faster moving electrons migrate away from the slower positive ions. Because of this effect, the number of electrons reaching the anode is independent of the type and the energy of incoming radiation, and is governed instead by the geometry and the gas pressure of the tube.

Ionization chambers. An ionization chamber detector is operated in the region bounded by V_1 and V_2 in Figure 16-4. The currents produced are minute (10^{-13} to 10^{-16} amp) and their measurement, therefore, demands high electronic amplification. The instruments employed in conjunction with a gas-ionization chamber either measure steady-state currents or detect current pulses resulting from individual ionization events. These instruments differ in the response time (the *time constant*) of their electrical circuits. If the response time is low compared with the rate at which ionizing particles enter the chamber, then a steady, direct current quickly results which can be amplified and measured. If, on the other hand, the electrical response is rapid compared with the number of ionizing events, a pulsed current results, and activity can be determined by counting the pulses. The magnitude of the pulse gives a measure of the energy of the individual nuclear particles. Only alpha radiation produces a sufficient number of electrons per particle for practical pulse counting; the pulse size provides a convenient way of distinguishing alpha particles of differing energies.

Beta and gamma radiation, as well as alpha radiation, can be conveniently measured by amplification of the steady-state current resulting from the ionization process. For alpha detection the gas in the ionization chamber is at or near atmospheric pressure; a very thin aluminum window is required because of the low penetrating power of the radiation. For beta detection a heavier window can be employed to absorb the alpha particles; a higher gas pressure is also needed to offset in part the lower ionizing capabilities of this type of radiation. High gas pressures are needed for the detection of gamma radiation in order to increase the likelihood of ion-pair formation; a lead window can be employed to filter out alpha and beta particles.

Geiger or Geiger-Müeller detectors. The Geiger tube is an ionization chamber operated in the V_4 to V_5 voltage range (Figure 16-4), where ion-current amplifications greater than 10^9 occur. Here, each nuclear particle or photon produces an avalanche of electrons and cations; the resulting currents are thus large and relatively easy to detect and measure.

The conduction of electricity through a chamber operated in the Geiger region (and in the proportional region also) is not continuous because the cation fragments move at a much slower rate than electrons. As a consequence, a space charge develops which effectively terminates the flow of electrons to the anode. The net effect is a momentary pulse of current followed by an interval during which the tube does not conduct. Before conduction can again occur, this space charge must be dissipated by migration of the cations to the walls of the chamber. During the *dead time* when the tube is nonconducting, response to radiation is impossible; the dead time thus represents an upper limit in the response capability of the tube. Typically, the dead time of a Geiger tube is in the range from 50 to 200 μsec.

Geiger tubes are usually filled with argon; a low concentration of an organic substance, often alcohol or methane (a *quench gas*), is also present to minimize the production of secondary electrons when the cations strike the chamber wall. The lifetime of a tube is limited to some 10^8 to 10^9 counts, by which time the quench gas has been depleted.

With a Geiger tube, radiation intensity is determined by a count of the pulses of current. The device is applicable to all types of nuclear and x-radiation. It does not, however, have the large counting range of other detectors because of its relatively long dead time.

Proportional counters. The proportional counter is an ionization chamber that is operated in the V_2 to V_3 voltage region of Figure 16-4. Here, the pulse produced by a nuclear particle or photon is amplified by a factor of 500 to 10,000, but the number of positive ions produced is small enough so that the dead time is only about one microsecond. In general, the pulses from a proportional counter tube must be amplified before being counted. In this respect, the proportional counter is similar to the Geiger tube.

The number of electrons per pulse (the *pulse height*) produced in the proportional region is directly dependent upon the energy of the incoming radiation. By employing a *pulse-height analyzer,* which counts a pulse only if its amplitude falls within certain limits, a proportional counter can be made to respond to beta ray and to x-ray frequencies within a certain range only. A pulse-height analyzer in effect permits electronic filtration of radiation; its function is analogous to a monochromator for visible radiation.

Scintillation Counters

The luminescence produced when radiation strikes a phosphor represents one of the oldest methods of detecting radioactivity, and one of the newest as well. In its earliest application, the technique involved the manual counting of the flashes that resulted when individual alpha particles struck a zinc sulfide screen. The tedium of counting individual flashes by eye led Geiger to the development of gas-ionization detectors that were not only more convenient and reliable but more responsive to beta and gamma radiation as well. The advent of the photomultiplier tube (p. 54) and better phosphors, however, has reversed this trend, and scintillation counting has again become the most important method for radiation detection.

The most widely used modern scintillation counter consists of a crystal of sodium iodide that has been activated by the introduction of perhaps one percent thallium. Often, the crystal is shaped as a cylinder, with a diameter and length of 3 or 4 in.; one of the plane surfaces then faces the cathode of a photomultiplier tube. As the incoming radiation traverses the crystal, its energy is first lost to the scintillator and then is subsequently released as photons of radiant energy. Several thousand photons with a wavelength of about 400 nm are produced by each primary particle or photon over a period of about 0.25 μsec (the *decay time*). Because the decay time is so brief, the dead time of a scintillation counter is short in comparison to the dead time of a gas-ionization detector.

The flashes of light produced in the scintillator crystal are reflected to the photocathode of the photomultiplier tube and are in turn converted to electrical pulses that can be amplified and counted. An important characteristic of scintillators is that the number of photons produced in each flash is approximately proportional to the energy of the incoming radiation. Thus, incorporation of a pulse-height analyzer to monitor the output of a scintillation counter forms the basis of a beta or a gamma ray spectrometer.

In addition to sodium iodide crystals a number of organic scintillators such as stilbene, anthracene, and terphenyl have been used. In crystalline form these compounds have decay times of 0.01 to 0.1 μsec. For counting such weak beta emitters as tritium and carbon-14, organic liquid scintillators are used to advantage because they exhibit less self-absorption of radiation than do the solids. Solutions of *p*-terphenyl in toluene find employment as liquid scintillators.

Semiconductor Counters

Crystals of silicon or germanium have also found use as detectors for radioactivity. Semiconductor detectors are employed in a manner similar to that of the gas-ionization chamber. When radiation is absorbed by the crystal, electrons and positive holes are formed which move toward opposite electrodes under the influence of an applied potential. The resulting current is proportional to the energy of the primary radiation.

Silicon and germanium crystals have been employed mainly for counting alpha and beta particles. Recently, a detector prepared by diffusing lithium ions into a germanium crystal has been developed for gamma ray spectrometers. The resolving power of an instrument equipped with this detector is greater than one employing a sodium iodide crystal. On the other hand, the germanium-lithium detector is less convenient since it must be operated at liquid-nitrogen temperature.

Auxiliary Equipment

From the foregoing discussion it is apparent that most nuclear radiation detectors deliver information in the form of electrical pulses. Thus, the measurement of decay rates requires an instrument for counting such pulses. For low counting rates (500 to 1000 cpm) a simple electromechanical device can be employed. For high counting rates, however, an electronic *scaler* is required. A scaler is an electronic amplifier circuit arranged so that its output terminals transmit only every other input signal. Several scalers may be arranged in series so that the pulses finally counted represent only $\frac{1}{2}$, $\frac{1}{4}$, $\frac{1}{8}$, and so on, of the number produced in the detector. In recent years, decade counters, which respond to pulses in factors of 10, have come into use.

Gamma ray spectrometers. A gamma ray spectrometer employs a sodium iodide crystal or a germanium-lithium semiconductor detector that produces electrical output pulses that are proportional in magnitude to the energy of the primary radiation. As shown in Figure 16-5 these output pulses are amplified by a linear amplifier and appear as voltage signals (in the 100-V range). These signals are fed into a linear pulse-height analyzer, the first stage of which consists of two discriminator circuits. Each discriminator can be set to reject any signal below a certain voltage. As shown in Figure 16-5, the lower discriminator rejects signal 1, which is smaller than V in voltage, but it transmits signals 2 and 3. The upper discriminator, on the other

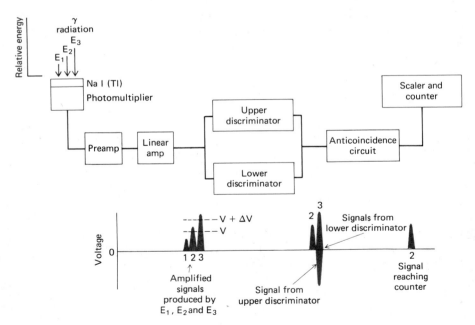

Figure 16-5. Schematic diagram of a single channel pulse height analyzer. The lower discriminator is set to remove voltage pulses smaller than V. The upper removes pulses smaller than V + ΔV, and reverses the polarity of transmitted pulses.

hand, is set to V + ΔV, and thus rejects all but signal 3. In addition, the upper circuit is so arranged that its output signal is reversed in polarity and thus cancels out signal 3 from the lower circuit in the anticoincidence circuit. As a consequence the only signal reaching the counter is one lying in the voltage range ΔV (ΔV is often referred to as the *window* of a pulse-height analyzer).

Pulse-height analyzers are either single- or multiple-channel devices. A single-channel analyzer typically has a voltage range of perhaps 100 V or more with a window of 5 to 10 V. The window can be manually or automatically adjusted to scan the entire voltage range, thus in effect providing data for a gamma ray spectrum.

Multichannel analyzers contain as few as two or as many as several hundred separate channels, each of which acts as a single channel set for a different voltage span, or width. Such an arrangement permits simultaneous counting and recording of an entire spectrum.

Figure 16-6 shows a gamma ray spectrum of a neutron-activated aluminum wire obtained with a 400-channel pulse-height analyzer.

NEUTRON ACTIVATION ANALYSIS

The basis for activation analysis is the measurement of the radioactivity induced in a sample as a result of irradiation by nuclear particles (usually thermal neutrons from a reactor). The single most important advantage of activation methods is high sensitivity, which for many elements

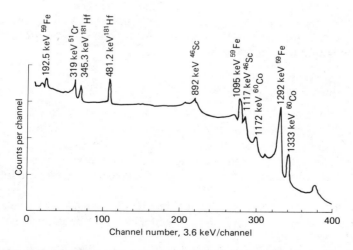

Figure 16-6. Gamma ray spectrum of aluminum wire after neutron activation. Taken from S. G. Prussin, J. A. Harris and J. H. Hollander, *Anal. Chem.*, **37**, 1127 (1965). (With permission of the American Chemical Society.)

exceeds that of other methods by a factor of 100 or more. Concentration determinations in the parts-per-billion range by the procedure are common.

Classification of Activation Methods

Activation methods can be classified in several ways. One is based on the type of radiation employed for excitation of the sample. Slow neutrons, fast neutrons, gamma rays, and various charged particles have been used. The majority of activation methods are based upon thermal neutrons although gamma excitation appears to be a field of growing interest. We shall confine our discussion to irradiation by thermal neutrons.

A second variable upon which to characterize activation methods is the type of emission measured in the final step of the analysis. Here both beta and gamma radiation have been monitored; the former often provides greater sensitivity but, on the other hand, frequently suffers from a lesser selectivity.

Finally, activation methods can be classified as being destructive or nondestructive of the sample. In destructive methods the sample, after irradiation, is dissolved and the element of interest is counted after it has been isolated from other species by suitable chemical or physical means; in this way, interferences from other species made radioactive by the irradiation are avoided. In the nondestructive procedure the activated sample is counted without separatory treatment; here the ability of a gamma ray spectrometer to discriminate between radiation of different energies is called upon to provide the required selectivity. When it can be used, the nondestructive method offers the advantage of great speed. On the other hand, the resolution of a gamma ray spectrometer may be insufficient to eliminate interferences. Also, it does not permit the use of beta emission for the analysis. For these reasons destructive methods have been the more widely used to date. Continued improvements in gamma ray spectrometers are, however, leading to a reversal in this trend.

Destructive Methods

The most common activation procedure involves solution of a known amount of the irradiated sample followed by separation of the element of interest from interferences; such separation procedures as chemical precipitation, solvent extraction, ion exchange, and chromatography are employed. The isolated material or a fraction thereof is then counted for its beta or gamma activity.

In principle it should be feasible to develop an absolute relationship between activity and concentration, given a constant neutron flux of known magnitude, a knowledge of the capture cross section of the nucleus of interest, and the decay constant for the irradiation product. It would also be necessary that the number of activated atoms be but a negligible fraction of the whole, that secondary reactions between neutrons and the product atom be absent, and that no self-shielding occur. The situations where all of these conditions are met are rare; as a result, it is more profitable to employ a comparison technique.

Conventional neutron activation analysis involves irradiation of a standard containing a known mass w_s of the element of interest simultaneously and in the same neutron flux as the sample. Insofar as the activity that results is proportional to mass, and provided also that the other components of the sample do not produce detectable radioactivity, the weight w_x of the element in the sample is given by

$$w_x = \frac{A_x}{A_s} w_s \qquad (16\text{-}17)$$

where A_x and A_s are the activities of the sample and standard. Generally, however, the neutron flux can be expected to generate activity in elements other than the one being determined. Thus, chemical isolation of the species of interest from a solution of the sample ordinarily precedes radio-assay. When the element of interest is present as a trace (as it usually is in a neutron activation analysis) its separation from the major constituents may be difficult and may be the source of large error. To avoid this problem a known weight W_x of the element is introduced to the sample solution as a nonactive *carrier* or *collector*. Separation of the carrier plus the irradiated element ($W_x + w_x$) is then accomplished by precipitation, extraction, or chromatographic means. A weighted quantity w_x' of the isolated material is counted, and the resulting activity a_x is related to the total activity of the original sample A_x by the relationship

$$a_x = A_x \frac{w_x'}{W_x + w_x} \qquad (16\text{-}18)$$

Ordinarily, the amount of nonactive element added is several orders of magnitude greater than the weight from the sample; that is, $w_x \ll W_x$. Equation (16-18) thus simplifies to

$$a_x = \frac{A_x w_x'}{W_x} \qquad (16\text{-}19)$$

The standard sample is treated in an identical way; thus, an analogous expression can be written:

$$a_s = \frac{A_s w_s'}{W_s} \qquad (16\text{-}20)$$

Substituting these expressions into equation (16-17) yields

$$w_x = w_s \frac{a_x W_x w_s'}{a_s W_s w_x'} \qquad (16\text{-}21)$$

In those instances where the condition $w_x \ll W_x$ is not satisfied, a more complex equation results.

The substoichiometric method.[4]   It is experimentally feasible to impose the further conditions that, at the time of assay, $W_x = W_s$ and $w'_x = w'_s$. Equation (16-21) then simplifies to

$$w_x = w_s \frac{a_x}{a_s} \qquad (16\text{-}22)$$

which forms the basis for the substoichiometric method. Where the mass of collector greatly exceeds that of the radioisotope, W_x and W_s are essentially identical if the same weight of collector is added to the sample and the standard. The requirement that the same total amount of species to be taken for radioassay (that is, that $w'_x = w'_s$) is analytically unique in that the same quantity must be taken from solutions of inherently dissimilar concentration. The problem is resolved by introducing a suitable reagent in an amount that is insufficient (substoichiometric) for the complete removal of the species of interest from either sample or standard. If the same amount of reagent is used for both sample and standard, the required equality in amounts assayed is realized.

Nondestructive Method

In the nondestructive method a gamma ray spectrometer is used to measure the activities of the sample and the standard immediately after irradiation. The weight of the element of interest is then calculated directly from equation (16-17).

Clearly, success of the nondestructive method requires that the spectrometer be able to isolate the signal of the gamma rays produced by the element of interest from those arising from the other components. Whether or not an adequate resolution is possible depends upon the complexity of the sample, the presence or absence of elements which produce gamma rays of about the same energy as that of the element of interest, and the resolving power of the spectrometer. Improvements in resolving power that have been made in the last few years have greatly broadened the scope of the nondestructive method. At the present time, however, the most selective and sensitive activation methods are still based upon isolation of the element of interest. The great advantage of the nondestructive approach is its simplicity in terms of sample handling and speed; to be sure, the required instrumentation is more complex.

Application of Neutron Activation

Figure 16-7 illustrates that neutron activation is potentially applicable to the determination of 69 elements. In addition, four of the inert gases form active isotopes with thermal neutrons and thus can also be determined.

[4]For a more detailed treatment see J. Ruzicka and J. Stary, *Substoichiometry in Radiochemical Analysis* (New York: Pergamon Press, Inc., 1968).

Figure 16-7. Estimated sensitivities of neutron activation methods. Upper numbers correspond to β sensitivities in micrograms; lower numbers to γ sensitivities in micrograms. In each case samples were irradiated for 1 hour or less in a thermal neutron flux of 1.8×10^{12} neutrons/cm²/sec. Data of V. P. Guinn and H. R. Lukens, Jr., in G. H. Morrison, Ed., *Trace Analysis: Physical Methods* (New York: Interscience Publishers, 1965), p. 345. (With permission.)

Element	β	γ
Na	5×10^{-3}	5×10^{-3}
Mg	5×10^{-1}	5×10^{-1}
K	5×10^{-2}	5×10^{-2}
Ca	1.0	5
Sc	1×10^{-2}	5×10^{-2}
Ti	5×10^{-1}	5×10^{-2}
V	5×10^{-1}	1×10^{-3}
Cr	—	1
Mn	5×10^{-5}	5×10^{-5}
Fe	50	200
Co	5×10^{-3}	1×10^{-1}
Ni	5×10^{-2}	5×10^{-1}
Cu	1×10^{-3}	1×10^{-3}
Zn	1×10^{-1}	1×10^{-1}
Al	1×10^{-1}	1×10^{-2}
Si	5×10^{-2}	500
P	5×10^{-1}	—
S	5	200
Cl	1×10^{-2}	1×10^{-1}
Ga	5×10^{-3}	5×10^{-3}
Ge	5×10^{-3}	5×10^{-2}
As	1×10^{-3}	5×10^{-3}
Se	—	5
Br	5×10^{-3}	5×10^{-3}
Rb	5×10^{-2}	5
Sr	5×10^{-3}	5×10^{-3}
Zr	1	1
Nb	5×10^{-3}	1
Mo	5×10^{-1}	1×10^{-1}
Ru	1×10^{-2}	5×10^{-2}
Rh	1×10^{-3}	5×10^{-4}
Pd	5×10^{-4}	5
Ag	5×10^{-3}	5×10^{-3}
Cd	5×10^{-2}	5×10^{-1}
In	5×10^{-5}	1×10^{-4}
Sn	5×10^{-1}	5×10^{-1}
Sb	5×10^{-3}	1×10^{-2}
Te	5×10^{-2}	5×10^{-2}
I	5×10^{-3}	1×10^{-2}
Cs	5×10^{-1}	5×10^{-1}
Ba	5×10^{-2}	1×10^{-1}
Hf	—	1
Ta	5×10^{-2}	5×10^{-1}
W	1×10^{-3}	5×10^{-3}
Re	5×10^{-4}	1×10^{-3}
Os	5×10^{-2}	—
Ir	1×10^{-4}	1×10^{-3}
Pt	5×10^{-2}	1×10^{-1}
Au	5×10^{-4}	5×10^{-4}
Hg	—	1×10^{-2}
La	1×10^{-3}	5×10^{-3}
Pb	10	—
Bi	5×10^{-1}	—
Ce	1×10^{-1}	1×10^{-1}
Pr	5×10^{-4}	1×10^{-2}
Nd	1×10^{-1}	1×10^{-1}
Sm	5×10^{-4}	5×10^{-3}
Eu	1×10^{-6}	5×10^{-4}
Gd	1×10^{-2}	5×10^{-1}
Tb	5×10^{-2}	1×10^{-1}
Dy	1×10^{-6}	5×10^{-6}
Ho	1×10^{-4}	1×10^{-4}
Er	1×10^{-3}	1×10^{-3}
Tm	1×10^{-2}	1×10^{-1}
Yb	1×10^{-3}	1×10^{-3}
Lu	5×10^{-5}	5×10^{-5}
Th	5×10^{-2}	5×10^{-2}
U	5×10^{-3}	5×10^{-3}

Finally, three additional elements (oxygen, nitrogen, and yttrium) can be activated with fast neutrons. A list of types of materials to which the method has been applied is impressive and includes metals, alloys, archeological objects, semiconductors, biological specimens, rocks, minerals, and water. Acceptance of evidence developed from activation analysis by courts of law has led to its use in forensic chemistry. Most applications have involved the determination of traces of the various elements.

Accuracy. The principal errors that arise in activation analyses are due to self-shielding, unequal neutron flux at sample and standard, counting uncertainties, and errors in counting due to scattering, absorption, and differences in geometry of the sample and the standard. In most instances it appears possible to reduce the error from these causes to less than 10 percent relative and often to the range of 1 to 3 percent.

Sensitivity. The most important characteristic of the neutron activation method is its remarkable sensitivity for many elements. Note in Figure 16-7, for example, that as little as 10^{-5} μg of several elements can be detected. Note also the wide variations in sensitivities among the elements; thus about 50 μg of iron are required for detection, in contrast to 10^{-6} μg of europium.

The sensitivity of the activation method for an element is a function of a number of variables. Some of these are associated with the properties of the particular nucleus. Others are related to the irradiation process; still others have to do with the efficiency of the counting apparatus.

The effect of a number of these variables on the activity A induced in a sample after irradiation for a time t is given by

$$A = N\sigma\phi \left[1 - \exp\left(-\frac{0.693\ t}{t_{1/2}} \right) \right] \qquad (16\text{-}23)$$

where A is given in counts per second. The quantity N refers to the number of target nuclei, and σ is their neutron capture cross section in cm^2 per nucleus. The neutron flux in units of neutrons per cm^2 second is given by ϕ while t and $t_{1/2}$ are the irradiation time and half-life of the isotope formed; the two times are expressed in the same units.

The neutron capture cross section is a measure of the probability that a given kind of nucleus will capture a neutron. This quantity depends in a complex way upon the energy of the neutron; typically one or more neutron energies exist that correspond to a very high probability for capture.

Figure 16-8 illustrates how the induced activity varies with neutron flux and with irradiation time. It is seen that irradiation in excess of the *saturation* time does not provide any further increase in activity; here the rate of decay and the rate of formation of the active species are the same. Equation (16-23) indicates that the irradiation time required to reach saturation increases with increasing half-life for the product nucleus.

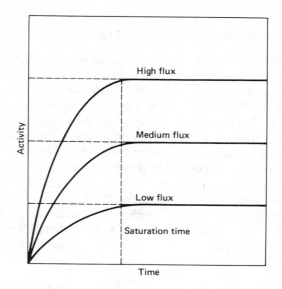

Figure 16-8. The effect of neutron flux upon the activity induced in a sample.

The efficiency of chemical recovery, if required prior to radioassay, may limit the sensitivity of an activation analysis. Other factors include the sensitivity of the detection equipment for the emitted radiation, the extent to which activity in the sample decays between irradiation and assay, the time available for counting, and the magnitude of the background count with respect to that for the sample. A high rate of decay is desirable from the standpoint of minimizing the duration of the counting period. Concomitant with high decay rates, however, is the need to establish with accuracy the time lapse between the cessation of irradiation and the commencement of counting. A further potential complication is associated with counting rates that exceed the resolving time of the detecting system; under these circumstances, a correction must be introduced to account for the difference between elapsed (clock) and live (real) counting times.

ISOTOPIC DILUTION METHODS

Isotopic dilution methods, which predate activation procedures, have been and still are extensively applied to problems in all branches of chemistry. These methods are among the most selective available to chemists.

Both stable and radioactive isotopes are employed in the isotopic dilution technique. The latter are the more convenient, however, because of the ease with which the concentration of the isotope can be determined. We shall limit this discussion to methods employing radioactive species.

Principles of the Isotopic Dilution Procedure

Isotopic dilution methods are based upon the introduction of a known activity of a radioactive species (either the active element or a compound of the element) into a weighed quantity of the sample. After treatment to assure homogeneity between the active species and the nonactive one in the sample, a part of the nonactive species and its active tracer are isolated chemically in the form of a purified compound of known composition. By counting a weighed portion of this product, the extent of dilution of the active material can be calculated and related to the amount of nonactive substance in the original sample. It is important to realize that quantitative recovery of the species is not required. Thus, in contrast to the typical analytical separation, steps can be employed to assure a highly pure product on which to base the analysis. It is this independence from the need for quantitative isolation that leads to the high selectivity of the isotopic dilution method.

Direct isotope dilution.   Assume that W_0 grams of a radioactive species having an activity of A_0 are mixed with a sample containing W_x grams of the inactive substance. After separation and purification, a weight W_r of the species is found to have an activity of A_r. We may then write

$$A_r = \frac{A_0\,W_r}{W_0 + W_x} \tag{16-24}$$

which rearranges to

$$W_x = \frac{A_0}{A_r}\,W_r - W_0 \tag{16-25}$$

Thus, the weight of the species originally present is obtained from the four measured quantities on the right-hand side of equation (16-25). If the activity of the tracer is large, the weight W_0 added can be kept small and equation (16-25) simplifies to

$$W_x = \frac{A_0}{A_r}\,W_r \tag{16-26}$$

Substoichiometric isotope dilution. A substoichiometric method, analogous to that for activation analysis, can also be used in isotopic dilution experiments. Here, identical amounts W_0 of the tracer are added to two solutions that are the same in every respect except that one contains the sample and the other does not. A suitable reagent is then added to isolate a quantity W_r of the species of interest from each; the amount of added reagent is, however, less than that required for complete removal of the species, so that W_r is identical for the two solutions. Equation (16-24) describes the activity of the product from the solution containing the sample; for the solution having no sample, W_x is zero, and equation (16-24) takes the form

$$A'_r = \frac{A_o W'_r}{W_o} \qquad (16\text{-}27)$$

Recall, however, that conditions have been chosen such that $W_r = W'_r$. As a consequence, by dividing equation (16-24) by equation (16-27) and rearranging we obtain

$$W_x = W_o\left(\frac{A'_r}{A_r} - 1\right) \qquad (16\text{-}28)$$

The substoichiometric procedure is advantageous when the amount recovered W_r is so small that its weight is difficult to assess.

Application of the Isotopic Dilution Method

The isotopic dilution technique has been employed for the determination of about thirty of the elements in a variety of matrix materials. Substoichiometric methods have proved useful for the determination of traces of several metallic elements. For example, fractions of a microgram of cadmium, copper, mercury, or zinc have been determined by a procedure in which the element is isolated for counting by extraction with a carbon tetrachloride solution of dithizone; a substoichiometric quantity of the dithizone is employed.

Isotopic dilution procedures have also proved useful for the determination of compounds that are of interest in organic chemistry and biochemistry. Thus, methods have been developed for the determination of such diverse substances as vitamin D, vitamin B_{12}, sucrose, insulin, penicillin, various amino acids, corticosterone, various alcohols, and thyroxine.

The isotopic dilution method has had less widespread application since the advent of activation methods. Continued use of the procedure can be expected, however, because of the relative simplicity of the equipment required. In addition, the procedure is often applicable where the activation method fails.

RADIOMETRIC METHODS[5]

Radiometric methods employ a radioactive reagent of known activity to isolate the substance of interest from the other components of a sample. After quantitative separation, the activity of the product is readily related to the amount of the species being determined. Methods for the

[5]For a review of radiometric methods see T. Brauer and J. Tölgyessy, *Radiometric Titrations* (New York: Pergamon Press, Inc., 1967).

radiometric determination of more than thirty of the common elements have been described. Examples include the determination of chromium by formation of active silver chromate with radioactive silver ion, precipitation of magnesium or zinc by phosphate containing phosphorus-32, and the determination of fluoride ion by precipitation with radioactive calcium.

Problems

1. Silicon-31 is a β-emitter with a half-life of 2.62 hr. Calculate the fraction of this isotope remaining in a sample after
   (a) 1 hr,
   (b) 3 hr,
   (c) 7 hr,
   (d) 12 hr.

2. Calculate the fraction that remains after 30 hr if the radioactive species in the sample is
   (a) ^{28}Mg ($t_{1/2} = 21.2$ hr),
   (b) ^{47}Ca ($t_{1/2} = 4.8$ days),
   (c) ^{130}I ($t_{1/2} = 12.6$ hr),
   (d) ^{101}Rh ($t_{1/2} = 4.3$ days).

3. A $BaSO_4$ sample contains 0.84 μ curie of barium-128 ($t_{1/2} = 2.4$ days). What storage period is needed to assure that its activity is less than 0.01 μ curie?

4. Estimate the standard deviation and the relative standard deviation associated with
   (a) 25 counts,
   (b) 250 counts,
   (c) 2500 counts,
   (d) 2.5×10^4 counts.

5. Estimate the standard deviation (in absolute and relative terms) associated with a counting rate of 125 cpm that is observed for
   (a) 0.50 min,
   (b) 1.0 min,
   (c) 5.0 min,
   (d) 10.0 min.

6. Estimate the absolute and relative uncertainty associated with a measurement involving 450 counts
   (a) at the 50-percent confidence level,
   (b) at the 90-percent confidence level,
   (c) at the 99-percent confidence level.

7. Estimate the absolute and relative uncertainty at the 90-percent confidence level for a measurement that involves a total of
   (a)   48 counts,
   (b) 170 counts,
   (c) 394 counts,
   (d) 925 counts.

8. Estimate the absolute and relative uncertainty at the 90-percent confidence

level associated with the corrected counting rate obtained from a total counting rate of 300 cpm for 10 min and a background count of
  (a) 8 cpm for 2 min,
  (b) 8 cpm for 10 min,
  (c) 14 cpm for 2 min,
  (d) 36 cpm for 2 min.

9. The background activity of a laboratory when measured for 3 min was found to be approximately 9 cpm. What total count should be taken in order to keep the relative uncertainty at the 90-percent confidence level smaller than 5.0 percent, given a total counting rate of about
  (a) 150 cpm,
  (b) 220 cpm,
  (c) 400 cpm.

10. In order to determine the mercury content of a specimen of animal tissue a 0.800-g sample of the tissue and a standard solution containing 0.120 μg of Hg as $HgCl_2$ were irradiated for three days in a thermal neutron flux of 10^{12} neutrons/cm^2 sec. After irradiation was complete, 20.0 mg of Hg as Hg_2Cl_2 was added to each and both were digested in a nitric acid–sulfuric acid mixture to oxidize organic material; suitable precautions were taken to avoid loss of mercury by volatilization. Hydrochloric acid was then added, and the $HgCl_2$ formed was distilled from the reaction mixtures. The mercury in each of the distillates was deposited electrolytically on gold foil electrodes, resulting in an increase in weight of 18.4 mg in the case of the sample and 17.9 mg for the standard. The γ activity due to ^{197}Hg was then determined. The sample was found to yield a count of 850 cpm and the standard 1120 cpm. Calculate the ppm Hg in the sample.

11. Identical electrolytic cells, fitted with silver anodes and platinum cathodes, were arranged in series. Exactly 1.00 ml of a solution containing 4.30×10^{-2} mg of KI labeled with ^{131}I (a β-emitter with a half-life of 8.0 days) and 5.00 ml of an HOAc/OAc$^-$ buffer were introduced to each cell. A 5.00-ml aliquot of an iodide-containing sample was added to one cell only. After passing a substoichiometric quantity of electricity, the anodes were removed and assayed for their β activity. Calculate the weight of I$^-$ in each milliliter of the sample solution if the activity (corrected for background) for the electrode from the cell containing the standard only was 6720 cpm, while that from the cell containing the unknown as well was 4940 cpm.

17 Introduction to Electroanalytical Chemistry

Electroanalytical chemistry comprises a group of methods for quantitative analysis based upon the behavior of a solution of a sample when it is made a part of an electrochemical cell. Three types of electroanalytical methods can be distinguished. A major subdivision includes methods dependent upon the direct relation between concentration and some electrical parameter such as potential, current, resistance (or conductance), capacitance, or quantity of electricity. A second group involves use of an electrical measurement to establish the end point in a volumetric analysis. A third category comprises methods in which a component of the sample is converted by an electrical current into a second phase which is subsequently measured gravimetrically or volumetrically.

Regardless of the type, the intelligent application of an electroanalytical method requires an understanding of the basic theory and the practical aspects of the operation of electrochemical cells. This chapter is devoted largely to these matters.

ELECTROCHEMICAL CELLS

Electrochemical cells can be conveniently classified as *galvanic* if they are employed to produce electrical energy and *electrolytic* when they consume electricity from an external source. Both types of cells are employed in electroanalytical chemistry. It is important to appreciate that many cells can be operated in either a galvanic or an electrolytic sense by variation of experimental conditions.

Cell Components

An electrochemical cell contains two conductors called *electrodes*, each immersed in a suitable electrolyte solution. For current to flow it is necessary (1) that the electrodes be connected externally by means of a metal conductor and (2) that the two electrolyte solutions be in contact so that movement of ions from one to the other can occur. The current in the external circuit then involves a net transfer of electrons from one electrode to the other; in the solutions, on the other hand, the passage of current entails a transfer of ions from one part of the cell to the second. Finally, current is carried across the solid-liquid interface between an electrode and its surrounding electrolyte by either an oxidation or a reduction process. *The electrode at which chemical reduction occurs is termed the cathode regardless of whether the cell is galvanic or electrolytic. The electrode at which oxidation takes place is always the anode.*

It is often convenient to focus attention on the events occurring at one electrode of a cell and to describe the oxidation or the reduction process occurring there in terms of a so-called *half-reaction*. It must always be borne in mind, however, that it is impossible to operate one half-cell in the absence of a second, nor can electrical measurements be performed on one electrode system without reference to a second.

Reactions at cathodes.   Some typical cathodic half-reactions are

$$Cu^{2+} + 2e \rightleftharpoons Cu$$
$$Fe^{3+} + e \rightleftharpoons Fe^{2+}$$
$$2H^+ + 2e \rightleftharpoons H_2(g)$$
$$AgCl(s) + e \rightleftharpoons Ag + Cl^-$$
$$IO_4^- + 2H^+ + 2e \rightleftharpoons IO_3^- + H_2O$$

For each of these processes, electrons are supplied from the external circuit via an electrode that does not participate directly in the chemical reaction. The first process involves deposition of copper on the electrode surface; in the second only a change in oxidation number of one of the components of the electrolyte occurs. The third reaction is frequently observed in aqueous solutions containing no easily reduced species.

The fourth half-reaction is of interest because it can be considered the result of a two-step process; that is,

$$AgCl(s) \rightleftharpoons Ag^+ + Cl^-$$
$$Ag^+ + e \rightleftharpoons Ag$$

Solution of the sparingly soluble precipitate occurs in the first step to provide the silver ions that are reduced in the second. The last half-reaction has been included to demonstrate that a cathodic reaction can involve anions as well as cations.

Reactions at anodes. Examples of typical anodic half-reactions include

$$Cu \rightleftharpoons Cu^{2+} + 2e$$
$$Fe^{2+} \rightleftharpoons Fe^{3+} + e$$
$$2Cl^- \rightleftharpoons Cl_2 + 2e$$
$$H_2 \text{ (g)} \rightleftharpoons 2H^+ + 2e$$
$$2H_2O \rightleftharpoons O_2 + 4H^+ + 4e$$

The first half-reaction requires a copper electrode to supply Cu^{2+} ions to the solution. The remainder can take place at any of a variety of metal surfaces. In order to cause the fourth reaction to occur it is necessary to bubble hydrogen over the surface of an inert metal such as platinum. The reaction can then be formulated as

$$H_2 \text{ (g)} \rightleftharpoons H_2 \text{ (sat'd)}$$
$$H_2 \text{ (sat'd)} \rightleftharpoons 2H^+ + 2e$$

The final reaction giving oxygen as a product is a common anodic process in aqueous solutions containing no easily oxidized species.

Liquid junctions. Electrochemical cells can also be classified by another characteristic. *Cells without liquid junctions* are those in which the two electrodes share a common electrolyte. Often, however, it is desirable or necessary to employ a cell in which the composition of the solution surrounding the cathode differs from that around the anode. Such a cell is called *a cell with liquid junction*, the junction referring to the interface between the two solutions. A variety of techniques are employed to minimize mechanical mixing of the solutions at the junction. One of these, involving separation of the two electrolytes with a fritted glass disk, is shown in Figure 17-1. Another method entails isolating one of the electrodes and its electrolyte in a porous paper or a ceramic cup; the cup then dips into the solution holding the second electrode.

Generally, cells with liquid junctions are employed to avoid direct reaction between the components of the two half-cells. Thus, for the cell shown in Figure 17-1 the desired half-reactions are

$$\text{anode} \qquad Zn \rightleftharpoons Zn^{2+} + 2e$$
$$\text{cathode} \qquad Cu^{2+} + 2e \rightleftharpoons Cu$$

and the flow of electrons in the external circuit occurs as shown. If the two electrolyte solutions are mixed, the cell efficiency would be reduced as a consequence of the direct deposition of the copper on the zinc electrode:

$$Cu^{2+} + Zn \rightarrow Cu + Zn^{2+}$$

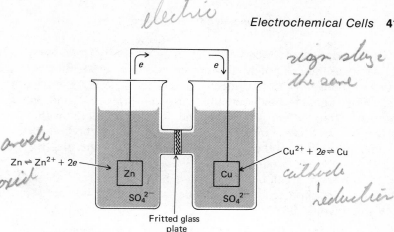

electric

zign slize
the same

anode
oxid

cathode
reduction

Figure 17-1. A cell with liquid junction.

Figure 17-2 shows a cell in which a liquid junction is unnecessary because the rate of oxidation of hydrogen by silver ions is negligibly slow.

The salt bridge. For reasons to be discussed later, electrochemical cells are often equipped with a *salt bridge* to separate the anode and cathode electrolytes. This device takes a variety of forms, one of which is shown in Figure 17-5. Here, the bridge consists of a U-shaped tube filled with a saturated solution of potassium chloride. Such a cell has two liquid junctions, one between the cathode electrolyte and the bridge liquid and the second between the anodic electrolyte and the other end of the bridge.

Schematic representation of cells. In order to simplify the description of cells, chemists often employ a shorthand notation. For example, the cells shown in Figures 17-1 and 17-2 can be described by

$$Zn \,|\, ZnSO_4 \,(xM) \,|\, CuSO_4 \,(yM) \,|\, Cu$$
$$Pt, H_2 \,(p = 1 \text{ atm}) \,|\, H^+ \,(0.1 \ M), \ Cl^- \,(0.1 \ M), \ AgCl \,(\text{sat'd}) \,|\, Ag$$

By convention, *the anode and information with respect to the solution with which it is in contact is listed on the left.* Single vertical lines represent

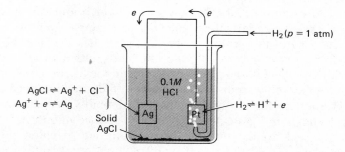

Figure 17-2. A cell without liquid junction.

phase boundaries at which potentials may develop. Thus, in the first example, a part of the cell potential is considered to be associated with the phase boundary between the zinc electrode and the zinc sulfate solution. As we shall show later, a small potential also develops at liquid junctions (between the zinc sulfate and the copper sulfate solutions in this cell); thus, another vertical line is inserted at this point. The cathode is then described symbolically with another vertical line separating the electrolyte solution from the copper electrode.

In the second cell only two phase boundaries are evident, the electrolyte being common to both electrodes. An equally correct representation of this cell would be

$$\text{Pt} \,|\, \text{H}_2 \text{ (sat'd)}, \text{HCl } (0.1 \text{ } M), \text{Ag}^+ \, (1.8 \times 10^{-9} \text{ } M) \,|\, \text{Ag}$$

Here, the hydrogen gas concentration is specified as that of a solution that has been saturated (a partial pressure of 1.00 atm being implicit); the molar silver ion concentration is that calculated from the solubility product for silver chloride.

The presence of a salt bridge in a cell is indicated by two vertical lines, implying that a potential difference is associated with each of the two interfaces. Thus, the cell shown in Figure 17-5 would be represented as

$$\text{M} \,|\, \text{M}^{2+} \, (xM) \,\|\, \text{H}^+ \, (1.00 \text{ } M) \,|\, \text{H}_2 \text{ (1 atm)}, \text{Pt}$$

Current Passage in a Cell

As noted earlier, current is transported within a cell by the migration of ions; in common with metallic conductors, Ohm's law is obeyed. That is,

$$I = E/R \tag{17-1}$$

where I is the current in amperes, E is the potential difference in volts responsible for movement of the ions, and R is the resistance in ohms of the electrolyte to current passage. The resistance is dependent upon the kinds and concentrations of ions in the solution.

It is found experimentally that the rate at which ions move in a solution under a fixed potential varies widely. For example, the rate of movement (or *mobility*) of the proton is about seven times that for the sodium ion and five times that for the chloride ion. Thus, although all of the ions in a solution participate in the passage of current, the fraction carried by one ion may differ markedly from that of another. This fraction depends upon the concentration of the ion and its inherent mobility. To illustrate, consider the cell shown in Figure 17-3. This cell is divided into three imaginary compartments, each containing six hydrogen ions and six chloride ions. Six electrons are then forced through the cell by a battery, resulting in formation of three molecules of hydrogen at the cathode and three of chlorine at the anode. To offset the resulting charge imbalance brought about by the removal of ions from the electrode compartments, migration occurs, positive ions moving toward the negative electrode and conversely. As we have mentioned, however, protons are about five times more mobile than the chloride ion; the consequence is a significant difference in concentrations in the outer electrode compartments after electrolysis is complete. In effect, five-sixths of the current has been carried through the solution by the hydrogen ions and one-sixth by the chloride ions.

It is important to appreciate that the current-carrying species need not be the electrode reactants. Thus, if we were to introduce, let us say, 100 potassium and nitrate ions into each of the three compartments of the cell under consideration, the charge imbalance resulting from electrolysis could be offset by migration of the added species as well as by the hydrogen and chloride ions. Since the added salt would represent an enormous excess, essentially all of the current would be carried by the potassium and nitrate ions rather than by the reactant ions.

Reversible and Irreversible Cells

A change in the direction of current flow in the cell shown in Figure 17-1 simply causes a reversal of the reactions that occur at the electrodes. A cell exhibiting this property is said to be electrochemically *reversible.*

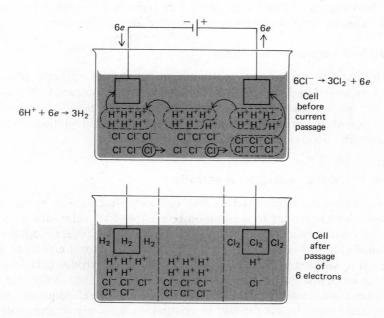

Figure 17-3. Illustration of changes brought about by current passage in a cell.

Cells in which reversal of current results in different reactions at one or both electrodes are called *irreversible*.

ELECTRODE POTENTIALS

It is useful to view the potential developed by a galvanic cell or required to operate an electrolytic cell as being the sum or difference of several potentials that arise at different parts of the cell. *Electrode, or half-cell, potentials* are produced by the chemical driving forces of the half-reactions taking place at the surfaces of the two electrodes.

Nature of Electrode Potentials

At the outset it should be realized that there is no way of determining an absolute value for the potential of a single electrode since all voltage-measuring devices determine only *differences* in potential. One lead of such a device is connected to the electrode in question; in order to measure a potential difference, however, the second lead must be brought in contact with the solution via another conductor. This latter contact inevitably involves a solid-solution interface and hence acts as a second half-cell at which a chemical reaction must also take place. Thus, we do not obtain an absolute measurement of the desired half-cell potential, but rather the resultant between the potential of interest and the half-cell potential for the contact between the voltage-measuring device and the solution.

Relative Half-cell Potentials

Our inability to measure absolute potentials for half-cell processes is not a serious handicap because relative half-cell potentials are just as useful for most purposes. These relative potentials can be combined to give cell potentials; in addition, they are useful for the calculation of equilibrium constants for oxidation-reduction processes.

The Standard Hydrogen Electrode

To obtain consistent relative half-cell potential data it is necessary to compare all electrodes to a common reference. This reference electrode should be relatively easy to construct, should exhibit reversible behavior, and should give constant and reproducible potentials for a given set of experimental conditions. The standard hydrogen electrode (SHE) meets these requirements and is used universally as the ultimate reference electrode. There are, in addition, a number of secondary reference electrodes that are more convenient for use in routine work; some of these are discussed later in this chapter.

Figure 17-4 illustrates a standard hydrogen electrode. It consists of a piece of platinum immersed in a solution with a hydrogen ion activity of one; hydrogen is bubbled across the surface of the platinum in an uninterrupted stream to assure that the electrode is continuously in contact with both the solution and the gas. In order to achieve the largest possible surface area the electrode is coated with a finely divided layer of platinum called *platinum black*. The partial pressure of hydrogen is maintained at one atmosphere above the liquid phase.

An electrode of this type is called a *gas electrode*. The platinum takes no part in the electrochemical reaction other than to serve as the site for the transfer of electrons. The half-cell reaction responsible for the transmission of current across the interface is

$$H_2 \text{ (gas)} \rightleftharpoons 2H^+ + 2e$$

The primary electron-transfer process undoubtedly involves hydrogen molecules dissolved in the solution. The continuous stream of gas at constant pressure assures a constant hydrogen-molecule concentration in the solution.

The hydrogen electrode may act as anode or cathode, depending upon the half-cell with which it is coupled. In the one case, oxidation of the gas to hydrogen ions occurs; in the other, the reverse reaction takes place. Under proper conditions, then, the hydrogen electrode is reversible in its behavior.

The potential of the hydrogen electrode is dependent upon temperature, the concentration of hydrogen ions in the solution, and the pressure of the hydrogen at the electrode surface. Values for these parameters must be carefully defined in order for the half-cell process to serve as a

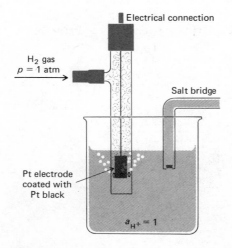

Figure 17-4. The standard hydrogen electrode.

reference. Specifications for the *standard* (known also as a *normal*) *hydrogen* electrode call for an activity of 1.00 mole per liter for the hydrogen ions and partial pressure of 1.00 atmosphere for the gas. For convenience, the potential of this half-cell is *assigned* the value of exactly zero volts at all temperatures.

Measurement of Electrode Potentials

Electrode potentials can be measured relative to the standard hydrogen electrode by means of a cell such as that shown in Figure 17-5. Here, one half of the cell consists of the standard hydrogen electrode and the other half is the electrode whose potential is to be determined. Connecting the two cells is a salt bridge, consisting of a tube containing a concentrated electrolyte solution—most often saturated potassium chloride. This bridge provides electrical contact between the two halves of the cell but prevents mixture of the contents of the half-cells; the passage of current takes place by ionic migration, as previously described. In general, the effect of a salt bridge upon the cell voltage is vanishingly small.

The left-hand half-cell in Figure 17-5 consists of a pure metal in contact with a solution of its ions. The electrode reaction thus is

$$M \rightleftharpoons M^{2+} + 2e$$

If the metal is cadmium and the solution is approximately one molar with respect to cadmium ions, the voltage indicated by the measuring device V will be about 0.4 V. Moreover, the metal electrode will behave as the anode; electrons thus tend to move spontaneously from the metal electrode to the hydrogen electrode via the external circuit. The half-cell reactions for this galvanic cell are

$$Cd \rightleftharpoons Cd^{2+} + 2e \qquad \text{anode}$$
$$2H^+ + 2e \rightleftharpoons H_2 \qquad \text{cathode}$$

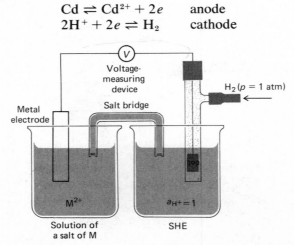

Figure 17-5. Schematic diagram of an arrangement for measurement of electrode potentials against the standard hydrogen electrode.

The overall reaction is the sum of these, or

$$Cd + 2H^+ \rightleftharpoons Cd^{2+} + H_2$$

If the cadmium electrode were replaced by a zinc electrode immersed in a solution that is 1 M in zinc ions, a potential of slightly less than 0.8 V would be observed. Again the metal electrode would behave as the anode. The larger voltage developed reflects the greater tendency of zinc to be oxidized; the difference between this potential and the one observed for cadmium gives a quantitative measure of the relative strengths of these two metals as reducing agents.

Since a value of zero is assigned as the potential for the standard hydrogen electrode, the potentials for the two half-cells thus become 0.4 and 0.8 V with respect to this reference. It must be understood, however, that these are actually the potentials of electrochemical cells in which the standard hydrogen electrode serves as a common reference.

If the half-cell in Figure 17-5 consisted of a copper electrode in a 1 M solution of copper(II) ions, the potential of the cell would be about 0.3 V. Here, however, copper would tend to plate out, and external electron flow, if allowed, would be from the hydrogen electrode to the copper electrode. Obviously, the spontaneous cell reaction is the reverse of the two cells considered earlier:

$$Cu^{2+} + H_2 \rightleftharpoons Cu + 2H^+$$

Thus, metallic copper is not only a much less effective reducing agent than either zinc or cadmium but also less effective than hydrogen. As before, the observed potential is a quantitative measure of this strength; in comparing this potential with those for the half-reactions of zinc and cadmium, however, it becomes necessary to indicate the differences in behavior of the electrode systems with respect to the reference electrode. This can be done conveniently by assigning a positive or a negative sign to the potentials, thereby making the sign of the potential for the copper half-reaction opposite to that for the other two. The choice as to which potential shall be positive and which shall be negative is purely arbitrary; whatever sign convention is chosen, however, must be used consistently.

Sign Conventions for Electrode Potentials

It is not surprising that the arbitrary nature of specifying signs has led to much controversy and confusion in the course of the development of electrochemistry. In 1953 the International Union of Pure and Applied Chemistry (IUPAC), meeting in Stockholm, attempted to resolve this controversy; the sign convention adopted at this meeting is sometimes called the IUPAC, or Stockholm, convention; there appears to be hope for its general adoption in years to come. The IUPAC sign convention will be used throughout this text.

The IUPAC sign convention for electrode potentials is designed to serve the needs of two groups of scientists: (1) the electrochemist, for whom the negative sign attached to the terminal of a galvanic cell is a practical indicator of the half-cell from which electrons tend to flow into the external circuit, and (2) the theoretical chemist, for whom the sign serves to indicate the direction in which the half-cell reaction tends to proceed.

The IUPAC convention defines an *electrode potential* (or more exactly, a *relative electrode potential*) as the electromotive force of a cell consisting of a standard hydrogen electrode and the half-cell of interest. The electrode potential is given the sign of the conductor attached to the half-cell of interest. In order to meet the needs of the two groups mentioned above it is necessary to designate a common direction for all half-reactions whose electrode potentials are being specified. Thus, the IUPAC convention states that the electrode potential and sign shall apply to half-reactions *written as reductions*, regardless of the actual direction in which the half-reaction will occur in the galvanic cell involving the standard hydrogen electrode.[1]

According to the IUPAC convention the electrode potentials for the three half-reactions we have considered are written as

$$Zn^{2+} + 2e \rightleftharpoons Zn \qquad E = -0.8 \text{ V}$$
$$Cd^{2+} + 2e \rightleftharpoons Cd \qquad E = -0.4 \text{ V}$$
$$Cu^{2+} + 2e \rightleftharpoons Cu \qquad E = +0.3 \text{ V}$$

Although all three half-reactions have been written as reductions, the signs of the potentials are determined by the actual direction of external electron flow occurring in the spontaneous cell reactions

$$Zn + 2H^+ \rightarrow Zn^{2+} + H_2$$
$$Cd + 2H^+ \rightarrow Cd^{2+} + H_2$$
$$Cu^{2+} + H_2 \rightarrow Cu + 2H^+$$

In the first two reactions, electrons tend to travel from the metal conductor to the external circuit, and the electrode is negative with respect to the standard hydrogen electrode; the electrode potentials for both are given a negative sign. With the copper electrode the travel is in the opposite direction, and the electrode is positive with respect to the standard hydrogen electrode; the electrode potential is thus given a positive sign.

A positive sign is always associated with half-reactions that proceed spontaneously in electrochemical cells containing the standard hydrogen electrode as the other electrode; a negative sign connotes a half-reaction that is nonspontaneous with respect to the same reference. Thus, in the preceding examples the reduction of Cu^{2+} occurs spontaneously and the electrode potential is positive; on the other hand, reduction of zinc ion and

---

[1]Thus, according to IUPAC nomenclature, the terms *electrode potential* and *reduction potential* are synonymous; there is no objection to the use of *oxidation potential* to connote processes occurring in the opposite sense, but an oxidation potential should no longer be called an "electrode potential."

cadmium ion by gaseous hydrogen is not spontaneous, and the electrode potentials bear negative signs.

The IUPAC convention was adopted in 1953, but electrode-potential data provided in many texts and reference works are not always in accord with it. For example, in the prime source of oxidation-potential data compiled by Latimer,[2] one finds

$$Zn \rightleftharpoons Zn^{2+} + 2e \qquad E = +0.76 \text{ V}$$
$$Cu \rightleftharpoons Cu^{2+} + 2e \qquad E = -0.34 \text{ V}$$

To convert these data to electrode potentials as defined by the IUPAC convention, one must mentally (1) express the half-reactions as reductions, and (2) change the signs of the potentials.

The sign convention employed in tables of electrode potentials is seldom explicitly stated; this information is readily determined, however, by referring to a half-reaction with which one is familiar and noting the direction of the reaction and the sign of the potential associated with it. Whatever changes, if any, are required to convert to the IUPAC convention are then applied to the remainder of the data in the table. For example, all one needs to remember is that under the IUPAC convention strong oxidizing agents such as chlorine have large positive electrode potentials; that is, the reaction

$$Cl_2 + 2e \rightleftharpoons 2Cl^- \qquad E = +1.36 \text{ V}$$

tends to occur spontaneously with respect to the hydrogen half-reaction. The sign and the direction of this reaction in a given table can then serve as a key to any changes needed to convert all data to the IUPAC convention.

Effect of Concentration on Electrode Potentials

The electrode potential, because it is a measure of the chemical driving force of a half-reaction, is affected by concentration. Thus, the tendency of copper(II) ions to be reduced to elemental copper is much greater from concentrated than from dilute solutions; therefore, the electrode potential for the reduction reaction is also greater in the more concentrated solution. In general, the concentrations of the reactants and products of a half-reaction have a marked effect on the electrode potential, and the quantitative aspects of this effect must now be considered.

Consider the generalized, reversible half-cell reaction

$$a\text{A} + b\text{B} + \cdots + ne \rightleftharpoons c\text{C} + d\text{D} + \cdots$$

where the capital roman letters represent formulas of reacting species (whether charged or uncharged), e represents the electron, and the re-

[2]W. M. Latimer, *The Oxidation States of the Elements and Their Potentials in Aqueous Solutions*, 2d ed. (Englewood Cliffs, N.J.: Prentice-Hall, Inc., 1952).

maining lower case italic letters indicate the number of moles of each species participating in the reaction. From thermodynamics it can be shown that the potential E for this electrode process is governed by the relation

$$E = E^0 - \frac{RT}{nF} \ln \frac{[C]^c[D]^d \cdots}{[A]^a[B]^b \cdots} \qquad (17\text{-}2)$$

where E^0 = a constant characteristic of a particular half-reaction
$\quad R$ = the gas constant = 8.314 volt coulombs/°K/mole
$\quad T$ = the absolute temperature
$\quad n$ = number of electrons participating in the reaction as defined by the equation describing the half-cell reaction
$\quad F$ = the faraday = 96,493 coulombs
$\quad \ln$ = the natural logarithm = 2.303 $\log_{10}$

Substituting numerical values for the various constants and converting to base-10 logarithms, equation (17-2) becomes, at 25°C

$$E = E^0 - \frac{0.059}{n} \log \frac{[C]^c[D]^d \cdots}{[A]^a[B]^b \cdots} \qquad (17\text{-}3)$$

The letters in brackets represent the activities of the reacting species. As with equilibrium-constant expressions, certain approximations for these activities are useful. Thus, where the substances are dissolved in a solvent,

$$[\ ] \cong \text{concentration in moles per liter}$$

If the reactant is a gas,

$$[\ ] \cong \text{partial pressure of gas, atmospheres}$$

If the reactant exists in a second phase as a pure solid or liquid, then by definition,[3]

$$[\ ] = 1$$

Finally, even though the solvent may be involved in the half-cell process, its concentration is ordinarily so large with respect to the other reactants that, for all practical purposes, it can be considered to remain unchanged; its contribution is customarily included in the constant E^0. Thus, in aqueous solutions, a term for water appearing in equation (17-3) simplifies to

$$[H_2O] = 1$$

Equation (17-2) is often called the *Nernst equation* in honor of a nineteenth-century electrochemist. Application of the Nernst equation, and

[3]The arbitrary assignment of unity to the activity of the second phase assumes that the concentration of the substance in this phase is constant and that the equilibrium is unaffected by the *quantity* present.

the approximations associated with it, is illustrated in the following examples:

(1) $\qquad Zn^{2+} + 2e \rightleftharpoons Zn \qquad E = E^0 - \dfrac{0.059}{2} \log \dfrac{1}{[Zn^{2+}]}$

Here, the activity of the elemental zinc is unity; the electrode potential varies with the logarithm of the molar concentration of zinc ions in the solution.

(2) $\qquad Fe^{3+} + e \rightleftharpoons Fe^{2+} \qquad E = E^0 - \dfrac{0.059}{1} \log \dfrac{[Fe^{2+}]}{[Fe^{3+}]}$

This electrode potential can be measured with an inert metal electrode immersed in a solution containing iron(II) and iron(III). The potential is dependent upon the ratio between the molar concentrations of these ions.

(3) $\qquad 2H^+ + 2e \rightleftharpoons H_2 \qquad E = E^0 - \dfrac{0.059}{2} \log \dfrac{p_{H_2}}{[H^+]^2}$

In this example p_{H_2} represents the partial pressure of hydrogen expressed in atmospheres at the surface of the electrode. Ordinarily, this is very close to atmospheric pressure.

(4) $\qquad Cr_2O_7^{2-} + 14H^+ + 6e \rightleftharpoons 2Cr^{3+} + 7H_2O$

$$E = E^0 - \dfrac{0.059}{6} \log \dfrac{[Cr^{3+}]^2}{[Cr_2O_7^{2-}] [H^+]^{14}}$$

Here, the potential is dependent not only on the concentration of the chromium(III) and dichromate ions, but also on the pH of the solution.

(5) $\qquad AgCl + e \rightleftharpoons Ag + Cl^- \qquad E = E^0 - \dfrac{0.059}{1} \log [Cl^-]$

Here, the activities of both metallic silver and silver chloride are equal to one by definition; the potential of the silver electrode will be dependent only upon the chloride ion concentration.

We need to make one additional remark with respect to the application of the Nernst equation. Although it is not obvious from our presentation, the quotient in the logarithm term is a unitless quantity. Each of the individual terms in the quotient is in fact a ratio of the present activity of the species compared with its activity in the standard state (which has been chosen as a unity). Thus, for the half-reaction in example 3,

$$E = E^0 - \dfrac{0.059}{2} \log \dfrac{p_{H_2}/(p_{H_2})_0}{[H^+]^2/[H^+]_0^2} = E^0 - \dfrac{0.059}{2} \log \dfrac{p_{H_2}/1.00}{[H^+]^2/(1.00)^2}$$

Here, $(p_{H_2})_0$ is the partial pressure of hydrogen in its standard state, one atmosphere. Similarly, $[H^+]_0$ represents the activity of hydrogen ion in its standard state, one mole per liter. Hence, the units cancel.

The Standard Electrode Potential E^0

An examination of equation (17-2) or equation (17-3) reveals that the constant E^0 is equal to the half-cell potential when the logarithmic term is zero. This condition occurs whenever the activity quotient is equal to unity, one instance being when the activities of all reactants and products are one; thus, the *standard electrode potential is defined as the potential of a half-cell reaction with respect to the standard hydrogen electrode when all reactants and products are at unit activity.*

The standard electrode potential is a fundamental physical constant that gives a quantitative description of the relative driving force of a half-cell reaction. Several facts regarding it should be kept in mind. First, a standard electrode potential is temperature-dependent; if it is to have significance, the temperature at which it is determined must be specified. Second, the standard electrode potential is a relative quantity in the sense that it is really the potential of a cell in which one of the electrodes is a carefully specified reference electrode — that is, the standard hydrogen electrode. Third, a standard electrode potential is given the sign of the conductor attached to the half-cell of interest under the specified conditions of activity. Finally, the value of a standard potential is a measure of the intensity of the driving force of a half-reaction. As such, it is independent of the notation employed to express the half-cell process. Thus, the potential for the reaction

$$Ag^+ + e \rightleftharpoons Ag \qquad E^0 = +0.799 \text{ V}$$

while dependent upon the *concentration* of silver ions, is the same regardless of whether we write the reaction as above or as

$$100Ag^+ + 100e \rightleftharpoons 100Ag \qquad E^0 = +0.799 \text{ V}$$

Standard electrode potentials are available for numerous half-reactions. Many of these have been determined directly from voltage measurements of cells in which the standard hydrogen electrode actually constitutes the other electrode. It is possible, however, to calculate E^0 values from equilibrium studies of oxidation-reduction systems and from thermochemical data relating to such reactions. Many of the values found in the literature were so obtained. W. M. Latimer[4] published a definitive work that is recommended as an authoritative source for standard electrode potential data.

A few standard electrode potentials are given in Table 17-1; a more comprehensive list is found in the Appendix. In these tabulations the species in the upper left part of the table are most easily reduced, as evidenced by the large positive E^0 values; they are therefore the most effective

[4]W. M. Latimer, *The Oxidation States of the Elements and Their Potentials in Aqueous Solutions,* 2d ed. (Englewood Cliffs, N.J.: Prentice-Hall, Inc., 1952).

Table 17-1 Standard Electrode Potentials[a]

Reaction	E^0 at 25°C, volts
$MnO_4^- + 8H^+ + 5e \rightleftharpoons Mn^{2+} + 4H_2O$	+1.51
$Cl_2 + 2e \rightleftharpoons 2Cl^-$	+1.359
$Ag^+ + e \rightleftharpoons Ag$	+0.799
$Fe^{3+} + e \rightleftharpoons Fe^{2+}$	+0.771
$I_3^- + 2e \rightleftharpoons 3I^-$	+0.536
$AgCl + e \rightleftharpoons Ag + Cl^-$	+0.222
$Ag(S_2O_3)_2^{3-} + e \rightleftharpoons Ag + 2S_2O_3^{2-}$	+0.010
$2H^+ + 2e \rightleftharpoons H_2$	0.000
$Cd^{2+} + 2e \rightleftharpoons Cd$	−0.403
$Cr^{3+} + e \rightleftharpoons Cr^{2+}$	−0.41
$Zn^{2+} + 2e \rightleftharpoons Zn$	−0.763

[a]See Appendix for a more extensive list and footnote 4 for source.

oxidizing agents. Proceeding down the table, each succeeding species is a less effective acceptor of electrons than the one above it. The half-cell reactions at the bottom of the table have little tendency to occur as written. On the other hand, they do tend to occur in the opposite sense, as oxidations; the most effective reducing agents, then, are those species that appear in the lower right-hand portion of a table of standard electrode potentials.

A compilation of standard potentials provides the chemist with a qualitative picture regarding the extent and direction of electron-transfer reactions between the tabulated species. On the basis of Table 17-1, for example, we see that zinc is more easily oxidized than cadmium and we conclude that a piece of zinc immersed in a solution containing cadmium ions causes the deposition of metallic cadmium. On the other hand, cadmium has little tendency to reduce zinc ions. Also from Table 17-1 we see that iron(III) is a better oxidizing agent than the triiodide ion. We may therefore predict that in a solution containing an equilibrium mixture of iron(III), iodide, iron(II), and triiodide ions, the latter two will predominate.

Calculations of Half-cell Potentials from E^0 Values

Application of the Nernst equation to the calculation of half-cell potentials is illustrated in the following example.

Example. What is the potential for the half-cell consisting of a cadmium electrode immersed in a solution that is 0.0100 F in Cd^{2+}?
From Table 17-1 we find

$$Cd^{2+} + 2e \rightleftharpoons Cd \qquad E^0 = -0.403 \text{ V}$$

and we write

$$E = E^0 - \frac{0.059}{2} \log \frac{1}{[Cd^{2+}]}$$

Substituting the Cd^{2+} concentration into the equation,

$$E = -0.403 - \frac{0.059}{2} \log \frac{1}{(0.0100)}$$

$$= -0.403 - \frac{0.059}{2} (+2.0)$$

$$= -0.462 \text{ V}$$

The sign for the potential simply indicates the direction of the half-reaction when this half-cell is coupled with the standard hydrogen electrode. The fact that it is negative shows that the reverse reaction,

$$Cd + 2H^+ \rightleftharpoons H_2 + Cd^{2+}$$

would occur spontaneously. Note that the calculated potential is a larger negative number than the standard electrode potential itself. This follows from equilibrium considerations, since the half-reaction, as written, has less tendency to occur with the lower cadmium ion concentration.

Electrode Potentials in Presence of Precipitation and Complex-forming Reagents

Reagents that react with the participants of an electrode process have a marked effect on the potential for that process. For example, the standard electrode potential for the reaction $Ag^+ + e \rightleftharpoons Ag$ is $+0.799$ V. Addition of chloride ions to such a solution materially alters the silver ion concentration and hence the electrode potential. This is illustrated in the following example.

Example. Calculate the potential of a silver electrode in a solution that is saturated with silver chloride and has a chloride ion activity of exactly 1.00.

$$Ag^+ + e \rightleftharpoons Ag \qquad E^0_{Ag^+ \rightarrow Ag} = +0.799 \text{ V}$$

$$E = 0.799 - 0.059 \log \frac{1}{[Ag^+]}$$

We may calculate $[Ag^+]$ from the solubility-product constant:

$$[Ag^+] = \frac{K_{sp}}{[Cl^-]}$$

Substituting this into the Nernst equation,

$$E = 0.799 - \frac{0.059}{1} \log \frac{[Cl^-]}{K_{sp}}$$

This may be rearranged to give

$$E = 0.799 + 0.059 \log K_{sp} - 0.059 \log [\text{Cl}^-] \qquad (17\text{-}4)$$

If we substitute 1.00 for [Cl$^-$] and use a value of 1.82×10^{-10} for K_{sp}, we obtain

$$E = +0.222 \text{ V}$$

This calculation shows that the half-cell potential for the reduction of silver ion becomes smaller in the presence of chloride ions. Qualitatively, this is what we would expect, since removal of silver ions decreases the tendency of the silver ions to be reduced.

Equation (17-4) relates the potential of a silver electrode to the chloride ion concentration of a solution that is also saturated with silver chloride. *When the chloride ion activity is unity*, the potential is the sum of two constants; this sum can be called the standard electrode potential for the half-reaction:

$$\text{AgCl} + e \rightleftharpoons \text{Ag} + \text{Cl}^- \qquad E^0_{\text{AgCl} \rightarrow \text{Ag}} = +0.222 \text{ V}$$

where

$$E^0_{\text{AgCl} \rightarrow \text{Ag}} = E^0_{\text{Ag}^+ \rightarrow \text{Ag}} + 0.059 \log K_{sp}$$

Thus, the potential of a silver electrode in contact with a solution saturated with silver chloride can be described either in terms of the silver ion concentration (using the standard electrode potential for the simple silver half-reaction) or in terms of the chloride ion concentration (using the standard potential for the silver-silver chloride half-reaction).

In an analogous fashion we can treat the behavior of a silver electrode in a solution of an ion that forms a soluble complex with the silver ion. For example, the half-reaction for a silver electrode in a solution containing silver and thiosulfate ions is

$$\text{Ag(S}_2\text{O}_3)_2^{3-} + e \rightleftharpoons \text{Ag} + 2\text{S}_2\text{O}_3^{2-}$$

The standard electrode potential for this half-reaction will be the electrode potential when both the complex and the complexing anion are at unit activity. Using the same approach as in the previous example, we find that

$$E^0_{\text{Ag(S}_2\text{O}_3)_2^{3-} \rightarrow \text{Ag}} = 0.799 + 0.059 \log \frac{1}{K_f}$$

where K_f is the formation constant[5] for the complex ion.

[5]$K_f = \dfrac{[\text{Ag(S}_2\text{O}_3)_2^{3-}]}{[\text{Ag}^+][\text{S}_2\text{O}_3^{2-}]^2}$

Data for the potential of the silver electrode in the presence of selected ions are given in the table of standard electrode potentials. Similar information is also provided for other electrode systems. Such data often simplify the calculation of half-cell potentials.

Some Limitations to the Use of Standard Electrode Potentials

We shall see that standard electrode potentials are of great importance in understanding electroanalytical processes. There are, however, certain inherent limitations to the use of these data that should be clearly understood.

Substitution of concentrations for activities. For expediency, molar concentrations of reactive species are generally employed in the Nernst equation instead of activities. Unfortunately, the assumption that these two quantities are identical is valid only in very dilute solutions; with increasing electrolyte concentrations, potentials calculated on the basis of molar concentrations can be expected to depart from those obtained by experiment.

To illustrate, the standard electrode potential for the half-reaction

$$Fe^{3+} + e \rightleftharpoons Fe^{2+}$$

is +0.771 V. Neglecting activities, we would predict that a platinum electrode immersed in a solution that was *one formal* in iron(II), iron(III), and perchloric acid would exhibit a potential numerically equal to this value relative to the standard hydrogen electrode. In fact, however, a potential of +0.732 V is observed experimentally. The reason for the discrepancy is seen if we write the Nernst equation in the form

$$E = E^0 - 0.059 \log \frac{C_{Fe^{2+}} f_{Fe^{2+}}}{C_{Fe^{3+}} f_{Fe^{3+}}}$$

where C is the molar concentration of each species and f is the activity coefficient. The activity coefficients of the two species are less than one in this system because of the high ionic strength imparted by the perchloric acid and the iron salts. More important, however, the activity coefficient of the iron(III) ion should be smaller than that of the iron(II) ion inasmuch as the effects of ionic strength on these coefficients increase with the charge on the ion. As a consequence, the ratio of the activity coefficients as they appear in the Nernst equation would be larger than one and the potential of half-cell would be smaller than the standard potential.

Activity coefficient data for ions in solutions of the types commonly encountered in oxidation-reduction titrations and electrochemical work are fairly limited; consequently, we are forced to use molar concentrations rather than activities in many calculations. Appreciable errors may result.

Effect of other equilibria. The application of standard electrode potential data to systems of interest to the analytical chemist is further complicated by the occurrence of solvolysis, dissociation, association, and complex-formation reactions involving the species in which he is interested. Often the equilibrium constants required to correct for these effects are not known. Lingane[6] cites the ferrocyanide-ferricyanide couple as an excellent example of this problem:

$$Fe(CN)_6^{3-} + e \rightleftharpoons Fe(CN)_6^{4-} \qquad E^0 = +0.356 \text{ V}$$

Although the hydrogen ion does not appear in this half-reaction, the experimentally measured potential is markedly affected by pH. Thus, instead of the expected value of $+0.356$ V, solutions containing equiformal concentrations of the two species yield potentials of $+0.71$, $+0.56$, and $+0.48$ V with respect to the standard hydrogen electrode when the measurements are made in media that are respectively $1.0\ F$, $0.1\ F$, and $0.01\ F$ in hydrochloric acid. The explanation for this variation is fairly simple. Both the ferrocyanide and ferricyanide ions are known to associate with hydrogen ions, and these processes occur to a degree in the presence of hydrochloric acid. The hydro-ferrocyanic acids, however, are weaker than the hydroferricyanic acids; thus, the concentration of the ferrocyanide ion is lowered more than that of the ferricyanide ion as the acid concentration increases. This effect tends in turn to shift the above equilibrium to the right and leads to more positive electrode potentials.

A somewhat analogous effect is encountered in the behavior of the potential of the iron(III)-iron(II) couple. As noted earlier, an equiformal mixture of these two ions in $1\ F$ perchloric acid exhibits a reduction potential of $+0.73$ V. Substitution of hydrochloric acid of the same concentration alters the observed potential to $+0.70$ V; a value of $+0.6$ V is observed in $1\ F$ phosphoric acid. These differences arise because the iron(III) ion forms more stable complexes with the chloride and the phosphate ions than does the iron(II) ion. As a result, the actual concentration of uncomplexed iron(III) ions in such solutions is less than that of uncomplexed iron(II), and the net effect is a shift in the observed potential.

Phenomena such as these can be taken into account in potential calculations only if the equilibria involved are known and constants for the processes are available. Often, however, such information is lacking; the chemist is then forced to neglect such effects and hope that they will not lead to serious errors in his calculations.

Formal potentials. In order to compensate partially for activity effects and errors resulting from side reactions, Swift[7] has proposed the use of a

[6] J. J. Lingane, *Electroanalytical Chemistry,* 2d ed. (New York: Interscience Publishers, Inc., 1958), p. 59.

[7] E. H. Swift, *A System of Chemical Analysis* (San Francisco: W. H. Freeman and Company, 1939), p. 50.

quantity called the *formal potential* in place of the standard electrode potential in oxidation-reduction calculations. The formal potential of a system is the potential of the half-cell with respect to the standard hydrogen electrode when the concentrations of reactants and products are 1 *F* and the concentrations of any other constituents of the solution are carefully specified. Thus, for example, the formal potential for the reduction of iron(III) is +0.731 V in 1 *F* perchloric acid and +0.700 V in 1 *F* hydrochloric acid; similarly, the formal potential for the reduction of ferricyanide ion would be +0.71 V in 1 *F* hydrochloric acid and +0.48 V in a 0.01 *F* solution of this acid. Use of these values in place of the standard electrode potential in the Nernst equation will yield better agreement between calculated and experimental potentials, provided the electrolyte concentration of the solution approximates that for which the formal potential was measured. Application of formal potentials to systems differing greatly as to kind and concentration of electrolyte can, however, lead to errors greater than those encountered when standard potentials are employed. The table in the Appendix contains selected formal potentials as well as standard potentials; in subsequent chapters we shall use whichever seems to be the more appropriate.

Reaction rates. While electrode potentials are of the greatest importance for calculation of cell potentials, it should be realized that the presence of a half-reaction in a table of electrode potentials does not necessarily imply that there is a real electrode whose potential will respond to the half-reaction. Many of the data in such tables have been obtained by calculations based upon equilibrium or thermal measurements rather than from the direct determination of a potential of an electrode system. For some, no suitable electrode is known; thus, the standard electrode potential for the process,

$$2CO_2 + 2H^+ + 2e \rightleftharpoons H_2C_2O_4 \qquad E^0 = -0.49 \text{ V}$$

has been arrived at indirectly. The reaction is not reversible and the rate at which carbon dioxide combines to give oxalic acid is negligibly slow. No electrode system is known whose potential varies with the ratio of activities of the reactants and products.

CELL POTENTIALS

The potential of a cell includes several components, only two of which are the electrode potentials of the anode and the cathode. We have shown that the magnitude of an electrode potential is readily obtained from the Nernst equation, which is based upon thermodynamics; as a consequence, that part of a cell potential that owes its origin to the electrodes is often called the *thermodynamic cell potential.* In addition, other potentials frequently contribute to the overall emf delivered by a galvanic cell or required for operation of an electrolytic cell. These additional potentials arise from liquid junctions, the internal cell resistance, and polarization effects.

Thermodynamic Cell Potential

One can obtain the electromotive force of a cell by combining half-cell potentials as follows:

$$E_{cell} = E_{cathode} - E_{anode} \qquad (17\text{-}5)$$

where E_{anode} and $E_{cathode}$ are the *electrode potentials* for the two half-reactions comprising the cell.

Consider the cell

$$Zn \mid ZnSO_4 \; (a_{Zn}{}^{2+} = 1.00) \parallel CuSO_4 \; (a_{Cu}{}^{2+} = 1.00) \mid Cu$$

The two half-reactions consist of the oxidation of zinc to zinc ions and the reduction of copper(II) ions to the metallic state. Standard electrode potentials for these reactions are

$$Zn^{2+} + 2e \rightleftharpoons Zn \qquad E^0 = -0.763 \text{ V}$$
$$Cu^{2+} + 2e \rightleftharpoons Cu \qquad E^0 = +0.337 \text{ V}$$

Since the activities of the two ions are specified as unity, the standard potentials are also the electrode potentials. The cell diagram also specifies that the zinc electrode is the anode; thus,

$$E_{cell} = +0.337 - (-0.763) = +1.100 \text{ V}$$

The positive sign for the cell potential indicates that the reaction

$$Zn + Cu^{2+} \rightarrow Zn^{2+} + Cu$$

occurs spontaneously and that this is a galvanic cell.

Diagraming the cell above as

$$Cu \mid Cu^{2+} \; (a_{Cu^{2+}} = 1.00) \parallel Zn^{2+} \; (a_{Zn^{2+}} = 1.00) \mid Zn$$

implies that the copper electrode is the anode; here

$$E_{cell} = -0.763 - (+0.337) = -1.100 \text{ V}$$

The negative sign indicates the nonspontaneity of the reaction

$$Cu + Zn^{2+} \rightarrow Cu^{2+} + Zn$$

To cause this reaction to occur would require the application of external potential of greater than 1.100 V.

Example. Calculate the theoretical potential of the cell

$$Pt, H_2(0.8 \text{ atm}) \mid HCl(0.20 \; F), AgCl(\text{sat'd}) \mid Ag$$

The two half-cell reactions and standard electrode potentials are

$$2H^+ + 2e \rightleftharpoons H_2 \qquad\qquad E^0 = 0.000 \text{ V}$$
$$AgCl + e \rightleftharpoons Ag + Cl^- \qquad E^0 = +0.222 \text{ V}$$

If we make the approximation that the activities of the various species are identical to their molar concentrations, then for the hydrogen electrode,

$$E = 0.000 - \frac{0.059}{2} \log \frac{0.8}{(0.2)^2} = -0.038 \text{ V}$$

and for the silver-silver chloride electrode,

$$E = +0.222 - 0.059 \log 0.20 = +0.263 \text{ V}$$

The cell diagram specified the hydrogen electrode as the anode and the silver electrode as the cathode. Therefore,

$$E_{cell} = +0.263 - (-0.038) = +0.301 \text{ V}$$

The positive sign indicates that the reaction

$$H_2 + 2AgCl \rightarrow 2H^+ + 2Ag + 2Cl^-$$

is occurring when current is drawn from this cell.

Example. Calculate the potential required to initiate deposition of copper from a solution that is 0.010 F in $CuSO_4$ and contains sufficient sulfuric acid to give a hydrogen ion concentration of 1.0×10^{-4} M.

Here, the cathode reaction is the deposition of copper. Since no easily oxidizable substances are present, the anode reaction will involve oxidation of H_2O to give O_2. From the table of standard potentials, we find

$$Cu^{2+} + 2e \rightleftharpoons Cu \qquad\qquad E^0 = +0.337 \text{ V}$$
$$\tfrac{1}{2}O_2 + 2H^+ + 2e \rightleftharpoons H_2O \qquad E^0 = +1.23 \text{ V}$$

Then, for the copper electrode,

$$E = +0.337 - \frac{0.059}{2} \log \frac{1}{0.01} = +0.278 \text{ V}$$

and for the oxygen electrode, assuming evolution of O_2 at 1.0 atmosphere,

$$E = +1.23 - \frac{0.059}{2} \log \frac{1}{(1 \times 10^{-4})^2 (1.0)^{1/2}} = +0.99 \text{ V}$$

The cell potential is then

$$E_{cell} = +0.278 - 0.99 = -0.71 \text{ V}$$

Thus, initiation of the reaction

$$Cu^{2+} + H_2O \rightarrow \tfrac{1}{2}O_2 + 2H^+ + Cu$$

would require the application of a potential greater than 0.71 V.

Liquid Junction Potential

When two electrolyte solutions of different composition are brought in contact with one another, a potential develops at the interface. This *junction potential*, as it is called, arises from an unequal distribution of cations and anions across the boundary, due to differences in rates with which these species migrate.

To take the very simplest case, consider events occurring at the interface between a 1 *F* and a 0.01 *F* hydrochloric acid solution. We may symbolize this interface as

$$HCl\ (1\ F)\ |\ HCl\ (0.01\ F)$$

Both hydrogen ions and chloride ions tend to diffuse across this boundary from the more concentrated to the more dilute solution, the driving force for this migration being proportional to the concentration difference. The rate at which various ions move under the influence of a fixed force varies considerably (that is, their *mobilities* are different); in the present example, hydrogen ions are several times more mobile than chloride ions. As a consequence, there is a tendency for the hydrogen ions to outstrip the chloride ions as the diffusion takes place; a separation of charge is the net result.

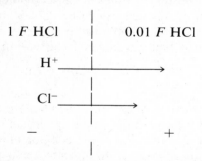

The more dilute side of the boundary becomes positively charged, owing to the more rapid migration of hydrogen ions; the concentrated side therefore acquires a negative charge from the slower-moving chloride ions. The charge that develops tends to counteract the differences in mobilities of the two ions; as a consequence, an equilibrium condition soon develops. The potential difference resulting from this charge separation may amount to several hundredths of a volt or more.

In a simple system such as this, where the behavior of only two ions need be considered, the magnitude of the junction potential can be calculated from a knowledge of the mobilities of the ions involved. However, it is seldom that a cell of analytical importance has a sufficiently simple composition to permit such a computation.

It is an experimental fact that the magnitude of the liquid-junction potential can be greatly decreased by interposition of a concentrated electrolyte solution (a *salt bridge*) between the two solutions. The effectiveness of this contrivance improves as the concentration of the salt in the bridge increases and as the mobilities of the ions of the salt approach one another in magnitude. A saturated potassium chloride solution is good from both standpoints, its concentration being somewhat greater than 4 *F* at room temperature, and the mobility of its ions differing by only 4 percent. With such a bridge the junction potential typically amounts to a few millivolts or less, a negligible quantity in most analytical measurements.

Ohmic Potential; *IR* Drop

Current flow affects the potential of either an electrolytic or a galvanic cell. This effect is due in part to the necessity of overcoming the resistance of the cell itself to the passage of current. The potential required to accomplish this is equal to the product of the current and the electrical resistance of the cell, and is frequently referred to as the *IR drop*.

To illustrate the effect of *IR* drop consider the behavior of the reversible cell

$$Cd \mid CdSO_4(1 \ F) \parallel CuSO_4(1 \ F) \mid Cu$$

during the passage of current. From the half-cell potentials we can readily calculate that the emf of this cell should be 0.74 V, provided the junction potential is kept small with a suitable salt bridge.

When it is operated as a galvanic cell, the overall chemical reaction is

$$Cd + Cu^{2+} \rightleftharpoons Cd^{2+} + Cu$$

Operation as an electrolytic cell would involve the reverse process, and would require the application of an emf somewhat greater than the theoretical potential of 0.74 V; this is shown in the upper half of Figure 17-6, where the current is plotted as a function of the cell potential. The potentials are given a negative sign in this region to indicate a nonspontaneous reaction.

The linear relationship between current and the applied potential extends over a considerable current range; within this region the potential required to overcome the resistance of the cell is directly proportional to the current. Thus, if the resistance of the cell is 4 ohms and the current is 0.02 amp, a potential of −0.08 V in excess of theoretical is required; operation of the cell will thus involve a total potential of −0.82 V.

The lower portion of Figure 17-6 describes the current-voltage behavior of the system as a galvanic cell. The current now passes in the opposite direction, and the potentials, by convention, are given positive values. The

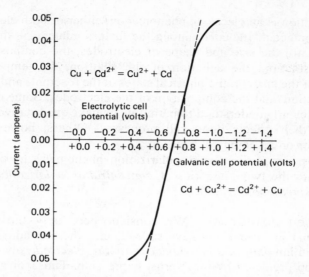

Figure 17-6. Current-voltage curve for the cell Cd|CdSO$_4$(1 F)‖CuSO$_4$(1 F)|Cu, (lower half of figure). Cell resistance is 4.0 ohms. Note that the direction of current flow for upper half of curve differs from that of lower half.

potential of the cell becomes smaller when current is drawn, the difference from the theoretical again being equal to IR over a considerable range. Thus, the potential of the cell in the preceding paragraph would be +0.66 V when a current of 0.02 amp is being drawn from it.

In general, the net effect of IR drop is to increase the potential required to operate an electrolytic cell and to decrease the potential of a galvanic cell. Therefore, the IR drop is always *subtracted* from the thermodynamic cell potential; that is,

$$E_{cell} = E_{cathode} - E_{anode} - IR$$

Polarization Effects

The term *polarization* refers to a condition in which cell or electrode potentials exhibit departures, during the passage of current, from values computed on the basis of standard-potential data and the IR drop. Its effects are illustrated in Figure 17-6, where deviations from linearity are observed for large currents. Thus, a polarized electrolytic cell requires application of potentials larger than theoretical for a given current flow; similarly, a polarized galvanic cell develops potentials that are smaller than predicted. Under some conditions the polarization of a cell may become considerably more extreme than is shown in Figure 17-6. As a matter of fact, the current is at times observed to become essentially independent of the voltage of the cell. Under these circumstances polarization is said to be complete.

Polarization is an electrode phenomenon; either or both electrodes in a cell can be affected. Included among the factors influencing the degree of polarization are the size and shape of electrodes, the composition of the electrolyte solution, the agitation of the solution, the temperature, the magnitude of the current, the physical states of the reactants and products of the cell reaction, and the composition of the electrodes. Some of these factors are sufficiently understood to permit us to make quantitative statements concerning their effects upon cell processes. Others, however, can be accounted for on an empirical basis only.

For purposes of discussion, polarization phenomena are conveniently classified into the two categories of *concentration polarization* and *overvoltage,* or *kinetic polarization.*

Concentration polarization. We consider here only those electrode processes that are rapid and reversible. Under these conditions the thin layer of solution *immediately adjacent to the electrode* always has the concentration predicted by the Nernst expression. Thus, for a silver electrode in contact with a solution of silver nitrate, we may write

$$E = 0.799 - 0.059 \log \frac{1}{[Ag^+]}$$

Since the process

$$Ag^+ + e \rightleftharpoons Ag$$

is rapid and reversible, the concentration of silver ion in the film of liquid surrounding the electrode is determined at any moment by the potential of the silver electrode at that moment. If the potential changes, there is an essentially instantaneous alteration of concentration in this film in accordance with the demands of the Nernst equation. This change involves deposition of silver or solution of the electrode, as appropriate.

In contrast to this substantially instantaneous surface process, the rate at which equilibrium is attained between the electrode and the bulk of the solution can be slow indeed, and depends upon the magnitude of the current passing through the cell.

With these facts in mind, consider again the cathodic reaction of the cell whose current-voltage behavior is depicted in the upper half of Figure 17-6; that is,

$$Cd^{2+} + 2e \rightleftharpoons Cd$$

If currents of the indicated magnitudes are to pass, the surface layer surrounding the cadmium electrode must be replenished with cadmium ions at a rate commensurate with the current demand. As an example, for a current of 0.01 amp the rate is given by

$$0.01 \text{ amp} = 0.01 \text{ coulomb/sec}$$

$$0.01 \text{ coulomb/sec} = \frac{0.01}{96,494}$$

$$\cong 10^{-7} \text{ faraday/sec}$$

Thus, to maintain this current, it is necessary that 10^{-7} equivalent of cadmium ion be transported to the cathode surface during each second of cell operation. If this rate of mass transfer cannot be met, concentration polarization sets in and lowered currents must result. This type of polarization, then, is the result of an inadequate rate of transfer of reactive species between the bulk of the solution and the electrode surface.

The transport of material through a solution can occur as a result of (1) diffusion forces, (2) electrostatic attractions and repulsions, or (3) mechanical or convection forces. We must therefore consider briefly the variables that influence these forces as they relate to electrode processes.

Whenever a concentration gradient develops in a solution, the process of diffusion tends to force molecules or ions from the more concentrated to the more dilute regions. The rate at which transfer occurs is proportional to the concentration difference. In an electrolysis a gradient is established as a result of the removal of ions from the film of solution next to the cathode. Diffusion then occurs, the rate being expressed by the relationship

$$\text{rate of diffusion to cathode surface} = k(C - C_0) \qquad (17\text{-}6)$$

where C is the reactant concentration in the bulk of the solution, C_0 is its equilibrium concentration at the cathode surface, and k is a proportionality constant. The value of C_0 is fixed by the potential of the electrode and can be calculated from the Nernst equation. As greater potentials are applied to the electrode, C_0 becomes smaller and smaller and the diffusion rate greater and greater.

Electrostatic forces also influence the rate at which an ionic reactant migrates toward or away from an electrode surface. Usually, the ion and the electrode are of opposite charge, which results in an attractive force between them. This condition does not always exist, however; iodate, periodate, and dichromate ions, for example, can be reduced at a negatively charged cathode. Owing to forces of repulsion between these species and the cathode, concentration polarization can be expected to occur more readily than in the reduction of cations. The electrostatic attraction (or repulsion) between a particular ionic species and the electrode becomes smaller as the total electrolyte concentration of the solution is increased and it may approach zero when the reactive species is but a small fraction of the total concentrations of ions of given charge.

Clearly, reactants can be transported to an electrode by mechanical means. Thus, stirring or agitation aids in preventing concentration polarization; convection currents due to temperature or density differences are also effective.

To summarize, then, concentration polarization occurs when the forces of diffusion, electrostatic attraction, and mechanical mixing are insufficient to transport the reactant to or from an electrode surface at a rate demanded by the theoretical current flow. Concentration polarization causes the potential of a galvanic cell to be smaller than the value predicted on the basis of the thermodynamic potential and the *IR* drop. In the case of an electrolytic cell a potential more negative than theoretical is required in order to maintain a given current.

Concentration polarization is important in several electroanalytical methods; in some applications effort is made to avoid it completely; in others, however, it is essential to the method, and every effort is made to promote its occurrence. Experimentally, the degree of concentration polarization can be influenced by (1) the reactant concentration, (2) the total electrolyte concentration, (3) mechanical agitation, and (4) the size of the electrodes; as the area toward which the reactant can be transported becomes greater, polarization effects become smaller.

Overvoltage effects.   Sometimes a voltage in excess of theoretical is required to cause the reaction in an electrolytic cell to occur at an appreciable rate even when conditions are such that concentration polarization is not likely. The output voltages of galvanic cells are also occasionally less than theoretical under similar circumstances. This voltage difference is called *overvoltage* and ordinarily results from the slow rate at which the electrochemical reaction occurs at one or both electrodes. The phenomenon is the result of the need for additional energy to overcome the energy barrier to the half-reaction. In contrast to concentration polarization, the current is controlled by the rate of the electrode process rather than by the rate of mass transfer.

While exceptions can be cited, some empirical generalizations can be made regarding the magnitude of overvoltage.

1. Overvoltage increases with current density (current density is defined as the amperes per square centimeter of electrode surface).

2. It usually decreases with increases in temperature.

3. Overvoltage varies with the chemical composition of the electrode, often being most pronounced with softer metals such as tin, lead, zinc, and particularly mercury.

4. Overvoltage is most marked for those electrode processes yielding gaseous products. It is frequently negligible where a metal is being deposited or where an ion is undergoing a change of oxidation state.

5. The magnitude of overvoltage in any given situation cannot be specified exactly because it is determined by a number of uncontrollable variables.

Overvoltages associated with the evolution of hydrogen and oxygen are of particular interest to the chemist. Table 17-2 presents data that depict the extent of the phenomenon under specific conditions. The difference between the overvoltage of the gases on smooth and on platinized platinum electrodes is of particular interest. This difference is primarily due to the much larger surface area associated with the platinum-black coating on the latter (see p. 415), which results in a smaller *real* current density than is apparent from the dimensions of the electrode. A platinized surface is always employed in construction of hydrogen reference electrodes to lower the current density to a point where the overvoltage is negligible.

The high overvoltage associated with the formation of hydrogen permits the electrolytic deposition of several metals that require potentials at which hydrogen would otherwise be expected to interfere. For example, it is readily shown from their standard potentials that rapid formation of hydrogen should occur well before a potential sufficient for the deposition of zinc from a neutral solution is reached. In fact, however, a quantitative deposition can be achieved, provided a mercury or a copper electrode is used; because of the high overvoltage of hydrogen on these metals, little or no gas is evolved during the process.

The magnitude of overvoltage can, at best, be only crudely approximated from empirical information available in the literature. Calculation of cell potentials in which overvoltage plays a part cannot be very accurate. As with IR drop, the overvoltage is subtracted from the theoretical cell potential.

Table 17-2 Overvoltage for Hydrogen and Oxygen Formation at Various Electrodes at 25°C[a]

Electrode Composition	Overvoltage (V), (Current Density 0.001 amp/cm²)		Overvoltage (V), (Current Density 0.01 amp/cm²)		Overvoltage (V), (Current Density 1 amp/cm²)	
	H_2	O_2	H_2	O_2	H_2	O_2
Smooth Pt	0.024	0.721	0.068	0.85	0.676	1.49
Platinized Pt	0.015	0.348	0.030	0.521	0.048	0.76
Au	0.241	0.673	0.391	0.963	0.798	1.63
Cu	0.479	0.422	0.584	0.580	1.269	0.793
Ni	0.563	0.353	0.747	0.519	1.241	0.853
Hg	0.9[b]		1.1[c]		1.1[d]	
Zn	0.716		0.746		1.229	
Sn	0.856		1.077		1.231	
Pb	0.52		1.090		1.262	
Bi	0.78		1.05		1.23	

[a]National Academy of Sciences, *International Critical Tables of Numerical Data*, Vol. 6 (New York: McGraw-Hill Book Company, Inc., 1929), pp. 339–340. (With permission.)
[b]0.556 V at 0.000077 amp/cm²; 0.929 V at 0.00154 amp/cm².
[c]1.063 V at 0.00769 amp/cm².
[d]1.126 V at 1.153 amp/cm².

SOME COMMON CELLS AND HALF-CELLS EMPLOYED IN ELECTROANALYTICAL CHEMISTRY

Reference Electrodes

In many electroanalytical methods it is desirable that the half-cell potential of one electrode should remain completely insensitive to changes in the composition of the solution during the analysis. An electrode that fits this description is called a *reference electrode*. Employed in conjunction with the reference electrode is an *indicator electrode,* whose response is dependent upon changes in solution environment.

A reference electrode should be easy to assemble, should provide reproducible potentials, and should be unchanged in potential with passage of small currents. Several electrode systems meet these requirements.

Calomel electrodes. Calomel half-cells may be represented as follows:

$$Hg_2Cl_2(\text{sat'd}), KCl\ (xF) \mid Hg$$

where x represents the formal concentration of potassium chloride in the solution. The electrode reaction is given by the equation

$$Hg_2Cl_2 + 2e \rightleftharpoons 2Hg + 2Cl^-\qquad E^0 = 0.2676\ V$$

The potential of this half-cell varies with the chloride concentration x, and this quantity must be specified in describing the electrode.

Table 17-3 lists the composition and the reduction potential for the three most commonly encountered calomel electrodes. Note that each solution is saturated with mercury(I) chloride and that the cells differ only with respect to the potassium chloride concentration. Note also that the potential of the normal calomel electrode is greater than the standard potential for the half-reaction because the chloride-ion *activity* in a 1 F solution of potassium chloride is significantly smaller than one. The last column in Table 17-3 gives expressions that permit calculations of the reduction potentials of calomel half-cells at temperatures other than 25°C.

The saturated calomel electrode (SCE) is most commonly used by the analytical chemist because of the ease with which it can be prepared.

Table 17-3 Specifications of Calomel Electrodes[a]

Name	Concentration of Hg_2Cl_2	KCl	*Reduction Potential (V) vs. Standard Hydrogen Electrode* $(Hg_2Cl_2 + 2e \rightleftharpoons 2Hg + 2Cl^-)$
Saturated	Saturated	Saturated	$+0.242 - 7.6 \times 10^{-4}(t - 25)$
Normal	Saturated	1.0 F	$+0.280 - 2.4 \times 10^{-4}(t - 25)$
Decinormal	Saturated	0.1 F	$+0.334 - 7 \times 10^{-5}(t - 25)$

[a]From *Fundamentals of Analytical Chemistry,* Second Edition, by Douglas A. Skoog and Donald M. West. Copyright © 1963 by Holt, Rinehart and Winston, Inc. Reprinted by permission of Holt, Rinehart and Winston, Inc.

Compared with the other two electrodes, its temperature coefficient is somewhat larger; this is a disadvantage only where substantial changes in temperature occur during the measurement process.

The form and shape of calomel electrodes vary tremendously according to the ingenuity and imagination of their designers. A simple, easily constructed, saturated electrode is shown in Figure 17-7. The container is a 2-oz widemouth bottle or small beaker equipped with a stopper that holds two tubes. Sealed into the end of one of these tubes is a small piece of platinum wire that provides electrical contact between the interior and the exterior of the cell. The other tube is a salt bridge that connects the calomel cell to the indicator half-cell; this bridge is ordinarily filled with saturated potassium chloride. To minimize siphoning of the cell liquid, a fine, fritted-glass disk is often sealed on one end of the tube. Alternatively, the tube may be filled with a solution prepared by heating agar or gelatin in a saturated potassium chloride solution; upon cooling a conducting gel is formed.

The bottom of the cell container is covered with a layer of pure mercury to a depth of approximately 5 mm. Upon this is placed a layer of paste, prepared by triturating pure mercury(I) chloride and a little mercury with saturated potassium chloride solution. A layer of solid potassium chloride follows; finally, the jar is nearly filled with a saturated solution of that salt. Several days are usually required for such a cell to come to equilibrium and develop a constant potential.

Several convenient calomel electrodes are available commercially. The one illustrated in Figure 17-8 is typical, and consists of a tube 5 to 15 cm in length and 0.5 to 1.0 cm in diameter. The mercury-mercury(I) chloride paste is contained in an inner tube that is connected to the saturated

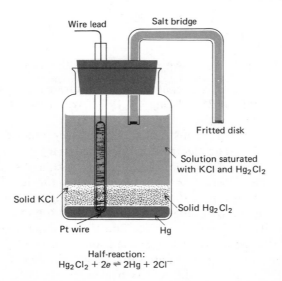

Half-reaction:
$$Hg_2Cl_2 + 2e \rightleftharpoons 2Hg + 2Cl^-$$

Figure 17-7. Diagram of a saturated calomel electrode.

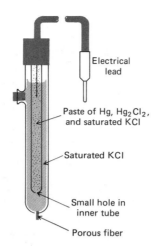

Figure 17-8. Diagram of a fiber-type saturated calomel electrode. (By courtesy, Beckman Instruments, Inc., Fullerton, California.)

potassium chloride solution in the outer tube by means of a small opening. Contact with the second half-cell is made by means of a porous fiber or a fritted-glass disk sealed in the end of the outer tubing. An electrode such as this has a relatively high resistance (2000 to 3000 ohms) and limited current-carrying capacity.

Silver-silver chloride electrodes. A reference electrode system analogous to the calomel electrode consists of a silver electrode immersed in a solution of potassium chloride that is also saturated with silver chloride:

$$AgCl(sat'd), KCl(xF) \mid Ag$$

The half-reaction is

$$AgCl + e \rightleftharpoons Ag + Cl^- \qquad E^0 = 0.222 \text{ V}$$

Normally, this electrode is prepared with a saturated potassium chloride solution, the potential at 25°C being +0.197 V with respect to the standard hydrogen electrode.

A simple and easily constructed silver-silver chloride electrode is shown in Figure 17-9. The electrode is contained in a Pyrex tube fitted with a 10-mm fritted-glass disk (tubing containing fritted disks is obtainable from Corning Glass Works). A plug of agar gel saturated with potassium chloride is formed on top of the disk to prevent loss of solution from the half-cell. The plug can be prepared by heating 4 to 6 g of pure agar in 100 ml of water until solution is complete and then adding about 35 g of potassium chloride. A portion of this suspension, while still warm, is poured into the tube; upon cooling, it solidifies to a gel with low electrical resistance. A layer of solid potassium chloride is placed on the gel, and the tube is

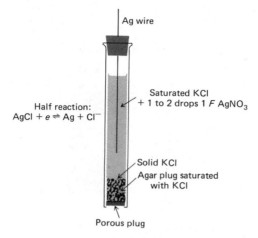

Ag wire

Saturated KCl
+ 1 to 2 drops 1 F AgNO₃

Half reaction:
AgCl + e ⇌ Ag + Cl⁻

Solid KCl
Agar plug saturated
with KCl

Porous plug

Figure 17-9. Diagram of a silver-silver chloride electrode.

filled with saturated solution of the salt. A drop or two of 1 F silver nitrate is then added and a heavy gauge (1- to 2-mm diameter) silver wire is inserted in the solution.

Standard Weston Cells

The accurate measurement of potentials for electrochemical work requires the frequent use of a cell whose emf is precisely known. The *Weston cell*, which is used almost universally for this purpose, can be represented as follows:

$$Cd(Hg) \mid CdSO_4 \cdot 8/3H_2O(sat'd), Hg_2SO_4(sat'd) \mid Hg$$

The half-reactions as they occur in the cell are

$$Cd(Hg) \rightarrow Cd^{2+} + Hg + 2e$$
$$Hg_2SO_4 + 2e \rightarrow 2Hg + SO_4^{2-}$$

Figure 17-10 shows a typical form of a Weston cell. Its potential at 25°C is 1.0183 V.

The emf of a Weston cell is governed by the activities of the cadmium and mercury(I) ions in the solution. At any given temperature, these quantities are invariant, being fixed by the solubilities of the cadmium sulfate and the mercury(I) sulfate. As a result, Weston cells remain constant in voltage for remarkably long periods of time if large currents are not drawn from them.

The Weston cell has a temperature coefficient of about −0.04 millivolt per °C, due primarily to changes in solubility of the cadmium and mercury(I) salts. A cell with a coefficient of about one-fourth this magnitude is obtained by using a solution of cadmium sulfate that has been saturated at 4°C; no

Cell reaction:
$$Cd(Hg) + Hg_2SO_4 \rightleftharpoons Cd^{2+} + 2Hg + SO_4^{2-}$$
$$E = 1.0183 \text{ volts}$$

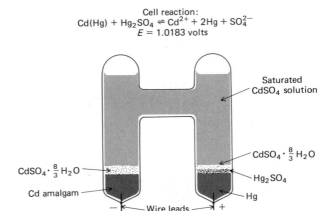

Figure 17-10. A Weston standard cell.

solid $CdSO_4 \cdot 8/3H_2O$ is incorporated in the cell itself. Most commercially available cells are of this type and are termed *unsaturated Weston cells*. Their potentials lie between 1.0185 and 1.0195 V.

Weston cells may be sent to the National Bureau of Standards for calibration and certification.

Problems

1. Calculate the electrode potentials of the following half-cells against the standard hydrogen electrode.
   (a) $Ag \mid Ag^+ (0.0100\ M)$
   (b) $Ni \mid Ni^{2+} (0.712\ M)$
   (c) $Pt, H_2 (1\ atm) \mid HCl (10^{-5}\ F)$
   (d) $Pt \mid Fe^{3+} (1.00 \times 10^{-4}\ M), Fe^{2+} (0.100\ M)$
   (e) $Ag \mid AgBr(sat'd), Br^- (3.00\ M)$

2. Calculate the electrode potentials of the following half-cells.
   (a) $Fe \mid Fe^{2+} (0.0100\ M)$
   (b) $Bi \mid BiO^+ (1.00 \times 10^{-3}\ M), H^+ (1.00 \times 10^{-2}\ M)$
   (c) $Pt, H_2 (1.00\ atm) \mid HCl (2.00\ F)$
   (d) $Pt, H_2 (1.00\ atm) \mid HOAc (0.100\ F)$
   (e) $Pt \mid IO_3^- (0.100\ M), I_2 (0.0100\ M), H^+ (1.00 \times 10^{-4}\ M)$
   (f) $Ag \mid Ag_2CrO_4 (sat'd), CrO_4^{2-} (0.0200\ M)$

3. Calculate the potential of a silver electrode in contact with the following.
   (a) $0.010\ M\ Ag^+$
   (b) $0.060\ M\ Ag^+$
   (c) a solution that is $0.050\ M$ in I^- and saturated with AgI
   (d) a solution that is $0.0050\ M$ in CN^- and $0.060\ M$ in $Ag(CN)_2^-$
   (e) the solution that results from mixing 25.0 ml of $0.050\ F$ KSCN with 20.0 ml of $0.080\ F\ Ag^+$
   (f) the solution that results from mixing 25.0 ml of $0.050\ F\ Ag^+$ with 20.0 ml of $0.080\ F$ KSCN

4. Calculate the electrode potentials for the following systems against the standard hydrogen electrode.
   (a) Pt | V^{3+} ($1.00 \times 10^{-3}\,M$), VO^{2+} ($1.00\,M$), HCl ($0.100\,F$)
   (b) Pt | MnO_4^- ($0.300\,M$), Mn^{2+} ($0.100\,M$), H^+ ($0.200\,M$)
5. Calculate the reduction potentials for the following systems.
   (a) Pt | $Cr_2O_7^{2-}$ ($1.00 \times 10^{-3}\,M$), Cr^{3+} ($1.00 \times 10^{-2}\,M$), H^+ ($0.100\,M$)
   (b) Pt | MnO_4^- ($0.200\,M$), Mn^{2+} ($0.100\,M$), H^+ ($0.300\,M$)
6. Calculate the theoretical potential of each of the following cells. Is the cell as written a galvanic or an electrolytic cell?
   (a) Pt | Cr^{3+}($1.0 \times 10^{-4}\,M$), Cr^{2+}($1.0 \times 10^{-1}\,M$) || Pb^{2+}($8.0 \times 10^{-2}\,M$) | Pb
   (b) Pt | Sn^{4+}($4.0 \times 10^{-4}\,M$), Sn^{2+}($2.0 \times 10^{-2}\,M$) || $Ag(CN)_2^-$($2.5 \times 10^{-3}\,M$), CN^-($5.0 \times 10^{-2}\,M$) | Ag
   (c) Hg | Hg_2^{2+}($1.00 \times 10^{-2}\,M$) || H^+($4.00 \times 10^{-2}\,M$), V^{3+}($8.0 \times 10^{-2}\,M$), VO^{2+}($5.0 \times 10^{-3}\,M$) | Pt
   (d) Pt | Fe^{3+}($4.0 \times 10^{-2}\,M$), Fe^{2+}($3.2 \times 10^{-5}\,M$) || Sn^{2+}($8.0 \times 10^{-2}\,M$), Sn^{4+}($2.0 \times 10^{-4}\,M$) | Pt
   (e) Mn | Mn^{2+}($6.0 \times 10^{-3}\,M$) || $Ag(CN)_2^-$($0.090\,M$), CN^-($0.030\,M$) | Ag
7. Calculate the theoretical potential of each of the following cells. Is the cell as written a galvanic or an electrolytic cell?
   (a) Pt | $Fe(CN)_6^{4-}$($8.0 \times 10^{-2}\,M$), $Fe(CN)_6^{3-}$($4.0 \times 10^{-4}\,M$) || I^-($1.00 \times 10^{-3}\,M$), I_2($4.0 \times 10^{-2}\,M$) | Pt
   (b) Pt | V^{3+}($8.0 \times 10^{-3}\,M$), V^{2+}($0.40\,M$) || Fe^{2+}($1.00 \times 10^{-4}\,M$) | Fe
   (c) Bi | BiO^+($0.080\,M$), H^+($1.00 \times 10^{-2}\,M$) || I^-($0.100\,M$), AgI(sat'd) | Ag
   (d) Zn | Zn^{2+}($5.0 \times 10^{-4}\,M$) || $Fe(CN)_6^{4-}$($2.00 \times 10^{-2}\,M$), $Fe(CN)_6^{3-}$($8.0 \times 10^{-2}\,M$) | Pt
   (e) Pt, H_2(0.100 atm) | HCl($2.00 \times 10^{-3}\,F$), AgCl(sat'd) | Ag
8. Calculate the potential of each of the following cells. Is the cell galvanic or electrolytic as written?
   (a) Cu | CuI(sat'd), I^-($0.100\,M$) || I^-($1.00 \times 10^{-3}\,M$), CuI(sat'd) | Cu
   (b) Pt, H_2(1.00 atm) | H^+($1.00 \times 10^{-6}\,M$) || H^+($1.00 \times 10^{-2}\,M$) | H_2(0.100 atm), Pt
9. Compute E^0 for the process

$$Cd(NH_3)_4^{2+} + 2e \rightleftharpoons Cd + 4NH_3$$

given that the formation constant for the complex is 1.33×10^7.
10. The solubility-product constant for TlI is 8.9×10^{-8} mole^2/liter^2 at 25°C. Calculate E^0 for the process

$$TlI + e \rightleftharpoons Tl + I^-$$

11. Calculate the standard potential for the half-reaction

$$Pd(OH)_2 + 2e \rightleftharpoons Pd + 2OH^-$$

given that K_{sp} for Pd(OH)$_2$ has a value of 1.0×10^{-31} mole^3/liter^3.
12. Calculate the standard potential for the half-reaction

$$Zn(C_2O_4)_2^{2-} + 2e \rightleftharpoons Zn + 2C_2O_4^{2-}$$

if the formation constant for the complex is 2.3×10^7 moles/liter.

13. Calculate the standard potential for the half-reaction

$$FeY^- + e \rightleftharpoons FeY^{2-}$$

if the formation constant for the EDTA complex of iron(III) is 1.3×10^{25} and that for the iron(II) complex is 2.1×10^{14}.

14. From the standard potentials

$$Tl^+ + e \rightleftharpoons Tl \qquad E^0 = -0.336 \text{ V}$$
$$TlCl + e \rightleftharpoons Tl + Cl^- \qquad E^0 = -0.557 \text{ V}$$

calculate the solubility-product constant for TlCl.

15. From the standard potentials

$$Ag_2CrO_4 + 2e \rightleftharpoons 2Ag + CrO_4^{2-} \qquad E^0 = 0.446 \text{ V}$$
$$Ag^+ + e \rightleftharpoons Ag \qquad E^0 = 0.799 \text{ V}$$

calculate the solubility-product constant for Ag_2CrO_4.

16. From the cell potentials shown on the right below, calculate the electrode potentials for the half-cells coupled to the reference electrode.
    (a) saturated calomel electrode $\|$ M^{+n} $|$ M $E = 0.672$ V
    (b) normal calomel electrode $\|$ X^{3+}, X^{2+} $|$ Pt $E = -0.713$ V
    (c) saturated, silver-silver chloride electrode $\|$ MA(sat'd), A^{2-} $|$ M
        $E = -0.272$ V

17. Convert each of the following electrode potentials to potentials versus the saturated calomel electrode.
    (a) $Cd^{2+} + 2e \rightleftharpoons Cd$ $E^0 = -0.403$ V
    (b) $Ce^{4+} + e \rightleftharpoons Ce^{3+}$ $E = 1.44$ V (in 1 F H_2SO_4)
    (c) $Tl^+ + e \rightleftharpoons Tl$ $E = -0.33$ V (in 1 F $HClO_4$)

18. A current of 0.03 amp is to be drawn from the cell

$$Pt \mid V^{3+} (1.0 \times 10^{-5} \text{ } M), V^{2+} (1.0 \times 10^{-1} \text{ } M) \| Br^- (2.0 \times 10^{-1} \text{ } M),$$
$$AgBr(sat'd) \mid Ag$$

As a consequence of its design, the cell has an internal resistance of 1.8 ohms. Calculate the potential to be expected initially.

19. The cell

$$Pt \mid TiO^{2+} (3.0 \times 10^{-4} \text{ } M), Ti^{3+} (7.2 \times 10^{-2} \text{ } M), H^+ (4.0 \times 10^{-3} \text{ } M) \| Cu^{2+}$$
$$(2.5 \times 10^{-2} \text{ } M) \mid Cu$$

has an internal resistance of 1.5 ohms.
    (a) What will be the initial potential if 0.040 amp is drawn from this cell?
    (b) Calculate the weight of Cu^{2+} involved in the reaction at the copper electrode during each second of operation at 0.040 amp.

20. The resistance of the galvanic cell

$$Pt \mid Fe(CN)_6^{4-} (4 \times 10^{-2} \text{ } M), Fe(CN)_6^{3-} (1 \times 10^{-3} \text{ } M) \| Ag^+ (2 \times 10^{-2} \text{ } M) \mid Ag$$

is 2.5 ohms. Each cell compartment contains 100 ml of solution.
    (a) Calculate the initial potential when 0.016 amp is drawn from this cell.
    (b) What will be the potential after a steady current of 0.016 amp has been drawn from this cell for 12.0 min? (Neglect any changes in resistance.)

21.   The cell

   Ni | Ni^{2+} (1.0 × 10^{-4} *M*) ‖ BiO^{+} (0.025 *M*), H^{+} (2.0 × 10^{-6} *M*) | Bi

has an internal resistance of 1.5 ohms. Briefly account for the experimental observation that an initial potential of 0.260 V is developed when a steady current of 0.020 amp is drawn from this cell.

18 Potentiometric Methods

As we have seen in Chapter 17, the potential of an electrode relative to a reference electrode varies in a predictable way with the activities of species which undergo oxidation or reduction reactions at the electrode surface. Thus, the measurement of cell potentials can be employed to determine the activity of electrode-active substances.

Potentiometric methods are of two types. In one, a suitable indicator electrode system is calibrated by measuring its potential with respect to a reference electrode in one or more standard solutions of the species to be determined; the calibration data then permit calculation of the concentration of the species from potential measurements involving solutions of samples. In the second method, called a *potentiometric titration,* the measurement of potential serves to locate the equivalence point for a titration.

It is important to appreciate that a direct potentiometric measurement provides equilibrium information regarding the system, whereas a potentiometric titration yields primarily stoichiometric information. To illustrate, consider equal volumes of two solutions containing respectively the same number of formula weights of a strong acid and a weak acid. Direct potentiometric measurement with a pH-sensitive electrode reveals the differences in the equilibrium hydronium ion activities of the two solutions; potentiometric titrations (using the same electrode) establish that both solutions contain the same number of titratable protons.

Speed and adaptability to continuous monitoring have resulted in the widespread application of the direct method to analytical problems. Because the measurement operation does not significantly alter the activity of the species under study, the direct potentiometric procedure is one of the most widely used and powerful tools for the determination of equilibrium constants. In comparison with potentiometric titrations, however, the direct procedure is inherently less accurate, requires a greater control of experimental variables, and greater precision of measurement.

The potentiometric end point has been applied to all types of reactions. In contrast to the visual end point, it can be used with colored or opaque solutions; it is less subjective and inherently more accurate; and it is readily adapted to instruments for automatic titration. Finally, the data from a potentiometric titration may reveal the presence in the solution of hitherto unsuspected species that are competing with the unknown for the titrant. Principal among the disadvantages is the likelihood that a potentiometric titration will be more time-consuming than the comparable indicator procedure; the necessity for special equipment is another disadvantage.

For potentiometric methods an *indicator* electrode whose response is sensitive to one or more of the species being determined is immersed in a solution of the sample. A reference electrode of known potential (ordinarily a calomel electrode) is then brought in contact with the sample, usually by means of a salt bridge. In addition to the two electrodes, a sensitive instrument for the determination of the cell potential is required. The characteristics of these instruments are considered on p. 463.

METALLIC INDICATOR ELECTRODES

Metals that undergo reversible oxidation reactions are useful indicator electrodes for the ions formed by their oxidation. Examples include silver, copper, mercury, lead, and cadmium. The potentials of these metals accurately reflect the activities of their ions in solution. In contrast, some of the harder and more brittle metals tend to develop potentials that are nonreproducible and that are dependent upon strains or crystal deformations in their structures as well as upon oxide coatings on their surfaces. Metals in this category include iron, nickel, cobalt, tungsten, and chromium. Such metals are not generally satisfactory for indicator electrodes.

A metal can serve as an indicator electrode not only for its cations but indirectly as well for those anions that form slightly soluble precipitates with its cations. For this application it is only necessary to saturate the solution under study with the sparingly soluble salt. For example, the potential of a silver electrode reflects the concentration of chloride ion in a solution that is saturated with silver chloride; under these circumstances we may describe the electrode behavior in terms of the half-reaction

$$AgCl + e \rightleftharpoons Ag + Cl^-$$

A silver electrode serving as an indicator for chloride ion is an example of an *electrode of the second order* because it measures the concentration of an ion not directly involved in the electron-transfer process. The same electrode, used as an indicator for silver ion, functions as an *electrode of the first order* because its potential is directly dependent upon a participant in the electrode process.

In an analogous way, a metallic electrode can function as a second-order electrode for anions that form stable complexes with the cation of the metal. An important example is the use of a mercury electrode for the detection and determination of the anion Y^{4-} of ethylenediaminetetraacetic acid (EDTA). The mercury(II) complex of this anion is unusually stable; thus, when a small concentration of mercury(II) is introduced into an EDTA solution, complexation is essentially complete. The potential-determining half-reaction is then

$$HgY^{2-} + 2e \rightleftharpoons Hg + Y^{4-}$$

and

$$E_{Hg} = E^0_{HgY^{2-} \to Hg} - \frac{0.059}{2} \log \frac{[Y^{4-}]}{[HgY^{2-}]}$$

If, however, the mercury(II) concentration is constant and small relative to $[Y^{4-}]$, then $[HgY^{2-}]$ is constant and

$$E = \text{constant} - \frac{0.059}{2} \log [Y^{4-}]$$

The mercury electrode is thus second order in Y^{4-}.

This technique can be carried one step further to measure the concentration of cations that form less stable complexes with EDTA than does mercury(II). Thus, if a small and constant amount of mercury(II) is introduced into a solution containing calcium and EDTA ions, in addition to the equilibrium shown above, we must consider the reaction

$$Ca^{2+} + Y^{4-} \rightleftharpoons CaY^{2-}$$

for which

$$K_{CaY^{2-}} = \frac{[CaY^{2-}]}{[Ca^{2+}][Y^{4-}]}$$

After solving this expression for $[Y^{4-}]$ and substituting into the Nernst equation, we obtain

$$E_{Hg} = E^0_{HgY^{2-} \to Hg} - \frac{0.059}{2} \log \frac{[CaY^{2-}]}{[HgY^{2-}][Ca^{2+}]K_{CaY^{2-}}} \tag{18-1}$$

Provided calcium ion is in excess in the solution and provided further that the formation constant of the metal-EDTA complex is reasonably favorable, $[CaY^{2-}]$ as well as $[HgY^{2-}]$ will remain approximately constant, and equation (18-1) reduces to

$$E_{Hg} = \text{constant} - \frac{0.059}{2} \log \frac{1}{[Ca^{2+}]}$$

Within these limitations the mercury electrode becomes a third-order indicator for calcium ion. This electrode system is important in potentiometric titrations involving EDTA.

Inert metals such as platinum or gold are frequently employed to measure the concentration ratio of ions in two oxidation states in a solution. For example, a platinum electrode serves as a convenient indicator electrode for equilibria such as

$$Fe^{3+} + e \rightleftharpoons Fe^{2+}$$

Metallic indicator electrodes are fabricated as coils of wire, as flat metal plates, or as heavy cylindrical billets. Generally, exposure of a large surface area to the solution assures rapid attainment of equilibrium. Thorough cleaning of the metal surface before use is often important; a brief dip in concentrated nitric acid followed by several rinsings with distilled water is satisfactory for many metals.

MEMBRANE ELECTRODES[1]

The development of membrane electrodes began with the empirical observation by Cremer,[2] shortly after the beginning of this century, of the potential that developed across a thin glass membrane interposed between solutions with different hydrogen ion concentrations. Further studies showed that with certain types of glass the magnitude of this potential was related to the difference in acidities of the two solutions and was independent of concentration differences of other components. Systematic investigations of this phenomenon led to development of the glass-membrane electrode, which has been widely used for the potentiometric determination of *p*H for nearly half a century. A remarkable and valuable property of the glass electrode is its highly selective response to hydrogen ions.

Recently, extensive research has resulted in a reasonably clear understanding of the *p*H sensitivity and selectivity of glass membranes, and the development of membranes that are selectively sensitive to other ions. As a consequence, membrane electrodes are now available for the direct potentiometric determination of such ions as K^+, Na^+, Li^+, F^-, Ca^{2+}, as well as others.[3]

It is convenient to divide membrane electrodes into three categories, based upon the membrane composition. These include (1) glass electrodes, (2) liquid-membrane electrodes, and (3) solid-state or precipitate electrodes.

---

[1]For further information on this topic see R. A. Durst, Ed., *Ion Selective Electrodes,* National Bureau of Standards Special Publication 314. (Washington, D.C.: U. S. Government Printing Office, 1969).

[2]M. Cremer, *Z. Biol.* **47**, 562 (1906).

[3]See G. A. Rechnitz, *Chem. Eng. News,* **45**, (25), 146 (1967).

The mechanism of potential development appears to be the same for the three classes. We shall consider the properties and behavior of the glass electrode in greater detail because of its pre-eminent importance at the present time.

The Glass Electrode for pH Measurement

The experimental observation fundamental to the development of this electrode is that a potential difference develops across a thin, conducting, glass membrane interposed between solutions of different pH. The potential can be detected by placing reference electrodes of known and constant potential in each solution, as shown schematically in Figure 18-1. With this arrangement the potential across the reference electrodes varies as a function of the ratio between the hydrogen ion activities a_1 and a_2. At 25°C the following relationship is found to hold

$$E_{obs} = k + 0.059 \log \frac{a_1}{a_2}$$

where k is a constant.

To measure the pH of an unknown solution, the hydrogen ion activity of one of the compartments (say, a_2) is kept constant. The solution to

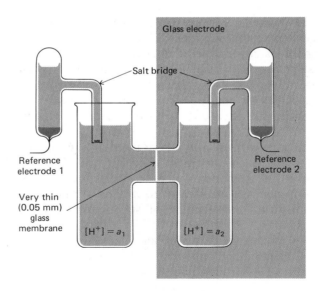

Figure 18-1. Schematic diagram of a cell for pH measurement. The glass electrode consists of reference electrode 2, the solution of known hydrogen ion activity a_2, and the glass membrane. The pH of the solution with hydrogen ion activity a_1 is being determined.

be measured is placed in the other compartment; if we denote its activity as a_1, the equation can be rewritten as

$$E_{obs} = K + 0.059 \log a_1$$
$$= K - 0.059 \, pH$$

where K is a new constant that contains the logarithmic function of a_2. The numerical value of K is obtained by measuring E for a solution whose hydrogen ion activity a_1 is known exactly.

It is important to emphasize that the potentials of the two reference electrodes E_1 and E_2 are quite independent of a_1 and a_2. Thus, any changes in potential between them must arise in some other part of the system — logically across the glass interface. As indicated in Figure 18-1, the glass electrode consists of a thin glass membrane, a solution of fixed pH, and one reference electrode. In making pH measurements, the glass electrode is coupled with a second reference electrode, usually a saturated calomel half-cell.

Figure 18-2 illustrates the construction of a typical, commercially available, glass electrode. The pH-sensitive membrane is blown or sealed onto the end of a heavy-walled glass tube; the resulting bulb contains a solution of fixed pH, often 0.1 F hydrochloric acid. As indicated, the inner reference electrode is a silver-silver chloride reference electrode (p. 440).

Composition of pH-sensitive glass membranes. Not all glass membranes show the pH response given by equation (18-2); indeed, quartz

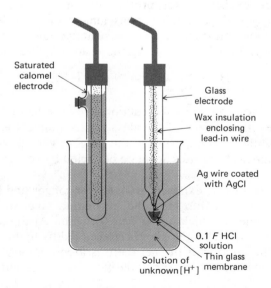

Figure 18-2. A typical electrode system for the potentiometric measurement of pH.

and Pyrex are virtually insensitive to pH variations. Much systematic investigation concerned with the effect of glass composition on sensitivity of membranes to protons and other cations has been carried out during the past quarter century, and a variety of compositions are now used commercially.[4] For many years Corning 015 glass, composed approximately of 22 percent Na_2O, 6 percent CaO, and 72 percent SiO_2, has been widely used. This glass shows an excellent specificity toward hydrogen ions up to a pH of about 9; at higher pH values, however, the membrane becomes sensitive to sodium and other alkali ions. As a consequence, a pH value calculated from equation (18-2) is likely to be low by several tenths of a unit in this region, particularly in solutions containing high concentrations of sodium ion. This effect, called the *alkaline error*, can be largely avoided by employing a membrane constructed of a glass in which the sodium has been replaced by lithium. The source of the alkaline error is considered on page 455.

Hygroscopicity of glass membranes. It has been shown that the surfaces of a glass membrane must be hydrated in order to function as a pH electrode. Nonhygroscopic glasses show no pH function. Even Corning 015 glass shows little pH response after dehydration by storage over a desiccant; however, its sensitivity is restored after standing for a few hours in water. This hydration involves absorption of approximately 50 mg of water per cubic centimeter of glass.

It has been demonstrated experimentally that hydration of a pH-sensitive glass membrane is accompanied by a reaction in which cations of the glass are exchanged for protons of the solution. The exchange process involves singly charged cations almost exclusively inasmuch as the di- and trivalent cations in the silicate structure are much more strongly bonded. Typically, the ion exchange reaction can be written as

$$\underset{\text{soln}}{H^+} + \underset{\text{solid}}{Na^+Gl^-} \rightleftharpoons \underset{\text{soln}}{Na^+} + \underset{\text{solid}}{H^+Gl^-} \tag{18-3}$$

The equilibrium constant for this reaction is highly favorable. Thus, after soaking in water, the surface of the glass consists almost entirely of silicic acid (H^+Gl^-). Only upon exposure of the glass to very basic solutions will a significant number of the sites again be occupied by sodium or other alkali metal ions.

As a glass surface is exposed to water for longer and longer periods, water penetrates farther into the solid, forming a layer of silicic acid gel that in time becomes as thick as 10^{-4} to 10^{-5} mm. At the outer surface of the gel, all fixed, singly charged sites are occupied by hydrogen ions. From the surface to the interior of the gel there is a continuous decrease in the number of protons and a corresponding increase in the number of sodium

[4]For a summary of this work see J. O. Isard, in G. Eisenman, *Glass Electrodes for Hydrogen and Other Cations* (New York: Marcel Dekker, Inc., 1967), Chap. 3.

[handwritten annotations at top:]
$H^+ + Na \cdot Gl^- \rightarrow H'Gl^- + Na^+$ *boundary potential*
$H^+Gl^- \rightleftharpoons H^+ + Gl^-$ *diffusion potential*

ions. A schematic representation of the two surfaces of a glass membrane is shown in Figure 18-3.

Resistance of glass membranes. The membrane in a typical commercial glass electrode ranges in thickness between 0.03 and 0.1 mm, and has an electrical resistance of 50 to 500 megohms. The problems associated with potential measurements in high-resistance cells are considered in later sections.

Current conduction through the dry glass region is ionic and involves movement of the alkali ions from one site to another. Within the two gel layers the current is carried by both alkali and hydrogen ions. Across each gel-solution interface, current passage involves the transfer of protons. The direction of migration is from gel to solution at the one interface and from solution to gel at the other. That is,

$$H^+Gl^- \rightleftharpoons H^+ + Gl^- \tag{18-4}$$
solid · soln solid

Theory of the glass-electrode potential.[5]   The potential across a glass membrane consists of a *boundary potential* and a *diffusion potential.* Under ideal conditions only the former is affected by pH.

The boundary potential for the membrane of a glass electrode contains two components, each being associated with one of the gel-solution interfaces. If V_1 is the component arising at the interface between the external solution and the gel (see Figure 18-3) and if V_2 is the corresponding potential at the inner surface, then the boundary potential E_b for the membrane is given by

$$E_b = V_1 - V_2 \tag{18-5}$$

[handwritten:] V_2 potential

The potential V_1 is determined by the hydrogen ion activities both in the external solution and upon the surface of the gel, and can be con-

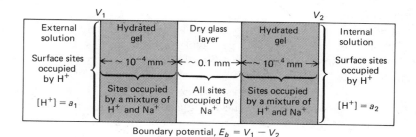

Boundary potential, $E_b = V_1 - V_2$

Figure 18-3.   Schematic representation of a well-soaked glass membrane.

[5]For a comprehensive discussion of this topic see G. Eisenman, *Glass Electrodes for Hydrogen and Other Cations* (New York: Marcel Dekker, Inc., 1967), Chaps. 4-6.

sidered a measure of the driving force of the reaction shown by equation (18-4). In the same way, the potential V_2 is related to the hydrogen ion activities in the internal solution and on the corresponding gel surface.

It can be demonstrated from thermodynamic considerations[6] that V_1 and V_2 are related to hydrogen ion activities at each interface as follows:

$$V_1 = k_1 + \frac{RT}{F} \ln \frac{a_1}{a_1'} \tag{18-6}$$

$$V_2 = k_2 + \frac{RT}{F} \ln \frac{a_2}{a_2'} \tag{18-7}$$

where R, T, and F have their usual meanings; a_1 and a_2 are activities of the hydrogen ion in the *solutions* on either side of the membrane; and a_1' and a_2' are the hydrogen ion activities in each of the *gel layers* contacting the two solutions. If the two gel surfaces have the same number of sites from which protons can leave, then the two constants k_1 and k_2 will be identical; so also will be the two activities a_1' and a_2' in the gel layers, provided that all original sodium ions on the surface have been replaced by protons (that is, insofar as the equilibrium shown in equation (18-3) lies far to the right). Assuming these equalities, substitution of equations (18-6) and (18-7) into (18-5) yields

$$E_b = V_1 - V_2 = \frac{RT}{F} \ln \frac{a_1}{a_2} \tag{18-8}$$

Thus, *provided the two gel surfaces are identical*, the boundary potential E_b depends only upon the activities of the hydrogen ion in the *solutions* on either side of the membrane. If one of these activities a_2 is kept constant, then equation (18-8) further simplifies to

$$E_b = \text{constant} + \frac{RT}{F} \ln a_1 \tag{18-9}$$

and the potential gives a measure of the activity of hydrogen ions in the external solution.

In addition to the boundary potential, a so-called *diffusion potential* also develops in each of the two gel layers. Its source lies in the difference in mobilities of hydrogen ions and the alkali metal ions in the membrane. We have pointed out that all occupied, singly charged, anionic sites available to cations at the gel-solution interface are populated by hydrogen ions. On the other hand, all sites at the dry glass-gel interface are occupied by alkali metal ions. Because of the resulting concentration differences, there is a tendency for hydrogen ions to move toward the dry glass surface and for the alkali ions to migrate toward the solution interface; the extent of this movement depends upon the inherent mobilities of the two ions in the

[6]G. Eisenman, *Biophys. J.*, **2** *Part 2*, 259 (1962).

gel. Because these mobilities differ, a charge separation (and thus a diffusion potential) develops in the gel; this potential is analogous to the liquid-junction potential discussed on page 431.

The overall magnitude of the diffusion potential in each of the gel layers is independent of the extent to which the diffusion has occurred, provided that all surface sites on the gel-solution interface are occupied solely by protons. Furthermore, the two diffusion potentials are equal and opposite in sign insofar as the two solution-gel interfaces are the same. Under these conditions, then, the net diffusion potential is zero, and the emf across the membrane depends only on the boundary potential as given by equation (18-8).

Asymmetry potential. If identical solutions and identical reference electrodes are placed on either side of the membrane shown in Figure 18-1, and if the two faces of the membrane have identical properties, then the measured potential of the cell should be zero. It is found, however, that a small potential, called the *asymmetry potential*, usually does develop when this experiment is performed. Moreover, the asymmetry potential associated with a given glass electrode changes slowly with time.

The causes of the asymmetry potential are obscure, but undoubtedly they include such factors as differences in strains established within the two surfaces during manufacture, mechanical and chemical attack of the external surface during use, and contamination of the outer surface by grease films and other adsorbed substances. The effect of the asymmetry potential on a pH measurement is eliminated by frequent calibration of the electrode against a standard buffer of known pH; the need for this calibration is considered further on page 472.

The alkaline error. In solutions containing very low hydrogen ion concentrations (pH > 9), some glass membranes respond not only to changes in hydrogen ion concentration but also to the concentration of alkali metal ions. The magnitude of the resulting error for four types of membranes is indicated on the right-hand side of Figure 18-4. In each of these curves, the sodium ion concentration was maintained at 1 M and the pH was varied. Note that the pH error is negative, suggesting that the electrode is responding to sodium ions as well as to protons. This idea is confirmed by data obtained at different sodium ion concentrations. Thus, at pH 12 the alkaline error for Corning 015 glass is about -0.7 pH when the sodium ion concentration is 1 M (Figure 18-4) and only about -0.3 pH in solutions that are 0.1 M in sodium ions.

All singly charged cations cause alkaline errors; their magnitudes vary according to the kind of metallic ion and the composition of the glass.

The alkaline error can be satisfactorily explained by assuming an

$H^+Cl^- + O^+ \rightleftarrows \textit{B} \, Cl^- + H^+$

exchange equilibrium between the hydrogen ions on the surface of the glass and the cations in the solution. This process can be formulated as

$$H^+Gl^- + B^+ \rightleftarrows B^+Gl^- + H^+$$

$$\text{gel} \qquad \text{soln} \qquad \text{gel} \qquad \text{soln}$$

where B^+ represents a singly charged cation, such as sodium ion. The equilibrium constant for this reaction is

$$K_{ex} = \frac{a_1 b_1'}{a_1' b_1} \tag{18-10}$$

where a_1 and b_1 are activities of H^+ and B^+ in solution, and a_1' and b_1' are the activities of the same ions in the surface of the gel. The constant K_{ex} depends upon the composition of the glass membrane and is ordinarily small. Thus, the fraction of the surface sites populated by cations other than hydrogen is small except when the hydrogen ion concentration is very low and the concentration of B^+ is high.

The effect of the presence of an alkali metal ion on the emf across a membrane can be described in quantitative terms by rewriting equation (18-10) in the form

$$\frac{a_1}{a_1'} = \frac{(a_1 + K_{ex}b_1)}{(a_1' + b_1')} \tag{18-11}$$

Substitution of equation (18-11) into equation (18-6) and subtraction of equation (18-7) gives

$$E_b = V_1 - V_2 = k_1 - k_2 + \frac{RT}{F} \ln \frac{(a_1 + K_{ex}b_1) \, a_2'}{(a_1' + b_1') \, a_2} \tag{18-12}$$

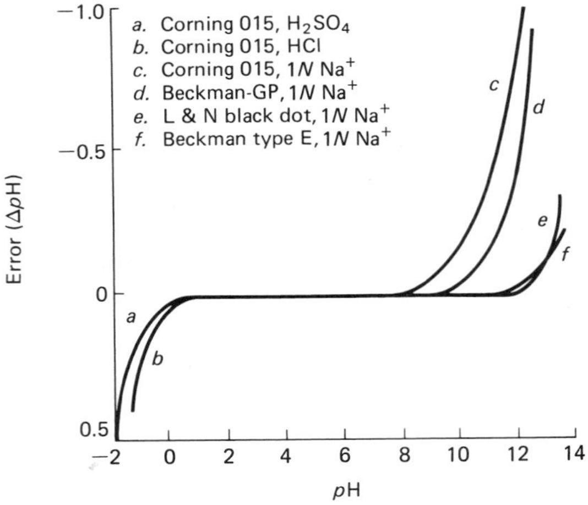

a. Corning 015, H_2SO_4
b. Corning 015, HCl
c. Corning 015, 1N Na^+
d. Beckman-GP, 1N Na^+
e. L & N black dot, 1N Na^+
f. Beckman type E, 1N Na^+

Figure 18-4. Acid and alkaline errors of selected glass electrodes at 25°C. From R. G. Bates, *Determination of pH: Theory and Practice* (New York: John Wiley & Sons, Inc., 1964), p. 316. (With permission.)

If the number of sites on the two sides of the membrane is the same, the activity of hydrogen ions on the internal surface is approximately equal to the sum of the activities of the two cations on the external surface. That is, $a_2' \cong (a_1' + b_1')$; further, $k_1 = k_2$, and so equation (18-12), reduces to

$$E_b = \frac{RT}{F} \ln \frac{(a_1 + K_{ex}b_1)}{a_2}$$

Finally, where a_2 is constant,

$$E_b = \text{constant} + \frac{RT}{F} \ln (a_1 + K_{ex}b_1) \qquad (18\text{-}13)$$

It has been shown that not only the boundary potential but also the diffusion potential on one side of the membrane is changed when some of the surface sites are occupied by cations other than hydrogen. As a consequence, the diffusion potentials do not subtract out as they did in the previous case (see p. 454), and E_b does not adequately describe the total emf that develops in the presence of another cation. It has also been shown that this effect can be accounted for by modifying equation (18-13) as follows:[7]

$$E = \text{constant} + \frac{RT}{F} \ln \left[a_1 + K_{ex}\left(\frac{U_B}{U_H}\right) b_1 \right] \qquad (18\text{-}14)$$

where U_B and U_H are measures of the mobilities of B^+ and H^+ in the gel. For many glasses the term $K_{ex}(U_B/U_H)b_1$ is small with respect to the hydrogen ion activity a_1 as long as the pH of the solution is less than about 9. Under these conditions, equation (18-14) simplifies to equation (18-9). With high concentrations of singly charged cations and at higher pH levels, however, this second term plays an increasingly important part in determining E. The composition of the glass membrane determines the magnitude of $K_{ex}(U_B/U_H)b_1$; this parameter is relatively small for glasses that have been designed for work in strongly basic solutions. The membrane of the Beckman Type E glass electrode, illustrated in Figure 18-4, is of this type.

The acid error.   As shown in Figure 18-4, the typical glass electrode exhibits an error, opposite in sign to the alkaline error, in solutions of pH less than one. As a consequence, pH readings tend to be too high in this region. The magnitude of the error is dependent upon a variety of factors and is generally not very reproducible. The causes of the acid error are not well understood.

[7]See G. Eisenman, *Glass Electrodes for Hydrogen and Other Cations* (New York: Marcel Dekker, Inc., 1967), Chaps. 4 and 5.

for determination of cations other than Hydrogen the ion activities a₁
is equation $E = k + R\frac{T}{F} \ln \left(a_1 + Kex \frac{UB}{UH} \right) b_1$

be negligible

458 *Potentiometric Methods*

Glass Electrodes for the Determination of Other Cations

As we have pointed out, the existence of the alkaline error in early glass electrodes led to studies of the effect of glass composition on the magnitude of this error. One consequence of this work has been the development of glasses for which the term $K_{ex}(U_B/U_H)b_1$ in equation (18-14) is small enough so that the alkaline error is negligible below a pH of about 12. Other studies have been directed toward finding glass compositions in which this term is greatly enhanced in the interests of developing glass electrodes for the determination of cations other than hydrogen. This application requires that the hydrogen ion activity a_1 in equation (18-14) be negligible with respect to the second term containing the activity b_1 of the other cation. Under these circumstances the potential of the electrode would be *independent* of pH, but would vary with pB in the typical way.

A number of investigators have demonstrated that the presence of Al_2O_3 or B_2O_3 in glass causes the desired effect. Eisenman and coworkers carried out a systematic study of glasses containing Na_2O, Al_2O_3, and SiO_2 in various proportions, and have demonstrated clearly that it is practical to prepare membranes that can be employed for the selective measurement of each of several cations in the presence of others.[8] Glass electrodes for potassium ion and sodium ion are now available commercially.

Figure 18-5 shows the response of two types of glass electrodes to pH changes in solutions that are 0.1 F in various alkali metal ions. The one electrode was fabricated from Corning 015 glass, a composition that contains no Al_2O_3; at pH levels below 9 the term $K_{ex}(U_B/U_H)b_1$ in equation (18-14) is small with respect to a_1. At higher pH values the second term becomes important; note that its magnitude depends upon the type of alkali ion present.

With the glass containing Al_2O_3, both potential-determining terms in equation (18-14) are clearly important at low pH values, but at levels above about pH 5 the potential becomes *independent* of pH. Thus, $a_1 \ll K_{ex}(U_B/U_H) b_1$ in this region, with the result that the potential can be expected to vary linearly with pNa, pK, or pLi. Experimentally, this prediction is verified; such a glass can be employed to measure the concentration of these ions.

Sources for the difference in behavior of the two glasses shown in Figure 18-5 (toward Na^+, for example) include the relative affinity of anionic surface sites of the gel for hydrogen ions as compared with sodium ions and (to a lesser extent) the mobility ratios of the ions in the two glasses. Above pH 4, where Al_2O_3-containing glass is no longer sensitive to hydrogen ions, these anionic sites are occupied entirely by the metallic ion rather than by the hydrogen ion, and the electrode behaves as a perfect sodium

---

[8]For a summary of this work see G. Eisenman in C. E. Reilley, Ed., *Advances in Analytical Chemistry and Instrumentation*, Vol. 4 (New York: John Wiley & Sons, Inc., 1965), p. 213.

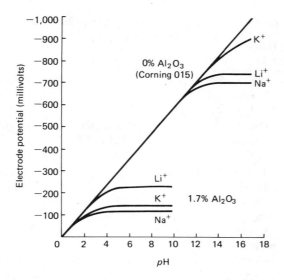

Figure 18-5. Response of two glass membranes in the presence of alkali metal ions (conc. 0.1 *M* in each case). Curves taken from G. Eisenman, in C. N. Reilley, Ed., *Advances in Analytical Chemistry and Instrumentation*, Vol. 4 (New York: John Wiley & Sons, Inc., 1965), p. 213. (With permission.)

electrode. With lithium ion, however, saturation occurs only at lower hydrogen ion concentrations ($pH \sim 6$). Thus, the surface has less affinity for lithium than for sodium ion.

Studies such as these have led to the development of glass membranes that are suitable for the direct potentiometric measurement of the concentrations of the following ions: Na^+, K^+, NH_4^+, Rb^+, Cs^+, Li^+, and Ag^+. Some of these glasses are reasonably selective, as is shown in Table 18-1. The selectivity constant K shown in the table is the ratio of concentration of the interfering ion to the ion of interest that produces the same potential response by the membrane.

The development of membranes that are selective for cations other than hydrogen has been rapid during the past decade, and much work in the field is currently in progress. Undoubtedly, new compositions with useful analytical properties can be expected in the near future.

Liquid-membrane Electrodes

Liquid membranes are a recent development. These electrodes owe their response to the potential that is established across the interface between the solution to be analyzed and an immiscible liquid that will selectively bond with the ion being determined. Liquid-membrane electrodes provide a means for the direct potentiometric determination of the activities of several polyvalent cations and certain anions as well.

Table 18-1 Properties of Certain Cation-Sensitive Glasses[a]

Principal Cation to be Measured	Glass Composition	Selectivity Characteristics	Remarks
Li^+	$15Li_2O$ $25Al_2O_3$ $60SiO_2$	$K_{Li^+/Na^+} \approx 3$, $K_{Li^+/K^+} > 1000$	Best for Li^+ in presence of H^+ and Na^+
Na^+	$11Na_2O$ $18Al_2O_3$ $71SiO_2$	$K_{Na^+/K^+} \approx 2800$ at pH 11 $K_{Na^+/K^+} \approx 300$ at pH 7	Nernst type of response to $\sim 10^{-5} M$ Na^+
	$10.4Li_2O$ $22.6Al_2O_3$ $67SiO_2$	$K_{Na^+/K^+} \approx 10^5$	Highly Na^+ selective, but very time-dependent
K^+	$27Na_2O$ $5Al_2O_3$ $68SiO_2$	$K_{K^+/Na^+} \approx 20$	Nernst type of response to $< 10^{-4} M$ K^+
Ag^+	$28.8Na_2O$ $19.1Al_2O_3$ $52.1SiO_2$	$K_{Ag^+/H^+} \approx 10^5$	Highly sensitive and selective to Ag^+, but poor stability
	$11Na_2O$ $18Al_2O_3$ $71SiO_2$	$K_{Ag^+/Na^+} > 1000$	Less selective for Ag^+ but more reliable

[a]Reprinted from *Chemical & Engineering News,* Vol. 45, June 12, 1967, page 149. Copyright 1967 by the American Chemical Society and reprinted by permission of the copyright owner.

Figure 18-6 is a schematic representation of a cell employing a liquid-membrane electrode. The membrane consists of a liquid ion exchanger supported between two porous glass or plastic disks that mechanically prevent mixing of the ion-exchange liquid and the aqueous solutions in contact with it. The ion-exchange liquid is a nonvolatile, water-immiscible, organic material that contains acidic, basic, or chelating functional groups. When in contact with an aqueous solution of a divalent cation, an exchange equilibrium is established that can be represented as

$$\underset{\substack{\text{organic} \\ \text{phase}}}{RH_{2x}} + \underset{\substack{\text{aqueous} \\ \text{phase}}}{xM^{2+}} \rightleftharpoons \underset{\substack{\text{organic} \\ \text{phase}}}{RM_x} + \underset{\substack{\text{aqueous} \\ \text{phase}}}{2xH^+}$$

By repeated treatment, the exchange liquid can be converted essentially completely to the cationic form, RM_x; it is this form that is employed in electrodes for the determination of M^{2+}.

Figure 18-7 shows construction details of a commercial liquid-membrane electrode that is selective for calcium ion. The ion exchanger is an organic phosphoric acid that bonds calcium ion more strongly than most other cations; the selective properties of the electrode are the result of this affinity for calcium ions. The internal solution in contact with the exchanger contains a fixed concentration of calcium chloride; a silver-

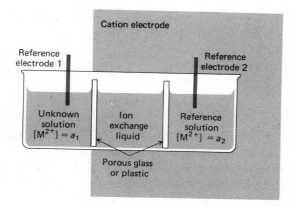

Figure 18-6. Schematic diagram of a cell employing a liquid-membrane electrode. Shaded portion represents the liquid-membrane electrode.

silver chloride reference electrode is immersed in this solution. When used for a calcium ion determination, the porous disk containing the ion-exchange liquid separates the solution to be analyzed and the reference calcium chloride solution; the equilibrium established at each interface can be represented as

$$RCa \rightleftharpoons R^{2-} + Ca^{2+}$$

organic   organic   aqueous

The emf developed at the interface with the test solution is given by

$$V_1 = k_1 + \frac{RT}{2F} \ln \frac{a_1}{a_1'}$$

and that resulting from the interface with the reference solution is

$$V_2 = k_2 + \frac{RT}{2F} \ln \frac{a_2}{a_2'}$$

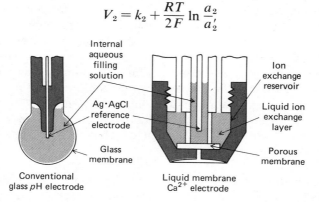

Figure 18-7. Comparison of a liquid membrane calcium ion electrode with a glass electrode. (Sketch courtesy of Orion Research Inc., Cambridge, Massachusetts.)

where a_1 and a_2 represent the calcium ion activities in the respective aqueous phases, and a'_1 and a'_2 are the activities for this ion in the ion-exchange liquid of the membrane. Normally, the latter two activities are equal; by a set of arguments similar to those given for a glass membrane (p. 454), the potential for the electrode is

$$E = \text{constant} + \frac{RT}{2F} \ln a_1$$

The performance of the calcium electrode just described is reported to be independent of pH in the range between 5.5 and 11. At lower pH levels, hydrogen ions undoubtedly exchange with calcium ions on the exchanger to a significant extent so that the electrode becomes pH sensitive as well as pCa dependent. As a device for measuring pCa, the electrode has a selectivity constant (p. 459) of 50 with respect to magnesium ion, and 1000 with respect to sodium and potassium. The electrode can be employed to measure calcium ion activities as low as $10^{-5}\ M$.

A number of ion-selective, liquid-membrane electrodes have appeared on the market. These include a copper(II) electrode, an electrode equally sensitive to calcium and magnesium for monitoring water hardness, and electrodes selective for perchlorate ion and nitrate ion. The latter pair involves use of liquid anion exchangers.

Solid-state, or Precipitate, Electrodes

Considerable work has been devoted recently to the development of solid membranes that are selective toward anions in the way that glass behaves toward certain cations. We have seen that selectivity of a glass membrane is the result of anionic sites on its surface that show special affinity toward certain positively charged ions. By analogy, a membrane having similar cationic sites might be expected to respond selectively toward anions. To exploit this possibility, attempts have been made to prepare membranes of salts containing the anion of interest and a cation that selectively precipitates that anion from aqueous solutions; for example, barium sulfate has been proposed for sulfate ion and silver halides for the various halide ions. The problem encountered in this approach has been in finding methods for fabricating membranes from the desired salt that have suitable physical strength, conductivity, and resistance to abrasion and corrosion. Two types have been developed: solid-state membranes and precipitate membranes.

Solid-state membranes. A solid-state electrode that is selective for fluoride ion has been described.[9] The membrane consists of a single crystal of lanthanum fluoride that has been doped with a rare earth to increase

---

[9]M. S. Frant and J. W. Ross, Jr., *Science,* **154,** 1553 (1966).

its electrical conductivity. The membrane, supported between a reference solution and the solution to be measured, shows the theoretical response to changes in fluoride ion activity (that is, $E = \text{constant} - 0.059 \log a_{F^-}$) to as low as 10^{-5} M. The electrode is reported to be as much as several orders of magnitude more selective for fluoride as for other common anions.

Membranes prepared from cast pellets of silver halides have been successfully employed in electrodes for the selective determination of chloride, bromide, and iodide ions. A sulfide electrode that employs a solid-state membrane is also available commercially.

Precipitate membranes. Electrode membranes have been prepared by suspending finely divided precipitates, such as barium sulfate or the silver halides, in a matrix material from which a membrane can be fabricated. A paraffin membrane containing barium sulfate has been employed for potentiometric titrations involving the sulfate ion. A silicone rubber matrix containing various slightly soluble salts has been recommended for the potentiometric determination of the halides, sulfate, and phosphate.[10]

INSTRUMENTS FOR MEASUREMENT OF CELL POTENTIALS

We noted in Chapter 17 that the potential of a galvanic cell is current-dependent because of the effects of IR drop and polarization. In addition, current flow causes changes in concentration of the electrode reactants and thus influences the electrode potentials. As a consequence, it is generally necessary to measure the potential of a galvanic cell with an instrument that requires essentially no current for its operation. Thus, an ordinary voltmeter is not satisfactory because of the appreciable electrical energy required to cause deflection of the needle.

Two voltage-measuring devices that draw infinitesimally small currents are the *potentiometer* and the direct-reading *electronic voltmeter*. Both are widely employed in potentiometric measurements.

Potentiometers

The potentiometer is the classical instrument for the accurate determination of the potentials of galvanic cells. It is a null-point instrument in which the unknown potential is balanced against an exactly known standard potential. At the null point no current is drawn from the galvanic cell being measured.

The voltage divider. Figure 18-8 shows a simple voltage divider, a device consisting of a resistance AB along which a sliding contact C can be moved.

[10]E. Pungor, J. Havas, and K. Toth, *Z. Chem.*, **5**, 9 (1965).

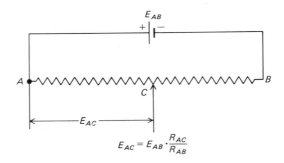

$$E_{AC} = E_{AB} \cdot \frac{R_{AC}}{R_{AB}}$$

Figure 18-8. Diagram of a simple voltage divider.

If a battery is placed across this resistance, a current I will flow; from Ohm's law we may write $E_{AB} = IR_{AB}$, where R_{AB} is the ohmic resistance of AB. Of interest here is the potential drop between the contact C and the end of the resistor A — that is, E_{AC}. The current passing through the portion of the resistor AC is identical to that passing through the entire resistor; it follows, then, that $E_{AC} = IR_{AC}$. Dividing this by the foregoing relationship we get

$$E_{AC} = E_{AB}\frac{R_{AC}}{R_{AB}} \tag{18-15}$$

Thus, we have a device that will supply a continuously variable voltage from zero (when contact C is at A) to the total output E_{AB} of the power supply (when the contact C has been moved to B). This very simple and useful contrivance is employed in many electroanalytical instruments.

Simple potentiometers. A simple instrument that permits the measurement of the emf of galvanic cells with passage of but negligible current is illustrated in Figure 18-9. This device consists of a voltage divider powered

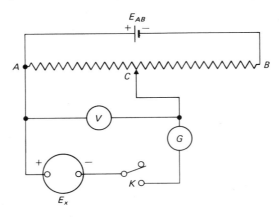

Figure 18-9. Diagram of a simple potentiometer.

by a battery whose emf is equal to or greater than the emf to be measured, E_x. The cell whose potential is to be measured is placed across AC so that its output *opposes* that of the working battery. In parallel with the galvanic cell is a tapping key K, which permits momentary closing of the circuit, and a galvanometer G, which serves as a current detector. The direct-current voltmeter V measures the potential drop across AC.

Now, if the potential across the voltage divider E_{AC} is greater than that of the galvanic cell E_x, electrons will be forced from right to left through the latter when K is closed. On the other hand, if E_{AC} is smaller than E_x, the electron flow will occur in the opposite direction. Finally, when E_x is equal to E_{AC}, no current will flow *in the circuit containing G, K, and the galvanic cell.*

An unknown voltage E_x is measured by moving C until the condition of zero current is indicated by the galvanometer G. This condition is achieved by repeated momentary closing of K, followed by suitable adjustment of C. The entire process draws only infinitesimal currents from the galvanic cell being measured. When balance is achieved, the voltage drop across AC is read on the voltmeter V and is equal, of course, to E_x. Thus, at balance, the voltmeter is powered by current supplied from the working battery and not from the unknown galvanic cell.

A potentiometer such as the one described is limited in accuracy by the precision of the voltmeter V; typically, the uncertainty ranges between about 0.005 and 0.01 V, which is entirely adequate for many purposes. Where greater precision is required, recourse must be made to a potentiometer with a linear voltage divider.

Potentiometer employing a linear voltage divider. To achieve the highest precision in potential measurements, an instrument such as that shown in Figure 18-10 is utilized. Here, the voltage divider contains an additional variable resistance R to allow adjustment of the voltage across AB. In addition, the resistance AB must now be *linear* in character. Thus, the resistance between one end A and any point C is directly proportional to the length AC of that portion of the resistor; then $R_{AC} = kAC$, where AC is expressed in convenient units of length and k is a proportionality constant. Similarly, $R_{AB} = kAB$. A scale, such as that shown in Figure 18-10, can be attached to the resistor to aid in measuring these lengths. Substituting the proportionality relationships into equation (18-15), we obtain

$$E_{AC} = E_{AB} \frac{AC}{AB} \tag{18-16}$$

The instrument in Figure 18-10 also contains a cell E_s whose potential is constant and known with great precision. Most commonly, a standard Weston cell is employed. Either the Weston cell or the unknown cell E_x can be placed in the circuit by means of the switch S.

As with the simple potentiometer, E_{AC} will equal the potential of the

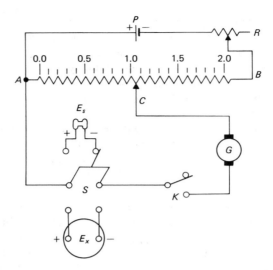

Figure 18-10. Diagram of a potentiometer with a linear voltage divider *AB*.

unknown or the standard cell when no current flow is indicated; with equation (18-16) we may write, then, that

$$E_x = E_{AC_x} = E_{AB} \frac{AC_x}{AB}$$

and

$$E_s = E_{AC_s} = E_{AB} \frac{AC_s}{AB}$$

where AC_x and AC_s represent the linear distances corresponding to balance when the unknown and standard cells are in the circuit. Dividing these equations, we obtain the relationship

$$E_x = E_s \frac{AC_x}{AC_s} \tag{18-17}$$

Thus, E_x may be obtained from the known emf of the Weston cell and the two measurable quantities AC_x and AC_s.

For convenience, the scale reading of AC_x is ordinarily made to correspond directly to the potential in volts by first switching the Weston cell into the circuit and positioning the contact so that AC_s corresponds numerically to the output of this standard cell (p. 441). The potential across *AB* is then adjusted by means of *R* until no current is indicated by the galvanometer. With the potentiometer adjusted thus, AC_s and E_s are numerically equal; similarly, then, AC_x and E_x will also be identical when balance is achieved with the unknown cell in the circuit.

The necessity for a working battery *P* may be questioned; in principle, there is nothing to prevent a direct measurement of E_x by replacing *P* with the standard cell E_s. It must be remembered, however, that current is being continuously drawn from P; a standard cell could not be expected to maintain a constant potential for long under such usage.

The accuracy of a voltage measurement with a potentiometer equipped with a linear voltage divider depends upon several factors. For example, it is necessary to assume that the voltage of the working battery P remains constant during the period of time required to balance the instrument against the standard cell and to measure the potential of the unknown cell. Ordinarily, this assumption does not lead to appreciable error if P consists of one or two heavy-duty dry cells in good condition or of a lead storage battery. Nevertheless, the instrument should be calibrated against the standard cell before each voltage measurement to compensate for possible changes in P.

The linearity of the resistance AB, as well as the precision with which distances along its length can be estimated, both contribute to the accuracy of the potentiometer. Ordinarily, however, the ultimate precision of a good-quality instrument is determined by the sensitivity of the galvanometer relative to the resistance of the circuit. Suppose, for example, that the electrical resistance of the galvanometer plus that of the unknown cell is 1000 ohms, a typical figure. Further, if we assume that the galvanometer is just capable of detecting a current of one microampere (10^{-6} amp), Ohm's law requires that the minimum distinguishable voltage difference will be $10^{-6} \times 1000 = 10^{-3}$ V, or 1 millivolt. By use of a galvanometer sensitive to 10^{-7} amp, a difference of 0.1 millivolt is detectable. A sensitivity of this order is found in an ordinary pointer-type galvanometer; more refined instruments with sensitivities of up to 10^{-10} amp are not uncommon.

A simple potentiometer is entirely satisfactory for measurement of the potential of cells with resistance of less than about 10,000 ohms, but it is totally unsuited for the voltage determination with systems that include a membrane electrode. As mentioned earlier, the resistance of such an electrode typically ranges from 50 to 500 megohms. Thus, the currents in the cell circuit corresponding to a potential difference of 0.1 millivolt range from 2×10^{-12} to 2×10^{-13} amp, a level that cannot be detected with even the best galvanometer. Here some kind of electronic current amplification is required, and several instruments with this feature have been developed for use with glass membrane electrodes. These instruments are generally called *pH meters* since their primary application has been for the determination of *p*H. The instruments, however, can be readily employed with other membrane electrodes and for the determination of potentials of low-resistance cells as well. It is of historical interest that the widespread use of the glass electrode for *p*H determination was delayed by a decade or more until convenient vacuum tube circuits became available to permit detection of very small currents. The first commercial *p*H meters were brought on the market in the early 1930s.

Potentiometers with electronic amplification; (potentiometric pH *meters).* In principle, a potentiometric pH meter differs from the ordinary potentiometer shown in Figure 18-10 only to the extent that the galvanometer is replaced by an electronic circuit that amplifies the current in the cell circuit by a factor of 10^9 or more. Figure 18-11 illustrates schematically how the replacement is made. Here, current amplification is accomplished by means of a single vacuum tube, and the out-of-balance signal is indicated by a simple rugged milliammeter.

The vacuum triode shown in Figure 18-11 contains three electrodes: a plate P, a grid G, and a heated cathode S. When the cathode is made negative with respect to the plate by means of the batteries B (45 to 90 V), electrons are emitted and pass through the wire-mesh grid to the plate. The magnitude of the resulting current depends upon the temperature of the cathode, the applied potential, and most important, the potential of the grid with respect to the cathode. The amplification properties of the tube depend upon the fact that under optimum conditions a small variation in this grid potential results in a large change in the current passing from the cathode to the plate. Thus, as the grid becomes more negative, electrons from the cathode are repelled and the number passing through the grid to the plate is reduced; a smaller current results. Conversely, as the grid becomes less negative, an increased current is observed.

It should be noted that the circuit diagram in Figure 18-11 is oversimplified in that the grid of a triode is usually kept negative with respect to the cathode by an additional battery between the grid and the cathode. The

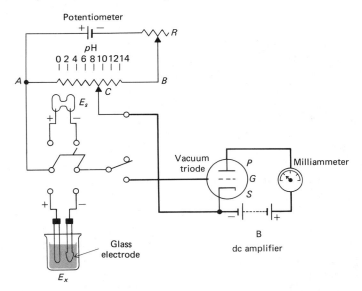

Figure 18-11. A potentiometer with a simple dc amplifier for null detection.

resulting grid-bias voltage permits the tube to be operated in its most favorable range. Thus, when E_{AC} (Figure 18-11) is exactly equal to E_x, the grid-bias voltage alone controls the current through the milliammeter at some suitable level. If, now, E_{AC} is made larger or smaller than E_x through the movement of contact C, the potential on the grid becomes correspondingly more or less negative. An increase or a decrease in the milliammeter current is observed. Thus, in balancing the potentiometer, the contact C is adjusted until the current indicated by the milliammeter is returned to the level corresponding to the original grid-bias voltage.

It should be appreciated that while the circuit shown in Figure 18-11 is adequate to illustrate the principles of the potentiometer pH meter, a modern instrument of this kind is considerably more complex. For example, the current amplification by a single triode is frequently insufficient, and additional tubes are required to provide currents large enough for ready detection. A second stage of amplification is readily accomplished by passage of the current from the first tube through a resistor; the potential drop across this resistor is then impressed across and controls the current of a second vacuum tube. In addition, commercial pH meters of this type also incorporate such features as potentiometer scales calibrated in both pH units and millivolts, compensators for the temperature coefficient and the asymmetry potential of the electrode system, and circuitry that permits standardization of the slide wire against a standard cell without having to reset the position of the sliding contact.

Direct-reading electronic pH *meters.* The amplifier circuit shown on the right in Figure 18-11 is a highly simplified version of a vacuum tube voltmeter in which the current indicated by the milliammeter varies regularly with the potential impressed across the grid and the cathode. According to this diagram, application of the output from a glass-calomel electrode system directly to the grid and cathode should yield a meter directly responsive to electrode potentials. By suitable calibration of the milliammeter scale, the meter could be made to provide pH values directly. Because of the high resistance to current flow between the grid and the cathode, the current drawn from the cell in this mode of operation is still small enough ($\sim 5 \times 10^{-11}$ amp) to be of no serious consequence. In general, however, this simple circuit lacks the stability needed for the highly accurate determination of pH.

Direct-reading pH meters of suitable stability and linear response have been designed by application of the negative or inverse feedback principle. In these circuits a fraction of the amplified output voltage is fed back in opposition to the input voltage to the amplifier. As a consequence, the amplification of the signal is reduced but, in exchange, a linear and highly stable net output signal is obtained. In order to understand the principle of negative feedback we must first consider the behavior of a typical amplifier.

Over small voltage ranges, the output voltage of such a device is approximately proportional to the input signal. That is,

$$E_0 = nE_{In} \qquad (18\text{-}18)$$

where E_0 and E_{In} are the output and input voltages, respectively, and n is a constant called the *amplification factor* for the amplifier. The value of n is determined by a number of variables, some of which are not readily controlled; it is the fluctuation in n brought about by these uncontrollable variables that causes the poor stability in the simple amplifier circuit.

A schematic diagram of a negative feedback circuit is shown in Figure 18-12. Here the output current i passes through a series of resistances and a milliammeter calibrated in pH units. Note that the resistance R is incorporated in both the input and the output circuits; furthermore, the potential drop $E = iR$ is so arranged that it opposes the output potential from the glass-calomel electrode system. Thus, the net input to the amplifier E_{In} is given by

$$E_{In} = E_s - E \qquad (18\text{-}19)$$

where E_s is the cell potential.

The magnitude of E depends upon the value of R relative to the total resistance of the output circuit. That is,

$$E = \frac{RE_O}{R + R_O} = \beta E_0 \qquad (18\text{-}20)$$

where R_O is the resistance of the ammeter and any other resistive compo-

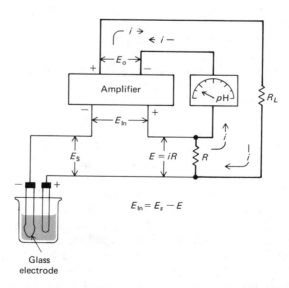

Figure 18-12. Direct-reading pH meter with negative feedback.

nents of the circuit. The quantity β represents the fraction of the output voltage that is employed in opposition to E_s.

Combining equations (18-18), (18-19), and (18-20) leads to the expression

$$E_0 = \frac{n}{1 + \beta n} E_s \qquad (18\text{-}21)$$

If conditions are so arranged that $\beta n \gg 1$, then

$$E_0 = \frac{E_s}{\beta} \qquad (18\text{-}22)$$

To the extent that the approximation is valid, the output signal E_0 is *independent of n*, the quantity that fluctuates with uncontrolled variables. Furthermore, the current i, registered by the meter, is directly proportional to E_0. That is,

$$i = \frac{E_0}{R + R_0}$$

Upon substituting into (18-22) and rearranging

$$E_s = i\beta(R + R_0)$$

But E_s is proportional to pH. Thus,

$$p\text{H} = ki$$

where k is an accumulation of constants.

Direct-reading pH meters of two types are manufactured commercially. Those with dc amplifiers are simpler and less expensive, but have lower precision (0.05 to 0.1 pH unit). In general, it is possible to design an ac amplifier with greater stability and less drift than its dc counterpart. Application of an ac amplifier to the measurement of cell potentials, however, requires that the signal from the cell first be converted to an ac signal before amplification; the amplified signal is then converted back to a dc signal for measurement. The complexities of circuitry associated with ac amplification add to the cost. Nevertheless, the best ac instruments approach the performance of a null-type instrument and are more convenient to use.

APPLICATIONS OF DIRECT POTENTIOMETRIC MEASUREMENT

Direct potentiometric measurements offer an attractive method for completing an analysis. The technique is simple and rapid. Preliminary separations are not ordinarily required because of the selectivity of the indicator electrode. Finally, the method is readily adapted to continuous and automatic recording of analytical data.

Techniques

The observed potential of a cell employed for a direct potentiometric measurement can be expressed in terms of the reference-electrode potential, the indicator-electrode potential, and a junction potential; that is,

$$E_{obs} = E_{ref} - E_{ind} + E_j \qquad (18\text{-}23)$$

The liquid-junction potential occurs between the electrolyte solution of the reference-electrode system and the solution being analyzed. Normally, this junction potential is minimized by the use of a salt bridge (p. 432), but its effect cannot be neglected in direct potentiometric measurements.

We have noted that the potential of the indicator electrode is generally related to the activity a_1 of the ion of interest M^{n+} by a Nernst-type relationship. Thus,

$$E_{ind} = K + \frac{RT}{nF} \ln a_1$$

where K is a constant. At 25°C

$$E_{ind} = K + \frac{0.059}{n} \log a_1 \qquad (18\text{-}24)$$

Combining equation (18-24) with equation (18-23) and rearranging gives

$$pM = -\log a_1 = \frac{E_{obs} - (E_{ref} + E_j - K)}{0.059/n} \qquad (18\text{-}25)$$

or

$$pM = -\log a_1 = \frac{E_{obs} - K'}{0.059/n} \qquad (18\text{-}26)$$

where the constant K' contains the three constants, at least one of which (the junction potential) cannot be evaluated from theoretical calculations. In some instances K is the known standard potential for the indicator electrode; with membrane electrodes, however, K may contain an indeterminate asymmetry potential (p. 455) that changes with time. Thus, the constant K' must be found *experimentally* before equation (18-26) can be employed for the determination of pM. Evaluation is accomplished by measuring E_{obs} for one or more standard solutions of known pM. The assumption is then made that K' does not change when the solutions to be analyzed replace the standards. Generally, this calibration operation should be performed at the time the unknowns are analyzed; recalibration may be desirable if the analysis extends over several hours.

Certain other general precautions should be observed in making direct potentiometric determinations. The solutions should be well stirred in contact with the electrode before measurements are made to insure that the surface film surrounding the electrode is homogeneous with the bulk of the solution. In addition, the response of some indicator electrodes is slow; with such electrodes, periodic measurements should be made with intervening stirring until constant potential values are observed.

Limitations to Direct Potentiometric Measurements

Inherent error in potentiometric measurements. As noted in the previous section it is necessary to assume that the constant K', determined by calibration, is unchanged when the potentiometric measurement of an unknown is performed. This assumption can seldom, if ever, be exactly true because the electrolyte composition of the unknown will almost inevitably differ from that of the solutions employed for calibration. The junction potential contained in K' is altered slightly as a consequence even when a salt bridge is used. Typically, this uncertainty is of the order of one millivolt; unfortunately, because of the nature of the potential-activity relationship, such an uncertainty has a significant effect on the inherent accuracy of the analysis. The magnitude of the uncertainty can be estimated by differentiating equation (18-26), holding E_{obs} constant

$$\frac{\log_{10} e}{a_1} \, da_1 = \frac{0.434 da_1}{a_1} = \frac{dK'}{0.059/n}$$

or

$$\frac{da_1}{a_1} = \frac{n \, dK'}{0.026}$$

Upon replacing da_1 and dK' with the finite increments Δa_1 and $\Delta K'$, and multiplying both sides of the equation by 100, we obtain

$$\frac{\Delta a_1}{a_1} \times 100 = \text{percent relative error} = 3900 \, n \, \Delta K'$$

The quantity $\Delta a_1/a_1$ is the *relative* error in a_1 associated with an *absolute* uncertainty in K' of $\Delta K'$. If, as is typical, $\Delta K'$ is ± 0.001 V, a relative error of about $\pm 4n$ percent can be expected. *It is important to appreciate that this uncertainty is characteristic of all direct potentiometric measurements of cells containing a salt bridge, and this error cannot be eliminated by even the most careful measurements of cell potentials;* indeed, it does not appear possible to devise a method for eliminating the uncertainty in K' that causes this error.

Effect of ionic strength. The potential of an indicator electrode changes linearly with the logarithm of the *activity* of the ion to which it is sensitive. Unless the ionic strength of the medium is held constant, however, a nonlinear relationship is obtained when the observed potential is plotted against the molar concentration. This difference is illustrated in Figure 18-13, where the lower curve depicts the change in potential of a calcium ion membrane electrode as a function of concentration of calcium chloride (note that the concentration scale is logarithmic). Nonlinearity in the curve results from the increases in ionic strength that accompany increases in the calcium chloride concentration. These cause decreases in the activity coefficient of the calcium ion. When concentrations are converted to activities, the upper curve is obtained; note that this straight line has the theoretical slope of 0.0295 (0.059/2). If this experiment were performed under conditions of constant ionic strength (by suitable additions of a salt to which the electrode was insensitive), a linear concentration-potential relationship would be observed. The resulting line would be parallel to, but displaced downward from, the straight activity line.

Changes in activity with ionic strength are most pronounced for ions that carry multiple charges. Thus, the effect shown in Figure 18-13 is less pronounced for electrodes that respond to H^+, Na^+, and other singly charged ions.

Because indicator electrodes respond to activities rather than concentrations, the results from a direct potentiometric analysis may differ significantly from an analysis that determines concentration. For example, a potentiometric calcium analysis based on the theoretical curve in Figure 18-13 would give lower results than a gravimetric analysis based upon precipitation of the oxalate.

An obvious way of correcting a potentiometric measurement to give results in terms of concentration is to prepare an empirical calibration curve, such as the lower line in Figure 18-13, and employ this for the analysis. For this technique to be successful, however, it is essential that the ionic strength of the standard closely approximate that of the samples to be analyzed.

In potentiometric pH measurements, the pH of the standard buffer employed for calibration is generally based on the activity of hydrogen ions. Thus, the resulting hydrogen ion analysis is also on an activity scale; if the

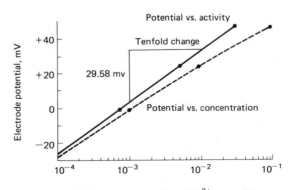

Figure 18-13. Response of a calcium ion electrode to variations in concentration and activity in solutions prepared from pure calcium chloride. (Courtesy of Orion Research Inc., Cambridge, Mass.)

unknown sample has a high ionic strength, the hydrogen ion concentration will differ appreciably from the activity measured.

Potentiometric *p*H Measurements[11]

The hydrogen ion concentration of aqueous solutions can vary over a tremendous range; yet relatively small changes in this quantity may greatly alter the chemical behavior and properties of a chemical system. For example, a tenfold change in hydrogen ion concentration may have a hundredfold or even a thousandfold effect on the solubility of a compound; indeed, such a change can be responsible for a millionfold alteration in the concentration of a participant in an oxidation-reduction equilibrium. As a consequence, a knowledge of hydrogen ion concentration is of vital importance to the chemist, and its measurement is a most important quantitative analytical process.

The hydrogen electrode. The potentiometric determination of *p*H can be accomplished with a hydrogen electrode (p. 414) in conjunction with a calomel reference electrode. Compared with the glass electrode, however, the hydrogen electrode lacks versatility and convenience, and is seldom used today. It cannot be used in solutions containing species capable of oxidizing hydrogen because *p*H changes result. Examples of such interferences include permanganate, iodine, iron(III), and easily reduced organic compounds. Titanium(II) and chromium(II) ions, whose oxidation by hydrogen ions is catalyzed by platinum, must also be absent. The platinum surface of the electrode is easily poisoned by other solutes, such as proteins and sulfides. These interferences, coupled with the inherent hazards associated with the use of hydrogen, have caused virtual abandonment of the hydrogen electrode for *p*H measurement.

The glass electrode. The glass electrode is unquestionably the most important indicator electrode for hydrogen ions and has almost completely displaced all other electrodes for *p*H measurements. It is convenient to use and is subject to few of the interferences affecting other electrodes.

Glass electrodes are available commercially at a relatively low cost and in a variety of shapes and sizes; one form is illustrated in Figure 18-2. As shown in this illustration, the second reference electrode is usually a commercial saturated-calomel electrode.

The glass-calomel electrode system is a remarkably versatile tool for the measurement of *p*H under many conditions. In contrast to other electrodes, the glass-calomel couple can be used without interference in solutions containing strong oxidants, reductants, proteins, and gases; the *p*H of viscous or even semisolid fluids can be determined. Electrodes for

[11]For a detailed discussion of potentiometric *p*H measurements see R. G. Bates, *Determination of pH, Theory and Practice* (New York: John Wiley & Sons, Inc., 1964).

special applications are available. Included among these are microelectrodes for the measurement of a drop (or less) of solution; systems for insertion in a flowing stream of liquid to provide the continuous monitoring of pH; a glass electrode small enough to study the pH in the cavity of a tooth; and a small glass electrode that can be swallowed to indicate the acidity of the stomach contents (the calomel electrode is kept in the mouth).

We have noted that water in the membrane of a glass electrode is essential to its proper performance as a pH indicator. As a consequence, conditions leading to the dehydration of the glass should be avoided whenever possible. Loss of water occurs upon prolonged exposure of the glass to the atmosphere or to such dehydrating solvents as alcohol or concentrated sulfuric acid. Fortunately, the water is easily restored if the electrode is soaked for several hours. To avoid dehydration, glass electrodes should be stored in distilled water.

A well-soaked glass electrode does retain sufficient moisture to permit evaluation of "pH numbers" of nonaqueous solutions and, more important, to indicate end points for acid-base titrations carried out in such solvents. While the theoretical interpretation of these numbers is not possible, their empirical application to analysis has proved useful.[12]

Summary of errors affecting pH measurements with the glass electrode.
The ubiquity of the pH meter and the general applicability of the glass electrode tends to lull the chemist into the idea that any reading obtained with such an instrument is surely correct. It is well to guard against this viewpoint, since there are distinct limitations to the electrode system. These have been discussed in earlier sections, and are summarized below.

1. *The alkaline error.* Ordinary glass electrodes become somewhat sensitive to the alkali metals at pH values greater than about 9. At a pH of 12 the error attributable to this effect is about -0.3 pH unit when the sodium ion concentration is 0.10 F; it approaches -0.7 unit in solutions that are 1.0 F in sodium ion. Electrodes designed for high pH values are available; with these the alkaline error is greatly reduced.

2. *The acid error.* At a pH less than zero, values obtained with a glass electrode tend to be somewhat high.

3. *Dehydration.* Dehydration of the electrode may lead to erratic response.

4. *Errors in unbuffered neutral solutions.* Equilibrium between the electrode surface layer and the solution is achieved only slowly in poorly buffered, approximately neutral solutions. Since the measured potential is determined by the surface layer of liquid, errors arise unless time is allowed for equilibrium to be established; this process may take several minutes. In determining the pH of poorly buffered solutions, the glass electrode should be thoroughly rinsed four or five times with fresh portions of dis-

---

[12]For example, see J. S. Fritz and G. S. Hammond, *Quantitative Organic Analysis* (New York: John Wiley & Sons, Inc., 1957), Chap. 3.

tilled water before use. Good stirring is also helpful, and several minutes should be allowed to obtain steady readings.

5. *Variation in junction potential.* This source of error was discussed on page 473. *The existence of this fundamental uncertainty in the measurement of pH for which a correction cannot be applied* should be reemphasized. Absolute values more reliable than 0.01 *p*H unit are generally unobtainable. On the other hand, with care, *p*H *differences* as small as 0.001 unit can be observed.

6. *Error in the pH of the buffer solution.* Since the glass electrode must be regularly calibrated, any inaccuracies in the preparation or any changes in composition of the buffer during storage are reflected as errors in *p*H measurements. Upon standing, many buffers suffer decomposition as a result of bacterial action.

APPLICATIONS OF POTENTIOMETRIC TITRATIONS

Figure 18-14 shows a typical apparatus for performing a potentiometric titration. Ordinarily, the titration involves measuring and recording the cell potential after each addition of reagent. The standard solution is added in large increments at the outset; as the end point is approached (as indicated by larger potential changes per addition), the increments are made smaller.

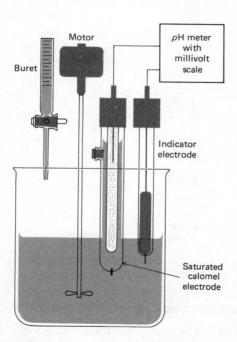

Figure 18-14.   Apparatus for a potentiometric titration.

For some purposes it is convenient to make small and *equal* additions near the equivalence point. Ordinarily, the titration is carried well beyond the end point.

Sufficient time must be allowed for the attainment of equilibrium after each addition of reagent; precipitation titrations may require several minutes for equilibration, particularly in the vicinity of the equivalence point. A close approach to equilibrium is indicated when the measured potential ceases to drift by more than a few millivolts. Good stirring is frequently effective in hastening the achievement of this condition.

The first two columns of Table 18-2 consist of a typical set of potentiometric-titration data obtained with the apparatus illustrated in Figure 18-14. These have been plotted in Figure 18-15(a). The experimental plot closely resembles the titration curves derived from theoretical considerations.

Table 18-2 Potentiometric Titration Data for 2.433 Milliequivalents of Chloride with 0.1000 N Silver Nitrate[a]

Vol AgNO₃ (ml)	E vs. SCE (V)	$\Delta E/\Delta V$ (V/ml)	$\Delta^2 E/\Delta V^2$
5.0	0.062		
		0.002	
15.0	0.085		
		0.004	
20.0	0.107		
		0.008	
22.0	0.123		
		0.015	
23.0	0.138		
		0.016	
23.50	0.146		
		0.050	
23.80	0.161		
		0.065	
24.00	0.174		
		0.09	
24.10	0.183		
		0.11	
24.20	0.194		0.28
		0.39	
24.30	0.233		0.44
		0.83	
24.40	0.316		−0.59
		0.24	
24.50	0.340		−0.13
		0.11	
24.60	0.351		−0.04
		0.07	
24.70	0.358		
		0.050	
25.00	0.373		
		0.024	
25.5	0.385		
		0.022	
26.0	0.396		
		0.015	
28.0	0.426		

[a]From *Fundamentals of Analytical Chemistry,* Second Edition, by Douglas A. Skoog and Donald M. West. Copyright © 1963 by Holt, Rinehart and Winston, Inc. Reprinted by permission of Holt, Rinehart and Winston, Inc.

End-point Determination

Graphical methods. The end point for a potentiometric titration may be established in any of several ways. The most straightforward method involves a visual estimation of the midpoint in the steeply rising portion of the

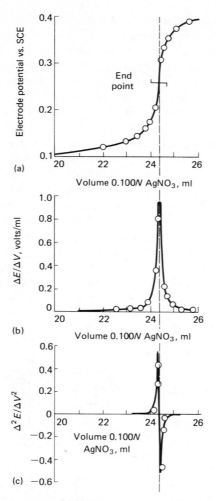

Figure 18-15. (a) Potentiometric titration curve for 2.433 meq of Cl⁻ with 0.100 N AgNO₃. (b) First derivative curve. (c) Second derivative curve.

titration curve. Various mechanical methods have been suggested for evaluation of this point; it seems doubtful, however, that these greatly improve the accuracy of the estimation.

Another graphical approach involves a plot of the change in potential per unit change in volume of reagent ($\Delta E/\Delta V$) as a function of the average volume of reagent added. The data from columns 1 and 2 of Table 18-2 have been used to calculate values for $\Delta E/\Delta V$ shown in column 3; these values are plotted in Figure 18-15(b). The end point is taken as the maximum in the curve and is obtained by extrapolation of the experimental points. Because of the uncertainty in the extrapolation procedure, it is debatable whether much is gained in terms of accuracy by such an approach despite the spectacular change in the parameter plotted; such a plot is certainly more troublesome to obtain.

With both techniques the assumption is made that the titration curve is symmetric about the true equivalence point, and that the inflection in the curve therefore corresponds to that point. This assumption is perfectly valid, provided the participants in the chemical process react with one another in an equimolar ratio, and also provided the electrode process is perfectly reversible. Where these provisions are not met, an asymmetric curve results, as shown in Figure 18-16. Note that the curve for the oxidation of iron(II) by cerium(IV) is symmetrical about the equivalence point. On the other hand, five moles of iron(II) are consumed by each mole of permanganate and a highly nonsymmetric titration curve results. Ordinarily, the change in potential within the equivalence-point region of these curves is large enough so that a negligible titration error is introduced if the midpoint of the steeply rising portion of the curve is chosen as the end point. Only when unusual accuracy is desired or where very dilute solutions are employed must account be taken of this source of uncertainty. In these instances a correction can be determined empirically by titration of a standard. Alternatively, when the error is due to a nonsymmetrical reaction, the correct position of the equivalence can be calculated from theoretical considerations.[13]

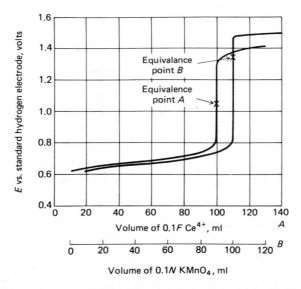

Figure 18-16. Curves depicting the titration of 100 ml of 0.100 N iron(II) solution with (A) cerium(IV) and with (B) permanganate. Note that the horizontal axes are displaced to permit comparison of the curves.

[13]See I. M. Kolthoff and N. H. Furman, *Potentiometric Titrations*, 2d ed. (New York: John Wiley & Sons, Inc., 1931), Chaps. 2 and 3.

Analytical determination of the end point. A less time-consuming method of determining the end point is to calculate the values for $\Delta E/\Delta V$ as the titration proceeds and record these data as shown in column 3 of Table 18-2. The assumption is then made that this function is a maximum at the equivalence point. Thus, in the data shown the end point is clearly between 24.3 and 24.4 ml; a value of 24.35 would be adequate for many purposes. Lingane[14] has shown that the volume can be fixed more exactly by estimating the point where the second derivative of the voltage with respect to volume (that is, $\Delta^2E/\Delta V^2$) becomes zero. This is easily done if equal increments of solution are added in the vicinity of the equivalence point. Values of $\Delta^2E/\Delta V^2$, calculated by subtracting the corresponding data for $\Delta E/\Delta V$, are given in the fourth column of Table 18-2. The function must become zero at some point between the two volumes where a change in sign occurs. The volume corresponding to this point can be obtained by interpolation; that is,

$$\text{at 24.30 ml} \quad \frac{\Delta^2E}{\Delta V^2} = +0.44$$

$$\text{at 24.40 ml} \quad \frac{\Delta^2E}{\Delta V^2} = -0.59$$

Therefore,

$$\text{end-point volume} = 24.30 + 0.1 \times \frac{0.44}{0.44 + 0.59}$$

$$= 24.34 \text{ ml}$$

It is clear that this method is also based upon the same assumption as the graphical methods. A plot of $\Delta^2E/\Delta V^2$ is shown in Figure 18-15(c).

Titration to a fixed potential. Another procedure consists of titrating to a predetermined end-point potential. This may be the theoretical equivalence-point potential calculated from formal potentials, or an empirical potential obtained by the titration of standards. Such a method demands that the equivalence-point behavior of the system be entirely reproducible.

Precipitation Titrations

Electrode systems. For a precipitation titration, the indicator electrode is often the metal from which the reacting cation is derived. Membrane electrodes that are sensitive to one of the ions involved in the titration process may also be employed. Occasionally, an inert platinum electrode can serve as the indicator. Thus, for example, the titration of zinc ion with ferro-cyanide may be followed with a platinum electrode, provided a quantity of

[14]J. J. Lingane, *Electroanalytical Chemistry*, 2d ed. (New York: Interscience Publishers, Inc., 1958), p. 93.

ferricyanide ion is also present. The potential developed by this electrode is determined by the half-reaction

$$Fe(CN)_6^{3-} + e \rightleftharpoons Fe(CN)_6^{4-}$$

With the first excess of reagent, the ferrocyanide concentration increases rapidly and causes a corresponding change in the potential of the electrode.

Titration curves. As the following examples demonstrate, a theoretical curve for a precipitation titration is readily derived from standard potential data.

Example. An electrode system consisting of a silver indicator electrode and a saturated calomel electrode was employed for the potentiometric titration of 100 ml of 0.0200 *F* chloride ion with 0.100 *F* silver nitrate. Calculate the cell potential at (a) 1.00 ml before the equivalence point, (b) the equivalence point, and (c) 1.00 ml beyond the equivalence point. The required electrode potential data are

$$AgCl + e \rightleftharpoons Ag + Cl^- \qquad\qquad E^0 = +0.222 \text{ V}$$
$$Hg_2Cl_2 + 2e \rightleftharpoons 2Hg + 2Cl^- \text{ (sat'd KCl)} \qquad E = +0.242 \text{ V}$$

(a) The titration will require 20.0 ml of $AgNO_3$. After 19.0 ml of the reagent have been added, the formal concentration of NaCl will be

$$F_{NaCl} = \frac{100 \times 0.0200 - 19.0 \times 0.100}{119} = 8.4 \times 10^{-4}$$

An additional source of chloride results from the finite solubility of the precipitate. Thus, the equilibrium chloride ion concentration is given by

$$[Cl^-] = 8.4 \times 10^{-4} + [Ag^+]$$

It is reasonable to assume that

$$[Ag^+] \ll 8.4 \times 10^{-4}$$

Thus, the potential of the indicator electrode will be given by

$$E_{Ag} = 0.222 - 0.059 \log 8.4 \times 10^{-4} = 0.403 \text{ V}$$

and treating the calomel electrode as the anode,

$$E_{cell} = 0.403 - 0.242 = 0.161 \text{ V}$$

(b) At the equivalence point

$$[Ag^+] = [Cl^-]$$

The solubility product for silver chloride is

$$K_{sp} = [Ag^+] [Cl^-] = 1.82 \times 10^{-10}$$

Thus,

$$[Cl^-] = \sqrt{1.82 \times 10^{-10}} = 1.35 \times 10^{-5}$$

Therefore,

$$E_{Ag} = 0.222 - 0.059 \log 1.35 \times 10^{-5} = 0.509$$

and

$$E_{cell} = 0.509 - 0.242 = 0.267 \text{ V}$$

(c) When the equivalence point has been exceeded by 1.00 ml, the concentration of excess Ag^+ is given by

$$F_{AgNO_3} = \frac{1.00 \times 0.100}{121} = 8.26 \times 10^{-4}$$

and

$$[Ag^+] = 8.26 \times 10^{-4} + [Cl^-] \cong 8.26 \times 10^{-4}$$

Here it is easier to describe the behavior of the silver electrode in terms of

$$Ag^+ + e \rightleftharpoons Ag \qquad E^0 = 0.799 \text{ V}$$

and

$$E_{Ag} = 0.799 - 0.059 \log \frac{1}{[Ag^+]}$$

$$= 0.799 - 0.059 \log \frac{1}{8.26 \times 10^{-4}} = 0.618 \text{ V}$$

$$E_{cell} = 0.618 - 0.242 = 0.376 \text{ V}$$

Effect of adsorption. Experimental curves for precipitation titrations usually correspond closely to the shape predicted by theory. Such discrepancies as are observed can be traced to adsorption of ions by the precipitate. This effect can be seen in the titration of the iodide ion with silver nitrate. Before the equivalence point has been reached, silver iodide strongly adsorbs iodide ions and thereby lowers their concentration in the solution. The effect of this phenomenon on the potential of a silver indicator electrode is particularly noticeable near the end point. An analogous situation exists in the region beyond the equivalence point; here, adsorption appreciably lowers the silver ion concentration in the solution and causes the experimentally measured potentials to differ somewhat from their calculated values. As may be seen in Figure 18-17, the net effect of adsorption phenomena is to produce less sharply defined end points. Ordinarily, this difficulty is not serious.

Titration curves for mixtures. Separate end points are often revealed by a potentiometric titration when more than one reacting species is present. The titration of halide mixtures with silver nitrate is an important example. To illustrate the derivation of a theoretical curve for such a system, consider the titration of 25.0 ml of a solution that is 0.100 F in both iodide and chloride ions with a standard 0.100 F silver nitrate solution. From solubility product calculations we conclude that the first additions of reagent will result in precipitation of silver iodide in preference to silver chloride. It is of interest to determine the extent to which this reaction will occur before appreciable chloride precipitation takes place. With the first appearance of silver chloride, both solubility-product relations are satisfied; thus, dividing the one by the other, we get

$$\frac{[\cancel{Ag^+}] [I^-]}{[\cancel{Ag^+}] [Cl^-]} = \frac{8.3 \times 10^{-17}}{1.8 \times 10^{-10}} = 4.6 \times 10^{-7}$$

After the first silver chloride forms, this ratio is maintained throughout the remainder of titration. The magnitude of the ratio suggests that nearly all the iodide will have precipitated before the appearance of any silver chloride. Thus, the chloride ion concentration, because of dilution, will be very close to half of its original concentration at the outset of its precipitation. With this assumption we can determine the iodide concentration when the chloride precipitate just appears; that is,

$$\frac{[I^-]}{0.1/2} \cong 4.6 \times 10^{-7}$$

$$[I^-] \cong 2.3 \times 10^{-8} \text{ mole/liter}$$

We can then calculate the percentage of iodide unprecipitated at this point in the titration:

$$\text{original no. mfw } I^- = 25 \times 0.1 = 2.5$$
$$\text{no. mfw } I^- \text{ at onset of } Cl^- \text{ precipitate} \cong 50 \times 2.3 \times 10^{-8} = 1.2 \times 10^{-6}$$
$$\text{percent } I^- \text{ unprecipitated} = \frac{1.2 \times 10^{-6}}{2.5} \times 100 = 0.00005$$

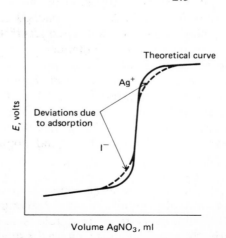

Figure 18-17. Potentiometric titration of iodide ion. Note the effect of adsorption upon the titration curve.

Thus, within 0.00005 percent of the first end point volume, the curve should be that of a simple iodide titration; the necessary data for plotting the theoretical curve may be calculated on this basis.

Example. Calculate the potential of a silver electrode (against a saturated calomel electrode) after the addition of 5 ml of 0.100 N AgNO$_3$ to the mixture under consideration.

$$[I^-] = \frac{25.00 \times 0.1 - 5.0 \times 0.1}{30.0} + [Ag^+] \cong \frac{1}{15}$$

We find in the table of standard potentials that

$$AgI + e \rightleftharpoons Ag + I^- \qquad E^0 = -0.151 \text{ V}$$

Thus,

$$E = -0.151 - 0.059 \log [I^-]$$

$$= -0.151 - 0.059 \log \frac{1}{15}$$

$$= -0.081 \text{ V}$$

Combining this with the potential for the saturated calomel electrode as an anode, we obtain

$$E_{cell} = -0.081 - 0.242$$
$$= -0.323 \text{ V}$$

The first half of the curve (solid line) shown in Figure 18-18 was obtained from data calculated in this manner. As expected, the curve rises steeply in the vicinity of the iodide equivalence point. The increase in potential, however, is terminated abruptly at the potential that corresponds to the first formation of silver chloride. This point is attained slightly before the iodide equivalence point.

Example. Calculate the cell potential at the onset of precipitation of AgCl.

We have established that at this point

$$[I^-] = 2.3 \times 10^{-8} \text{ mole/liter}$$

Thus,

$$E = -0.151 - 0.059 \log 2.3 \times 10^{-8}$$

$$= 0.301 \text{ V}$$

and

$$E_{cell} = 0.301 - 0.242 = 0.059 \text{ V}$$

After precipitation of chloride has commenced, the potential can be most conveniently calculated by means of the standard potential for the AgCl-Ag couple. The remainder of the curve is thus essentially identical to the titration for chloride by itself; points defining this portion are calculated as described previously.

Figure 18-18 illustrates that the theoretical curve for a mixture is a synthesis of the two curves for the individual ions and that marked potential changes are associated with each equivalence point. It is to be expected that the first equivalence point in the titration of a mixture will become less well-defined as the solubilities of the two precipitates approach one another. This effect is shown by the dotted line in Figure 18-18, which represents the theoretical curve for the titration of bromide ion in the presence of chloride. Even here an end point can be distinguished fairly readily. As a matter of fact, the solubilities of the three silver halides differ sufficiently for a mixture of the three to give a titration curve with three distinct end points.

One obvious conclusion from Figure 18-18 is that the potentiometric method should make possible the analysis of the individual components in halide mixtures. These curves, of course, are theoretical; the predicted sharp discontinuities are considerably rounded in actuality. More important,

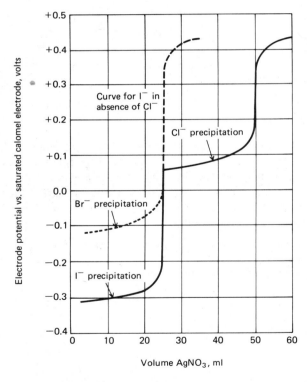

Figure 18-18. Potentiometric titration of halide mixtures. The solid curve represents titration of 25.0 ml of a solution that is 0.100 *F* with respect to both Cl^- and I^-. Titrant is 0.100 *F* Ag^+. The dashed extension past the first end point indicates the course of the I^- titration if Cl^- were not present. Shown also is the curve for titration of 25.0 ml of solution that is 0.100 *F* with respect to both Br^- and Cl^-.

the volume of silver nitrate required to reach the first end point is generally somewhat greater than the amount calculated by theoretical means; the total volume, however, approaches the correct amount. These observations apply to the titration of chloride-bromide, chloride-iodide, and bromide-iodide mixtures, and can be explained by assuming that coprecipitation of the more soluble silver halide occurs during formation of the less soluble compound. This would account for the overconsumption of reagent in the first part of the titration.

Despite the coprecipitation error, the potentiometric method is useful for the analysis of halide mixtures. When approximately equal quantities are present, relative errors can be kept to about 1 to 2 percent.[15]

Complex-formation Titrations

Both metal electrodes and membrane electrodes have been applied to the detection of end points in reactions that involve formation of a soluble complex. The mercury electrode, illustrated in Figure 18-19, is particularly useful for EDTA titrations. It functions as an indicator electrode for cations whose EDTA complexes are less stable than HgY^{2-}. For example, in the titration of calcium ion, a constant amount of HgY^{2-} can be introduced into the solution to be analyzed. Under these conditions, equation (18-1) on page 448 reduces to

$$E_{Hg} = (\text{constant})' - \frac{0.059}{2} \log \frac{[CaY^{2-}]}{[Ca^{2+}]}$$

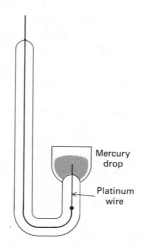

Mercury drop

Platinum wire

Figure 18-19. A typical mercury electrode.

[15]For further details see I. M. Kolthoff and N. H. Furman, *Potentiometric Titrations,* 2d ed. (New York: John Wiley & Sons, Inc., 1931), pp. 154–158.

Since the concentration ratio between the cation titrated and its complex undergoes a profound change in the end-point region, it follows that the mercury electrode will suffer a corresponding change in potential and fulfill its role as an indicator electrode for the titration.

Reilley and Schmid[16] have carried out a systematic study of this end point and have described conditions that are necessary for various EDTA titrations.

Neutralization Titrations

Potentiometric acid-base titrations are frequently employed where the sample solutions are colored or turbid. They are particularly useful for the analysis of mixtures of acids or polyprotic acids (or bases), since individual end points are frequently discernible. A numerical value for the dissociation constant of the reacting species can also be estimated from potentiometric titration curves. In theory, this can be obtained from any point along the curve; as a practical matter, it is most easily evaluated from the pH at the point of half-neutralization. For example, in the titration of the weak acid HA we may often assume that at the midpoint

$$[HA] \cong [A^-]$$

and therefore

$$K_a = \frac{[H^+][\cancel{A^-}]}{[\cancel{HA}]} = [H^+]$$

or

$$pK_a = pH$$

The assumption that $[HA] = [A^-]$ at the midpoint in the titration leads to significant errors if the ionization constant for the acid is greater than about 10^{-4} or if the formal concentrations of the salt and the acid are less than about 0.01. It is readily shown that under these circumstances

$$K_a = \frac{[H^+](F_{A^-} + [H^+])}{(F_{HA} - [H^+])}$$

where F_{A^-} and F_{HA} refer to the formal concentrations of the salt formed by the titration and the acid remaining.

A value of the dissociation constant and the equivalent weight of a pure sample of an unknown acid can be obtained from a single potentiometric titration; this information is frequently sufficient to identify the acid.[17]

Titrations in nonaqueous solvents. Acid-base titrations in aqueous solvents are limited to substances with acidic or basic dissociation constants that are greater than about 10^{-8}. For weaker acids or bases the

---

[16]C. N. Reilley and R. W. Schmid, *Anal. Chem.*, **30**, 947 (1958).

[17]For a more complete discussion of the theory and applications of neutralization titrations see D. A. Skoog and D. M. West, *Fundamentals of Analytical Chemistry*, 2d ed. (New York: Holt, Rinehart and Winston, Inc., 1969), Chaps. 11-13.

reactions with the reagent are so incomplete that the change in pH at the equivalence point is not sufficient to permit accurate establishment of the end point. Often, however, analysis becomes feasible if a nonaqueous solvent is employed.[18] For example, aqueous solutions of aniline ($K_b \sim 10^{-10}$) are so weakly basic that titration with standard perchloric acid is not satisfactory. On the other hand, if glacial acetic acid solutions of aniline and perchloric acid are employed, the reaction between the two is nearly complete at the equivalence point and a large change in pH occurs as a consequence. Similarly, the reaction of phenol ($K_a \sim 10^{-10}$) with a base in a nonaqueous solvent is sufficiently favorable to make titration feasible. In water the end point for this titration is not satisfactory.

The potentiometric method has proved particularly useful for signaling end points for titrations in nonaqueous solvents. The ordinary glass-calomel electrode system can be used; the electrodes must be stored in water between titrations to prevent dehydration of the glass and precipitation of potassium chloride in the salt bridge. Ordinarily, the millivolt scale rather than the pH scale of the potentiometer should be employed because the potentials in nonaqueous solvents may exceed the pH scale. Furthermore, when standardized against an aqueous buffer, the pH scale has no significance in a nonaqueous environment. Even though the titration curves are empirical, they provide a useful and satisfactory means of end-point detection.

Oxidation-reduction Titrations

A platinum electrode responds rapidly to many important oxidation-reduction couples and develops a potential that is dependent upon the concentration (strictly, activity) ratio of the reactants and the products of such half-reactions. For example, a platinum electrode can be employed for the titration of iron(II) with cerium(IV); the two half-reactions are

$$Ce^{4+} + e \rightleftharpoons Ce^{3+} \qquad E^f = 1.44 \text{ V } (1 \text{ } F \text{ } H_2SO_4)$$
$$Fe^{3+} + e \rightleftharpoons Fe^{2+} \qquad E^f = 0.68 \text{ V } (1 \text{ } F \text{ } H_2SO_4)$$

where the superscript f indicates formal, rather than standard, potentials. The electrode is sensitive to both half-reactions so that its potential can be calculated from the ratio of either $[Ce^{3+}]/[Ce^{4+}]$ or $[Fe^{2+}]/[Fe^{3+}]$. That is, interaction among the species occurs until the half-cell potentials are identical; the electrode potential can then be thought of as arising from either of the two couples. The following examples illustrate the derivation of points for theoretical titration curves.

Example. A platinum-saturated calomel electrode system was employed for a titration of 80.0 ml of 0.0250 F Fe^{2+} with 0.100 Ce^{4+}. Both

---

[18]For a more complete discussion of acid-base titrations in nonaqueous solvents see D. A. Skoog and D. M. West, *Fundamentals of Analytical Chemistry,* 2d ed. (New York: Holt, Rinehart and Winston, Inc., 1969), Chap. 14.

solutions were 1.0 F in H_2SO_4. Calculate the potential of the electrode system after (a) 19.9, (b) 20.0, and (c) 20.1 ml of reagent has been added.

(a) The formal concentration of unreacted Fe^{2+} is given by

$$F_{Fe^{2+}} = \frac{80.0 \times 0.0250 - 19.9 \times 0.100}{99.9} \cong \frac{0.0100}{99.9}$$

and the concentration of the Fe^{3+} formed is

$$F_{Fe^{3+}} = \frac{19.9 \times 0.100}{99.9} = \frac{1.99}{99.9}$$

The Ce^{3+} concentration is also 1.99/99.9, but the Ce^{4+} concentration can only be calculated from the equilibrium constant for the oxidation-reduction reaction. Therefore, it is more convenient to employ the formal potential for Fe(II)–Fe(III) than for Ce(III)–Ce(IV). Assuming that the reaction is nearly complete, we write

$$[Fe^{3+}] = 1.99/99.9$$
$$[Fe^{2+}] = 0.0100/99.9$$

Thus,

$$E_{Pt} = E^f - 0.059 \log \frac{[Fe^{2+}]}{[Fe^{3+}]}$$

$$E_{Pt} = 0.68 - 0.059 \log \frac{0.0100/99.9}{1.99/99.9} = 0.81 \text{ V}$$

In order to obtain a positive potential (that of a galvanic cell) we treat the platinum electrode as the cathode, and obtain

$$E_{cell} = 0.81 \text{ V} - E_{SCE} = 0.81 - 0.242$$
$$= 0.57 \text{ V}$$

Note that the same potential would have been obtained had we used the potential for Ce(IV)–Ce(III) and the appropriate concentrations for these species.

(b) The equivalence point occurs when 20.0 ml of Ce(IV) have been added. Here the equilibrium concentration of both Ce^{4+} and Fe^{2+} are small and can only be calculated from the equilibrium constant for the reaction. Advantage can be taken of the fact, however, that at the equivalence point

$$[Ce^{4+}] = [Fe^{2+}]$$
$$[Ce^{3+}] = [Fe^{3+}]$$

We may also write

$$E_{Pt} = 1.44 - 0.059 \log \frac{[Ce^{3+}]}{[Ce^{4+}]}$$

$$E_{Pt} = 0.68 - 0.059 \log \frac{[Fe^{2+}]}{[Fe^{3+}]}$$

Adding the two equations,

Thus,

$$2E_{Pt} = 2.12 - 0.059 \log \frac{[Ce^{3+}] [Fe^{2+}]}{[Ce^{4+}] [Fe^{3+}]}$$

$$E_{Pt} = \frac{2.12}{2} = 1.06 \text{ V}$$

and

$$E_{cell} = 1.06 - 0.242 = 0.82 \text{ V}$$

(c) After 20.1 ml of Ce^{4+} have been added we find

$$F_{Ce^{4+}} = \frac{20.1 \times 0.100 - 80.0 \times 0.0250}{100.1} = \frac{0.0100}{100.1}$$

$$F_{Ce^{3+}} = \frac{20.0 \times 0.100}{100.1} = \frac{2.00}{100.1}$$

Here, $F_{Fe^{3+}} \cong 2.00/100.1$, but the concentration of Fe^{2+} is not directly obtainable. Thus, we employ the formal potential for Ce(IV)-Ce(III) and write

$$E_{Pt} = 1.44 - 0.059 \log \frac{2.00/100.1}{0.01/100.1} = 1.30 \text{ V}$$

$$E_{cell} = 1.30 - 0.242 = 1.06 \text{ V}$$

Example. Calculate the potential for a platinum-SCE system at the equivalence point in the titration of 0.100 *N* U^{4+} with 0.100 *N* $KMnO_4$. Assume that $[H^+] = 1.00$ at this point.
The reaction is

$$5U^{4+} + 2MnO_4^- + 2H_2O \rightleftharpoons 5UO_2^{2+} + 2Mn^{2+} + 4H^+$$

The half-reactions are

$$UO_2^{2+} + 4H^+ + 2e \rightleftharpoons U^{4+} + 2H_2O \qquad E^0 = 0.334 \text{ V}$$
$$MnO_4^- + 8H^+ + 5e \rightleftharpoons Mn^{2+} + 4H_2O \qquad E^0 = 1.51 \text{ V}$$

and we may write

$$E_{Pt} = 0.334 - \frac{0.059}{2} \log \frac{[U^{4+}]}{[UO_2^{2+}] [H^+]^4}$$

$$E_{Pt} = 1.51 - \frac{0.059}{5} \log \frac{[Mn^{2+}]}{[MnO_4^-] [H^+]^8}$$

Here we have employed standard potentials because the formal potential for the uranium couple under the specified conditions is not available.
In order to combine the log terms we multiply the first equation through by 2 and the second by 5. Then, upon adding the two, we obtain

$$2E_{Pt} + 5E_{Pt} = 2 \times 0.334 + 5 \times 1.51 - 0.059 \log \frac{[U^{4+}] [Mn^{2+}]}{[UO_2^{2+}] [MnO_4^-] [H^+]^{12}}$$

At the equivalence point in this titration

$$[UO_2^{2+}] = \frac{5}{2} [Mn^{2+}]$$

$$[U^{4+}] = \frac{5}{2} [MnO_4^-]$$

Substituting these equalities into the log term gives

$$E_{Pt} = \frac{0.668 + 7.75}{7} - \frac{0.059}{7} \log \frac{1}{[H^+]^{12}}$$

and since $[H^+] = 1.00$,

$$E_{Pt} = 1.20 \text{ V}$$

or

$$E_{cell} = 1.20 - 0.24 = 0.96 \text{ V}$$

Differential Titrations

We have seen that a derivative curve generated from the data of a conventional potentiometric titration (Figure 18-15b) contains a distinct maximum in the vicinity of the equivalence point. It is also possible to acquire such derivative data directly by means of a differential titration. The technique, as first proposed by Cox[19], was extraordinarily cumbersome; fortunately, however, derivative data can be obtained in a more convenient manner with rather simple equipment.

In essence, a differential titration requires the use of two identical indicator electrodes, one of which is well shielded from the bulk of the solution. Figure 18-20 illustrates one such arrangement. Here, one of the electrodes is contained in a sidearm test tube. As an alternative for oxidation-reduction reactions, a piece of platinum wire sealed into a small medicine dropper can be employed in conjunction with an unshielded wire of the same metal. Contact with the bulk of the solution is made through a small ($\sim$ 1 mm) hole in the bottom of the tube. Because of this restricted access, the composition of the solution surrounding the shielded electrode is not materially affected by an addition of titrant to the major portion of the solution; the resulting difference in solution composition gives rise to the difference in potential ΔE between the electrodes. After each potential measurement the solution is homogenized as the rubber bulb is squeezed several times, whereupon ΔE again becomes zero. In common with the calculated derivative curve and subject to the same limitations, ΔE for a fixed volume of titrant attains a maximum value at the equivalence point. If the volume of solution in the tube that shields the electrode is kept small (say 1 to 5 ml), the error arising from failure of the final addition of reagent to react with this portion of the solution can be shown to be negligibly small.

[19]D. C. Cox, *J. Am. Chem. Soc.*, **47**, 2138 (1925).

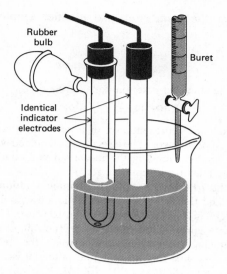

Figure 18-20. Apparatus for differential potentiometric titrations.

The main advantage of a differential method is the elimination of the reference electrode and the salt bridge. The end points are ordinarily sharply defined.

Automatic Titrations

In recent years several automatic titrators based on the potentiometric principle have come on the market. These are useful where a large number of routine analyses are to be carried out. Such instruments cannot yield more accurate results than those obtained by manual potentiometric techniques; however, they do decrease the time needed to perform the titrations and thus may offer some economic advantages.

Basically, two types of automatic titrators are available. The first yields a titration curve of potential versus reagent volume or, in some instances, $\Delta E/\Delta V$ or $\Delta^2 E/\Delta V^2$ against volume. The end point is then obtained from the curve by inspection. In the second type, the titration is stopped automatically when the potential of the electrode system reaches a predetermined value; the volume of reagent delivered is then read at the operator's convenience.

Automatic titrators normally employ a buret system with a solenoid operator valve to control the flow, or alternatively a syringe, the plunger of which is activated by a motor-driven micrometer screw. With both types the rate of addition of reagent must be either very slow throughout to prevent overrunning the end point, or some means must be provided to add reagent in smaller and smaller increments as the end point is approached. The latter method is to be preferred because of the shorter time required for titration. A number of instruments have been designed that anticipate the end point and add reagent in the same stepwise manner employed by a human operator.[20]

Problems

1. (a) Calculate the standard potential for the reaction

$$CuBr + e \rightleftharpoons Cu + Br^-$$

   (b) Give a schematic representation of a cell with a copper indicator electrode and a SCE that would be a second-order electrode for Br^-.

   (c) Derive an equation relating the measured potential of the cell in (b) to pBr (assume junction potential is zero).

   (d) Calculate the pBr of a bromide-containing solution that was saturated with CuBr and then employed in conjunction with a copper electrode in the type of cell in (b) if the resulting potential was 0.076 V.

2. Give a schematic representation for each of the following cells. Also derive an equation showing the relationship between the cell potential and the desired quantity. Assume that the junction potential is negligible and specify any necessary concentrations as $1.00 \times 10^{-4}\,M$.

   (a) a cell with a silver indicator electrode for the determination of pIO_3

   (b) a cell with a silver indicator electrode for the determination of pC_2O_4

   (c) a cell with a silver indicator electrode for the determination of pS_2O_3

   (d) a cell with a platinum indicator electrode for the determination of pI_2

   (e) a cell with a platinum electrode for the determination of $pFe(III)$

3. The following cell was employed for the determination of $pCrO_4$:

$$Ag \,|\, Ag_2CrO_4 \text{ (sat'd), } CrO_4^{2-} \text{ } (xM) \,\|\, SCE$$

   (a) Assuming a negligible junction potential, derive an equation relating the cell potential to $pCrO_4$.

   (b) Calculate $pCrO_4$ when the cell potential was found to be -0.285 V.

4. The formation constant for the soluble mercury(II) acetate is

$$Hg^{2+} + 2OAc^- \rightleftharpoons Hg(OAc)_2 \qquad K_{form} = 2.5 \times 10^8$$

   (a) Calculate the standard potential for the reaction

$$Hg(OAc)_2 + 2e \rightleftharpoons Hg + 2OAc^-$$

   (b) If a small quantity of $Hg(OAc)_2$ were introduced into a solution of sodium acetate, what would be the minimum acetate ion concentration needed to keep the Hg^{2+} concentration less than 1 percent of the concentration of the complex?

   (c) In theory, a mercury electrode could be made to function as a second-order electrode for the determination of the acetate ion by the introduction of a small known quantity of $Hg(OAc)_2$ into the solution to be analyzed, provided the acetate ion concentration was greater than the minimum calculated in part (b). The cell

$$Hg \,|\, Hg(OAc)_2 \text{ } (1.00 \times 10^{-4}\,M), OAc^- \text{ } (xM) \,\|\, SCE$$

[20]For a description of such instruments see J. J. Lingane, *Electroanalytical Chemistry*, 2d ed. (New York: Interscience Publishers, Inc., 1958), Chap. 8.

was found to have a potential of -0.375 V. What was the pOAc of the solution in the anode compartment (junction potential negligible)?

5. The following cell was found to have a potential of 0.209 V when the solution in the left compartment was a buffer of pH 4.00:

$$\text{glass electrode} \,|\, H^+ \,(a = x) \,\|\, \text{SCE}$$

The following potentials were obtained when the buffered solution was replaced with unknowns. Calculate the pH for each unknown.
   (a)   0.312 V
   (b)   0.088 V
   (c) -0.017 V

6. The following cell was found to have a potential of 0.275 V:

$$\text{membrane electrode for } Mg^{2+} \,|\, Mg^{2+} \,(a = 1.15 \times 10^{-2} \ M) \,\|\, \text{SCE}$$

   (a)   When the solution of known magnesium activity was replaced with an unknown solution, the potential was found to be 0.412 V. What was the pMg of this unknown solution?
   (b)   Assuming an uncertainty of $\pm$ 0.002 V in the junction potential, what is the range of Mg^{2+} activities within which the true value might be expected?

7. The following cell was found to have a potential of 1.034 V:

$$Cd \,|\, CdX_2(\text{sat'd}), X^- (0.0100 \ M) \,\|\, \text{SCE}$$

Calculate the solubility product of CdX_2 (neglect junction potential).

8. The following cell was found to have a potential of 0.672 V:

$$Pt, H_2(1.00 \text{ atm}) \,|\, HA(0.200 \ F), NaA(0.300 \ F) \,\|\, \text{SCE}$$

Calculate the dissociation constant of HA, neglecting the junction potential.

9. Zinc ion is known to form a single, relatively stable complex with the multidentate ligand Z^{3-}:

$$Zn^{2+} + Z^{3-} \rightleftharpoons ZnZ^-$$

In order to measure the formation constant of the complex, the following cell was constructed:

$$Zn \,|\, Zn(NO_3)_2(1.00 \times 10^{-4} \ F), Na_3Z(0.500 \ F) \,\|\, \text{SCE}$$

The potential of this cell was found to be 1.597 V. Calculate the formation constant, assuming the junction potential to be zero.

10. The response associated with the glass electrode is related to the pH of its environment by the expression

$$E = K + 0.059 \, p\text{H}$$

What will be the uncertainty in pH that results from an uncertainty of 0.001 V in the measurement of E?

11. Calculate the silver ion concentration after the addition of 5.00, 15.0, 25.0, 30.0, 35.0, 39.0, 40.0, 41.0, 45.0, and 50.0 ml of 0.100 F AgNO$_3$ to 50.0 ml of 0.080 F KSCN. Construct a titration curve from these data, plotting the potential of a silver cathode (vs. SCE) as a function of titrant volume.

12. A 40.0-ml aliquot of 0.050 F U^{4+} is diluted to 75.0 ml and titrated with 0.080 F Ce^{4+}. The pH of the solution is maintained at 1.00 throughout the titration.

    (a)   Calculate the potential of the indicator electrode with respect to a saturated calomel reference electrode after the addition of 5.00, 10.0, 15.0, 25.0, 40.0, 49.0, 50.0, 51.0, 55.0, and 60.0 ml of cerium(IV).

    (b)   Draw a titration curve for these data.

13.   Calculate the equivalence-point cell potential for each of the following potentiometric titrations. Treat the indicator electrode system as the cathode; the reference electrode is a saturated calomel electrode. In each instance assume that the solutions of the reagent and the substance titrated are 0.100 N at the outset.

    (a)   the titration of Cl^- with a standard solution of $Hg_2(NO_3)_2$ (for Hg_2Cl_2, $K_{sp} = 1.3 \times 10^{-18}$), employing a mercury indicator electrode

    (b)   the titration of U^{4+} with I_3^-, employing a platinum indicator electrode (assume $[I^-] = 0.500\ M$ and $[H^+] = 0.100\ M$ at the equivalence point)

    (c)   the titration of I^- with $AgNO_3$, using a silver electrode

    (d)   the titration of Mn^{2+} with MnO_4^- to give MnO_2, employing a platinum electrode ($[H^+] = 1.00 \times 10^{-8}\ M$ at equivalence point) For the reaction

$$MnO_4^- + 4H^+ + 3e \rightleftharpoons MnO_2 + 2H_2O \qquad E^0 = 1.695\ V$$

14.   Quinhydrone is an equimolar mixture of quinone (Q) and hydroquinone (H_2Q). These two compounds react reversibly at a platinum electrode

$$Q + 2H^+ + 2e \rightleftharpoons H_2Q \qquad E^0 = 0.699\ V$$

The pH of a solution can be determined by saturating the solution with quinhydrone and making it a part of the cell

$$Pt\,|\,quinhydrone(sat'd),H^+(xM)\,\|\,SCE$$

Such a cell was found to have a potential of -0.235 V. What was the pH of the solution? (Assume junction potential was zero.)

15.   The potential of the following cell was found to be 0.873 V:

$$Cd\,|\,Cd(CN)_4^{2-}(8.0 \times 10^{-2}\ M),CN^-(0.100\ M)\,\|\,SHE$$

Calculate the formation constant for $Cd(CN)_4^{2-}$.

16.   The potential of the following cell was found to be 0.981 V:

$$Zn\,|\,Zn^{2+}(5.0 \times 10^{-3}\ F),NH_3(0.120\ F)\,\|\,SHE$$

Calculate the formation constant for the reaction

$$Zn^{2+} + 4NH_3 \rightleftharpoons Zn(NH_3)_4^{2+}$$

17.   The potential of the following cell was found to be 0.245 V:

$$SHE\,\|\,IO_3^-(0.0100\ F),Cu(IO_3)_2(sat'd)\,|\,Cu$$

Calculate the K_{sp} for $Cu(IO_3)_2$.

18.   The potential of the following cell was found to be 0.584 V:

$$SHE\,\|\,C_2O_4^{2-}(1.00 \times 10^{-3}\ M),Ag_2C_2O_4(sat'd)\,|\,Ag$$

Calculate the K_{sp} for $Ag_2C_2O_4$.

19.   In order to determine the formation constant for the EDTA complex of Cu(II), the following cell was constructed:

$$Cu\,|\,CuY^{2-}(1.00 \times 10^{-4}\ M),Y^{4-}(1.00 \times 10^{-2}\ M)\,\|\,SHE$$

The measured potential was 0.277 V. Calculate the formation constant.

20. The following cell was employed for the determination of the dissociation constant of the weak acid HX:

$$Pt, H_2(1.00 \text{ atm}) \,|\, NaX(0.300 \; F), HX(0.200 \; F) \,\|\, SHE$$

If the potential was 0.262 V, what was K_a?

21. The following cell was used to determine the basic ionization constant for the organic amine RNH_2:

$$Pt, H_2(1.00 \text{ atm}) \,|\, RNH_2(0.100 \; F), RNH_3Cl(0.050 \; F) \,\|\, SHE$$

where RNH_3Cl is the chloride salt. The potential was 0.490 V; calculate K_b.

19 Conductometric Methods

Conduction of an electric current through an electrolyte solution involves migration of positively charged species toward the cathode and negatively charged ones toward the anode. The *conductance,* which is a measure of the current that results from the application of a given electrical force, is directly dependent upon the number of charged particles in the solution. All ions contribute to the conduction process, but the fraction of current carried by any given species is determined by its relative concentration and its inherent mobility in the medium.

The application of direct conductance measurements to analysis is limited because of the nonselective nature of this property. The principal uses of direct measurements have been confined to the analysis of binary water-electrolyte mixtures and to the determination of total electrolyte concentration. The latter measurement is particularly useful as a criterion of purity for distilled water.

On the other hand, *conductometric titrations,* in which conductance measurements are used for end-point detection, can be applied to the determination of numerous substances.[1]

The principal advantage to the conductometric end point is its applica-

[1]For further discussion of conductometric methods, see the chapter by T. Shedlovsky in A. Weissberger, Ed., *Physical Methods of Organic Chemistry,* 3d ed., Vol. 1 (New York: Interscience Publishers, Inc., 1960), Chap. 4, pp. 3011-3048; also see J. W. Loveland in I. M. Kolthoff and P. J. Elving, Eds., *Treatise on Analytical Chemistry*, Part I, Vol. 4 (New York: Interscience Publishers, Inc., 1963), Chap. 51.

bility to the titration of very dilute solutions and to systems in which reaction is relatively incomplete. Thus, for example, the conductometric titration of an aqueous phenol ($K_a \approx 10^{-10}$) solution is feasible even though the change in $p\text{H}$ at the equivalence point is insufficient for either a potentiometric or an indicator end point.

The technique has its limitations. In particular, it becomes less accurate and less satisfactory with increasing total electrolyte concentration. Indeed, the change in conductance due to the addition of titrant can become largely masked by high salt concentrations in the solution being titrated; under these circumstances the method cannot be used.

ELECTROLYTIC CONDUCTANCE

Under the influence of an applied potential, the ions in a solution are essentially instantly accelerated toward the electrode of opposite charge. The rate at which they migrate, however, is limited by the frictional forces generated by their motion. As in a metallic conductor, the velocity of movement of the particles is linearly related to the applied field; Ohm's law is thus obeyed by electrolyte solutions.

Some Important Relationships

Conductance L. The conductance of a solution is the reciprocal of the electrical resistance and has the units of ohm^{-1}. That is,

$$L = \frac{1}{R} \tag{19-1}$$

where R is the resistance in ohms.

Specific conductance k. Conductance is directly proportional to the cross sectional area A and inversely proportional to the length l of a uniform conductor; thus,

$$L = k\frac{A}{l} \tag{19-2}$$

where k is a proportionality constant called the *specific conductance*. Clearly, it is the conductance when A and l are numerically equal. If these parameters are based upon the centimeter, k is the conductance of a cube of liquid one centimeter on a side. The dimensions of specific conductance are then ohm^{-1} cm^{-1}.

Equivalent conductance, Λ. The equivalent conductance is defined as the conductance of one gram equivalent of solute contained between electrodes spaced one centimeter apart.[2] Neither the volume of the solution nor the area of the electrodes is specified; these vary to satisfy the conditions of the definition. For example, a 1.0 N solution (1.0 gram equivalent per liter) would require electrodes with individual surface areas of 1000 cm²; a 0.1 N solution would need 10,000-cm² electrodes. The direct measurement of equivalent conductance is thus seldom, if ever, undertaken because of the experimental inconvenience associated with such relatively large electrodes. Instead, this quantity is determined indirectly from specific conductance data. By definition, Λ will be equal to L when one gram equivalent of solute is contained between electrodes spaced one centimeter apart. The volume V of the solution (cm³) that will contain one gram equivalent of solute is given by

$$V = \frac{1000}{C}$$

where C is the concentration in equivalents per liter. This volume can also be expressed in terms of the dimensions of the cell

$$V = l\,A$$

With l fixed by definition at one centimeter,

$$V = A = \frac{1000}{C}$$

Substitution into equation (19-2) thus gives

$$\Lambda = \frac{1000k}{C} \tag{19-3}$$

Equation (19-3) permits calculation of the equivalent conductance from the experimental value of k for a solution of known concentration C.

Equivalent conductance at infinite dilution. The mobility of an ion in solution is governed by four forces. An *electrical force*, equal to the product of the potential of the electrode and the charge of the ion, tends to move the particle toward one of the electrodes. This effect is partially balanced by a *frictional force*, which is a characteristic property for each ion. In dilute solutions only these two effects play a significant role in determining the conductivity; thus, under these circumstances the equivalent conductance of a salt is independent of its concentration.

At finite concentrations, two other factors, the *electrophoretic* effect and the *relaxation* effect, become important and cause the equivalent conductance of a substance to decrease with concentration increases. The behavior of a sodium chloride solution is typical and is shown in the following table.

[2]In the context of electrolytic conductance the equivalent is defined in terms of the number of charges carried by an ion and not upon its fate in a specific reaction.

Concentration of NaCl, eq/liter	Λ
0.1	106.7
0.01	118.5
0.001	123.7
Infinite dilution	126.4(Λ_0)

The electrophoretic effect stems from the motion of the oppositely charged ions surrounding the ion of interest. These ions carry with them molecules of solvent; the motion of the primary particle is thus retarded by the flow of solvent in the opposite direction. The relaxation effect also owes its genesis to movement of the ionic atmosphere surrounding a given particle. Here, however, the ion is slowed by the charge of opposite sign that builds up behind the moving particle.

For strong electrolytes a linear relationship exists between equivalent conductance and square root of the concentration. Extrapolation of this straight-line relationship to zero concentrations yields a value for the equivalent conductance at infinite dilution Λ_0. A similar plot for a weak electrolyte is nonlinear, and direct evaluation of Λ_0 is difficult.

At infinite dilution, interionic attractions become nil; the over-all conductance of the solution then consists of the sum of the individual equivalent ionic conductances

$$\Lambda_0 = \lambda_+^0 + \lambda_-^0$$

where λ_+^0 and λ_-^0 are the equivalent ionic conductances of the anion and the cation of the salt at infinite dilution. Individual ionic conductances can be determined from other electrolytic measurements; values for a number of common ions are shown in Table 19-1. Note that symbols such as

Table 19-1 Equivalent Ionic Conductances at 25°C

Cation	λ_+^0	Anion	λ_-^0
H_3O^+	349.8	OH^-	199.
Li^+	38.7	Cl^-	76.3
Na^+	50.1	Br^-	78.1
K^+	73.5	I^-	76.8
NH_4^+	73.4	NO_3^-	71.4
Ag^+	61.9	ClO_4^-	67.3
$\frac{1}{2}Mg^{2+}$	53.1	$C_2H_3O_2^-$	40.9
$\frac{1}{2}Ca^{2+}$	59.5	$\frac{1}{2}SO_4^{2-}$	80.0
$\frac{1}{2}Ba^{2+}$	63.6	$\frac{1}{2}CO_3^{2-}$	69.3
$\frac{1}{2}Pb^{2+}$	69.5	$\frac{1}{2}C_2O_4^{2-}$	74.2
$\frac{1}{3}Fe^{3+}$	68.	$\frac{1}{4}Fe(CN)_6^{4-}$	110.5
$\frac{1}{3}La^{3+}$	69.6	—	—

$\frac{1}{2}Mg^{2+}$, $\frac{1}{3}Fe^{3+}$, $\frac{1}{2}SO_4^{2-}$, are used to emphasize that the concentration units are in *equivalents* per liter.

The differences that exist in the equivalent ionic conductance of various species (Table 19-1) arise primarily from differences in their size and the degree of their hydration.

The equivalent ionic conductance is a measure of the mobility of an ion under the influence of an electric force field and is thus a gauge of its current-carrying capacity. For example, the ionic conductance of a potassium ion is nearly the same as that for a chloride ion; therefore, a current passed through a potassium chloride solution is carried nearly equally by the two species. The situation is quite different with hydrochloric acid; because of the greater mobility of the hydronium ion, a greater fraction of the current $[350/(350 + 76)]$ is carried by that species in an electrolysis.

Ionic conductance data permit comparison of the relative conductivity of various solutes. Thus, we are justified in saying that 0.01 F hydrochloric acid will have a greater conductivity than 0.01 F sodium chloride because of the very large ionic conductance of the hydronium ion. Such conclusions are important in predicting the course of a conductometric titration.

Alternating Current Conduction

Conduction of a direct current through a cell requires an oxidation reaction at the anode and a reduction at the cathode. The term *faradaic* is sometimes used to denote such currents and processes. When alternating current is passed through a cell, conduction may occur by nonfaradaic as well as by faradaic processes. Because the changes associated with faradaic conduction can materially alter the electrical characteristics of a cell, conductometric measurements are advantageously based upon nonfaradaic processes.

Flow of current across the electrode-solution interface. Nonfaradaic currents involve the formation of an *electrical double layer* at the electrode-solution interface. When a potential is applied to an electrode immersed in an electrolyte, a momentary surge of current creates an excess (or a deficiency) of negative charge at the surface of the metal. As a consequence of ionic mobility, however, the layer of solution immediately adjacent to the electrode acquires an opposing charge. This effect is shown in Figure 19-1(a). This charged layer consists of two parts: (1) a compact inner layer, in which the potential decreases linearly with distance from the electrode surface and (2) a more diffuse layer, in which the decrease is exponential; see Figure 19-1(b). This assemblage of charge inhomogeneities is termed the electrical double layer.

With direct current, formation of the double layer involves the passage of a momentary current which then drops to zero (that is, the electrode

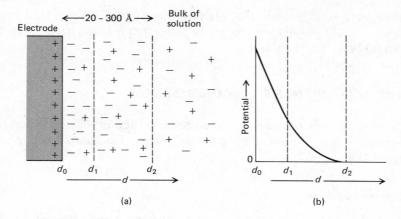

Figure 19-1. Electric double layer formed at electrode surface as a result of an applied potential.

becomes polarized) unless some faradaic process occurs. With an alternating current, however, reversal of the charge relationship occurs at each half-cycle as first negative and then positive ions are attracted to the electrode surface. Electrical energy is consumed and converted to frictional heat from this ion movement. Thus, each electrode surface behaves as a capacitor, the capacitance of which may be remarkably large (several hundred to several thousand microfarads per cm^2). The capacitance current increases with frequency and with electrode size; by controlling these variables it is possible to arrange conditions so that essentially all of the alternating current flowing through a cell is carried across the electrode interface by this nonfaradaic process.

Flow of current through an electrolyte solution. For frequencies up to a few thousand cycles per second, alternating current is carried almost exclusively by ion movement. The direction of this motion, of course, reverses with each half-cycle. At very high frequencies a significant fraction of the current is carried by a second mechanism that results from electrical polarization of the dielectric medium. Here, the voltage gradient causes *induced polarization* and *orientation polarization* of the *molecules* in the medium. In the former, distortion of the electron cloud surrounding the nucleus of a molecule causes a temporary polarized condition; in the latter, molecules with a permanent dipole moment become aligned with the electrical field. Regardless of the mechanism, current is carried by the periodic alteration in these processes as a result of the alternating voltage.

The dielectric current is dependent upon the dielectric constant of the medium and is directly proportional to frequency; only at radio frequencies ($\sim 10^6$ Hz) does this current become important. For ordinary conductance measurements the frequencies used are so low that these conductance mechanisms are largely avoided. *Oscillometry*, which is described briefly at the end of this chapter, makes use of radio frequencies; here the dielectric current becomes important.

THE MEASUREMENT OF CONDUCTANCE

A conductance measurement requires a source of electrical power, a cell to contain the solution, and a suitable bridge to measure the resistance of the solution.

Power Sources

Use of an alternating current source eliminates the effect of faradaic currents. There are, however, both upper and lower limits to the frequencies that can be employed; audio oscillators that produce signals of about 1000 Hz are the most satisfactory.

For less refined work an ordinary 60-cycle current, stepped down from 110 to perhaps 10 V, is also used. With such a source, however, faradaic processes often limit the accuracy of the conductance measurements. On the other hand, the convenience and ready availability of 60-cycle power justify its use for many purposes.

Power sources with frequencies much greater than 1000 Hz create problems in conductance measurements with a bridge. Here, the cell capacitance and stray capacitances in other parts of the circuit cause phase changes in the current that are difficult to compensate for.

Resistance Bridges

The Wheatstone bridge arrangement, shown in Figure 19-2, is typical of the apparatus used for conductance measurements. The power source *S* provides an alternating current in the frequency range of 60 to 1000 Hz at a potential of 6 to 10 V. The resistance R_{AC} and R_{BC} can be calculated from the position of *C*. The cell, of unknown resistance R_x, is placed in

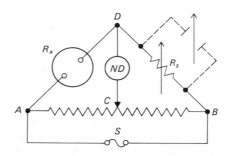

Figure 19-2. Diagram of a bridge circuit for resistance measurements.

the upper left arm of the bridge and a precision variable resistance R_s is placed in the right-hand side. A null detector *ND* is used to indicate the condition of no-current flow between *D* and *C*. The detector may consist of a pair of ordinary earphones, since the ear is responsive to frequencies in the 1000 Hz range; alternatively, it may be a "magic eye" tube that, with appropriate circuitry, indicates the current minimum.

An alternating-current device such as this suffers a loss in sensitivity when very high resistances are measured as a result of capacitance effects within R_x; in practice, a variable capacitor (shown in dotted lines) across R_s compensates for this effect.

The unknown resistance R_x is measured by adjusting the contact *C* until no current can be detected through *CD*. At this point E_{AC} must be equal to E_{AD}. We may then use equation (18-15) for a voltage divider to express E_{AC} as

$$E_{AC} = E_{AB} \frac{R_{AC}}{R_{AC} + R_{BC}}$$

The circuit *ADB* may also be treated in this way; thus

$$E_{AD} = E_{AB} \frac{R_x}{R_x + R_s}$$

Equating these two expressions and rearranging gives

$$R_x = \frac{R_{AC} R_s}{R_{BC}} \tag{19-4}$$

Figure 19-3 shows a circuit diagram for a commercial multirange

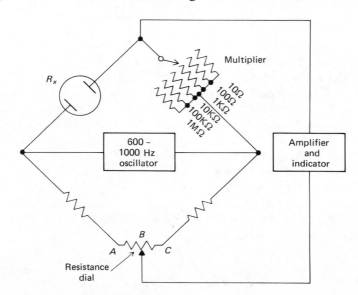

Figure 19-3. Conductance bridge.

conductance bridge. By choosing the proper multiplier position, the resistance scale AC can be made to cover several ranges. The detector consists of an electron-ray tube powered by a simple amplifier system.

Cells

Figure 19-4 depicts two common arrangements for the measurement of conductivity. The fundamental requirement is for a pair of electrodes that are firmly located in a constant geometry with respect to one another. These electrodes are ordinarily platinized to increase their effective surface and thus their capacitances; faradaic currents are minimized as a result.

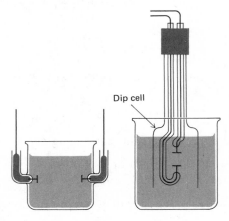

Dip cell

Figure 19-4. Typical cells for conductometric titrations.

Determination of the cell constant. The specific conductance k is the quantity determined in direct conductometric measurements. According to equation (19-2), this quantity is related to the measured conductance L by the ratio of the distance separating the electrodes to their surface area. This ratio has a fixed and constant value in any given cell, and is known as the *cell constant*. Its value is seldom determined directly; instead, it is evaluated by measuring the conductance L of a solution whose specific conductance is reliably known. Solutions of potassium chloride are commonly chosen for this purpose.[3] Typical data are shown on page 507.

Having once determined the value of the cell constant, conductivity data obtained with the cell can be easily converted to terms of specific conductance with the aid of equation (19-2).

Temperature control. The temperature coefficient for conductance measurements is about 2 percent per degree Celsius; as a consequence,

[3]G. Jones and B. C. Bradshaw, *J. Am. Chem. Soc.*, **55**, 1780 (1933).

Grams KCl per 1000 g of Solution in Vacuum	Specific Conductance at 25°C, ohm^{-1} cm^{-1}
71.1352	0.111342
7.41913	0.0128560
0.745263	0.00140877

some temperature control is ordinarily required during a conductometric measurement. For many purposes it is sufficient to immerse the cell in a reasonably large bath of water or oil maintained at about room temperature. Although a constant temperature is necessary, control at some specific value is not required for a successful conductometric titration.

CONDUCTOMETRIC TITRATIONS

Conductometric measurements provide a convenient means for the location of end points in titrations. To establish a conductometric end point, sufficient measurements are needed to define the titration curve. After correcting for volume change, the conductance data are plotted as a function of titrant volume. The two linear portions are then extrapolated, the point of intersection being taken as the equivalence point.

Because reactions fail to proceed to absolute completion, conductometric titration curves invariably show departures from strict linearity in the region of the equivalence point. Curved regions become more pronounced as the reaction in question becomes less favorable and as the solution becomes more dilute. The linear portions of the curve are best defined by measurements sufficiently removed from the equivalence point so that the common ion effect forces the reaction more nearly to completion. It is in this respect that the conductometric technique appears to best advantage. In contrast to potentiometric or indicator methods that depend upon observations under conditions where reaction is least complete, a conductometric analysis can be employed successfully for titrations based upon relatively unfavorable equilibria.

The conductometric end point is completely nonspecific. Although the method is potentially adaptable to all types of volumetric reactions, the number of useful applications to oxidation-reduction systems is limited; the substantial excess of hydronium ion typically needed for such reactions tends to mask conductivity changes associated with the volumetric reaction.

Correction for Volume Change

Throughout a titration the volume of the solution is always increasing; unless the conductance is corrected for this effect, nonlinear titration curves result. The correction can be accomplished by multiplying the observed conductance by the factor $(V + v)/V$, where V is the initial volume of solution and v is the total volume of reagent added. The correction presupposes that the conductivity is a linear function of dilution; this is true only to a first approximation. In the interests of keeping v small, the reagent for a conductometric titration is ordinarily several times more concentrated than the solution being titrated. A microburet may then be used for delivery of titrant.

Acid-base Titrations

Neutralization titrations are particularly well adapted to the conductometric end point because of the very high conductance of the hydronium and hydroxide ions compared with the conductance of the reaction products.

Titration of strong acids or bases.   The solid line in Figure 19-5 represents a curve (corrected for volume change) obtained when hydrochloric acid is titrated with sodium hydroxide. Also plotted are the calculated contributions of the individual ions to the conductance of the solution. During neutralization, hydronium ions are replaced by an equivalent number of less mobile sodium ions, and the conductance changes to lower values as a result of this substitution. At the equivalence point the concentrations of hydronium and hydroxide ions are at a minimum and the solution exhibits its lowest conductance. A reversal of slope occurs past the end point as the sodium ion and the hydroxide ion concentrations increase. With the excep-

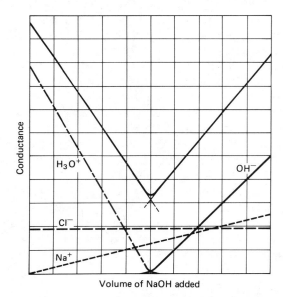

Volume of NaOH added

Figure 19-5.   Conductometric titration of a strong acid with a strong base. The solid line represents the titration curve, corrected for volume change. Broken lines indicate the contribution of the individual species, also corrected for volume change, to the conductance of the solution.

tion of the immediate equivalence-point region, there is an excellent linearity between conductance and the volume of base added; as a result, only two or three observations on each side of the equivalence point are needed for an analysis.

The percentage change in conductivity during the course of the titration of a strong acid or base is the same regardless of the concentration of the solution. Thus, very dilute solutions can be analyzed with an accuracy comparable to more concentrated ones.

Titration of weak acids or bases. Figure 19-6(a) illustrates application of the conductometric end point to the titration of boric acid ($K_a = 6 \times 10^{-10}$) with strong base. This reaction is so incomplete that a potentiometric or an indicator end point would be quite unsatisfactory. In the early stages of the titration, a buffer is rapidly established that imparts to the solution a relatively

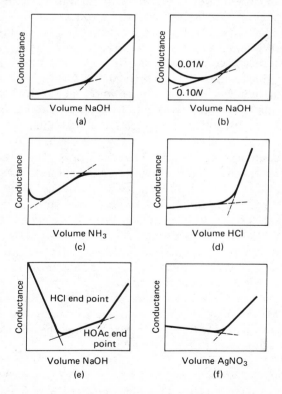

Figure 19-6. Typical conductometric titration curves. Titration of (a) a very weak acid ($K_a \approx 10^{-10}$) with sodium hydroxide, (b) a weak acid ($K_a \approx 10^{-5}$) with sodium hydroxide (Note that for 0.01 N solutions, conductance $\times$ 10 is plotted.), (c) a weak acid ($K_a \approx 10^{-5}$) with aqueous ammonia, (d) the salt of a weak acid, (e) a mixture of hydrochloric and acetic acids with sodium hydroxide, and (f) chloride ion with silver nitrate.

small and nearly constant hydronium ion concentration. The added hydroxide ions are consumed by this buffer and thus do not directly contribute to the conductivity. A gradual increase in conductance does result, however, owing to the increase in concentration of sodium and of borate ions. With attainment of the equivalence point, no further borate is produced; further additions of base cause a more rapid increase in conductance due to the increase in concentration of the mobile hydroxide ion.

Figure 19-6(b) illustrates the titration of a moderately weak acid, such as acetic acid ($K_a \cong 10^{-5}$), with sodium hydroxide. Nonlinearity in the early portions of the titration curve creates problems in establishing the end point; with concentrated solutions, however, the titration is feasible. As before, we can interpret this curve in light of the changes in composition that occur. Here, the solution initially has a moderate concentration of hydronium ions ($\approx 10^{-3} M$). Addition of base results in the establishment of a buffer system and a consequent diminution in the hydronium ion concentration. Concurrent with this decrease is the increase in the concentration of sodium ion as well as the conjugate base of the acid. These two factors act in opposition to one another. At first the decrease in hydronium ion concentration predominates and a decrease in conductance is observed. As the titration progresses, however, the pH becomes stabilized (in the buffer region); the increase in the salt content then becomes the more important factor, and a linear increase in conductance finally results. Beyond the equivalence point the curve steepens because of the greater ionic conductance of hydroxide ion.

In principle, all titration curves for weak acids or bases contain the general features of Figure 19-6(b). The ionization of very weak species is so slight, however, that little or no curvature occurs with the establishment of the buffer region [see Figure 19-6(a), for example]. As the strength of the acid (or base) becomes greater, so also does the extent of the curvature in the early portions of the titration curve. For weak acids or bases with dissociation constants greater than about 10^{-5}, the curvature becomes so pronounced that an end point cannot be distinguished.

Figure 19-6(c) illustrates the titration of the same weak acid as in Figure 19-6(b), but with aqueous ammonia instead of sodium hydroxide. Here, because the titrant is a weak electrolyte, the curve is essentially horizontal past the equivalence point. Use of ammonia as titrant actually provides a curve that is extrapolated with less uncertainty than the corresponding curve based upon titration with sodium hydroxide.

Titration of salts of weak acids or bases. Figure 19-6(d) represents the titration curve for a weak base, such as acetate ion, with a standard solution of hydrochloric acid. The addition of strong acid results in formation of sodium chloride and undissociated acetic acid. The net effect is a slight rise in conductance due to the greater mobility of the chloride ion over that of the acetate ion it replaces. After the end point has been passed, a sharp rise in conductance attends the addition of excess hydronium ions. The conductometric method is convenient for the titration of salts whose acidic or basic character is too weak to give satisfactory end points with indicators.

Figure 19-6(e) is typical of the titration of a mixture of two acids that differ in degree of dissociation. The conductometric titration of such mixtures is frequently more accurate than a potentiometric method.

Precipitation and complex-formation titrations. Figure 19-6(f) illustrates the titration of sodium chloride with silver nitrate. The initial additions of reagent in effect cause a substitution of chloride ions by the somewhat less mobile nitrate ions of the reagent; a slight decrease in conductance results. After the reaction is complete a rapid increase occurs, owing to the addition of excess silver nitrate.

The curve in Figure 19-6(f) is fairly typical for a precipitation titration. The slope of the initial portion of the curve may be either downward or upward, however, depending upon the relative conductance of the ion being determined and the ion of like charge in the reagent that replaces it. A downward-sloping line produces a V-shaped titration curve that provides a sharper definition of the end point. It is therefore preferable, where possible, to choose a reagent in which the ionic conductance of the nonreactive ion is less than that of the ion being titrated. According to the data in Table 19-1, then, we may predict that lithium chloride would be preferable to potassium chloride as a precipitating agent for silver ion.

Conductometric methods based upon precipitation or complex-formation reactions are not as useful as those involving neutralization processes. Conductance changes during these titrations are seldom as large as those observed with acid-base reactions because no other ion approaches the conductance of either the hydronium or the hydroxide ion. Such factors as slowness of reaction and coprecipitation represent further sources of difficulty with precipitation reactions.

APPLICATIONS OF DIRECT CONDUCTANCE MEASUREMENTS

Direct conductometric measurements suffer from a lack of selectivity, any charged species contributing to the total conductance of a solution. On the other hand, the high sensitivity of the procedure makes it an important analytical tool for certain applications. As we have noted, an important use of the method has been for the estimation of the purity of distilled or deionized water. The specific conductance of pure water is only about 5×10^{-8} ohm^{-1} cm^{-1}, and traces of an ionic impurity will increase the conductance by an order of magnitude or more.

Conductance measurements are also employed for the determination of the concentration of solutions containing a single strong electrolyte, such as solutions of the common alkalis or acids. A nearly linear increase in conductance with concentration is observed for solutions containing as much as 20 percent by weight of solute. Analyses are based upon calibration curves.

Conductance measurements are also useful for the completion of certain types of elemental analysis. For example, hydrocarbons can be analyzed for their sulfur content by combustion of the sample followed by absorption of the sulfur dioxide in hydrogen peroxide. The increase in conductance resulting from the sulfuric acid that is produced can be related to the sulfur concentration. Similarly, small amounts of nitrogen in biological materials can be determined by an ordinary Kjeldahl digestion with sulfuric acid followed by distillation of the ammonia into a boric acid solution. The conductance of the resulting ammonium borate solution can then be related to the percentage of nitrogen in the sample.

Conductance measurements are also widely employed to measure the salinity of sea water in oceanographic work.

Finally, conductance measurements yield much information about association and dissociation equilibria in aqueous solutions provided, of course, that one or more of the reacting species is ionic.

OSCILLOMETRY

As we have mentioned earlier, the conduction of an alternating current of radio frequency (10^5 to 10^7 Hz) is a complex process that involves not only ionic motion but also induced and orientation polarization of the molecules in the medium. As a result of the various conduction mechanisms, the impedance of a solution to the flow of current varies in an intricate fashion with the dielectric constant, the electrolyte concentration, and the frequency. Interpretation of the relationships among these variables is difficult, and the use of high-frequency currents could hardly be justified were it not for the fact that the experimental measurement does not require direct contact between the electrodes and the solution.

Instruments

As shown in Figure 19-7, a typical cell for high-frequency measurements consists of a glass container designed to fit snugly between a pair of cylindrical metal electrodes. The electrode and cell are then made part of the resonant circuit of a sine-wave oscillator; the behavior of this resonant circuit can be interpreted by assuming that the solution in the cell acts like a parallel capacitor and resistance. The capacititive reactance is dependent upon frequency and the dielectric constant of the solution; the resistive impedance is also frequency dependent and is related to the number and kinds of ions in the solution as well. At high frequencies the resistance impedance is generally lower than that measured with currents in the 1000 cycle range.

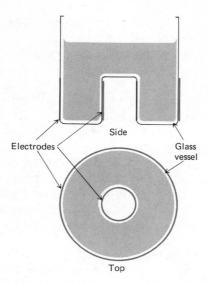

Side

Electrodes

Glass
vessel

Top

Figure 19-7. Sample cell and electrodes for high-frequency measurements.

Instruments are available commercially which are based either upon the change in frequency or the change in oscillator circuit current (at constant frequency) brought about by the presence of the sample and cell. In the case of the former, one or more variable capacitors are wired in parallel with the cell and the circuit is tuned to its resonance frequency in the absence of sample. The sample is then introduced and the capacity that must be subtracted by means of the calibrated capacitor to restore resonance is determined. In the case of instruments based on the resistive impedance of the sample, the change in oscillator current resulting from the introduction of sample is measured.

Because of the limited scope of oscillometry, we shall not dwell further on the somewhat complicated instrumentation and the complex interrelations associated with the technique; the reader is referred to the treatment of the subject by Reilley[4] and Pungor.[5]

Applications

Analysis of binary mixtures. Oscillometric measurements have been employed for the determination of binary mixtures of nonionic species. Such analyses are based upon the dielectric behavior of the mixtures and require that the dielectric constants for the two components differ significantly. For this application an instrument based on frequency changes is employed.

[4]C. N. Reilley, in P. Delahay, *New Instrumental Methods in Electrochemistry* (New York: Interscience Publishers, 1954), Chap. 15.

[5]E. Pungor, *Oscillometry and Conductometry* (New York: Pergamon Press, 1965).

Empirical calibration curves permit conversion of instrument readings to concentration ratios. Examples of this type of application include the analysis of mixtures of ethanol and nitrobenzene, benzene and chlorobenzene, alcohol and water, and ortho and para xylene.

Titrations. Oscillometric measurements have been employed to locate the end point in various types of titrations. The methods used are similar to those for the more conventional low-frequency conductometric titrations. The appearance of the titration curves is often quite different, however; sometimes V-shaped curves result, but on other occasions inverted V-type end points are found. Curvature is also frequently encountered and data must therefore be collected in the end-point region. For a given reaction the titration curve depends upon the concentration of the substance titrated as well as upon the parameter measured (that is, the change in frequency or the oscillator current).

The high-frequency method does not appear to yield titration data of higher accuracy than the classical method and suffers the drawbacks of requiring measurements near the end point and of greater empiricism. Thus, except where the presence of electrodes interferes with a titration, oscillometry offers little advantage.

Problems

For each of the following problems derive an approximate titration curve of specific conductance versus volume of reagent, employing the assumptions that

    (a)   100.0 ml of $1.00 \times 10^{-3} F$ solution is being titrated with $1.00 \times 10^{-2} F$ reagent;

    (b)   the volume change occurring as the titration proceeds can be neglected;

    (c)   the equivalent ionic conductances of the various ions are not significantly different from their equivalent ionic conductances at infinite dilution (Table 19-1).

Calculate theoretical specific conductances of the mixtures after the following additions of reagent: 0.00, 2.00, 4.00, 6.00, 8.00, 10.0, 12.0, 14.0, 16.0 ml.

1. NaOH with HCl
2. Phenol with NaOH (use $K_a = 1.0 \times 10^{-10}$ and $\lambda^\circ = 30$)
3. Sodium acetate with HCl (for acetic acid use $K_a = 1.0 \times 10^{-5}$)
4. Acetic acid with NaOH (for acetic acid use $K_a = 1.0 \times 10^{-5}$)
5. Acetic acid with NH$_3$ (for acetic acid use $K_a = 1.0 \times 10^{-5}$; for NH$_3$ use $K_b = 1.0 \times 10^{-5}$)
6. AgNO$_3$ with NaCl
7. AgNO$_3$ with LiCl

20 Electrogravimetric Methods

Electrolytic precipitation has been used for over a century for the gravimetric determination of metals. In most applications the metal is deposited on a weighed platinum cathode and the increase in weight is determined. Important exceptions to this procedure include the deposition of lead as its dioxide on a platinum anode and chloride as silver chloride on a silver electrode.

In most of the early applications electrodeposition was carried out by forcing a sufficiently large current through the cell to complete the reduction in a brief period. Often the applied potential had to be increased as the electrolysis proceeded to maintain the current flow. It has become apparent, however, that much of the potential selectivity of the electrodeposition procedure is lost when a constant current is employed, and that by holding the working electrode at a fixed, predetermined emf many useful separations can be performed. Both constant current and constant cathode potential methods are considered in the pages that follow, as well as a procedure in which no external source of power is employed.

CURRENT-VOLTAGE RELATIONSHIP DURING AN ELECTROLYSIS

It is useful to consider the relationship between current, voltage, and time for an electrolytic cell when it is operated in three different ways throughout an electrolysis: (1) the applied potential is held constant, (2) the cell current is kept constant, and (3) the potential of one of the electrodes (the working electrode) is held constant.

Operation of a Cell at a Fixed Applied Potential

To illustrate current-voltage relationships during an electrolysis at fixed potential, consider a cell consisting of two platinum electrodes, each with a surface area of 100 cm², immersed in a solution that is 0.100 M with respect to copper(II) ion and 1.00 M with respect to hydrogen ion. The cell has a resistance of 0.50 ohm. When current is forced through the cell, copper is deposited upon the cathode and oxygen is evolved at a partial pressure of 1.00 atm at the anode. The over-all cell reaction can be expressed as

$$Cu^{2+} + H_2O \rightarrow Cu + \frac{1}{2} O_2 + 2H^+$$

Decomposition potential. From standard potential data for the half-reactions

$$Cu^{2+} + 2e \rightleftharpoons Cu \qquad E^0 = 0.34 \text{ V}$$

$$\frac{1}{2} O_2 + 2H^+ + 2e \rightleftharpoons H_2O \qquad E^0 = 1.23 \text{ V}$$

a value of −0.92 V is obtained for the theoretical decomposition potential. No current flow would be expected at less negative potentials; at greater applied potentials, the current would be determined by the magnitude of the cell resistance.

The dotted lines in Figure 20-1 illustrate the theoretical current-voltage behavior of the cell; the decomposition potential is seen to be the intersection of two straight lines. In actuality, the current-voltage relationship for this system will more closely resemble the solid curve in Figure 20-1. Here, the oxygen overvoltage at the anode has the effect of displacing the curve to the right. Moreover, a small current begins to pass as soon as a potential is applied. There are two causes for this flow. First, most solutions contain small concentrations of easily reduced impurities that can diffuse to the cathode surface; dissolved oxygen and iron(III) are two common examples. In addition, part of this initial current flow is due to the reduction of copper (II) itself. This apparent anomaly results from the assumption of unit activity for copper in the calculation for the theoretical decomposition potential. This assumption becomes correct as soon as the electrode has acquired a uniform copper coating. At the outset of the electrolysis, however, the cathode is initially platinum and does not become a copper electrode until deposition has occurred. Experiments have shown that the activity of a metal in a deposit that only partially covers a platinum surface is less than one.[1] Initially, then, the numerator of the logarithmic term in the Nernst

[1]See L. B. Rogers, *et al., J. Electrochem. Soc.*, **95**, 25, 33, 129 (1949); **98**, 447, 452, 457 (1951).

equation for the reduction of copper(II) is infinitely small and approaches a value of one only after a continuous copper film has formed on the platinum surface. As a consequence, deposition does in fact take place at potentials more positive than the one based on the assumption of unit activity for copper. Quite generally, deposition of a metal on a platinum surface does not start suddenly when the decomposition potential is attained; instead, it increases gradually, becoming far more rapid in the region of, and beyond, this potential.

Current-changes with time.   In order for the cell reaction to occur at an appreciable rate it is necessary to impress a potential considerably in excess of the calculated theoretical value of -0.92 V. For example, operation at a current level of 1.0 amp requires application of an additional 0.5 V in order to overcome the 0.50 ohm resistance of the cell. Moreover, account must be taken of the overvoltage of oxygen on the platinum anode. Since the anode area is 100 cm², a current density of 0.010 amp per cm² will be involved under the proposed conditions of operation. From Table 17-2 we find that an overvoltage of about 0.85 V can be expected. Thus, a reasonable estimate of the potential required for the operation of this cell would be

$$E_{cell} = E_{cathode} - E_{anode} - IR - E_{overvoltage} \qquad (20\text{-}1)$$
$$= -0.92 - 0.5 - 0.85$$
$$= -2.3 \text{ V}$$

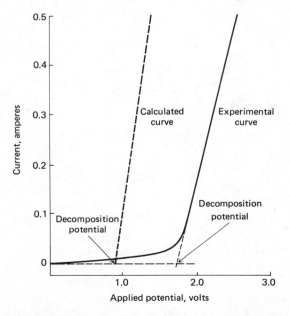

Figure 20-1. Current-voltage curve for the electrolysis of a copper(II) solution.

Let us now consider the changes that occur as the electrolysis proceeds at this fixed applied potential. As a consequence of the cell reaction there will be a decrease in the copper concentration and a corresponding increase in the hydrogen ion concentration. These changes will cause both E_{anode} and $E_{cathode}$ to become less positive, or more negative. Thus, when the copper concentration has been lowered to 10^{-6} M, the theoretical cathode potential will have decreased from $+0.31$ to $+0.16$ V; the anode potential will have changed only slightly to -1.22 V (the hydrogen ion concentration will have changed from 1.0 to 1.2 M as a result of the anode reaction). The sum of these potentials is -1.06 V as compared with the initial value of -0.92 V. Since the cell is to be operated at a fixed applied potential, it is apparent from equation (20-1) that changes in $E_{cathode}$ and E_{anode} must be offset by corresponding decreases in IR and $E_{overvoltage}$. Long before the copper(II) concentration has been diminished to 10^{-6} M, however, the cathode will have been affected by concentration polarization; the onset of this phenomenon will have an even more profound effect on IR and $E_{overvoltage}$.

Concentration polarization occurs when copper(II) ions can no longer be brought to the electrode surface at a sufficient rate to carry the theoretical current. As a result, the current must fall, and thus decrease the magnitude of IR in equation (20-1). The overvoltage term will also fall off, since it is dependent upon current density. Figure 20-2 illustrates these changes in the cell being considered. The current drops off very rapidly after a few minutes and eventually approaches zero as the electrolysis nears completion.

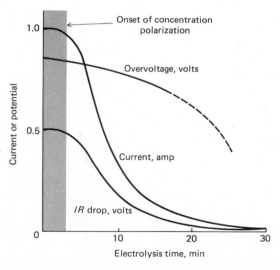

Figure 20-2. Hypothetical cell behavior during electrolysis. The applied potential is constant at -2.3 V. The cell has a resistance of 0.5 ohm. The area of each electrode is 100 cm². Initial concentrations: $[Cu^{2+}] = 0.100$ M, $[H^+] = 1.0$ M.

Cathode potential changes. Even more important than the changes in current, *IR* drop, and overvoltage are the changes that these phenomena induce in the cathode potential. To understand this, again consider equation (20-1) for the overall cell potential. Here, $E_{applied}$ is fixed at −2.3 V; at the onset of concentration polarization $E_{overvoltage}$ and *IR* drop decrease. Therefore, E_{anode}, $E_{cathode}$, or both, must become more negative. Now the potential of the anode is stabilized at the equilibrium potential for the oxidation of water because this reactant is present in plentiful supply at the electrode surface. Consequently, it is the cathode potential that must change to more negative values as the *IR* drop and the overvoltage decrease. The effect is shown graphically in Figure 20-3.

The sharp change in cathode potential accompanying concentration polarization may have several consequences. First, from a prior calculation (p. 518) it was established that a cathode potential of +0.16 V (vs. the standard hydrogen electrode) was required for quantitative deposition of copper (that is, to lower the copper concentration to 10^{-6} M); from Figure 20-3 it is obvious that a value considerably more negative than this is assured. Given sufficient time, then, essentially complete removal of copper (II) ion may be expected. Second, there is the possibility of additional electrode reactions taking place at the more negative cathode potentials resulting from electrode polarization. Certainly, if other ions that are reduced in the region between zero and −0.5 V are also present, codeposition must be expected. Cobalt(II) ion, with a reduction potential of −0.25 V, and cadmium ion, which has a standard potential of −0.4 V, are examples of these. Another possible electrode process at the more negative potential is

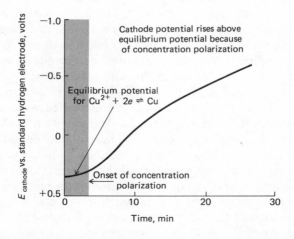

Figure 20-3. Variation in cathode potential during the electrodeposition of copper. Data are computed from Figure 20-2, assuming a fixed potential of −2.3 V and an anode potential of −1.2 V.

the formation of hydrogen. Under the specified conditions this process would commence at a cathode potential of about zero volt were it not for the high overvoltage of hydrogen on the copper-plated cathode. According to Table 17-2, reduction of hydrogen ions may be expected at about -0.5 V; therefore, formation of elemental hydrogen might well be observed near the end of this electrolysis. Gas formation during an electrodeposition is often an undesirable phenomenon.

From this discussion we can obtain a clear understanding of the limitations associated with an analytical electrolysis carried out at a fixed applied potential. It is mainly the shift in cathode potential toward more negative values that prevents the method from being very specific. The magnitude of this change can be minimized by decreasing the initial applied potential; this is done, however, at the expense of a diminution in the initial current and a concomitant increase in the time required for completion of the electrolysis. At best, then, electrolysis with a constant applied potential is effective in depositing metals with reduction potentials appreciably less negative than that at which hydrogen evolution occurs. The only feasible separations are from metals that are reduced with difficulty.

Constant-current Electrolysis

An analytical electrodeposition can be carried out by maintaining the current, rather than the applied potential, at a more or less constant level. This procedure requires periodic increases in the applied emf as the electrolysis proceeds.

We have shown in the preceding section that the onset of concentration polarization is accompanied by a decrease in current. To offset this effect, the applied potential can be made more negative; the resulting increase in electrostatic attraction causes the copper(II) ions to migrate to the cathode surface at a rate sufficient to maintain the desired current flow. A situation rapidly develops, however, in which the solution has been so depleted of copper(II) ion that reduction of this species by itself is inadequate to maintain a constant current. Further increases in the applied voltage then result in a rapid change in the cathode potential (see equation 20-1) to a point where codeposition of hydrogen occurs. The cathode potential then becomes stabilized at a level fixed by the standard potential and the overvoltage for the new electrode reaction; further large increases in the cell potential will no longer be necessary to maintain a constant current. Copper continues to deposit as copper(II) ions reach the electrode surface; however, the contribution of this electrode process to the total current becomes smaller and smaller as the deposition becomes more and more nearly complete. Hydrogen evolution soon predominates. The changes in cathode potential under conditions of constant current are shown in Figure 20-4.

In electroanalytical work, the vigorous evolution of hydrogen results in deposits with poor physical qualities. For this reason electrodeposition at

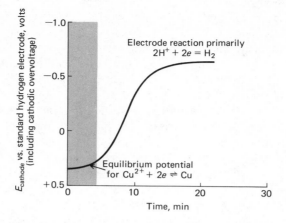

Figure 20-4. Changes in cathode potential during the deposition of copper with a constant current of 1.0 amp.

constant current is seldom, if ever, employed. Instead, the current is permitted to drop in preference to gas formation. Alternatively, the electrolysis is performed in the presence of a substance that is more easily reduced than hydrogen and whose concurrent reduction does not harm the analytical deposit. In a copper analysis, for example, the nitrate ion functions in this fashion, its reduction to an ammonium ion occurring more readily than hydrogen evolution:

$$NO_3^- + 10 \ H^+ + 8e \rightleftharpoons NH_4^+ + 3H_2O$$

Constant Cathode-potential Electrolysis

From the Nernst equation it is seen that a tenfold decrease in the concentration of an ion being deposited requires a negative shift in potential of only $0.059/n$ V. Thus, an ion can be quantitatively removed from solution with a relatively small change in cathode potential. As a consequence, electrogravimetric methods are potentially quite selective. In the present example, the copper concentration is decreased from $0.1 \ M$ to $10^{-6} \ M$ as the cathode potential changes from an initial value of $+0.31$ to $+0.16$ V. In theory, then, it should be feasible to separate copper from any element that is not deposited within this 0.15-V potential range; species that precipitate at potentials more positive than $+0.31$ V could be removed by preliminary deposition, while ions depositing at potentials smaller than $+0.16$ V should not be reduced until copper deposition is complete. To summarize, if we are willing to accept a hundred-thousandfold lowering of concentration as a quantitative separation, it follows that univalent ions differing in standard potentials by 0.3 V or greater can, theoretically, be separated quantitatively by electrodeposition provided their initial concentrations are about the same. Correspondingly, 0.15- and 0.1-V differences are required for divalent and trivalent ions.

An approach to these theoretical separation values, within a reasonable electrolysis period, requires a more sophisticated technique than we have thus far discussed because the onset of concentration polarization at the cathode will, if unchecked, prevent all but the crudest of separations. The change in cathode potential is governed by the decrease in *IR* drop. Thus, where relatively large currents are passed at the onset, the change in cathode potential can be expected to be large. On the other hand, if the cell is operated at low-current levels, the time required for completion of the deposition may become prohibitively long. An obvious answer to this dilemma is to initiate the electrolysis with an applied cell potential that is sufficiently high to ensure a reasonable current flow; as concentration polarization sets in, the applied potential is then continuously reduced to keep the cathode potential at the level necessary to accomplish the desired separation. Unfortunately, it is not feasible to set up such a program of applied potential changes on a theoretical basis because of uncertainties in variables affecting the deposition, such as overvoltage effects and conductivity changes. Nor, indeed, does it help to measure the potential across the working electrodes, since this measures only the over-all cell potential. The alternative is to measure the cathode potential against a third electrode whose potential in the solution is known and constant — that is, a reference electrode. The potential impressed across the working electrodes can then be adjusted to the level that will impart the desired potential to the cathode with respect to the reference. This technique is called *controlled cathode electrolysis.*

Experimental details for performing a controlled cathode-potential electrolysis are presented in a later section. For the present it is sufficient to note that the potential difference between the reference electrode and the cathode is measured with a potentiometer. The potential applied between the working electrodes is controlled with a voltage divider so that the cathode potential is maintained at a level suitable for the separation. Figure 20-5 shows a diagram of a typical apparatus. Calculation of the approximate cathode potential required for a separation with such an apparatus is illustrated in the following example.

Example. A solution is approximately 0.1 *F* in both zinc and cadmium ions. Calculate the cathode potential (with respect to a saturated calomel electrode) that should be used in order to separate these ions by electrodeposition.

From the table of standard electrode potentials, we find

$$Zn^{2+} + 2e \rightleftharpoons Zn \qquad\qquad E^0 = -0.76 \text{ V}$$
$$Cd^{2+} + 2e \rightleftharpoons Cd \qquad\qquad E^0 = -0.40 \text{ V}$$
$$Hg_2Cl_2 + 2e \rightleftharpoons 2Hg + 2Cl^- \text{ (sat'd KCl)} \qquad E = +0.24 \text{ V}$$

If we consider that quantitative removal has been accomplished when $[Cd^{2+}] = 10^{-6}$ *M*, then the cathode potential needed to achieve this condition will be

$$E = -0.40 - \frac{0.059}{2} \log \frac{1}{10^{-6}}$$
$$= -0.58 \text{ V}$$

Precipitation of Zn requires a cathode potential of

$$E = -0.76 - \frac{0.059}{2} \log \frac{1}{0.1}$$
$$= -0.79 \text{ V}$$

Thus, if the cathode is maintained between −0.58 and −0.79 V (with respect to the standard hydrogen electrode), a quantitative separation of cadmium should occur. Suppose we choose a potential of −0.70 V; against the saturated calomel electrode this is

$$E_{\text{vs. SCE}} = -0.70 - (+0.24) = -0.94 \text{ V}$$

To maintain this cathode potential a considerably larger emf is required across *AC* in Figure 20-5; its magnitude will depend upon the potential of the anode, the resistance of the solution, and overvoltage effects, if any.

An apparatus of the type shown in Figure 20-5 can be operated at relatively high initial applied potentials to give high currents. As the elec-

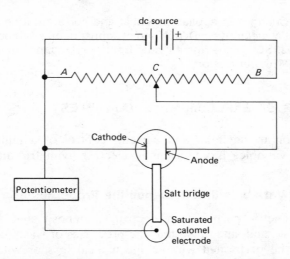

Figure 20-5. Apparatus for electrolysis at a controlled cathode potential. Contact *C* is continuously adjusted to maintain the cathode potential at the desired level.

trolysis progresses, however, a lowering of the applied potential across *AC* is required. This decrease, in turn, diminishes the current flow. Completion of the electrolysis will be indicated by the current approaching zero. The changes that occur in a typical constant cathode-potential electrolysis are depicted in Figure 20-6. In contrast to the electrolytic methods described earlier, this technique demands constant attention during operation. Unless some provision is made for automatic control, the operator-time required represents a major disadvantage to the controlled cathode-potential method.

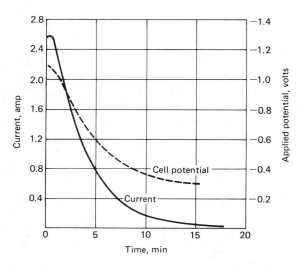

Figure 20-6. Changes in applied potential and current during a controlled-cathode potential electrolysis. Deposition of copper upon a cathode maintained at −0.36 volt vs. SCE. Experimental data from J. J. Lingane, *Anal. Chim. Acta,* **2**, 589 (1949). (With permission.)

EFFECTS OF EXPERIMENTAL VARIABLES

In addition to potential and current, control of a number of other experimental variables is important in electrogravimetric analysis.

Physical Variables that Influence the Properties of a Deposit

For gravimetric purposes an electrolytic deposit should be strongly adherent, dense, and smooth so that the processes of washing, drying, and weighing can be performed without mechanical loss or without reaction with the atmosphere. Good metallic deposits are fine-grained and have a metallic luster; spongy, powdery, or flaky precipitates are likely to be less pure and less adherent.

The principal factors that influence the physical characteristics of deposits include competing electrode processes, current density, temperature, and stirring.

Gas evolution.   If a gas is formed during an electrodeposition, a spongy and irregular deposit is usually obtained. In cathodic reactions the usual offender is hydrogen, and care must always be taken to prevent its formation by control of the cathode potential or by the addition of a so-called *depolarizer.* We have noted that the nitrate ion functions as a depolarizer in a copper analysis (p. 521).

Current density.   Electrolytic precipitates resemble chemical precipitates in that their crystal size decreases as their rate of formation increases — that is, as the current density increases. Here, however, small crystal size is a desirable characteristic; metallic deposits that are smooth, strong, and adherent contain very fine crystals.

While moderately high current densities generally give more satisfactory deposits, extremes should be avoided; very high current densities often lead to irregular precipitates with little physical strength, which develop as "treelike" structures from a few spots on the electrode. In addition, very high currents also lead to concentration polarization and concomitant gas formation. Ordinarily, a current density between 0.01 and 0.1 amp per cm^2 is suitable for electroanalytical work.

Stirring.   Stirring tends to reduce concentration polarization, and is ordinarily desirable in an electrolysis.

Temperature.   Although temperature may play an important part in determining the characteristics of a deposit, prediction of its effect is seldom possible. On the positive side, higher temperatures tend to inhibit concentration polarization by increasing the mobility of the ions and reducing the viscosity of the solvent. At the same time, elevated temperatures also tend to minimize overvoltage effects; increased gas formation may be observed under these circumstances. Thus, the best temperature for a given electrolysis can be determined only by experiment.

Chemical Variables

The success or failure of an electrolytic determination is often influenced by the chemical environment from which deposition occurs. The pH of the medium and the presence of complexing agents deserve particular mention.

Effect of pH.   Whether or not a given metal can be deposited completely frequently depends upon the pH of the solution. No problem is encountered with such easily reduced species as copper(II) ion or silver ion; these can be quantitatively removed from quite acidic media without inter-ference. On the other hand, less readily reduced elements cannot be de-posited from acid solution, owing to the simultaneous evolution of hydrogen; thus, for example, neutral or alkaline media are required for the electrolytic deposition of nickel or cadmium.

Proper pH control sometimes makes feasible the quantitative separation of cations. For example, copper is readily separated electrolytically from nickel, cadmium, or zinc in acidic solutions. Even if extreme concentration polarization occurs during the deposition of copper, the resulting change in cathode potential cannot become great enough to cause codeposition of the other metals. Hydrogen evolution occurs first; this process stabilizes the cathode potential at a value less negative than that required to initiate deposition of these metals.

Effect of complexing agents.   It is found empirically that many metals form smoother and more adherent films when deposited from solutions in which their ions exist primarily as complexes. The best metallic surfaces are frequently produced from solutions containing large amounts of cyanide or ammonia. The reasons for this effect are not obvious.

Deposition of a metal from a solution in which its ion exists as a complex requires a higher applied potential than in the absence of the complexing reagent. The magnitude of this potential shift is readily cal-culated, provided the formation constant for the complex ion is known. The data in Table 20-1 show that these effects can be large and must be taken into account when considering the feasibility of an electrolytic determination or separation. Thus, while copper is readily separated from zinc or cadmium in acidic solution, simultaneous deposition of all three occurs in the presence of appreciable quantities of cyanide ion. The greater potential shifts for silver and copper can be directly attributed to the higher stability of their cyanide complexes.

Occasionally, electrolytic separation of ions that would ordinarily codeposit can be achieved by selective complex formation. For example, the copper in a steel sample can be deposited electrolytically from a solution

Table 20-1   Effect of Cyanide Concentration on the Cathode Potential Required for the Deposition of Certain Metals from 0.1 F Solutions[a]

	Calculated Equilibrium Potential		
Ion	*No* CN^- *Present*	*In 0.1 M* CN^-	*In 1 M* CN^-
Zn^{2+}	-0.79	-1.16	-1.28
Cd^{2+}	-0.43	-0.81	-0.93
Cu^{2+}	$+0.31$	-0.99	-1.15
Ag^+	$+0.74$	-0.38	-0.50

[a]From *Fundamentals of Analytical Chemistry*, Second Edition, by Douglas A. Skoog and Donald M. West. Copyright © 1963 by Holt, Rinehart and Winston, Inc. Reprinted by permission of Holt, Rinehart and Winston, Inc.

containing phosphate or fluoride ions. Even though a large amount of iron is present, reduction of iron(III) ion does not occur because of the great stability of its complexes with these anions.

Anodic Deposits

Most electrogravimetric methods involve reduction of a metallic ion at the cathode. Occasionally, however, precipitates formed on an anode can also be used for analytical purposes. For example, lead is frequently oxidized to lead dioxide in acid solution:

$$Pb^{2+} + 2H_2O \rightleftharpoons PbO_2 + 4H^+ + 2e$$

The physical properties of the deposit make it a suitable weighing form for lead; a gravimetric determination is thus possible. Similarly, cobalt can be isolated and weighed as Co_2O_3.

INSTRUMENTATION

The apparatus for an analytical electrodeposition consists of a suitable cell and a direct-current power supply.

Cells

Figure 20-7 shows a typical cell employed for the deposition of a metal on a solid electrode. Tall-form beakers are ordinarily employed and mechanical stirring is provided to minimize concentration polarization; frequently the anode is rotated with an electric motor.

Electrodes. Electrodes are usually constructed of platinum, although copper, brass, and other metals find occasional use. Platinum electrodes have the advantage of being relatively nonreactive; moreover, they can be ignited to remove any grease, organic matter, or gases that could have a deleterious effect upon the physical properties of the deposit. Certain metals (notably bismuth, zinc, and gallium) cannot be deposited directly onto platinum without causing permanent damage to the electrode. A protective coating of copper should always be deposited on a platinum electrode before the electrolysis of these metals is undertaken.

Platinum should not be used as an anode in solutions containing high concentrations of chloride ion because chlorine may be evolved instead of oxygen; oxidation of the electrode itself may then occur. If a platinum anode must be used in these circumstances, it should be protected from attack by means of a depolarizer. Hydrazine works well in this capacity; it is preferentially oxidized to nitrogen:

$$H_2NNH_2 \rightarrow N_2 + 5H^+ + 4e$$

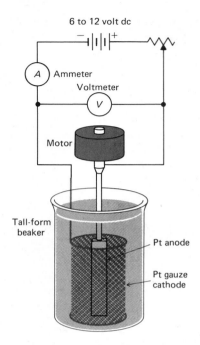

6 to 12 volt dc

Ammeter

Voltmeter

Motor

Tall-form beaker

Pt anode

Pt gauze cathode

Figure 20-7. Apparatus for electrodeposition of metals.

Cathodes are usually formed as gauze cylinders 2 to 3 cm in diameter and perhaps 6 cm in length. This construction minimizes polarization effects by providing a large surface area to which the solution can freely circulate. The anode may also be a gauze cylinder of somewhat smaller diameter so that it can be fitted inside the cathode; alternatively, it may take the form of a heavy wire spiral or a solid paddle.

The mercury cathode. For certain applications, electrolytic reductions are advantageously performed with a mercury cathode. This electrode is particularly useful in removing easily reduced elements as a preliminary step in an analysis. For example, copper, nickel, cobalt, silver, and cadmium are readily separated from ions such as aluminum, titanium, phosphates, and the alkali metals. The precipitated elements dissolve in the mercury; little hydrogen evolution occurs even at high applied potentials because of large overvoltage effects. Ordinarily, no attempt is made to determine the elements deposited in the mercury, the goal being simply their removal from solution. A cell such as that shown in Figure 20-8 is used.

Power Supplies

For deposition without cathode-potential control. The apparatus shown in Figure 20-8 is typical of that employed for most electrolytic analyses.

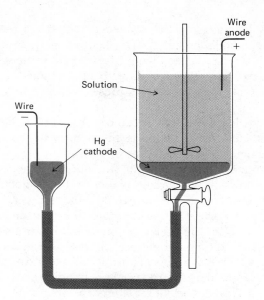

Figure 20-8. Mercury cathode for the electrolytic removal of metal ions from solution.

The direct-current power supply may consist of a storage battery, a generator, or an alternating-current rectifier. A rheostat is used to control the applied potential; an ammeter and a voltmeter are provided to indicate the approximate current and applied voltage. An entirely adequate electrolysis unit can be assembled from components found in most laboratories; more elaborate equipment is commercially available.

Apparatus for controlled electrode-potential electrolysis. An apparatus required for a controlled potential electrolysis need not be elaborate. A schematic diagram of the essential components is given in Figure 20-5. The cathode potential is measured against a saturated calomel or silver-silver chloride reference electrode with a simple potentiometer or vacuum-tube voltmeter. An ordinary moving-coil voltmeter is not satisfactory because it draws sufficient current to introduce an error in the voltage measurement.

The power supply can be a storage battery or a rectifier unit with a well-filtered, direct-current output. The voltage divider *AB* should be of high current rating and have a resistance of no more than 20 or 30 ohms. A tubular type of rheostat, found in most laboratories, is quite suitable. The working electrodes can be similar to those described in previous sections.

Employment of a simple manual device, such as this, demands the constant attention of the operator. Initially, the applied potential can be quite high and the currents, therefore, large. As the electrolysis proceeds, however, continuous decreases in the applied potential are required to maintain a constant cathode potential. During this period the chemist can do little else but adjust the equipment. Fortunately, however, automatic instruments, called *potentiostats*, have been designed to maintain the cathode potential at some constant, preset value throughout the electrolysis.[2]

In principle, at least, the conversion of a manual instrument to a potentiostat is not difficult. An example is the device shown in Figure 20-9. Here, a potentiometer is connected in opposition to the cathode-reference electrode system and the output of the potentiometer is set at the desired emf. No current flows in this circuit as long as the cathode remains at the proper potential. On the other hand, when cathode potential departs from the desired value, a current passes through the galvanometer relay; this in turn activates an electronic relay in one direction or another depending upon the direction of the potentiometer current. The electronic relay controls a reversible motor which is mechanically coupled to an autotransformer that supplies power to the electrodes via a rectifier unit. Thus, when the electronic relay is closed, the motor alters the output of the transformer and thereby changes the dc potential applied to the cell. When the cathode potential again reaches the desired level, current ceases to flow through the galvanometer relay and the motor stops.

Many electronic circuits have been designed that provide electrode potential control without mechanical parts. Several potentiostats are available commercially.

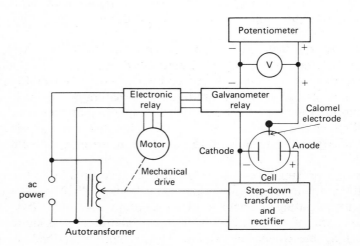

Figure 20-9. Schematic diagram of a potentiostat.

[2]For details of such instruments see J. J. Lingane, *Electroanalytical Chemistry,* 2d ed. (New York: Interscience Publishers, Inc., 1958), pp. 308–339.

APPLICATIONS

Constant Current Methods

Without control of the cathode potential, electrolytic methods suffer from a lack of specificity. Despite this limitation, several applications of practical importance make use of this relatively unrefined technique. In general, the species being determined must be the only component in the solution that is more readily reduced than the hydrogen ion. Should more than one such element be present, some preliminary treatment is required, involving chemical precipitation of the potential interference or the addition of some ligand that will prevent deposition of the offender without appreciably influencing the behavior of the element of interest.

This type of electrodeposition is also useful in removing easily reduced ions from solution prior to completion of an analysis by some other method. The deposition of interfering heavy metals prior to the quantitative determination of the alkali metals provides an example of this application.

The equipment necessary for constant current electrolysis is simple when contrasted to that required for constant cathode-potential methods.

Table 20-2 lists the common elements that can be determined by electrogravimetric procedures for which control over the cathode potential is not required.

Table 20-2 Common Elements That Can Be Determined by Electrogravimetric Methods[a]

Ion	Weighed as	Conditions
Cd^{2+}	Cd	Alkaline cyanide solution
Co^{2+}	Co	Ammoniacal sulfate solution
Cu^{2+}	Cu	HNO_3-H_2SO_4 solution
Fe^{3+}	Fe	$(NH_4)_2CO_3$ solution
Pb^{2+}	PbO_2	HNO_3 solution
Ni^{2+}	Ni	Ammoniacal sulfate solution
Ag^+	Ag	Cyanide solution
Sn^{2+}	Sn	$(NH_4)_2C_2O_4$-$H_2C_2O_4$ solution
Zn^{2+}	Zn	Ammoniacal or strong NaOH solution

[a]From *Fundamentals of Analytical Chemistry*, Second Edition, by Douglas A. Skoog and Donald M. West. Copyright © 1963 by Holt, Rinehart and Winston, Inc. Reprinted by permission of Holt, Rinehart and Winston, Inc.

Controlled Electrode-potential Methods[3]

The controlled cathode-potential method is a potent tool for the direct analysis of solutions containing a mixture of the metallic elements.

[3]This method was first suggested by H. J. S. Sand, *Trans. Chem. Soc.*, **91**, 373 (1907). For many of its applications, see H. J. S. Sand, *Electrochemistry and Electrochemical Analysis*, Vol. 2 (Glasgow: Blackie & Son, Ltd., 1940). An excellent discussion of applications of automatic control to the method can be found in J. J. Lingane, *Electroanalytical Chemistry*, 2d ed. (New York: Interscience Publishers, Inc., 1958), Chaps. 13–16. See also G. A. Rechnitz, *Controlled-Potential Analysis* (New York: The Macmillan Company, 1963).

Such control permits quantitative separation of elements with standard potentials that differ by only a few tenths of a volt. For example, Lingane and Jones[4] developed a method for the successive determination of copper, bismuth, lead, and tin. The first three can be deposited from a nearly neutral tartrate solution. Copper is first reduced quantitatively by maintaining the cathode potential at -0.2 V with respect to a saturated calomel electrode. After weighing, the copper-plated cathode is returned to the solution and bismuth is removed at a potential of -0.4 V. Lead is then plated out quantitatively by increasing the cathode potential to -0.6 V. Throughout these depositions the tin is retained in solution as a very stable tartrate complex. Acidification of the solution after deposition of the lead is sufficient to decompose the complex by converting tartrate ion to the undissociated acid; tin can then be readily deposited at a potential of -0.65 V. This method can be extended to include zinc and cadmium also. Here, the solution is made ammoniacal after removal of the copper, bismuth, and lead. Cadmium and zinc are then successively plated out and weighed. Finally, the tin is determined after acidification, as before.

A procedure such as this is particularly attractive for use with a potentiostat because of the small operator-time required for the complete analysis.

Table 20-3 indicates other separations that have been performed by the controlled-cathode method.

Table 20-3 Some Applications of Controlled Cathode-potential Electrolysis[a]

Element Determined	*Other Elements That May Be Present*
Ag	Cu and base metals
Cu	Bi, Sb, Pb, Sn, Ni, Cd, Zn
Bi	Cu, Pb, Zn, Sb, Cd, Sn
Sb	Pb, Sn
Sn	Cd, Zn, Mn, Fe
Pb	Cd, Sn, Ni, Zn, Mn, Al, Fe
Cd	Zn
Ni	Zn, Al, Fe

[a]From *Fundamentals of Analytical Chemistry*, Second Edition, by Douglas A. Skoog and Donald M. West. Copyright © 1963 by Holt, Rinehart and Winston, Inc. Reprinted by permission of Holt, Rinehart and Winston, Inc.

SPONTANEOUS OR INTERNAL ELECTROLYSIS

An electrogravimetric analysis can occasionally be accomplished within a short-circuited galvanic cell. Under these circumstances no external power is required; the deposition takes place as a consequence of the energetics of the cell reaction. For example, copper(II) ions in a solution will quantitatively deposit upon a platinum cathode when external contact is made with a zinc anode immersed in a solution of zinc ions. The cell

[4]J. J. Lingane and S. L. Jones, *Anal. Chem.,* **23**, 1798 (1951).

reaction may be represented as

$$Zn + Cu^{2+} \rightleftharpoons Zn^{2+} + Cu$$

If the reaction is allowed to proceed to equilibrium, substantially all copper(II) ions are removed from the solution.

This technique is termed *internal electrolysis* or, perhaps more aptly, *spontaneous electrolysis*. Aside from simplicity insofar as equipment is concerned, it enjoys the advantage of being somewhat more selective than an ordinary electrolysis without cathode-potential control; through suitable choice of anode system the codeposition of many elements can be eliminated. Thus, for example, employment of a lead anode for the deposition of copper prevents interference from all species with more negative potentials than the lead ion-lead couple.

Apparatus

Figure 20-10 illustrates a typical arrangement for the spontaneous electrolytic determination of copper. The element is deposited on a weighed platinum gauze cathode. A stirrer is employed to circulate the solution around the cathode.

The anode is a piece of zinc immersed in a zinc sulfate solution. This electrode must be isolated from the solution being analyzed, to prevent the deposition of copper directly on the zinc. The anode may be conveniently isolated with a paper or a porous ceramic cup. A solution of zinc sulfate or some other electrolyte is placed in the cup.

Electrolysis is initiated by connecting the platinum and zinc electrodes with an external wire, and is continued until removal of the copper is judged complete.

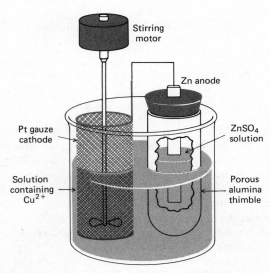

Figure 20-10. Apparatus for the spontaneous electrogravimetric determination of copper.

A constant source of concern in a spontaneous electrolysis is the internal resistance of the cell, since this variable controls the rate at which deposition occurs. If the resistance becomes very high, inordinate lengths of time are required for completion of the reaction. This problem does not arise in an ordinary electrolysis, where the effects of a high-cell resistance can be readily overcome by increasing the applied potential. With spontaneous deposition, large currents can be obtained only if R is kept small. Therefore, the apparatus for this method must always be designed with the view of minimizing the ohmic drop through the use of large electrodes, good stirring, and reasonably high concentrations of electrolyte. Depositions can be completed in less than an hour by this method under ideal conditions; often, however, several hours are required. Long deposition times are not necessarily a serious handicap, since the cell requires no attention during operation.

Applications

Table 20-4 lists some of the applications of the internal-electrolysis method.

Table 20-4 Application of the Internal-electrolysis Method[a]

Element Determined	Anode	Noninterfering Elements
Ag	Cu, $CuSO_4$	Cu, Fe, Ni, Zn
Cu	Zn, $ZnCl_2$	Ni, Zn
Bi	Mg, $MgCl_2$	—
Pb	Zn, $ZnCl_2$	Zn
Ni	Mg, $MgSO_4$	—
Co	Mg, NH_4Cl, HCl	—
Cd	Zn, $ZnCl_2$	Zn
Zn	Mg, NH_4Cl, HCl	—

[a]From *Fundamentals of Analytical Chemistry*, Second Edition, by Douglas A. Skoog and Donald M. West. Copyright © 1963 by Holt, Rinehart and Winston, Inc. Reprinted by permission of Holt, Rinehart and Winston, Inc.

Problems

1. A current of 0.50 amp is to be passed through a cell in which oxygen is deposited ($p = 1.00$ atm) at a platinum anode having a surface area of 100 cm²; lead is to be plated out at the cathode. The solution is 0.200 M in Pb^{2+} and is buffered to a pH of 5.00. The resistance of the cell is 0.80 ohm.
    (a) Calculate the theoretical potential (zero current potential) of the cell.
    (b) Calculate the IR drop.
    (c) Estimate the oxygen overvoltage (see Table 17-2).
    (d) Estimate the total applied potential required to begin operation of the cell at the specified conditions.
    (e) If the solution volume were exactly 100 ml and the current were maintained at 0.500 amp, how long would it take to reduce the Pb^{2+} concentration to 0.0100 M?

    (f)   What total applied potential would be required to operate the cell under the specified conditions when the Pb^{2+} concentration is 0.0100 *M*?

2. The indium in a 0.200-*M* solution of In^{3+} is to be isolated by electrodeposition. The solution is buffered to *p*H 4.00 and a current of 0.200 amp is passed. Oxygen is evolved (1.00 atm) at a platinum anode which has a surface area of 20 cm²; resistance of the cell is 2.50 ohms. For indium (III),

$$In^{3+} + 3e \rightleftharpoons In \qquad E^0 = -0.342 \text{ V}$$

    (a)   Calculate the theoretical potential (zero current potential) of the cell.

    (b)   Calculate the *IR* drop.

    (c)   Estimate the oxygen overvoltage (see Table 17-2).

    (d)   Estimate the total applied potential required to begin operation of the cell at the specified conditions.

    (e)   If the solution volume were exactly 100 ml and the current were maintained at 0.200 amp, how long would it take to reduce the In^{3+} concentration to 0.0100 *M*?

    (f)   What total applied potential would be required to operate the cell under the specified conditions when the In^{3+} concentration was 0.0100 *M*?

3. Electrodeposition has been proposed as a means for separating the cations of a solution that is 0.080 *F* with respect to Zn^{2+} and 0.060 *F* with respect to Co^{2+}.

    (a)   Assuming for the moment that such a separation is feasible, indicate the cation that will be deposited and the cation that will remain in solution.

    (b)   Taking a residual concentration of 1.0×10^{-6} *M* as a reasonable estimate of quantitative separation, calculate the range (if such exists) with respect to a saturated calomel electrode within which the potential of the cathode must be maintained in order to achieve such a separation.

4. (a)   Determine whether it would be feasible to attempt an electrogravimetric separation of lead from nickel in a solution that is 5.0×10^{-2} *M* with respect to Pb^{2+} and 4.0×10^{-2} *M* with respect to Ni^{2+} through control of the potential of the cathode. Use 1.0×10^{-6} *M* as the criterion for quantitative removal.

    (b)   If separation is feasible, evaluate the range (vs. SCE) within which the cathode potential should be controlled.

    (c)   If separation is not feasible, propose a modification that would increase the possibilities of success with this separation.

5. Halide ions can be deposited at a silver anode, the reaction being

$$Ag + X^- \rightarrow AgX + e$$

    (a)   Determine whether or not it is theoretically feasible to separate Cl^- and Br^- ions from a solution that is 0.0500 *M* in each ion by controlling the silver anode potential. Take 1.00×10^{-6} *M* as the criterion of quantitative removal of one ion.

    (b)   Is a separation of Cl^- and I^- theoretically feasible?

    (c)   If a separation is feasible in either (a) or (b), what range of anode potentials (vs. the saturated calomel electrode) should be employed?

6. What cathode potential (vs. the saturated calomel electrode) would be required to reduce the total silver concentration of the following solutions to 1×10^{-5} *F*?

(a)   water containing $AgNO_3$
(b)   a silver nitrate solution in which the equilibrium thiosulfate concentration is $1.00 \times 10^{-3}\ M$

$$Ag^+ + 2S_2O_3^{2-} \rightleftharpoons Ag(S_2O_3)_2^{3-} \qquad K_f = 2.9 \times 10^{13}$$

(c)   a silver nitrate solution in which the equilibrium CN^- concentration is $1.00 \times 10^{-2}\ M$

$$Ag^+ + 2CN^- \rightleftharpoons Ag(CN)_2^- \qquad K_f = 1.3 \times 10^{21}$$

(d)   a silver nitrate solution in which the equilibrium NH_3 concentration is $0.100\ M$

$$Ag^+ + 2NH_3 \rightleftharpoons Ag(NH_3)_2^+ \qquad K_f = 1.4 \times 10^7$$

7.   At what potentials (vs. the saturated calomel electrode) would copper begin to deposit from each of the following solutions in which the formal Cu^{2+} concentration is 0.0100?
(a)   water
(b)   a solution having an equilibrium cyanide ion concentration of $0.100\ M$

$$Cu(CN)_4^{2-} + 2e \rightleftharpoons Cu + 4CN^- \qquad E^0 = -0.43\ V$$

(c)   a solution having an equilibrium NH_3 concentration of $0.100\ M$

$$Cu^{2+} + 4NH_3 \rightleftharpoons Cu(NH_3)_4^{2+} \qquad K_f = 1.1 \times 10^{12}$$

(d)   a solution having an equilibrium Y^{4-} (EDTA) concentration of $1.00 \times 10^{-3}\ M$

$$Cu^{2+} + Y^{4-} \rightleftharpoons CuY^{2-} \qquad K_f = 6.3 \times 10^{18}$$

8.   An internal-electrolysis cell such as that shown in Figure 20-10 contained 100.0 ml of $0.100\ F\ Cu^{2+}$ and a copper electrode as the cathode compartment; a zinc electrode immersed in 20.0 ml of $1.00 \times 10^{-3}\ F\ Zn^{2+}$ served as the anode. The cell resistance was found to be 7.5 ohms.
(a)   What was the initial potential of this cell if no current was drawn?
(b)   What would be the initial current if the electrodes were short-circuited with a conductor?
(c)   What would be the cell potential when the Cu^{2+} concentration had been reduced to $1.00 \times 10^{-5}\ M$?
(d)   What would be the theoretical current when the Cu^{2+} concentration was $1.00 \times 10^{-5}\ M$ if the cell resistance did not change? Is it likely that the observed current would be this large? Explain.

9.   Answer Problem 8, assuming an identical cell except that the cathode compartment now consists of a cadmium electrode in 100 ml of $0.100\ F Cd^{2+}$.

10.   An internal-electrolysis cell consisted of a silver cathode immersed in 50.0 ml of $0.200\ M\ AgNO_3$ and a copper anode in 20 ml of $1.00 \times 10^{-4}\ FCu^{2+}$. The resistance of the cell was 8.0 ohms.
(a)   What was the initial potential of the cell if no current was drawn?
(b)   What was the potential after the silver ion concentration was reduced to $1.00 \times 10^{-3}\ M$?
(c)   What was the initial current when the electrodes were short-circuited?
(d)   What was the theoretical current when the Ag^+ concentration had been reduced to $1.00 \times 10^{-3}\ M$?

21 Coulometric Methods of Analysis

More than a century ago Michael Faraday discovered the direct proportionality between quantity of electricity and amount of oxidation and reduction that occurs upon passage of current through an electrochemical cell. His observations led to development of the chemical coulometer, a device for measuring quantity of electricity by determining the amount of chemical change brought about by the current. The reverse application of Faraday's laws, in which the amount of a substance is determined from the measurement of quantity of electricity, is of much more recent origin. Such *coulometric methods* of analysis date from about 1940.[1]

Coulometric methods embrace two general techniques. The first involves maintaining the potential of the working electrode at some fixed value until the approach to zero current serves to indicate completion of the reaction. The total quantity of electricity required for the electrolysis is evaluated with a chemical coulometer or by integration of the current-time curve for the reaction. The second technique makes use of a constant cur-

[1]For summaries of coulometric methods see H. J. Kies, *J. Electroanal. Chem.*, **4**, 257 (1962); J. J. Lingane, *Electroanalytical Chemistry*, 2d ed. (New York: Interscience Publishers, Inc., 1958), Chaps. 19-21; D. D. DeFord in I. M. Kolthoff and P. J. Elving, Eds., *Treatise on Analytical Chemistry*, Part 1, Vol. 4 (New York: Interscience Publishers, Inc., 1963), Chap. 49.

rent that is passed until an indicator signals completion of the reaction. The quantity of electricity required to attain the end point is then calculated from the magnitude of the current and the time of its passage. This method has a wider variety of applications than the former; it is frequently called a *coulometric titration.*

A fundamental requirement of all coulometric methods is that the species determined interact with 100 percent current efficiency. Thus, each faraday of electricity must bring about a chemical change corresponding to one equivalent of the substance of interest. This requirement does not, however, imply that the species must necessarily participate directly in the electron-transfer process at the electrode. Indeed, more often than not, the substance being determined is involved wholly or in part in a reaction that is secondary to the electrode reaction. For example, at the outset of the oxidation of iron(II) at a platinum anode, all current transfer results from the reaction

$$Fe^{2+} \rightarrow Fe^{3+} + e$$

As the concentration of iron(II) decreases, however, concentration polarization may cause the anode potential to rise until decomposition of water occurs as a competing process. That is,

$$2H_2O \rightarrow O_2 + 4H^+ + 4e$$

The current required to complete the oxidation of iron(II) would then exceed that demanded by theory. To avoid the consequent error, an excess of cerium(III) can be introduced at the start of the electrolysis. This ion is oxidized at a lower anode potential than that for water:

$$Ce^{3+} \rightarrow Ce^{4+} + e$$

The cerium(IV) produced diffuses rapidly from the electrode surface, where it can then oxidize an equivalent amount of iron(II):

$$Ce^{4+} + Fe^{2+} \rightarrow Ce^{3+} + Fe^{3+}$$

The net effect is an electrochemical oxidation of iron(II) with 100 percent current efficiency even though only a fraction of the iron(II) ions are directly oxidized at the electrode surface.

The coulometric determination of chloride provides another example of an indirect process. Here, a silver electrode serves as the anode and produces silver ions when current is passed. These cations diffuse into the solution and precipitate the chloride. A current efficiency of 100 percent with respect to the chloride ion is achieved even though this species is neither oxidized nor reduced in the cell.

COULOMETRIC METHODS AT CONSTANT ELECTRODE POTENTIAL

Coulometric methods employing a controlled potential were first suggested by Hickling[2] in 1942 and have been further investigated by Lingane[3] and others. The techniques are similar to electrogravimetric methods using potential control (see Chapter 20); they differ only in that a quantity of electricity is measured rather than a weight of deposit. In contrast to a coulometric titration, a single reaction at the working electrode is required, although the species being determined need not react directly.

Apparatus and Methods

A controlled-potential coulometric analysis requires a potentiostat such as that described on page 529. In addition, a chemical coulometer, placed in series with the working electrode, is needed to determine the quantity of electricity used. An example of a hydrogen–oxygen coulometer, designed by Lingane[4], is illustrated in Figure 21-1. It consists of a tube equipped with a stopcock and a pair of platinum electrodes. The tube is connected to a buret by a rubber tubing; both are filled with 0.5 F K_2SO_4. Passage of current through this device liberates hydrogen at the cathode and oxygen at the anode. Both gases are collected, and their total volume is measured by determining the volume of liquid displaced. The water jacket and thermometer provide a means of ascertaining the gas temperature.

The total quantity of electricity can also be determined graphically by careful measurement of the current passing through the cell at known time intervals. The area under the curve that relates these two variables yields the desired quantity. This procedure is more time-consuming and generally less accurate than the procedure using a chemical coulometer.

A number of mechanical and electronic integrators have been developed that can be adapted to the evaluation of the current-time integral in a coulometric electrolysis.[5] These are more convenient to use than the chemical coulometer.

Applications of Controlled-potential Coulometric Methods

A coulometric analysis with controlled potential possesses all advantages of a controlled-cathode potential electrogravimetric method (see p. 521)

[2]A. Hickling, *Trans. Faraday Soc.*, **38**, 27 (1942).

[3]For a good discussion of this method see J. J. Lingane, *Electroanalytical Chemistry*, 2d ed. (New York: Interscience Publishers, Inc., 1958), pp. 450-483.

[4]J. J. Lingane, *J. Am. Chem. Soc.*, **67**, 1916 (1945).

[5]For a description of several of these see J. J. Lingane, *Electroanalytical Chemistry*, 2d ed. (New York: Interscience Publishers, Inc., 1958), pp. 340-350.

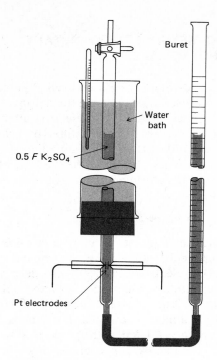

Figure 21-1. A hydrogen-oxygen coulometer.

and, in addition, is not subject to the limitation imposed by the need for a weighable product. The technique can therefore be applied to systems that yield deposits with poor physical properties, as well as to reactions in which no solid product is formed. For example, arsenic may be determined coulometrically by electrolytic oxidation of arsenious acid (H_3AsO_3) to arsenic acid (H_3AsO_4) at a platinum anode. Similarly, the analytical conversion of iron(II) to iron(III) can be accomplished with suitable control of the anode potential. Other metallic ions having more than one stable oxidation state can also be analyzed in this way.

The coulometric method has been advantageously applied to the deposition of metals at a mercury cathode (see p. 529); with this electrode a gravimetric completion of the analysis is inconvenient. Excellent methods have been described for the analysis of lead in the presence of cadmium, copper in the presence of bismuth, and nickel in the presence of cobalt.

The controlled-potential coulometric procedure also offers possibilities for the electrolytic determination of organic compounds. For example, Meites and Meites[6] have demonstrated that trichloroacetic acid and picric

[6]T. Meites and L. Meites, *Anal. Chem.,* **27**, 1531 (1955); **28**, 103 (1956).

acid are quantitatively reduced at a mercury cathode whose potential is suitably controlled.

$$Cl_3CCOO^- + H^+ + 2e \rightleftharpoons Cl_2HCCOO^- + Cl^-$$

Coulometric measurements make possible the estimation of these compounds with an accuracy of a few tenths of a percent.

COULOMETRIC TITRATIONS

A coulometric titration involves the electrolytic generation of a reagent that reacts with the species to be determined. The electrode reaction may involve only reagent preparation, as in the formation of silver ion for the precipitation of halides. In other titrations the substance being determined may also be directly involved at the generator electrode; the coulometric oxidation of iron(II) — in part by electrolytically generated cerium(IV) and in part by direct electrode reaction — is an example. Under any circumstances, the net process must approach 100 percent current efficiency with respect to the substance being determined.

In contrast to the controlled-potential method, the current during a coulometric titration is carefully maintained at a constant and accurately known level; the product of this current in amperes and the time in seconds required to reach the equivalence point for the reaction yields the number of coulombs and thus the number of equivalents involved in the electrolysis. The constant-current aspect of this operation precludes the quantitative oxidation or reduction of the unknown species entirely at the generator electrode; as the solution is depleted of the substance being determined, concentration polarization is inevitable. The electrode potential must then rise if a constant current is to be maintained. Unless this potential rise produces a reagent that can react with the species of interest, current efficiencies that are less than 100 percent result. Thus, at least part (and often all) of the analytical reaction does not occur at the surface of the working electrode.

A coulometric titration, in common with the more conventional titration, requires some means of detecting the point of chemical equivalence. Most of the end points applicable to volumetric analysis are equally satisfactory here; the color change of indicators, potentiometric, amperometric (p. 579), and conductance measurements have all been successfully applied.

The analogy between a volumetric and a coulometric titration extends well beyond the common requirement of an observable end point. In both, the amount of unknown is determined through evaluation of its combining capacity—in the one case for a standard solution and in the other for a quantity of electricity. Similar demands are made of the reactions; that is, they must be rapid, essentially complete, and free of side reactions.

It is of interest to compare volumetric and coulometric analyses from the standpoint of equipment and methods. Figure 21-2 shows a block diagram for a coulometric apparatus; included is a constant-current source, an electric timer, a switch that simultaneously activates the stopclock and the generator circuit, and a device for measuring current. The direct-current source and its magnitude can be considered analogous to the titrant and its normality. The electric clock and switch correspond to the buret, the switch performing the same function as a stopcock. During the early phases of a coulometric titration the switch is kept closed for extended periods; as the end point is approached, however, small additions of "reagent" are achieved by closing the switch for shorter and shorter intervals. The similarity to the operation of a buret is obvious.

Some real advantages can be claimed for a coulometric titration when it is compared with the classical volumetric process. Principal among these is the elimination of problems associated with the preparation, the standardization, and the storage of standard solutions. This advantage is particularly important with labile reagents such as chlorine, bromine, or titanium(II) ion; owing to their instability, these substances are inconvenient as volumetric reagents. Their utilization in coulometric analysis is straightforward, however, since they undergo reaction almost immediately after generation.

Where small quantities of reagent are required, a coulometric titration offers a considerable advantage. By proper choice of current, micro quantities of a substance can be introduced with ease and accuracy, whereas

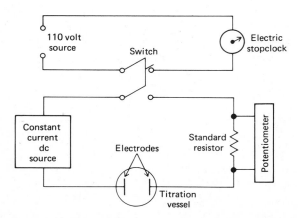

Figure 21-2. Schematic diagram of a coulometric titration apparatus.

the equivalent volumetric process would require the employment of very dilute solutions, a recourse that is always difficult.

A single constant-current source can be employed to generate precipitation, oxidation-reduction, or neutralization reagents. Furthermore, the coulometric method is readily adapted to automatic titrations, since current control is easily accomplished.

Coulometric titrations are subject to five potential sources of error: (1) variation in the current during electrolysis, (2) departure of the process from 100 percent current efficiency, (3) error in the measurement of current, (4) error in the measurement of time, and (5) titration error due to the difference between the equivalence point and the end point. The last of these difficulties is common to volumetric methods as well; where the indicator error is the limiting factor, the two methods are likely to be comparable in reliability.

With simple instrumentation, currents constant to 0.2 to 0.5 percent relative are easily achieved; with somewhat more sophisticated apparatus, control to 0.01 percent is obtainable. In general, then, errors due to fluctuations in current need not be serious.

Although generalizations concerning the magnitude of uncertainty associated with the electrode process are difficult, current efficiency does not appear to be the factor limiting the accuracy of many coulometric titrations.

Errors in measurement of current can be kept small. It is not difficult to determine the magnitude of even the smallest currents to 0.01 percent or better. Error in the measurement of time, however, can represent the limiting factor in the accuracy of a coulometric titration. With a good-quality electric stopclock, relative time errors of 0.1 percent or smaller can be achieved.

To summarize, then, the current-time measurements required for a coulometric titration are inherently as accurate or more accurate than the comparable volume-normality measurements of classical volumetric analysis, particularly where small quantities of reagent are involved. Often, however, the accuracy of a titration is not limited by these measurements, but instead by the sensitivity of the end point; in this respect the two procedures are equivalent.

Apparatus and Methods

The apparatus for a coulometric titration can be relatively simple in comparison to the instrumentation required for the controlled-potential method. The basic components are shown in Figure 21-2 and are discussed in the sections that follow.

Constant-current sources. Many constant-current sources for coulometric titrations have been described in the literature. These vary considerably in their complexity and in their performance characteristics. We shall consider only the simplest type; it is capable of delivering currents of about 20 mamp that are constant to approximately 0.5 percent. Devices yielding currents of an ampere or greater and which vary no more than 0.01 percent over extended periods of time are considerably more complex.[7]

A diagram for a constant-current source is shown in Figure 21-3. The power supply consists of two or more high-capacity, 45-V B-batteries, the current from which passes through a calibrated standard resistance R_1. A potentiometer is connected across R_1 to permit accurate measurement of the potential drop, from which the current is calculated by Ohm's law. The resistance of R_1 should be chosen so that IR_1 is about one volt; with this arrangement a precise determination of I can be obtained with even a relatively simple potentiometer. The variable resistance R_2 has a maximum value of about 20,000 ohms.

When the circuit is completed by throwing the switch to the number 2 position, the current passing through the cell is

$$I = \frac{E_B + E_{cell}}{R_1 + R_2 + R_B + R_{cell}}$$

where E_B is the potential of the B-batteries and E_{cell} comprises the cathode and anode potentials of the titration cell plus any overvoltage or junction potentials associated with its operation. The resistance of the batteries and cell are symbolized by R_B and R_{cell}, respectively.

The potential of dry cells remains reasonably constant for short periods of time, provided the current drawn is not too large; it is safe to

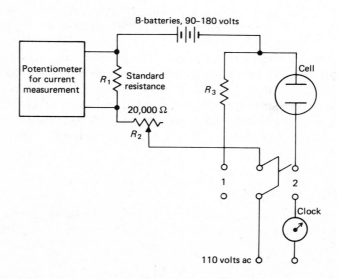

Figure 21-3. A simple apparatus for coulometric titrations.

[7]For a description of constant-current sources see J. J. Lingane, *Electroanalytical Chemistry*, 2d ed. (New York: Interscience Publishers, Inc., 1958), pp. 499–511.

assume, therefore, that E_B as well as R_B will remain unchanged during any given titration. Variations in I, then, arise only from changes in E_{cell} and R_{cell}. Ordinarily, however, R_{cell} will be on the order of 10 to 20 ohms compared with R_2, which is perhaps 10,000 ohms. Thus, even if R_{cell} were to change by as much as 10 ohms, which is highly unlikely, the effect on the current would be less than one part in a thousand.

Changes in E_{cell} during a titration usually have a greater effect on the current, for the cell potential may be altered by as much as 0.5 V during the electrolysis. This change will cause a variation of 0.5 to 0.6 percent in I if E_B is 90 V; the same change would cause a variation of only about 0.3 percent if E_B were 180 V. Experience has shown these to be fairly realistic figures for a simple power source of this kind, *provided current is drawn more or less continuously from the battery.* Accordingly, a resistance R_3 is employed that has about the same magnitude as R_{cell}. The switching arrangement shown allows imposition of R_3 in the circuit whenever current is not being passed through the cell.

Measurement of time. The electrolysis time during a titration is best measured with an electric stopclock actuated by the same switch used to operate the cell. Typically, a titration involves a time period on the order of 100 to 500 sec; the clock should therefore be accurate to a few tenths of a second. An ordinary electric stopclock is not very satisfactory because the motor tends to coast when the current is shut off and also to lag when started. While the error resulting from one start-stop sequence may be small, a coulometric titration involves many such operations, and the accumulated error from this source can become appreciable. Stopclocks with solenoid-operated brakes eliminate this problem, but unfortunately they are more expensive than the simple laboratory timer.

Another error that may arise in connection with electric timing is caused by variations in the frequency of the 110-V power supply used to operate the clock. Ordinarily, these variations are less than 0.2 percent and need be considered only when accuracies greater than this are sought.

Cells for coulometric titrations. A typical coulometric titration cell is shown in Figure 21-4. It consists of a generator electrode at which the reagent is formed and a second electrode to complete the circuit. The generator electrode, which should have a relatively large surface area, is often formed in the shape of a rectangular strip or a wire coil of platinum; a gauze electrode such as that shown on page 528 can also be employed.

The products formed at the second electrode frequently represent potential sources of interference. For example, the anodic generation of oxidizing agents is often accompanied by the evolution of hydrogen from the cathode; unless this gas is allowed to escape from the solution, reaction with the oxidizing agent becomes a likelihood. To eliminate this type of difficulty the second electrode is isolated by a sintered disk or some other porous medium.

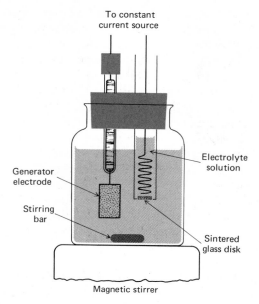

To constant
current source

Generator
electrode

Stirring
bar

Electrolyte
solution

Sintered
glass disk

Magnetic stirrer

Figure 21-4. A typical coulometric titration cell.

External generation of reagent. Occasionally a coulometric titration cannot be used because of side reactions between the generator electrode and some other constituent in the solution. For example, the coulometric titration of acids involves the formation of base at the cathode

$$2e + 2H_2O \rightleftharpoons H_2 + 2OH^-$$

In the presence of easily reduced substances, a reaction other than the desired one can occur at the generator electrode; departures from the required 100 percent current efficiency thus result. To overcome this problem several ingenious devices for external reagent generation have been developed; Figure 21-5 shows the essential features of a design by DeFord, Pitts, and Johns.[8] During electrolysis, an electrolyte solution such as sodium sulfate is fed through the tubing at a rate of about 0.2 ml per sec. The hydrogen ions formed at the anode are washed down one arm of the T-tube (along with an equivalent number of sulfate ions), while the hydroxide ions produced at the cathode are transported through the other. The apparatus is so arranged that flow of the electrolyte is discontinued whenever the electrolysis current is shut off, and a flush-out system is provided to rinse the residual reagent from the tube into the titration vessel.

Both electrode reactions shown in the illustration proceed with 100 percent current efficiency; thus, the solution emerging from the left arm of the apparatus can be used for the titration of acids, and that from the right

[8]D. D. DeFord, J. N. Pitts, and C. J. Johns, *Anal. Chem.,* **23**, 938 (1951).

arm for the titration of bases. A current of about 250 mamp is satisfactory for the titration of various acids or bases in the range between 0.2 and 2 milliequivalents; end points are detected with a glass-calomel electrode system. This apparatus has also been used for the generation of iodine by electrolysis of an iodide solution.

Applications of Coulometric Titrations[9]

Coulometric titrations have been developed for all types of volumetric reactions. Typical applications are described in the following paragraphs.

Neutralization titrations. Both weak and strong acids can be titrated with a high degree of accuracy, using electrogenerated hydroxide ions. The most convenient method, where applicable, involves generation of hydroxide ion at a platinum cathode within the solution. In this application the platinum anode must be isolated by some sort of diaphragm (see Figure 21-4) to eliminate potential interference from the hydrogen ions produced. A convenient alternative involves the addition of chloride or bromide ions to the solution to be analyzed and the use of a silver wire as the anode; the reaction at this electrode then becomes

$$Ag + Br^- \rightleftharpoons AgBr + e$$

Clearly, this anode product will not interfere with the neutralization reaction.

Both potentiometric and indicator end points can be employed for these titrations; the problems associated with the estimation of the equivalence point are identical with those encountered in the corresponding

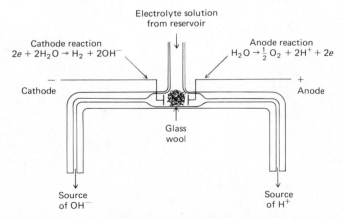

Figure 21-5. Cell for the external generation of acid and base.

[9]Applications of the coulometric procedure are summarized in J. J. Lingane, *Electroanalytical Chemistry,* 2d ed. (New York: Interscience Publishers, Inc., 1958) pp. 536-613; and in H. L. Kies, *J. Electroanal. Chem.* **4**, 257 (1962).

volumetric analysis. A real advantage to the coulometric method is that the carbonate problem is far less troublesome; it is necessary only to eliminate carbon dioxide from the solution to be analyzed, by aeration with a carbon dioxide-free gas, before beginning the analysis. Thus, the problems associated with the preparation and storage of a carbonate-free solution of standard base are avoided.

Coulometric titration of strong and weak bases can be performed with hydrogen ions generated at a platinum anode:

$$H_2O \rightleftharpoons \frac{1}{2} O_2 + 2H^+ + 2e$$

The generation may be carried out either internally or externally; again, if the former method is used, the cathode must be isolated from the solution, to prevent interference from the hydroxide ions produced at that electrode.

Precipitation and complex-formation titrations. Numerous coulometric precipitation titrations are based upon anodically generated silver ions (see Table 21-1). A cell such as that shown in Figure 21-4 can be employed with a generator electrode constructed from a piece of heavy silver wire. Adsorption indicators or the potentiometric method provide end-point detection. Similar applications employing mercury(I) ions formed at a mercury anode have been described.

The coulometric method has also been applied to the titration of several cations by means of ethylenediaminetetraacetate ion (HY^{3-}) generated at a mercury cathode.[10] In this application an excess of the mercury(II) complex of EDTA is introduced into an ammoniacal solution of the sample.

Table 21-1 Typical Applications of Coulometric Titrations Involving Neutralization, Precipitation, and Complex-formation Reactions[a]

Species Determined	Generator Electrode Reaction	Secondary Analytical Reaction
Acids	$2H_2O + 2e \rightleftharpoons 2OH^- + H_2$	$OH^- + H^+ \rightleftharpoons H_2O$
Bases	$H_2O \rightleftharpoons 2H^+ + \frac{1}{2}O_2 + 2e$	$H^+ + OH^- \rightleftharpoons H_2O$
Cl^-, Br^-, I^-	$Ag \rightleftharpoons Ag^+ + e$	$Ag^+ + Cl^- \rightleftharpoons AgCl$ etc.
Mercaptans	$Ag \rightleftharpoons Ag^+ + e$	$Ag^+ + RSH \rightleftharpoons AgSR + H^+$
Cl^-, Br^-, I^-	$2Hg \rightleftharpoons Hg_2^{2+} + 2e$	$Hg_2^{2+} + 2Cl^- \rightleftharpoons Hg_2Cl_2$ etc.
Zn^{2+}	$Fe(CN)_6^{3-} + e \rightleftharpoons Fe(CN)_6^{4-}$	$3Zn^{2+} + 2K^+ + 2Fe(CN)_6^{4-} \rightleftharpoons K_2Zn_3[Fe(CN)_6]_2$
Ca^{2+}, Cu^{2+}, Zn^{2+} and Pb^{2+}	$HgNH_3Y^{2-} + NH_4^+ + 2e \rightleftharpoons Hg + 2NH_3 + HY^{3-}$ (where Y^{4-} is ethylenediaminetetraacetate ion)	$HY^{3-} + Ca^{2+} \rightleftharpoons CaY^{2-} + H^+$, etc.

[a]From *Fundamentals of Analytical Chemistry,* Second Edition, by Douglas A. Skoog and Donald M. West. Copyright © 1963 by Holt, Rinehart and Winston, Inc. Reprinted by permission of Holt, Rinehart and Winston, Inc.

[10]C. N. Reilley and W. W. Porterfield, *Anal. Chem.,* **28**, 443 (1956).

The EDTA anion is then released by electrochemical reduction of the mercury(II):

$$HgNH_3Y^{2-} + NH_4^+ + 2e \rightarrow Hg + 2NH_3 + HY^{3-}$$

The liberated HY^{3-} then reacts with the cation being determined. For example, for Ca^{2+} (and most other divalent cations) the reaction is

$$Ca^{2+} + HY^{3-} + NH_3 \rightarrow CaY^{2-} + NH_4^+$$

Because the mercury chelate is more stable than the corresponding complexes with calcium, zinc, lead, or copper, complexation of these ions cannot occur until the electrode process frees the complexing agent.

Oxidation-reduction titrations. Table 21-2 lists oxidizing and reducing agents that can be generated by the coulometric procedure and the analyses to which they have been applied. Electrogenerated bromine has proved to be particularly useful among the oxidizing agents, and the development of many interesting methods has been based upon this substance. Of importance also are some of the unusual reagents not encountered in volumetric titrations because of the instability of their solutions; these include dipositive silver ion, tripositive manganese, and unipositive copper as the chloride complex.

Table 21-2 Typical Applications of Coulometric Titrations Involving Oxidation-reduction Reactions[a]

Reagent	Generator Electrode Reaction	Substance Determined
Br_2	$2Br^- \rightleftharpoons Br_2 + 2e$	As(III), Sb(III), U(IV), Tl(I), I^-, SCN^-, NH_3, N_2H_4, NH_2OH, phenol, aniline, mustard gas, 8-hydroxyquinoline
Cl_2	$2Cl^- \rightleftharpoons Cl_2 + 2e$	As(III), I^-
I_2	$2I^- \rightleftharpoons I_2 + 2e$	As(III), Sb(III), $S_2O_3^{2-}$, H_2S
Ce^{4+}	$Ce^{3+} \rightleftharpoons Ce^{4+} + e$	Fe(II), Ti(III), U(IV), As(III), I^-, $Fe(CN)_6^{4-}$
Mn^{3+}	$Mn^{2+} \rightleftharpoons Mn^{3+} + e$	$H_2C_2O_4$, Fe(II), As(III)
Ag^{2+}	$Ag^+ \rightleftharpoons Ag^{2+} + e$	Ce(III), V(IV), $H_2C_2O_4$, As(III)
Fe^{2+}	$Fe^{3+} + e \rightleftharpoons Fe^{2+}$	Cr(VI), Mn(VII), V(V), Ce(IV)
Ti^{3+}	$TiO^{2+} + 2H^+ + e \rightleftharpoons Ti^{3+} + H_2O$	Fe(III), V(V), Ce(IV), U(VI)
$CuCl_3^{2-}$	$Cu^{2+} + 3Cl^- + e \rightleftharpoons CuCl_3^{2-}$	V(V), Cr(VI), IO_3^-
U^{4+}	$UO_2^{2+} + 4H^+ + 2e \rightleftharpoons U^{4+} + 2H_2O$	Cr(VI), Ce(IV)

[a]From *Fundamentals of Analytical Chemistry*, Second Edition, by Douglas A. Skoog and Donald M. West. Copyright © 1963 by Holt, Rinehart and Winston, Inc. Reprinted by permission of Holt, Rinehart and Winston, Inc.

Problems

1. Nickel and cobalt can be separated by deposition at a mercury cathode from an aqueous pyridine solution. At -0.95 V (vs. the SCE) the nickel

is deposited without interference from the cobalt. By raising the cathode potential to -1.20 V, cobalt can then be deposited quantitatively.

A sample of an ore weighing 1.32 g was dissolved and treated in such a way as to give Co^{2+} and Ni^{2+} ions. The sample was then electrolyzed with a mercury cathode that was maintained at -0.95 V until current ceased. The volume of gas produced in a hydrogen-oxygen coulometer, in series with the electrolysis cell, was 47.4 ml (corrected for water vapor) at a temperature of 19.0°C and a barometric pressure of 748 mm of Hg. The electrolysis was then continued at -1.2 V, and 11.3 ml of gas were collected.

Calculate the percent Co and Ni in the sample.

2. A chloride-iodide sample weighing 2.10 g was dissolved, made ammoniacal, and introduced into a coulometric cell equipped with a silver anode. By maintaining the anode at -0.060 V (vs. the SCE), iodide was quantitatively precipitated as AgI without interference from chloride. The volume of hydrogen and oxygen produced in a coulometer in series with the cell was 38.1 ml (corrected for water vapor) at 23.5°C and at a pressure of 755 mm of Hg.

Upon completion of the iodide analysis, the solution was acidified and the potential of the anode was maintained at $+0.25$ V (vs. the SCE), whereupon quantitative deposition of AgCl occurred. The volume of gas formed in the coulometer during this process was 44.6 ml.

Calculate the percent $BaCl_2$ and BaI_2 in the sample.

3. The iron in a 0.737-g sample of an ore was converted to the Fe(II) state by suitable treatment and then oxidized quantitatively at a Pt anode maintained at -1.0 V (vs. the SCE). The quantity of electricity required to complete the oxidation was determined with a chemical coulometer equipped with a platinum anode immersed in an excess of iodide ion. The iodine liberated by the passage of current required 27.2 ml of 0.0217 N sodium thiosulfate to reach a starch end point. What was the percent Fe_3O_4 in the sample?

4. An apparatus similar to that shown in Figure 21-3 was employed for the coulometric determination of Fe(III) in 50.0 ml of a solution that was 0.100 N in HCl. Initially, the Fe(III) was reduced directly to Fe(II) at the platinum cathode of the cell. An excess of TiO^{2+} was added to the solution, however, to provide Ti^{3+} to complete the reduction when the Fe(III) concentration had become small and concentration polarization likely. The anode of the cell was a silver electrode ($Ag + Cl^- \rightleftharpoons AgCl + e$). The resistance of the various components of the apparatus were as follows: $R_1 = 9.84$ ohms; $R_{cell} = 19$ ohms; $R_B + R_2 = x$ ohms. The power supply consisted of three 45-V dry cells in series.

    (a) If 532 sec were required to complete the titration and if the measured potential across R_1 was 91.4 mV, what was the concentration of Fe^{3+} in the original solution?

    (b) Calculate the theoretical potential of the cell when 1.0 percent of the Fe^{3+} had been reduced; when 99 percent had been reduced. Assume no concentration polarization effects.

    (c) In order to obtain the current indicated in (a), what should be the approximate value of $x = (R_2 + R_B)$?

    (d) What would be the percentage change in the current that would result from the potential change found in (b)?

(e) If the cell resistance changed by 1.0 ohm in the course of the analysis, what would be the resulting percentage change in the current?

5. A 7.20-g sample of an ant-control preparation was decomposed by wet ashing with H_2SO_4 and HNO_3. The arsenic in the residue was reduced to the trivalent state with hydrazine. After the excess reducing agent was removed, the arsenic(III) was oxidized with electrolytically generated I_2 in a faintly alkaline medium:

$$HAsO_3^{2-} + I_2 + 2HCO_3^- \rightarrow HAsO_4^{2-} + 2I^- + 2CO_2 + H_2O$$

The titration was complete after a constant current of 120 mamp had been passed for 12.40 min. Express the results of this analysis in terms of the percentage As_2O_3 in the original sample.

6. The chromium deposited on a 10.0-cm² test plate was dissolved by treatment with HCl and oxidized to the +6 state with ammonium peroxodisulfate:

$$3S_2O_8^{2-} + 2Cr^{3+} + 7H_2O = 6SO_4^{2-} + Cr_2O_7^{2-} + 14H^+$$

The solution was boiled to remove the excess peroxodisulfate, was cooled, and was then subjected to coulometric titration with Cu(I) generated from 50 ml of 0.10 F Cu^{2+}. Calculate the weight of chromium that was deposited on each square centimeter of the test plate if the titration required a steady current of 0.0500 amp for a period of 8.40 min.

7. The H_2S content of a water sample was assayed with electrolytically generated iodine. After 3.00 g of potassium iodide had been introduced to a 50.0-ml portion of the water, titration required a constant current of 0.0731 amp for a total of 9.2 min. Reaction:

$$H_2S + I_2 \rightarrow S + 2H^+ + 2I^-$$

Express the concentration of H_2S in terms of milligrams per liter of sample.

8. The phenol content of water downstream from a coking mill was determined by coulometric means. A 100-ml sample was rendered slightly acidic and an excess of KBr was introduced. To produce Br_2 for the reaction

$$C_6H_5OH + 3Br_2 \rightarrow Br_3C_6H_2OH + 2HBr$$

a steady current of 0.0208 amp for 580 sec was required. Express the results of this analysis in terms of parts phenol per million parts of water.

9. The odorant concentration of household gas can be monitored by passage of a fraction of the gas stream through a solution containing an excess of bromide ion; electrogenerated bromine reacts rapidly with the mercaptan odorant

$$RSH + Br_2 \rightarrow RSSR + 2H^+ + 2Br^-$$

Continuous analysis is made possible by means of an electrode system that signals the need for additional Br_2 to oxidize the mercaptan. Ordinarily, the current required to keep the unreacted odorant level at zero is automatically plotted as a function of time. Suppose the following data were obtained from the smooth curve from the recorder when a gas stream was being sampled at the rate of 10.0 liters/min.

Time (P.M.)	Current (mamp)
2:32	1.65
2:33	1.55
2:34	1.50
2:35	1.50
2:36	1.53
2:37	1.58
2:38	1.65
2:39	1.70
2:40	1.73
2:41	1.74

(a) Calculate the minimum concentration of odorant (as C_2H_5SH) in parts per million during the 10-min interval. The gas density was 2.00×10^{-3} g/ml.

(b) What was the average odorant concentration during the 10-min period?

10. The calcium content of a water sample was determined by introducing an excess of a solution of $HgNH_3Y^{2-}$ into a 50.0-ml sample. The anion of EDTA was then generated at a mercury cathode (see Table 21-1). A constant current of 0.0180 amp was employed to reach an end point after 3 min and 32 sec. Calculate the milligrams of $CaCO_3$ per milliliter of sample.

11. The equivalent weight of an organic acid was obtained by dissolving 0.0231 g of the purified compound in an alcohol-water mixture and titrating with coulometrically generated hydroxide ions. With a current of 0.0427 amp, 402 sec were required to reach a phenolphthalein end point. Calculate the equivalent weight of the compound.

12. The cyanide concentration of plating solution was determined by titration of 10.0 ml to a methyl-orange end point with electrogenerated hydrogen ions. A color change occurred after 2 min and 57 sec with a current of 0.0391 amp. Calculate the grams of NaCN per liter of solution.

13. Ascorbic acid is oxidized to dehydroascorbic acid with Br_2

A vitamin C tablet was dissolved in sufficient water to give exactly 250 ml of solution. A 50.0-ml aliquot was then mixed with an equal volume of 0.100 F KBr. Calculate the weight of ascorbic acid in the tablet if the bromine generated by a steady current of 0.050 amp for a total of 7.53 min was required for the titration.

22 Voltammetry

Voltammetry comprises a group of electroanalytical procedures that are based upon the potential-current behavior of a polarizable electrode in the solution being analyzed. In order to ensure polarization of this electrode, its dimensions generally are made small.

Historically, voltammetry developed from the discovery of *polarography* by the Czechoslovakian chemist Jaroslav Heyrovsky[1] in the early 1920s. Later in this same decade Heyrovsky and coworkers adapted the principles of polarography to the detection of end points in volumetric analyses; such methods have come to be known as *amperometric titrations.*[2]

Recent years have seen the development of numerous modifications of the original polarographic method as well as the development of several methods closely related to polarography. Some of these more recent developments are considered near the end of this chapter, but the emphasis of our discussion is focused on polarography, not only for historical reasons, but also because polarography is still the most widely used of the voltammetric procedures.

POLAROGRAPHY[3]

Virtually every element, in one form or another, is amenable to polarographic analysis. In addition, the method can be extended to the determina-

[1]J. Heyrovsky, *Chem. listy,* **16**, 256 (1922). Heyrovsky was awarded the 1959 Nobel prize in chemistry for his discovery and development of polarography.

[2]J. Heyrovsky and S. Berezicky, *Collection Czechoslov. Chem. Commum.,* **1**, 19 (1929).

[3]The principles and applications of polarography are considered in detail in a number of monographs; see, for example, I. M. Kolthoff and J. J. Lingane, *Polarography,* 2d ed. (New York: Interscience Publishers, Inc., 1952); L. Meites, *Polarographic Techniques,* 2d ed. (New York: Interscience Publishers, Inc., 1965); J. Heyrovsky in W. G. Berl, Ed., *Physical Methods in Chemical Analysis,* Vol. 2 (New York: Academic Press, Inc., 1951); P. Zuman, *Organic Polarographic Analysis* (Oxford: Pergamon Press, Ltd., 1964); P. Zuman, *The Elucidation of Organic Electrode Processes* (New York: Academic Press, Inc., 1969).

tion of several organic functional groups. Because the polarographic behavior of any species is unique for a given set of experimental conditions, the technique offers attractive possibilities for selective analysis.

Most polarographic analyses are performed in aqueous solution, but if necessary other solvent systems may be used instead. For quantitative analyses the optimum concentration range lies between 10^{-2} and 10^{-4} M; the minimum figure can, however, frequently be diminished by yet another factor of 10. An analysis can be easily performed on 1 to 2 ml of solution, and with a little effort, a volume as small as one drop is sufficient. The polarographic method is thus particularly useful for the determination of quantities in the milligram to microgram range.

Relative errors ranging between 2 and 3 percent are to be expected in routine polarographic work. This order of uncertainty is comparable with or superior to errors affecting other methods for the analysis of small quantities.

A Brief Description of Polarographic Measurements

Polarographic data are obtained by measuring current as a function of the potential applied to a special type of electrolytic cell. A plot of the data gives current-voltage curves, called *polarograms*; these provide both qualitative and quantitative information about the composition of the solution in which the electrodes are immersed.

Polarographic cells. A polarographic cell consists of a small, easily polarized *microelectrode*, a large nonpolarizable reference electrode, and the solution to be analyzed. The microelectrode at which the analytical reaction occurs is an inert metal surface with an area of a few square millimeters. The *dropping mercury electrode,* shown in Figure 22-1, is the most common type of microelectrode. Here, mercury is forced by gravity through a very fine capillary to provide a continuous stream of identical droplets, each having a maximum diameter of between 0.5 and 1 mm. The lifetime of a drop is typically 2 to 6 sec. We shall see that the dropping mercury electrode has properties that make it particularly well-suited for polarographic work. Other microelectrodes, consisting of small-diameter wires or disks of platinum and other metals, can be used instead.

The reference electrode in a polarographic cell should be massive relative to the microelectrode so that its behavior remains essentially constant with the passage of small currents; that is, it should remain unpolarized during the analysis. A saturated calomel electrode and salt bridge, arranged in the manner shown in Figure 22-1, is frequently employed; another common reference electrode consists simply of a large pool of mercury.

Polarograms. A polarogram is a plot of current as a function of the potential applied to a polarographic cell. In most analyses the microelectrode is

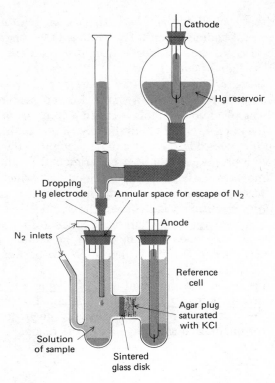

Cathode

Hg reservoir

Dropping
Hg electrode

Annular space for escape of N$_2$

N$_2$ inlets

Anode

Reference
cell

Agar plug
saturated
with KCl

Solution
of sample

Sintered
glass disk

Figure 22-1. A dropping mercury electrode and cell. From J. J. Lingane and H. A. Laitinen, *Ind. Eng. Chem., Anal. Ed.,* **11**, 504 (1939). (With permission of the American Chemical Society.)

connected to the negative terminal of the power supply; *by convention* the applied potential is given a negative sign under these circumstances. By convention also, the currents are designated as positive when the flow of electrons is from the power supply into the microelectrode—that is, when that electrode behaves as a cathode.

Figure 22-2 shows two polarograms, the lower one being for a solution that is 1.0 *F* in potassium chloride and the upper one for a solution that is additionally 1 × 10⁻³ *F* in cadmium chloride. An S-shaped current-voltage curve, called a *polarographic wave,* is produced as result of the reaction

$$Cd^{2+} + 2e + Hg \rightleftharpoons Cd(Hg)$$

In both plots a sharp rise in current occurs at about −2 V; this increase is associated with reduction of potassium ions to give a potassium amalgam.

For reasons to be considered presently, a polarographic wave suitable for analysis is obtained only in the presence of a large excess of a *supporting electrolyte*; potassium chloride serves this function in the present example. Examination of the polarogram for the supporting electrolyte alone reveals that a small current, called the *residual current*, passes through the cell even in the absence of cadmium ions. The voltage at which the polarogram for the electrode-reactive species departs from the residual-current curve is called the *decomposition potential*.

One of the characteristic features of a polarographic wave is the region in which the current levels off after a sharp rise and becomes essentially independent of the applied voltage; this is called a *limiting current*. We shall see that the limiting current is the result of a restriction in the rate at which the participant in the electrode process can be brought to the surface of the microelectrode; with proper control over experimental conditions, this rate is determined almost exclusively by the velocity at which the reactant diffuses. Under these circumstances the limiting current is given a special name, the *diffusion current,* and is assigned the symbol i_d. Ordinarily, the diffusion current is directly proportional to the concentration of the reactive constituent and is thus of prime importance from the standpoint of analysis. In Figure 22-2 the diffusion current is the difference between the limiting and the residual currents.

We must define one other important term, the *half-wave potential*; this is the potential corresponding to a current that is equal to one-half the diffusion current. The half-wave potential is usually given the symbol $E_{1/2}$ and is important for qualitative identification of the reactant.

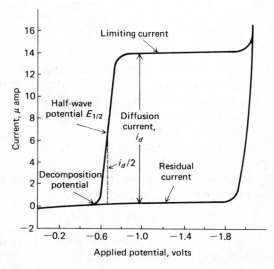

Figure 22-2.   Polarogram for cadmium ion. The upper curve is for a solution that is 1×10^{-3} F with respect to Cd^{2+} and 1 F with respect to KCl. The lower curve is for a solution that is 1 F in KCl only.

Interpretation of Polarographic Waves

A typical polarographic reduction can be represented by the half-cell process:

$$\text{Ox} + ne \rightleftharpoons \text{Red} \qquad E^0 = -0.30 \text{ V}$$

where "Ox" represents the substance undergoing reduction at the micro-electrode, and "Red" stands for the reduction product. No restriction is placed upon the character of the latter; it may consist of a metal amalgam, a metallic deposit on the surface of a solid microelectrode, or a soluble ion or molecule. It is only necessary to stipulate that the electrode process is both rapid and reversible. A standard potential of -0.30 V for the reduction has been assumed for convenience.

We must first consider the chemical changes that occur in the film of liquid immediately surrounding the surface of the microelectrode. The layer under consideration is so thin that its constituents may be considered to be in *instantaneous* equilibrium with the electrode surface; reactant concentrations in this layer are thus determined by the potential of the electrode alone and can be calculated with the Nernst equation. Hence,

$$E_{\text{cathode}} = E^0 - \frac{0.059}{n} \log \frac{[\text{Red}]_0}{[\text{Ox}]_0}$$

where the subscripts signify that the activities apply only to the film immediately adjacent to the cathode. If sufficient time were allowed, equilibrium would be ultimately established, and these activities would then be applicable for the entire solution. At first, however, equilibrium will exist only in the surface film in which we are interested.

On the basis of the discussion in earlier chapters, we may express the applied potential as

$$E_{\text{applied}} = E_{\text{cathode}} - E_{\text{anode}}$$

We are assuming that the junction potential in the cell is negligibly small and that there are no overvoltage effects. Furthermore, we have not taken account of the *IR* drop, inasmuch as the resistance of the cell is ordinarily small and the currents are minute.

Typically, the anode of a polarographic cell is a saturated calomel electrode; the preceding equation, when applied to our example, becomes

$$E_{\text{applied}} = E^0 - \frac{0.059}{n} \log \frac{[\text{Red}]_0}{[\text{Ox}]_0} - E_{\text{SCE}}$$

which in turn can be rearranged to give

$$\frac{n(E_{\text{applied}} + E_{\text{SCE}} - E^0)}{0.059} = \log \frac{[\text{Ox}]_0}{[\text{Red}]_0}$$

This expression permits calculation of the ratio between the activities of reactant and product *at the electrode surface* for any applied potential. For example, at -0.60 V, $[Ox]_0'/[Red]_0'$ will be about 0.1 if n is one and a value of 0.242 is employed for the potential of the calomel electrode. Upon application of -0.60 V, then, an instantaneous current sufficient to establish an activity ratio of 0.1 in the surface film will result. This current would diminish rapidly to zero were it not for the fact that more Ox ions or molecules migrate into the surface film from the bulk of the solution. If the electrochemical reaction is essentially instantaneous, the current will be directly dependent upon the rate at which Ox particles are transported into the surface layer; as a result, we may write

$$i' = k' \times \text{(rate of movement of Ox into surface film)}$$

where k' is a proportionality constant and i' is the current at the applied potential of -0.60 V.

From an earlier discussion (p. 435) we know that transport of ions or molecules in a cell can be brought about by diffusion, by thermal or mechanical convection, and by electrostatic attraction. In polarography every effort is made to eliminate the last two effects; vibration and stirring of the cell solution are avoided and a large excess of a nonreactive supporting electrolyte is employed to minimize the effects of electrostatic attraction between the electrode and the reactant ions. Thus, the rate at which Ox particles are brought to the depleted surface film depends upon the rate at which these species diffuse through the solution. Since the rate of diffusion of a substance is directly proportional to the concentration difference between the parts of the solution through which this process occurs, we may write

$$\text{rate of diffusion of Ox} = k'' \, ([Ox] - [Ox]_0')$$

where $[Ox]$ represents the concentration in the bulk of the solution and $[Ox]_0'$ is the concentration in the film adjacent to the electrode; k'' is a proportionality constant. Furthermore, if diffusion is the *only* process responsible for movement of Ox particles to the electrode surface, then

$$i' = k' \times \text{(rate of diffusion of Ox)}$$

and

$$i' = k([Ox] - [Ox]_0')$$

Therefore, the variable which determines the magnitude of the current is the concentration of the reactant at the electrode surface, which in turn is dependent upon the electrode potential.

Now consider the situation when the applied potential has been increased to -0.70 V. The activity ratio at the surface of the electrode will now be 2×10^{-3}. As a consequence, the surface concentration of Ox at this potential will have decreased to a new value $[Ox]_0''$, and we may write

$$i'' = k([Ox] - [Ox]_0'')$$

The decrease in the surface concentration of Ox results in an increased diffusion rate and thus an increase in the current.

At higher potentials, calculation indicates that the ratio $[Ox]_0/[Red]_0$ continues to diminish; it has a value of 1×10^{-6} at -0.9 V, and 2×10^{-8} at -1.0 V. With $[Ox]_0$ becoming smaller and smaller the difference $([Ox] - [Ox]_0)$ assumes an essentially constant value. Thus, as the concentration of Ox at the surface of the electrode approaches zero, the expression for the current in the cell simplifies to

$$i_d = k[Ox] \qquad \text{when} \qquad [Ox]_0 \ll [Ox]$$

At large applied potentials, then, the diffusion rate is constant and therefore the current is also constant. Note that *the magnitude of the diffusion current is directly proportional to the concentration of the reactant in the bulk of the solution.* Quantitative polarography is based upon this fact.

When a limiting current is achieved as a consequence of the limitation of the rate at which a reactant can be brought to the surface of an electrode, a state of *complete concentration polarization* is said to exist. With a micro-electrode, the current required to reach this condition is small (typically 3 to 10 μamp for a 10^{-3} M solution). Such current levels do not significantly alter the reactant concentration, as shown by the following example.

Example. The diffusion current for a 1.00×10^{-3} M solution of Zn^{2+} was found to be 8.4 μamp. Calculate the percent decrease in the concentration of Zn^{2+} after this current had been allowed to flow for 8.0 min through 10.0 ml of the solution.

$$Q = 8.0 \text{ min} \times 60 \, \frac{\text{sec}}{\text{min}} \times 8.4 \times 10^{-6} \text{ amp}$$
$$= 4.03 \times 10^{-3} \text{ coulombs}$$

$$\begin{array}{ll} \text{no. of meq.} \\ \text{Zn}^{2+} \text{ consumed} \end{array} = \frac{4.03 \times 10^{-3}}{96,494} \times 10^3 = 4.18 \times 10^{-5}$$

$$\begin{array}{ll} \text{no. of millimoles} \\ \text{Zn}^{2+} \text{ consumed} \end{array} = 2.09 \times 10^{-5}$$

$$\begin{array}{ll} \text{percent decrease in} \\ \text{Zn}^{2+} \text{ concentration} \end{array} = \frac{2.09 \times 10^{-5}}{1.00 \times 10^{-3} \times 10} \times 100 = 0.21$$

Equation for the Polarographic Wave

To derive an equation for a typical polarographic wave, we again write the Nernst expression for the process we have been considering as

$$E_{\text{applied}} + E_{\text{SCE}} = E^0 - \frac{0.059}{n} \log \frac{[Red]_0}{[Ox]_0} \qquad (22\text{-}1)$$

We have seen that at any applied potential,

$$i = k([Ox] - [Ox]_0) \qquad (22\text{-}2)$$

and that

$$i_d = k[\text{Ox}] \tag{22-3}$$

Subtracting equation (22-2) from equation (22-3) gives

$$[\text{Ox}]_0 = \frac{i_d - i}{k}$$

Now, if Red is a soluble substance, its concentration at the surface of the electrode will also be proportional to the current, and we may write

$$i = k_r\,[\text{Red}]_0$$

Thus,

$$[\text{Red}]_0 = \frac{i}{k_r}$$

Substituting these quantities into equation (22-1) gives

$$E_{\text{applied}} + E_{\text{SCE}} = E^0 - \frac{0.059}{n} \log \frac{i}{i_d - i} \cdot \frac{k}{k_r} \qquad \frac{\frac{i_d}{2}}{i_d - \frac{i_d}{2}} = 1$$

which can then be written as

$$E_{\text{applied}} = E^0 - E_{\text{SCE}} - \frac{0.059}{n} \log \frac{k}{k_r} - \frac{0.059}{n} \log \frac{i}{i_d - i} \tag{22-4}$$

According to the definition of half-wave potential, when $i = i_d/2$,

$$E_{\text{applied}} = E_{1/2}$$

Substituting into equation (22-4), we obtain

$$E_{1/2} = E^0 - E_{\text{SCE}} - \frac{0.059}{n} \log \frac{k}{k_r} \tag{22-5}$$

and we see that the half-wave potential is a constant that is related to the standard potential for the half-reaction and is *independent* of the reactant concentration. Equation (22-4) can be simplified by substituting $E_{1/2}$ for the several equivalent terms; thus,

$$E_{\text{applied}} = E_{1/2} - \frac{0.059}{n} \log \frac{i}{i_d - i} \tag{22-6}$$

This equation defines the relationship between current and applied potential for a reversible reaction involving the formation of a soluble product. A similar expression results when the half-reaction involves reduction of an ion to a metal that is soluble in the mercury drop.

Half-wave potential. An examination of equation (22-5) reveals the value of the half-wave potential as a reference point on a polarographic wave; it is independent of the reactant concentration, but is directly related to the standard potential for the reaction. In practice, the half-wave potential can be a useful quantity for identification of the species responsible for a given polarographic wave.

It is important to note that the half-wave potential may vary considerably with concentration for electrode reactions that are not rapid and reversible; in addition, equation (22-4) is inadequate for the description of the wave in such cases.

Effect of complex formation on polarographic waves. We have already seen (p. 526) that the potential for the oxidation or reduction of a metallic ion is greatly affected by the presence of species that form complexes with that ion. It is not surprising therefore, that similar effects are observed with polarographic half-wave potentials. The data in Table 22-1 indicate that the half-wave potential for the reduction of a metal complex is generally more negative than that for reduction of the corresponding simple metal ion. Lingane[4] has shown that the magnitude of this shift in half-wave potential is related to the stability of the complex as well as to the concentration of the complexing reagent; as shown in the accompanying example, formation constants can often be estimated from measurement of such shifts.

Table 22-1 Effect of Complexing Agents on Polarographic Half-wave Potentials at the Dropping Mercury Electrode[a]

Ion	Noncomplexing Media	*1 F KCN*	*1 F KCl*	*1 F NH_3, 1 F NH_4Cl*
Cd^{2+}	−0.59	−1.18	−0.64	−0.81
Zn^{2+}	−1.00	NR^b	−1.00	−1.35
Pb^{2+}	−0.40	−0.72	−0.44	−0.67
Ni^{2+}	−	−1.36	−1.20	−1.10
Co^{2+}	−	−1.45	−1.20	−1.29
Cu^{2+}	+0.02	NR^a	+0.04 and −0.22	−0.24 and −0.51

[a]From *Fundamentals of Analytical Chemistry*, Second Edition, by Douglas A. Skoog and Donald M. West. Copyright © 1963 by Holt, Rinehart and Winston, Inc. Reprinted by permission of Holt, Rinehart and Winston, Inc.
[b]NR: no reduction before decomposition of the supporting electrolyte.

Example. Consider the reversible reaction at a dropping electrode:

$$MA_x^{(n-x)+} + ne + Hg \rightleftharpoons M(Hg) + xA^-$$

Here the product is an amalgam of the metal M(Hg). An equation for the polarographic wave can be derived if we consider the reaction as a two-step process, involving first the dissociation of the complex

$$MA_x^{(n-x)+} \rightleftharpoons M^{n+} + xA^-$$

[4]J. J. Lingane, *Chem. Rev.*, **29**, 1 (1941).

followed by reduction of M^{n+}:

$$M^{n+} + ne + Hg \rightleftharpoons M(Hg)$$

It is not necessary that the actual mechanism follow this sequence. Equation (22-1) will serve to describe events at the electrode surface

$$E_{\text{applied}} = E^0 - E_{\text{SCE}} - \frac{0.059}{n} \log \frac{[M(Hg)]_0}{[M^{n+}]_0} \tag{22-7}$$

where $[M(Hg)]_0$ refers to the activity of the metal M at the surface of the mercury drop and E^0 is the standard potential for formation of the amalgam from M^{n+}.

The formation constant for the complex K_f allows us to express $[M^{n+}]_0$ in terms of the concentration of the complex *at the electrode surface;* that is,

$$[M^{n+}]_0 = \frac{[MA_x^{(n-x)+}]_0}{K_f[A^-]_0^x}$$

If the solution contains an excess of the complexing agent, two sources contribute to $[A^-]_0$. First and foremost is the excess of A^- in the solution, given by its formal concentration F_A. Of much smaller magnitude is the A^- that is released as a result of the electrode reaction. Ordinarily, this second source is negligibly small with respect to the first; that is, $[A^-]_0 \cong F_A$. Equation (22-1) can thus be written as

$$E_{\text{applied}} = E^0 - E_{\text{SCE}} - \frac{0.059}{n} \log F_A^x - \frac{0.059}{n} \log K_f - \frac{0.059}{n} \log \frac{[M(Hg)]_0}{[MA_x^{(n-x)+}]_0} \tag{22-8}$$

Now, as before,

$$i = k_r[M(Hg)]_0$$

$$= k_c([MA_x^{(n-x)+}] - [MA_x^{(n-x)+}]_0)$$

$$i_d = k_c[MA_x^{(n-x)+}]$$

and

$$[M(Hg)]_0 = \frac{i}{k_r}$$

$$[MA_x^{(n-x)+}]_0 = \frac{i_d - i}{k_c}$$

Substitution in equation (22-8) yields

$$E_{\text{applied}} = E^0 - E_{\text{SCE}} - \frac{0.059}{n} \log F_A^x - \frac{0.059}{n} \log K_f$$

$$- \frac{0.059}{n} \log \frac{k_c}{k_r} - \frac{0.059}{n} \log \frac{i}{i_d - i} \tag{22-9}$$

Again, $E_{\text{applied}} = E_{1/2}(c)$ when $i = i_d/2$. Thus,

$$E_{1/2}(c) = E^0 - E_{\text{SCE}} - \frac{0.059}{n} \log F_A^x - \frac{0.059}{n} \log K_f - \frac{0.059}{n} \log \frac{k_c}{k_r}$$

$$(22\text{-}10)$$

Notice that $E_{1/2}(c)$ is dependent upon both K_f and F_A, and becomes more negative as both increase.

Ordinarily the proportionality constant k for the simple ion is not greatly different from k_c for the complex; thus, subtraction of equation (22-5) from equation (22-10) reveals

$$E_{1/2}(c) - E_{1/2} \cong -\frac{0.059}{n} \log K_f - \frac{0.059}{n} \log F_A^x \qquad (22\text{-}11)$$

Equation (22-10) can be employed for determination of the formula of the complex. Thus, a plot of the half-wave potential against log F_A for several formal concentrations of the complexing agent, gives a straight line, the slope of which is $-0.059x/n$. If n is known, the combining ratio of ligand to metal ion is thus obtained. If necessary, n can be found from a plot of log $i/(i_d - i)$ against E_{applied} at a fixed formal concentration of A^- (equation 22-9).

Equation (22-11) can be employed to determine the formation constant of the complex once the value for x has been established.

Because of the effect of complexing reagents on half-wave potentials, the electrolyte content of the solution should be carefully controlled whenever polarographic data are used for the qualitative identification of the constituents of a solution.

Polarograms for irreversible reactions. Many polarographic electrode processes, particularly those associated with organic systems, are irreversible; as shown in Figure 22-7, drawn-out and less well-defined waves result. The quantitative description of such waves requires an additional term in equation (22-6) to account for the kinetics of the electrode reaction. Although half-wave potentials for irreversible reactions ordinarily show a dependence upon concentration, diffusion currents remain linearly related to this variable, and such processes are readily adapted to quantitative analysis.

The Dropping Mercury Electrode

Most polarographic work has been performed with a dropping mercury electrode; it is, therefore, of interest to consider some of the unique features of this device.

Current variations during the lifetime of a drop. The current passing through a cell containing a dropping electrode undergoes periodic fluctuations corresponding in frequency to the drop rate. As a drop breaks, the current falls to zero; it then increases rapidly as the electrode area grows because of the greater surface to which diffusion can occur. For convenience, a well-damped galvanometer is generally employed for current measurement. As shown in Figure 22-3 the oscillations under these circumstances are limited to a reasonable magnitude and the average current is readily determined provided the drop rate is reproducible. Note the effect of irregular drops in the center of the limiting current region; these were probably caused by vibration of the apparatus.

Advantages and limitations of the dropping mercury electrode. The dropping mercury electrode offers several advantages over other types of microelectrodes. The first arises from the large overvoltage for the formation of hydrogen from hydrogen ions. As a consequence, the reduction of many substances from acidic solutions can be studied without interference. Second, because a new metal surface is continuously generated, the behavior of the electrode is independent of its past history. Thus, reproducible current-voltage curves are obtained regardless of how the electrode has been used previously. The third unique feature of a dropping electrode is that reproducible average currents are immediately achieved at any given applied potential. In contrast, when a stationary solid microelectrode is set at a new potential, a large current is first observed which decays to a constant and reproducible value only after several minutes. As a consequence,

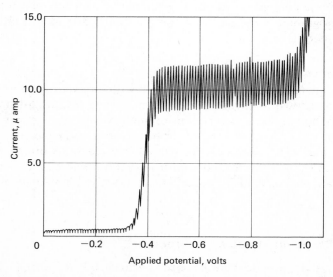

Figure 22-3. Typical polarogram produced by an instrument that continuously and automatically records current-voltage data.

a good deal more time is required to obtain a polarogram with this type of electrode system.

The large initial currents with a solid microelectrode are associated with the formation of a concentration gradient that finally extends several tenths of a millimeter from the electrode surface. Within this gradient the reactant concentration increases continuously from the equilibrium value at the electrode surface to that of the bulk of the solution. Constant currents that are applicable for analysis are obtained only when the slope of the relationship between concentration and distance from the electrode surface becomes essentially constant; ordinarily two to five minutes are required to reach this condition. As we have mentioned, current changes are also observed with a dropping electrode. Here, however, the solution becomes homogenized by local stirring as a drop breaks, and for each new drop conditions are identical to those encountered by the previous drop. Thus, the current fluctuation is perfectly reproducible and the average current is independent of when it is measured.

The most serious limitation to the dropping mercury electrode is the ease with which it is oxidized; this property severely restricts the use of mercury as an anode. At applied potentials much above $+ 0.4$ V (vs. the saturated calomel electrode), formation of mercury(I) occurs; the resulting current masks the polarographic waves of other oxidizable species in the solution. Thus, the dropping mercury electrode can be employed only for the analysis of reducible or very easily oxidizable substances.

Polarographic Diffusion Currents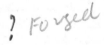

The Ilkovic equation. In 1934 D. Ilkovic[5] derived a fundamental equation relating the various parameters that determine the magnitude of the diffusion currents obtained with a dropping mercury electrode. He showed that at a temperature of 25°C

$$i_d = 607nD^{1/2} \ m^{2/3} \ t^{1/6} \ C \qquad (22\text{-}12)$$

Here, i_d is the time-average diffusion current (in microamperes) during the lifetime of a drop; n is the number of faradays of electricity per mole of reactant; D is the diffusion coefficient for the reactive species expressed in units of square centimeters per second; m is the rate of mercury flow in milligrams per second; t is the drop time in seconds; and C is the concentration of the reactant in millimoles per liter. The quantity 607 represents the combination of several constants.

The Ilkovic equation contains certain assumptions that cause discrepancies of a few percent between calculated and experimental diffusion currents. Corrections to the equation have been derived that yield better cor-

[5]D. Ilkovic, *Collection Czechoslov. Chem. Commum.*, **6**, 498 (1934).

respondence[6]; for most purposes, however, the simple equation gives a satisfactory accounting of the factors that influence the current.

Capillary characteristics. The product $m^{2/3} t^{1/6}$ in the Ilkovic equation, called the *capillary constant,* describes the influence of dropping-electrode characteristics upon the diffusion current; since both m and t are readily evaluated experimentally, comparison of diffusion currents from different capillaries is thus possible.

Two factors, other than the geometry of the capillary itself, play a part in determining the magnitude of the capillary constant. The height of the head that forces the mercury through the capillary influences both m and t such that the diffusion current is directly proportional to the square root of the mercury height. The drop time t of a given electrode is also affected by the applied potential, since the interfacial tension between the mercury and the solution varies with the charge on the drop. Generally t passes through a maximum at about -0.4 V (vs. the saturated calomel electrode) and then falls off fairly rapidly; at -2.0 V, t may be only half of its maximum value. Fortunately, however, the diffusion current varies only as the one-sixth power of the drop time so that, over small potential ranges, the decrease in current due to this variation is negligibly small.

Diffusion coefficient. The Ilkovic equation indicates that the diffusion current for a particular reactant varies as the square root of its diffusion coefficient D. This quantity is a measure of the rate at which the species diffuses through a unit concentration gradient and is dependent upon such factors as the size of the ion or molecule, its charge, if any, and the viscosity and composition of the solvent. The coefficient for a simple, hydrated metal ion is often different from that of its complexes; as a result, the diffusion current as well as the half-wave potential may be affected by the presence of a complexing reagent.

Temperature. Temperature affects several of the variables that govern the diffusion current for a given species, and its overall influence is thus complex. The most temperature-sensitive of the factors in the Ilkovic equation is the diffusion coefficient, which ordinarily can be expected to change by about 2.5 percent per degree. As a consequence, temperature control to a few tenths of a degree is needed for accurate polarographic analysis.

Polarograms for Mixtures of Reactants

Ordinarily, the reactants of a mixture will behave independently of one another at a microelectrode so that a polarogram for a mixture is simply the summation of the waves of the individual components. Figure 22-4 shows

[6]See J. J. Lingane and B. A. Loveridge, *J. Am. Chem. Soc.,* **72**, 438 (1950).

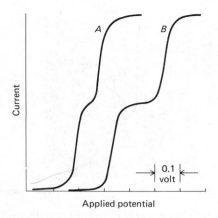

Figure 22-4. Polarograms for two-component mixtures. Half-wave potentials differ by 0.1 V in curve *A*, and by 0.2 V in curve *B*.

polarograms for a pair of two-component mixtures. The half-wave potentials of the two reactants differ by about 0.1 V in curve *A* and by about 0.2 V in curve *B*. Figure 22-4 suggests the possibility of using a single polarogram to analyze for several components in a mixture. Success depends upon having a sufficient separation in half-wave potentials to permit evaluation of the individual diffusion currents. Approximately 0.2 V must separate the half-wave potential for a two-electron reduction from that for a succeeding reaction; a minimum of between 0.2 and 0.3 V is needed if the first reduction is a one-electron process. The analysis of mixtures is considered further on page 574.

Anodic Waves and Mixed Anodic-cathodic Waves

Anodic waves as well as cathodic waves are encountered in polarography. The former are the less common because of the relatively small range of anodic potentials that can be covered with the dropping mercury electrode before oxidation of the electrode itself commences. An example of an anodic wave is illustrated in curve 1 of Figure 22-5, where the electrode reaction involves the oxidation of iron(II) to iron(III) in the presence of citrate ion. A diffusion current is obtained at zero volt (vs. the saturated calomel electrode), which is due to the half-reaction

$$Fe^{2+} \rightleftharpoons Fe^{3+} + e$$

As the potential is made more negative, a decrease in the anodic current occurs; at about -0.2 V the current becomes zero because the oxidation of iron(II) ion has ceased.

Curve 3 represents the polarogram for a solution of iron(III) in the same medium. Here, a cathodic wave results from the reduction of the iron(III) to

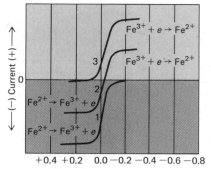

Applied emf vs. saturated calomel electrode, volts

Figure 22-5. Polarographic behavior of iron(II) and iron(III) in a citrate medium. (1) Anodic wave for a solution in which $[Fe^{2+}] = 1 \times 10^{-3}$ *F*. (2) Anodic-cathodic wave for a solution in which $[Fe^{2+}] = [Fe^{3+}] = 0.5 \times 10^{-3}$ *F*. (3) Cathodic wave for a solution in which $[Fe^{3+}] = 1 \times 10^{-3}$ *F*.

the divalent state. The half-wave potential is identical with that of the anodic wave, indicating that the oxidation and reduction of the two iron species are perfectly reversible at the dropping electrode.

Curve 2 is the polarogram of an equiformal mixture of iron(II) and iron(III). The portion of the curve below the zero-current line corresponds to the oxidation of the iron(II); this reaction ceases at an applied potential equal to the half-wave potential. The upper portion of the curve is due to the reduction of iron(III).

Current Maxima

The shapes of polarograms are frequently distorted by so-called current maxima (see Figure 22-6). These are troublesome because they interfere with the accurate evaluation of diffusion currents and half-wave potentials. Although the cause or causes of maxima are not fully understood, considerable empirical knowledge of methods for their elimination exists. These gen-

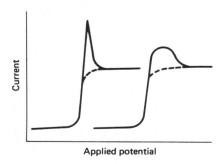

Applied potential

Figure 22-6. Typical current maxima.

erally involve the addition of traces of such high molecular-weight substances as gelatin, Triton X-100 (a commercial surface-active agent), methyl red, other dyes, or carpenter's glue. The first two of these additives are particularly useful.

Residual Current

The typical residual current curve, such as that shown in Figure 22-2, has two sources. One of these is the reduction of trace impurities that are almost inevitably present in the blank solution; contributors here include small amounts of dissolved oxygen, heavy-metal ions from the distilled water, and impurities present in the salt used as the supporting electrolyte. Generally, the concentrations of these contaminants are not great enough to produce a distinct wave; their presence does, however, affect the residual current. The purity of the salt employed as a supporting electrolyte is particularly important. For example, if the solution to be studied is made 1.0 *F* in potassium nitrate, as little as 0.001 percent of reducible impurity in that salt may contribute appreciably to the residual current.

A second component of the residual current is the so-called *charging,* or *condenser, current* arising from a flow of electrons that charges the mercury droplets with respect to the solution; this current may be either negative or positive. At potentials more negative than −0.4 V (vs. the saturated calomel electrode) an excess of electrons provides the surface of each droplet with a negative charge. These excess electrons are carried down with the drop as it breaks; since each new drop is charged as it forms, a small but steady current results. At applied potentials smaller than about −0.4 V the mercury tends to be positive with respect to the solution; thus, as each drop is formed, electrons are repelled from the surface toward the bulk of mercury and a negative current is the result. At about −0.4 V the mercury surface is uncharged and the condenser current is zero.

Ultimately, the accuracy and the sensitivity of the polarographic method depends upon the size of the residual current and the accuracy with which a correction for its effect can be determined. Two procedures for correction are employed. One requires a polarogram of a blank solution that is as nearly as possible identical to the solution being analyzed except for the component of interest; the diffusion current is then taken as the arithmetic difference between the current for the sample and the blank at an identical potential. In the second procedure the correction is determined from a linear extrapolation of the polarogram of the sample in the region short of the decomposition potential (see Figure 22-9). The first method is usually the more accurate.

Supporting Electrolyte

An electrode process is diffusion-controlled only if the solution contains a sufficient concentration of a supporting electrolyte. The data in Table

Table 22-2 Effect of Supporting Electrolyte Concentration on Polarographic Currents for Lead Ion[a,b]

Potassium Nitrate Concentration (F)	Limiting Current (μamp)
0	17.6
0.0001	16.2
0.001	12.0
0.005	9.8
0.10	8.45
1.00	8.45

[a]Data From J. J. Lingane and I. M. Kolthoff, *J. Am. Chem. Soc.*, **61**, 1045 (1939). With permission of the American Chemical Society.

[b]Solution: $9.5 \times 10^{-4}\ F$ in $PbCl_2$.

22-2 illustrate the effect of the supporting electrolyte concentration upon the limiting current. It is seen that the limiting current for lead ion decreases markedly with the addition of potassium nitrate, and becomes constant only in the presence of high concentrations of that salt. In solutions of low electrolyte concentration the portion of the limiting current that is due to electrostatic forces is sometimes called the *migration current;* for the first entry in Table 22-2 the migration current associated with a $9.5 \times 10^{-4}\ F$ solution of lead ion is about (17.6-8.45) 9.2 μamp.

It is of interest to note that the limiting current for the reduction of anions (such as iodate or chromate) becomes larger as the supporting electrolyte concentration increases, since the electrostatic forces are repulsive rather than attractive.

The migration current can quite generally be eliminated by the employment of a supporting electrolyte whose concentration exceeds that of the reactive species by a factor of 50 to 100. Under these circumstances the fraction of the current carried through the solution by the species of interest is negligibly small because of the large excess of other particles of the same charge. The limiting current is then a diffusion current which is independent of electrolyte concentration.

Oxygen Waves

Dissolved oxygen is readily reduced at the dropping mercury electrode; an aqueous solution saturated with air exhibits two distinct waves attributable to this element (see Figure 22-7). The first results from the reduction of oxygen to peroxide:

$$O_2 + 2H^+ + 2e \rightleftharpoons H_2O_2$$

The second corresponds to the further reduction of the hydrogen peroxide:

$$H_2O_2 + 2H^+ + 2e \rightleftharpoons 2H_2O$$

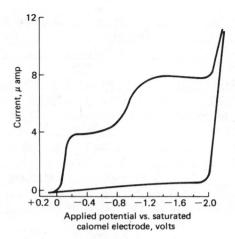

Figure 22-7. Polarogram for the reduction of oxygen in an air-saturated 0.1 *F* KCl solution. The lower curve is for 0.1 *F* KCl alone.

As would be expected from stoichiometric considerations, the two wave heights are equal.

While the polarographic waves for oxygen are convenient for the determination of the concentration of the dissolved gas in solutions, the presence of this element often interferes with the accurate determination of other species. Thus, oxygen removal is ordinarily the first step in polarographic analysis. Aeration of the solution for several minutes with an inert gas accomplishes this end; a stream of the same gas, usually nitrogen, is passed over the surface during the analysis to prevent reabsorption.

Kinetic and Catalytic Currents

There are instances where polarographic limiting currents are controlled not only by the diffusion rate of the reactive species but also by the rate of some *chemical reaction* related to the electrode process. These currents are no longer predictable by the Ilkovic equation and are abnormally influenced by temperature, capillary characteristics, and solution composition. They are known as *kinetic currents*. The polarographic reduction of formaldehyde is an example of this behavior. In aqueous solution two forms of this compound are in equilibrium:

$$CH_2(OH)_2 \rightleftharpoons HCHO + H_2O$$

The hydrated form predominates, but only the unhydrated form is reduced at a dropping electrode. Thus, when a suitable potential is applied, the concentration of the latter approaches zero at the electrode surface. This results in a shift in the equilibrium to the right and the formation of more unhydrated formaldehyde which then can react. Here, however, the rate of the equilibrium shift is slow; thus, the supply of available reactant is controlled by a reaction rate rather than by a diffusion rate. The net effect is a smaller limiting current than would be observed if the process were totally diffusion controlled.

A *catalytic current* represents another type of limiting current that depends upon the rate of a chemical reaction. Here, the reactive species is regenerated by a chemical reaction involving some other component of the solution. The reduction of iron(III) in the presence of hydrogen peroxide provides an example of this phenomenon. The magnitude of the limiting current for iron in the presence of this compound is greatly enhanced even when the applied potential is kept well below the value required to reduce the peroxide. This effect is readily explained by assuming that the following reaction occurs in the surface layer following the electrolytic formation of iron(II) ion:

$$2Fe^{2+} + H_2O_2 \rightarrow 2Fe^{3+} + 2OH^-$$

The current is then controlled in part by the rate of this reaction.

Both kinetic and catalytic currents can be used for analytical purposes. The latter are particularly useful for determining very small concentrations of certain species. Both are quite sensitive to the variables that influence the rates of chemical reactions.

APPLICATIONS OF POLAROGRAPHY

Apparatus

Cells.   A typical general-purpose cell for polarographic analysis is shown in Figure 22-1. Separation of the solution to be analyzed from the calomel electrode is accomplished by means of a sintered disk backed by an agar plug that has been rendered conducting by the addition of potassium chloride. Such a bridge is readily prepared and lasts for extended periods, provided a potassium chloride solution is kept in the reaction compartment when the cell is not in use. A capillary sidearm is provided to permit the passage of nitrogen or other gas through the solution. Provision is also made for blanketing the solution with the gas during the analysis.

In a simpler arrangement, a mercury pool in the bottom of the sample container can be used as the nonpolarizable electrode. Here, the observed half-wave potentials differ from published values, which are based upon a saturated calomel-reference electrode.

Dropping electrodes.   A dropping electrode, such as that shown in Figure 22-1, can be purchased from commercial sources.[7] A 10-cm length ordinarily

[7]For example: Sargent-Welch Scientific Co., Skokie, Ill.; Fisher Scientific Co., Pittsburgh, Pa.

has a drop time of 3 to 6 sec under a mercury head of about 50 cm. The tip of the capillary should be as nearly square as possible, and care should be taken to assure a vertical mounting of the electrode; otherwise erratic and nonreproducible drop times and sizes will be observed.

With reasonable care, a capillary can be used for several months or even years. To ensure such performance, scrupulously clean mercury must be used, and a mercury head, no matter how slight, must be maintained at all times. If the head is ever diminished to the point where solution comes in contact with the inner surface of the tip, malfunction of the electrode is to be expected. For this reason, the head of mercury should always be increased to provide a good flow before the tip is immersed in a solution.

Storage of an electrode always presents a problem. One method is to rinse the electrode thoroughly with water, to dry it, and then carefully to reduce the head until the flow of mercury in air just ceases. Care must be taken to avoid lowering the mercury too far. Before use, the head is increased, the tip in immersed in 1:1 nitric acid for a minute or so, and then washed with distilled water.

Electrical apparatus. To make polarographic measurements it is necessary to have the means for applying a voltage that can be varied continuously over the range from 0 to −2.5 V; the applied potential should usually be known to about 0.01 V. In addition, it must be possible to measure the cell currents over the range between 0.01 and perhaps 100 μamp with an accuracy of about 0.01 μamp. A manual instrument that meets these requirements is easily constructed from equipment available in most laboratories. More elaborate devices for recording polarograms automatically are commercially available.

Figure 22-8 shows a circuit diagram of a simple instrument for polarographic work. Two 1.5-V batteries provide a voltage across the 100-ohm potential divider R_1, by means of which the potential applied to the cell can be varied. The magnitude of this voltage can be determined by means of the potentiometer with the double-pole, double-throw switch in position 2. The current is measured by determining the potential drop across the precision 10,000-ohm resistance R_2 with the same potentiometer and the switch in position 1. The current oscillations associated with the dropping electrode require that the null-detecting galvanometer be damped with a suitable resistance.

The apparatus shown in Figure 22-8 provides data that are as good as or better than the most sophisticated and expensive polarographic equipment available on the market. It is, furthermore, just as convenient and rapid for routine quantitative work, since in such applications current measurements at only two voltages are needed to define i_d (one preceding the decomposition potential and one in the limiting current region). On the other hand, where the entire polarogram is required, the point-by-point determination of the curve with a manual instrument is tedious and time-consuming; for this type of work automatic recording is of great value.

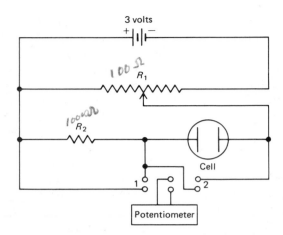

Figure 22-8. A simple circuit for polarographic measurements. From J. J. Lingane, *Anal. Chem.,* **21**, 45 (1949). (With permission of the American Chemical Society.)

Treatment of Data

In measuring currents obtained with the dropping electrode it is common practice to use the *average* value of the galvanometer or recorder oscillations rather than the maximum or minimum values; the measurement is thus less dependent upon the damping employed.

Determination of diffusion currents. For analytical work, limiting currents must always be corrected for the residual current. A residual-current curve is experimentally determined along with the curve for the sample. The difference between the two can be then taken at some potential in the limiting-current region. Because the residual current usually increases nearly linearly with applied voltage, it is often possible to correct for the residual-current curve by extrapolation; this technique is illustrated in Figure 22-9.

Analysis of mixtures. The quantitative determination of each species in a multicomponent mixture from a single polarogram is theoretically feasible, provided the half-wave potentials of the various species are sufficiently different (see Figure 22-4). Where a major constituent is more easily reduced than a minor one, however, the accuracy with which the latter can be determined may be poor because its diffusion current can occupy only a small fraction of the current scale of the instrument. Under such circumstances a small error in current measurement leads to a large relative error in the analysis. This problem does not arise when the minor constituent is the more easily reduced component. Here, its diffusion current can be determined at high current sensitivity; then the sensitivity can be decreased to allow measurement of the major species.

Several ways exist for treating mixtures containing species in unfavor-

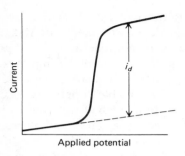

Applied potential

Figure 22-9. Determination of the diffusion current by extrapolation of the residual current.

able concentration ratios. The best method is to cause the wave of the minor component to appear first through alteration of the supporting electrolyte; with the variety of complexing agents available, this technique is often feasible. Alternatively, a preliminary chemical separation can be employed. Finally, there is the so-called *compensation technique*. Here, the current due to the major constituent is lowered to zero (or a very small value) by application of a counter emf in the current-measuring circuit. The current sensitivity can then be increased to give a satisfactory signal for the reduction of the minor component. Most modern polarographs are equipped with such a compensating device.

Concentration determination. The best and most straightforward method for quantitative polarographic analysis involves preliminary calibration with a series of standard solutions; as nearly as possible, these standards should be identical with the samples being analyzed and should cover a concentration range within which the unknown samples will likely fall. The linearity of the current-concentration relationship can be assessed from such data; if the relationship is nonlinear, the analysis can be based upon the calibration curve.

Another useful technique is the standard addition method. The polarogram of an exactly known volume of the sample solution is obtained. Then a carefully measured volume of a standard solution of the substance of interest is added and the polarogram is again obtained. From the increase in wave height and the quantity of standard added, the concentration of the original solution can be calculated; the analyst must assume here that the concentration-current relationship is linear. This procedure is particularly effective when the diffusion current is sensitive to other components of the solution that are introduced with the sample.

Inorganic Polarographic Analysis

The polarographic method is generally applicable to the analysis of inorganic substances. Most metallic cations, for example, are reduced at the dropping electrode to form a metal amalgam or a lower oxidation state. Even the alkali metals and alkaline-earth metals are reducible, provided the supporting electrolyte used does not react at the high potentials required; here the tetraalkyl ammonium halides are often employed.

The successful polarographic analysis of cations frequently depends upon the employment of a suitable supporting electrolyte. To aid in this selection, tabular compilations of half-wave potential data should be consulted.[8] The judicious choice of anion often enhances the selectivity of the method. For example, with potassium chloride as a supporting electrolyte, the waves for iron(III) and copper interfere with one another; in a fluoride medium, however, the half-wave potential of the former is shifted by about -0.5 V, while that for the latter is altered by only a few hundredths of a volt. The presence of fluoride thus results in the appearance of separate waves for the two ions.

The polarographic method is also applicable to the analysis of such inorganic anions as bromate, iodate, dichromate, vanadate, selenite, and nitrite. In general, polarograms for these substances are affected by the pH of the solution because the hydrogen ion is a participant in the reduction process. As a consequence, strong buffering of the solutions to some fixed pH is necessary to obtain reproducible data.

Certain inorganic anions that form complexes or precipitates with the ions of mercury are responsible for anodic waves that occur in the region of zero volt (vs. the saturated calomel electrode). Here, the electrode reaction involves oxidation of the electrode; for example,

$$2Hg + 2Cl^- \rightleftharpoons Hg_2Cl_2 + 2e$$
$$Hg + 2S_2O_3^{2-} \rightleftharpoons Hg(S_2O_3)_2^{2-} + 2e$$

Bromide, iodide, thiocyanate, and cyanide act similarly. The magnitude of such diffusion currents is controlled by the rate of diffusion of the anions to the electrode surface; as a consequence, a linear relationship between current and concentration is observed.

The polarographic method can be used for the analysis of a few inorganic substances that exist as uncharged molecules in the solvent used. The determination of oxygen in gases, biological fluids, and water is a most important example of this application. Other neutral substances that react at the dropping electrode include hydrogen peroxide, hydrazine, cyanogen, elemental sulfur, and sulfur dioxide.

For a discussion of further applications of polarography to inorganic analysis the reader is referred to the monographs by Kolthoff and Lingane and by Meites.[9]

---

[8] See, for example, the references cited on page 553; another extensive source is L. Meites, Ed. *Handbook of Analytical Chemistry* (New York: McGraw-Hill Book Company, Inc., 1963).

[9] I. M. Kolthoff and J. J. Lingane, *Polarography*, 2d ed., Vol. 2, (New York: Interscience Publishers, Inc., 1952); L. Meites, *Polarographic Techniques*, (New York: Interscience Publishers, Inc., 1965).

Organic Polarographic Analysis

Almost from its inception the polarographic method has been used for the study and analysis of organic compounds, and a large number of papers have been devoted to this subject. Several common functional groups are oxidized or reduced at the dropping electrode, and compounds containing these groups are thus subject to analysis by the polarographic technique.

In general, the reactions of organic compounds at a microelectrode are slower and more complex than those of inorganic cations. Thus, theoretical interpretation of the polarographic data is more difficult or even impossible; furthermore, a much stricter adherence to detail is required for quantitative work. Despite these handicaps, organic polarography has proved fruitful for the determination of structure, for the qualitative identification of compounds, and for the quantitative analysis of mixtures.

Effect of pH on polarograms. Organic electrode processes ordinarily involve hydrogen ions, the most common reaction being represented as

$$R + nH^+ + ne = RH_n$$

where R and RH_n are the oxidized and reduced forms of the organic molecule. Half-wave potentials for organic compounds are therefore markedly pH-dependent. Furthermore, alteration of the pH often results in a change in the reaction products. For example, when benzaldehyde is reduced in a basic solution, a wave is obtained at about -1.4 V, attributable to the formation of benzyl alcohol

$$C_6H_5CHO + 2H^+ + 2e \rightleftharpoons C_6H_5CH_2OH$$

If the pH is less than 2, however, a wave occurs at about -1.0 V that is just half the size of the foregoing one; here, the reaction consists of the production of hydrobenzoin

$$2C_6H_5CHO + 2H^+ + 2e \rightleftharpoons C_6H_5CHOHCHOHC_6H_5$$

At intermediate pH values two waves are observed, indicating the occurrence of both reactions.

It should be emphasized that an electrode process consuming or producing hydrogen ions tends to alter the pH of the solution *at the electrode surface*; unless the solution is well-buffered, marked changes in pH can occur in the surface film as the electrolysis proceeds. These changes affect the reduction potential of the reaction and lead to drawn-out and poorly defined waves. Moreover, where the electrode process is altered by the pH, nonlinearity in the diffusion current-concentration relationship must also be expected. Thus, in organic polarography good buffering is vital for the generation of reproducible half-wave potentials and diffusion currents.

Solvents for organic polarography. In organic polarography solubility considerations frequently demand the use of some solvent other than pure water; aqueous mixtures containing varying amounts of such miscible solvents as glycols, dioxane, alcohols, Cellosolve, or glacial acetic acid have been employed. Anhydrous media of acetic acid, formamide, and ethylene glycol have also been investigated. Supporting electrolytes are often lithium salts or tetraalkyl ammonium salts.

Irreversibility of electrode reactions. Few organic electrode reactions are reversible; as a consequence, equation (22-6) does not adequately describe the polarographic waves of organic compounds. Because non-reversibility results in drawn-out waves, greater differences in half-wave potentials are required to permit discrimination between substances undergoing consecutive reduction.

In general, the Ilkovic equation applies to the diffusion currents for electrode reactions that are nonreversible. Thus, the quantitative aspects of organic and inorganic polarography are similar.

Reactive functional groups. Organic compounds containing any of the following functional groups can be expected to react at the dropping mercury electrode and thus produce one or more polarographic waves.

1. *The carbonyl group,* including aldehydes, ketones, and quinones, produce polarographic waves. In general, aldehydes are reduced at lower potentials than ketones; conjugation of the carbonyl double bond also leads to lower half-wave potentials.

2. *Certain carboxylic acids* are reduced polarographically, although simple aliphatic and aromatic monocarboxylic acids are not. Dicarboxylic acids such as fumaric, maleic, or phthalic acid, in which the carboxyl groups are conjugated with one another, give characteristic polarograms; the same is true of certain keto and aldehydo acids.

3. *Most peroxides and epoxides* yield polarograms.

4. *Nitro, nitroso, amine oxide, and azo groups* are generally reduced at the dropping electrode.

5. *Most organic halogen groups* produce a polarographic wave as a result of replacement of the halogen group with an atom of hydrogen.

6. *The carbon–carbon double bond* is reduced when it is conjugated with another double bond, an aromatic ring, or an unsaturated group.

7. *Hydroquinones and mercaptans* produce anodic waves.

In addition, a number of other organic groups cause catalytic hydrogen waves that can be used for analysis. These include amines, mercaptans, acids, and heterocyclic nitrogen compounds. Numerous applications to biological systems have been reported.[10]

---

[10]M. Brezina and P. Zuman, *Polarography in Medicine, Biochemistry and Pharmacy* (New York: Interscience Publishers, Inc., 1958); P. Zuman, *Organic Polarographic Analysis* (Oxford: Pergamon Press, Ltd., 1964).

AMPEROMETRIC TITRATIONS

The polarographic method can be employed for the estimation of the equivalence point, provided at least one of the participants or products of the titration is oxidized or reduced at a microelectrode. Here the current passing through a polarographic cell at some fixed potential is measured as a function of the volume of reagent (or of time if the reagent is generated by a constant-current coulometric process). Plots of the data on either side of the equivalence point are straight lines with differing slopes so that the end point can be fixed by extrapolation to their intersection.

The amperometric method is inherently more accurate than the polarographic method and is less dependent upon the characteristics of the capillary and the supporting electrolyte. Furthermore, the temperature need not be fixed accurately, although it must be kept constant during the titration. Finally, the substance being determined need not be reactive at the electrode; a reactive reagent or product is equally satisfactory.

Amperometric and conductometric titrations are similar in the respect that the data for each are collected well away from the equivalence point. Therefore, reactions that are relatively incomplete can be employed. In contrast to the conductometric procedure, however, the presence of a high-electrolyte concentration in no way affects the accuracy of amperometric titrations, provided, of course, the electrolyte does not react at the microelectrode. Very dilute solutions can be titrated by the amperometric method because the indicator electrode is sensitive to minute quantities of the reactants.

Titration Curves

Amperometric titration curves typically take one of the forms shown in Figure 22-10. Figure 22-10(a) represents a titration in which the substance being analyzed reacts at the electrode while the reagent does not. The

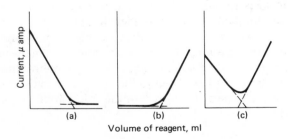

Figure 22-10. Typical amperometric titration curves. (a) Substance analyzed is reduced, reagent is not. (b) Reagent is reduced, substance analyzed is not. (c) Both reagent and substance analyzed are reduced.

titration of lead with sulfate or oxalate ions may be cited as an example. Here, a sufficiently high potential is applied to give a diffusion current for lead; a linear decrease in current is observed as lead ions are removed from the solution by precipitation. The curvature near the equivalence point reflects the incompleteness of the analytical reaction in this region. The end point is obtained by extrapolation of the linear portions, as shown.

The curve of Figure 22-10(b) is typical of a titration in which the reagent reacts at the microelectrode and the substance being analyzed does not. An example of this would be the titration of magnesium with 8-hydroxyquinoline. A diffusion current for the latter is obtained at -1.6 V (vs. the saturated calomel electrode), whereas magnesium ion is inert at this potential.

The curve of Figure 22-10(c) corresponds to the titration of lead ion with a chromate solution at an applied potential greater than -1.0 V. Both lead and chromate ions give diffusion currents, and a minimum in the curve signals the end point. This system yields a curve resembling Figure 22-10(b) with zero applied potential, since only chromate ions are reduced under these conditions.

Apparatus and Techniques

With relatively simple apparatus, accurate results can be obtained by the amperometric method.

Cells. Figure 22-11 shows a typical cell for an amperometric titration. A calomel half-cell is usually employed as the nonpolarizable electrode; the indicator electrode may be a dropping mercury electrode or a microwire electrode, as shown. The cell should have a capacity of 75 to 100 ml.

Volume measurements. In order to obtain linear plots for fixing the end point of the titration, it is necessary to correct for volume changes due to the added titrant. The diffusion current can be multiplied by $(V + v)/V$, where V is the original volume and v is the volume of reagent; thus, all measured currents are corrected back to the original volume. An alternative, which is often satisfactory, is to use a reagent that is 20 or more times as concentrated as the solution being titrated. Under these circumstances v is so small with respect to V that the correction is negligible. This approach, however, does require the use of a microburet so that a total reagent volume of 1 or 2 ml can be measured with a suitable accuracy. The microburet should be so arranged that its tip can be touched to the surface of the solution after each addition of reagent, to permit removal of the fraction of a drop that tends to remain attached.

Electrical measurements. For amperometric titrations a simple manual polarograph is entirely adequate. An appropriate voltage is applied to the cell by means of the linear potential divider; the current is measured by a damped galvanometer (which need not be calibrated) or a low-resistance microammeter. Ordinarily, the applied voltage does not have to be known any closer than about ±0.05 V, since it is only necessary to select a potential within the diffusion-current region of at least one of the participants in the titration.

Microelectrodes; the rotating platinum electrode. Many amperometric titrations can be carried out conveniently with a dropping mercury electrode. For reactions involving oxidizing agents that attack mercury (bromine, silver ion, iron(III), among others), a rotating platinum electrode is preferable. This microelectrode consists of a short length of platinum wire sealed into the side of a glass tube; mercury inside the tube provides electrical contact between the wire and the lead to the polarograph. The tube is held in the hollow chuck of a synchronous motor and is rotated at a constant speed in excess of 600 rpm. Commercial models of the rotating electrode are available. A typical apparatus is shown in Figure 22-11.

Polarographic waves, similar in appearance to those observed with the dropping electrode, can be obtained with the rotating platinum electrode. Here, however, the reactive species is brought to the electrode surface not only by diffusion but also by mechanical mixing. As a consequence, the limiting currents are as much as 20 times larger than those obtained with a microelectrode that is supplied by diffusion only. With a rotating electrode,

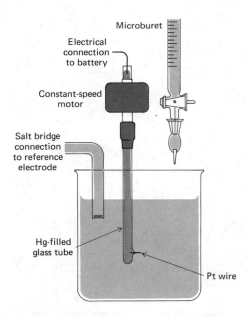

Figure 22-11. Typical cell arrangement for amperometric titrations with a rotating platinum electrode.

steady currents are instantaneously obtained. This behavior is in distinct contrast to the behavior of a solid microelectrode without stirring (p. 565).

Several limitations restrict the widespread application of the rotating platinum electrode to polarography. The low hydrogen overvoltage prevents its use as a cathode in acidic solutions. In addition, the high currents obtained with the electrode make it particularly sensitive to traces of oxygen in the solution. These two factors have largely confined its employment to anodic reactions. Limiting currents from a rotating electrode are often influenced by the previous history of the electrode and are seldom as reproducible as the diffusion currents obtained with a dropping electrode. These limitations, however, do not seriously restrict the use of the rotating electrode for amperometric titrations.

Application of Amperometric Titrations

The amperometric end point has been largely confined to titrations in which a slightly soluble precipitate is the reaction product. Selected applications are listed in Table 22-3. A variety of organic precipitants that are reducible at the dropping electrode appear in the table.

A few applications of the amperometric method to oxidation-reduction reactions can be found. For example, the technique has been applied to various titrations involving iodine and bromine (in the form of bromate) as reagents.

Table 22-3 Some Precipitation Titrations Employing the Amperometric End Point[a]

Electrode	Reagent	Substance Determined
Dropping mercury	K_2CrO_4	Pb^{2+}, Ba^{2+}
	$Pb(NO_3)_2$	SO_4^{2-}, MoO_4^{2-}, F^-, Cl^-
	8-Hydroxyquinoline	Mg^{2+}, Zn^{2+}, Cu^{2+}, Cd^{2+}, Al^{3+}, Bi^{3+}, Fe^{3+}
	Cupferron	Cu^{2+}, Fe^{3+}
	Dimethylglyoxime	Ni^{2+}
	α-Nitroso-β-napthol	Co^{2+}, Cu^{2+}, Pd^{2+}
	$K_4Fe(CN)_6$	Zn^{2+}
Rotating platinum	$AgNO_3$	Cl^-, Br^-, I^-, CN^-, RSH

[a]From *Fundamentals of Analytical Chemistry*, Second Edition, by Douglas A. Skoog and Donald M. West. Copyright © 1963 by Holt, Rinehart and Winston, Inc. Reprinted by permission of Holt, Rinehart and Winston, Inc.

AMPEROMETRIC TITRATIONS WITH TWO POLARIZED MICROELECTRODES

A convenient modification of the amperometric method involves the use of two stationary microelectrodes immersed in a well-stirred solution of the sample. A small potential (say, 0.1 to 0.2 V) is applied between these electrodes and such current that flows is followed as a function of the volume of added reagent. The end point is marked by a sudden current rise from zero, a decrease in the current to zero, or a minimum (at zero) in a V-shaped curve.

Although the use of two polarized electrodes for end-point detection was first proposed before 1900, almost 30 years passed before chemists came to appreciate the potentialities of the method.[11] The name *dead-stop end point* was used to describe the technique, and this term is still occasionally used. It was not until about 1950 that a clear interpretation of dead-stop titration curves was made.[12]

Oxidation-reduction Titrations

Twin-polarized platinum microelectrodes are conveniently used for end-point detection for oxidation-reduction titrations. Figure 22-12 illustrates three types of titration curves that are commonly encountered. The curve in Figure 22-12(a) is observed when both reactant systems behave reversibly with respect to the electrodes. The curves in Figure 22-12(b) and 22-12(c) are obtained when only one of the reactants exhibits reversible behavior.

Titration curves when both systems are reversible. A typical reversible system is the reaction between iron(II) and cerium(IV) in a sulfuric acid medium. Figure 22-12(a) illustrates the titration curve for this system when twin-platinum electrodes, maintained at a potential difference of 0.2 V, are employed. In order to analyze this curve it is first helpful to consider the *current-voltage* curves shown in Figure 22-13; each of the five plots corresponds to the solution at a different stage in the titration. Note that the data refer to a cell consisting of just *one of the microelectrodes coupled with a standard hydrogen reference electrode*.

Figure 22-13(a) is a current-voltage curve for the original solution that contains only iron(II) ions and sulfuric acid. An anodic wave at about 0.6 V is observed which results from oxidation of iron(II) to iron(III) at the plat-

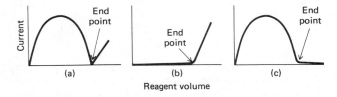

Figure 22-12. End points for amperometric oxidation-reduction titrations with twin polarized electrodes. Curve (a): both reactants behave reversibly at the electrode. Curve (b): only reagent behaves reversibly. Curve (c): only species titrated behaves reversibly.

[11]C. W. Foulk and A. T. Bawden, *J. Am. Chem. Soc.,* **48**, 2045 (1926).

[12]For an excellent analysis of this type of end point see J. J. Lingane, *Electroanalytical Chemistry,* 2d ed. (New York: Interscience Publishers, Inc., 1958), pp. 280-294.

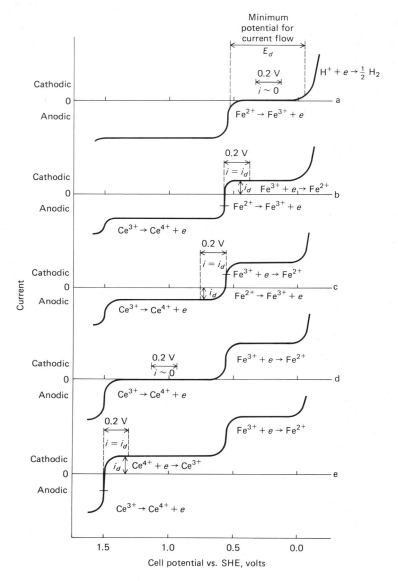

Figure 22-13. Current-voltage curves for a single platinum microelectrode vs. the standard hydrogen electrode. Curve a, original solution before addition of cerium(IV). Curve b, solution after one-third of the iron(II) is oxidized. Curve c, solution after two-thirds of the iron(II) is oxidized. Curve d, solution at the equivalence point. Curve e, solution containing excess of cerium(IV) reagent. Adapted from J. J. Lingane, *Electroanalytical Chemistry,* 2d ed. (New York: Interscience Publishers, Inc., 1958), p. 288. (With permission.)

584

inum electrode. The rapid increase in the cathodic current at about -0.1 V is due to the reduction of hydrogen ions at this same electrode. From this curve we can now predict the conditions required to cause a current flow in a cell consisting of *two platinum electrodes*. Here, the potential applied must be sufficient to provide both an anodic and a cathodic current. As indicated in the figure a minimum potential of about 0.7 V is necessary to cause both oxidation of iron(II) at the anode and reduction of iron(III) at the cathode. The potential specified for the titration was 0.2 V, however; thus, no current flow can occur.

Figure 22-13(b) corresponds to the system after sufficient cerium(IV) has been added to oxidize one-third of the iron(II) to iron(III). Now, both an anodic and a cathodic current are observed in the region of 0.6 V (vs. the SHE), the former resulting from the oxidation of iron(II) and the latter from the reduction of iron(III) produced by the addition of reagent. In addition, an anodic wave for the oxidation of cerium(III) appears at about 1.5 V. In contrast to the previous situation, application of a potential of 0.2 V across a pair of platinum electrodes should produce a current since the presence of both iron(II) and iron(III) ions permit an anodic and a cathodic current within this voltage span. The magnitude of the current is determined by the limiting current i_d associated with the reduction of iron(III) ions which are present in the lesser amount. That is, the cathode is the more polarized of the two electrodes; as a consequence, it serves as the indicator electrode at this point in the titration.

As shown by Figure 22-13(c), the situation becomes reversed when two-thirds of the iron(II) has been oxidized by the cerium(IV). Here, the current flow i_d between platinum electrodes at an applied potential of 0.2 V is limited by the iron(II) concentration because it is now smaller than that of iron(III). Thus, the anode is now the indicator electrode. At some stage of the titration between 22-13(b) and 22-13(c), the limiting currents for the oxidation and reduction processes are equal, and at this point the curve that expresses current as a function of volume of reagent exhibits a maximum [see Figure 22-12(a)].

Figure 22-13(d) is the current-voltage curve corresponding to the equivalence point in the titration. Here the iron(II) and cerium(IV) concentrations are negligible, and amperometric waves for only the reduction of iron(III) and the oxidation of cerium(III) are found. In order to produce a current between twin microelectrodes a potential of nearly a volt would be needed. At 0.2 V polarization is complete and no current flows.

Figure 22-13(e) refers to the system after a small excess of cerium(IV) has been added. Here, both an anodic and a cathodic current occur in the region of 1.5 V (vs. the SHE), since cerium(IV) ions are now present to carry the cathodic current. With twin platinum electrodes at an applied potential of 0.2 V, a current i_d limited by the cerium(IV) concentration is observed. Thus, beyond the equivalence point the titration curve in Figure 22-12(a) shows a linear increase in current with added cerium(IV) until such a time as the cerium(IV) concentration exceeds the cerium(III). The current is then limited by the cerium(III) concentration and becomes independent of the amount of reagent added (neglecting dilution effects).

Titration curves when only one system is reversible. Let us now consider the behavior of a twin-electrode system in the presence of species that are oxidized or reduced only slowly at an electrode (that is, a nonreversible system). An example is the arsenic(III)-arsenic(V) couple. Possible electrode reactions are

$$\text{cathode} \qquad H_3AsO_4 + 2H^+ + 2e \rightarrow H_3AsO_3 + H_2O$$
$$\text{anode} \qquad H_3AsO_3 + H_2O \rightarrow H_3AsO_4 + 2H^+ + 2e$$

These processes are slow at a platinum surface and occur only by applying an overpotential of several tenths of a volt. At an applied potential of 0.1 V little reaction occurs because of this kinetic polarization, and essentially no current is observed.

Upon addition of iodine to a solution of arsenious acid, a mixture of the two arsenic species and iodide ion results. While iodide can serve as an anode reactant, no cathode reactant is available because of the slow rate at which arsenic acid is reduced at a platinum surface; thus, at an applied potential of 0.1 V no current passes [see Figure 22-12(b)]. Beyond the equivalence point, depolarization of the cell can occur at 0.1 V by virtue of the reactions

$$\text{cathode} \qquad I_2 + 2e \rightarrow 2I^-$$
$$\text{anode} \qquad 2I^- \rightarrow I_2 + 2e$$

Here, the current is dependent upon the concentration of iodine as long as it remains in excess.

The curve in Figure 22-12(c) shows the titration curve of a dilute solution of iodine with the thiosulfate ion. In the initial stages, iodine and iodide are present, and both react reversibly at the electrodes; the current is dependent upon the concentration of the species present in lesser amount. Thiosulfate does not behave reversibly at the electrodes and therefore the current remains at zero beyond the equivalence point.

Precipitation Titrations

Twin silver microelectrodes have been employed for end-point detection in titrations involving the silver ion as a participant. For example, in the titration of the silver ion with a standard solution of the chloride ion, currents proportional to the metal ion concentration would result from the reactions

$$\text{cathode} \qquad Ag^+ + e \rightarrow Ag$$
$$\text{anode} \qquad Ag \rightarrow Ag^+ + e$$

With the effective removal of silver ion by the analytical reaction, cathodic polarization would occur and the current would approach zero at the end point.

Applications

Amperometric methods with two microelectrodes have not been fully exploited, and there are undoubtedly a number of oxidation-reduction systems to which this end-point technique could be applied with advantage. Most of the published data have been concerned with titrations involving iodine, although a few applications with reagents such as bromine, titanium(III), and cerium(IV) have been reported. An important use is in the titration of water with the Karl Fischer reagent. The technique has also found a number of applications for detecting end points in coulometric titrations.

The principal advantage of the twin microelectrode procedure is its simplicity. One can dispense with a reference electrode; the only instrumentation needed is a simple voltage divider, powered by a dry cell, and a galvanometer or microammeter for current detection.

METHODS RELATED TO CLASSICAL POLAROGRAPHY

In addition to amperometric titrations, several other methods that are closely related to classical polarography have been developed. None of these has had as widespread use, however, either for reasons of instrumental complexity or because of inherent limitations in scope or accuracy. On the other hand, these methods can provide useful information under circumstances where standard polarographic techniques are inapplicable.

Oscillographic (Rapid-scan) Polarography

In oscillographic polarography the applied potential is swept over a range of perhaps 0.5 V during the lifetime (or part of the lifetime) of a single mercury drop. An oscilloscope is required to display the current-voltage relationship; for a permanent record the image on the oscilloscope screen may be photographed.

The most successful oscillographic procedure appears to be one in which the voltage scan occurs over the final two to three seconds of a drop that has a total lifetime of six to seven seconds. During this period the surface area of the drop increases only slightly; as a consequence the charging or residual current, which depends directly upon the change in electrode area (p. 569), is small when compared with the residual currents found in ordinary polarography.

As will be seen from Figure 22-14, an oscillographic polarogram is quite different from the conventional wave form, the current increasing to a maximum (a *summit*) and then decaying to an equilibrium value. To account for this different behavior, let us consider the events that must occur in the surface layer around an electrode as the potential is quickly shifted from zero to some value beyond the half-wave potential for a reactive species. If the electron transfer process is rapid, a relatively large current flows momentarily; in this way the concentration of the reactive substance is adjusted to that demanded by the Nernst expression. Once equilibrium is established, however, the current decreases to a level just sufficient to counteract the effect of diffusion. This large momentary current is not observed in conventional polarography because the time-scale of the measurement is long relative to the lifetime of the momentary current. In rapid-scan polarography, on the other hand, an essentially instantaneous current-voltage relationship is followed; thus, the summit current consists not only of a diffusion current but also of the much larger current required for initial depletion of the surface layer. By the end of the scan, however, the initial current has largely decayed, and the true diffusion current is approached.

It can be shown that the *summit potential E_s* for a reversible process is related to the half-wave potential as follows:

$$E_s = E_{1/2} - 1.1 \frac{RT}{nF} \qquad (22\text{-}13)$$

The current i_s at the summit is given by

$$i_s = 2.72 \times 10^5 \, n^{3/2} \, A \, D^{1/2} \, C \, \nu^{1/2} \qquad (22\text{-}14)$$

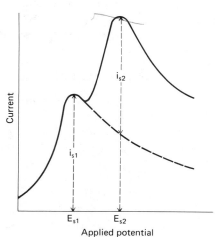

Figure 22-14. Oscillographic polarogram of a two-component mixture. Taken from H. Schmidt and M. von Stackelberg, *Modern Polarographic Methods* (New York: Academic Press, 1963), p. 27. (With permission.)

where A is the area of the electrode in cm^2, ν is the rate of scan in volts per second, and the remaining terms are the same as in the Ilkovic equation (p. 565). Generally, a summit current exceeds a diffusion current by a factor of ten or more.

Of the several advantages that can be cited for oscillographic polarography, perhaps the most important is the enhanced sensitivity. Not only are summit currents larger than diffusion currents, but as we have mentioned, the residual currents are smaller; therefore, the signal-to-background ratio is considerably more favorable than in normal polarography. Quantitative analysis thus becomes possible for solutions as dilute as 10^{-6} to 10^{-7} M. In addition, the shapes of oscillographic polarograms are such that accurate current measurements are possible even when the half-wave potentials of two species differ by as little as 0.1 V.

Not surprisingly, oscillographic equipment is more complex and more expensive than that required for ordinary polarography. A commercial instrument for this purpose is available.

Alternating Current Polarography

In this modification, a constant ac potential (sine wave) of a few millivolts is superimposed upon the dc potential normally employed in polarography. As in the regular procedure, the dc potential is varied over its usual range; here, however, the *ac current is measured.*

Consider the application of this technique to a solution containing a single, easily reduced species Ox which, at a suitable potential, reacts rapidly and reversibly as follows:

$$\text{Ox} + ne \rightleftharpoons \text{Red}$$

Assume also that an excess of supporting electrolyte is present.

We shall first focus our attention on the passage of the *faradaic* ac current through this cell; that is, the current that is carried across the interface by an oxidation-reduction process. In order for such a current to exist it is clearly necessary to have both an oxidizable and a reducible species at the microelectrode surface. Significant concentrations of Red occur, however, only when the *applied dc potential* corresponds to the steeply rising portion of the dc polarogram. Thus, before the decomposition potential, no faradaic ac current will flow. Beyond the decomposition potential but short of the half-wave potential, the following relationship exists between the *surface* concentration of Ox and Red

$$0 < [\text{Red}]_0 < [\text{Ox}]_0$$

and the magnitude of the ac current will be determined by $[\text{Red}]_0$. At applied dc potentials slightly greater than the half-wave potential

$$[\text{Red}]_0 > [\text{Ox}]_0 > 0$$

oxidizable species is reduced

and here the magnitude of the ac current is controlled by $[Ox]_0$. At the half-wave potential

$$[Red]_0 = [Ox]_0$$

and the faradaic current will have reached its maximum value. When a potential corresponding to the diffusion current region is reached, $[Ox]_0 \rightarrow 0$ and the ac current will, for lack of a reducible species, again approach zero. Thus, a typical ac polarogram consists of one or more peaks, the maxima occurring at the half-wave potentials for each of the reactive species present. The relationship between ac and dc polarograms is shown in Figure 22-15.

As noted in Chapter 19, ac currents are also carried by a nonfaradaic process that involves charging of the double layer surrounding an electrode. Thus, ac polarographic peaks are superimposed upon a base line of nonfaradaic current. The double layer structure is, however, affected by the dc component; as a consequence, the base line of an ac polarogram is not horizontal. Fortunately, over a small potential range the nonfaradaic current is nearly linear, and a satisfactory correction can be made by linear extrapolation between the base lines on either side of a peak.

The height of an ac polarographic peak is proportional to bulk concentration and, for a reversible reaction, to the square root of frequency. For reversible electrode reactions, ac polarography is more sensitive than the dc procedure and permits analysis of solutions as dilute as $10^{-6}\ M$.

For a totally nonreversible couple (that is, where the rate of the half-reaction is very slow in òne direction), no ac peak results since the product

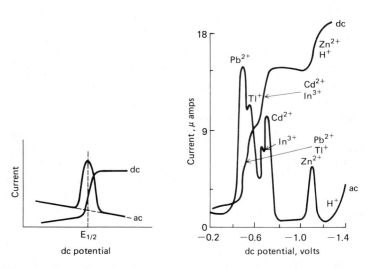

Figure 22-15. Comparison of ac and dc polarograms. The right-hand plot is taken from B. Breyer, F. Gutmann, and S. Hacobian, *Australian J. Sci. Research*, **A3**, 567 (1950). (With permission.)

generated by the dc current does not provide a reactant for transport of the oxidation half cycle of the ac current. For those instances where the rate of one or the other half-reaction is less than instantaneous but still appreciable, the magnitude of the ac current becomes dependent upon the frequency of the ac source. As the frequency is increased, a condition is reached in which the time lapse of a half cycle is not sufficient for completion of the slow half reaction; the ac current will then be limited by the reaction rate, and the current maximum will be smaller than for an entirely reversible process. Thus, ac polarography provides a useful means for kinetic studies of electrode reactions.

AC polarography with current having a square wave rather than a sinusoidal form has also been investigated (*square-wave polarography*). The advantage of the square wave form arises from the fact that the nonfaradaic current decays exponentially with time and approaches zero near the end of each half cycle. By intermittent measurement of the current at the end of each half cycle only, the base-line current becomes vanishingly small.

AC polarography has the advantage of enhanced sensitivity, and in contrast to dc polarography, permits the ready determination of a trace of a reducible species in the presence of a much larger concentration of a substance that gives a wave at lower potentials. The technique is also useful for the study of electrode kinetics. The disadvantage of the procedure is, of course, the complex nature of the equipment required.

Chronopotentiometry

In chronopotentiometry the potential of a working electrode is measured as a function of time while a *constant current* is passed through a cell containing the solution under study. A potential-time curve, which can be employed for analysis, is thus obtained. Chronopotentiometry is not a true voltammetric method; we have chosen to include the procedure here because, like polarography, the phenomenon measured is based upon diffusion-controlled processes.

Instrumentation. Figure 22-16 is a schematic diagram of an instrument for obtaining potential-time curves. Note the close resemblance to the instrument for coulometric titration shown in Figure 21-2. As in the latter, the constant current source can be a series of simple dry cells with a large series resistor; alternatively, a more sophisticated electronic source can be employed. The current is measured periodically by means of a potentiometer across a standard resistance. The reaction under study takes place at a so-called working electrode which may be a mercury pool or a platinum plate with a surface area of perhaps 1 cm². The potential of this electrode is measured relative to a calomel electrode. The third electrode, through which the constant current flows, is usually of platinum.

Many of the precautions required in conventional polarography are

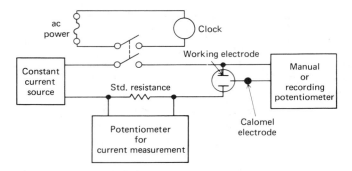

Figure 22-16. Apparatus for chronopotentiometry.

important for chronopotentiometry as well. Thus, an excess of supporting electrolyte must be used, removal of oxygen may be necessary, and care must be taken to avoid vibrations and consequent stirring of the solution. All of these measures, of course, are needed to assure a diffusion-controlled electrode process.

The time-potential relationship can be determined in a number of ways. For example, a high-impedance electronic pH meter has been used to measure the potential across the reference and working electrode. In this method the source and an electric stopclock are activated simultaneously, the potential scale of the pH meter is read at timed intervals, and the data are plotted manually. It is more satisfactory, however, to replace the clock and direct-reading meter with a recording potentiometer or an oscilloscope. If the observed phenomena occur in less than 20 to 30 sec, a recorder with a fast response time must be employed; for changes occurring in a second or less, only an oscilloscope is satisfactory.

Theory. Figure 22-17 shows a typical potential-time plot for a solution containing low concentrations of both cadmium and lead ions as well as an excess of a supporting electrolyte; a mercury pool served as the working electrode whose potential was measured relative to a silver-silver chloride reference. The two *transition times*, labeled τ_1 and τ_2 in the figure, are of analytical interest because they are related to the concentrations of lead and cadmium ions, respectively.

Upon activation of the power supply, the mercury cathode immediately assumes the negative potential A required to deposit the more easily reduced lead ions thus permitting the current flow. Formation of lead amalgam results; as before, the cathode potential at any instant is given by the relationship

$$E_{\text{cathode}} = E^0_{Pb^{2+} \rightarrow Pb(Hg)} - \frac{0.059}{2} \log \frac{(a_{Pb})_0}{(a_{Pb^{2+}})_0}$$

where $(a_{Pb})_0$ is the activity of lead atoms in the surface layer of the mercury and $(a_{Pb^{2+}})_0$ is the activity of lead ions in the layer of solution adjacent to the electrode surface. Initially, the cathode potential changes rapidly because of the large relative increase of $(a_{Pb})_0$ from zero to a value that begins to approach $(a_{Pb^{2+}})_0$. Further passage of current results in a continuous but more gradual alteration of the ratio of activities and thus correspondingly smaller changes in the cathode potential. In the middle portion of the curve between A and B, the removal of lead ions from the surface layer is partially offset by diffusion from the bulk of the solution. It is clear from the negative shift in potential, however, that the rate of deposition exceeds the rate of diffusion and that $(a_{Pb^{2+}})_0$ must be approaching zero by the time point B is reached; that is, a state of complete concentration polarization with respect to lead ions is being approached. As $(a_{Pb^{2+}})_0$ approaches zero the ratio of activities becomes very large and the rate of change of the cathode potential accelerates until a potential is reached at which the current can be carried by reduction of another species — in this example, the cadmium ions. The second portion of the curve (B to C) then reflects the time required for the cathode to become totally polarized with respect to this species. If the electrolysis were continued further, the cathode potential would rapidly fall until formation of hydrogen again stabilized the system.

The transition times τ_1 and τ_2 then are the times required for complete concentration polarization to be achieved. Clearly, these quantities are dependent upon the same variables that influence the size of the diffusion cur-

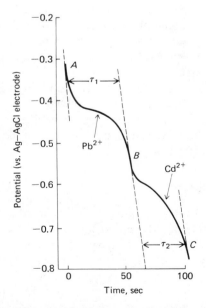

Figure 22-17. Potential-time curve for a solution that was 1.5×10^{-5} *M* in Pb^{2+} and 8×10^{-6} *M* in Cd^{2+}. Taken from C. N. Reilley, G. W. Everett and R. H. Johns, *Anal. Chem.*, **27**, 483 (1955). (With permission of the American Chemical Society.)

rent in polarography. When a single reducible species is present, it can be readily shown that

$$\tau^{1/2} = \frac{\pi^{1/2} \, n \, F \, A \, D^{1/2} \, c}{2i} \tag{22-15}$$

where A is the surface area in cm^2, i is the value of the constant current in amperes, and c is the concentration of the active species in moles per cm^3. The rest of the terms have been defined earlier in this chapter (p. 565).

If two or more reducible substances are present in the solution, the transition time τ_1 for the more easily reduced is given by equation (22-15); expressions for the times for the remaining species are more complex functions because the preceding processes continue as each new one begins. Thus, for the two-component system shown in Figure 22-17, τ_1 is defined by equation (22-15), but τ_2 is given by

$$(\tau_1 + \tau_2)^{1/2} - \tau_1^{1/2} = \frac{\pi^{1/2} \, n_2 \, F \, A \, D^{1/2} \, c_2}{2i} \tag{22-16}$$

Application. Chronopotentiometry can be employed for the quantitative determination of species in the concentration range of 10^{-1} to 10^{-5} M. Potentially it may be possible to obtain somewhat higher precision with potential-time measurements than with the potential-current measurements of polarography, but this advantage is offset by the larger effect of traces of impurities. The presence of small amounts of oxygen has, in general, a more pronounced effect on transition times than on diffusion currents of species that are reduced at potentials more negative than that for oxygen.

Chronopotentiometry with a platinum electrode may prove useful for studying anodic processes, and a number of investigations of this kind have been reported. In addition, chronopotentiometry has been employed for the analysis of molten salt solutions. For these studies a eutectic mixture of potassium and lithium chloride (m.p. $\sim 450°C$) has been employed as a solvent; both the reduction of metal ions and the oxidation of metals have been investigated with platinum electrodes.

Stripping Methods

Stripping analysis encompasses a variety of electrochemical methods having a common, characteristic initial step. In all these procedures, the substance of interest is first concentrated by electrodeposition at a mercury or a solid electrode and then redissolved (*stripped*) from the electrode to produce a solution that is more concentrated than the original solution. The analysis is finally based either upon electrical measurements made during the stripping process itself or upon other types of electroanalytical studies of the concentrated solution produced by the stripping process.

Stripping methods are of prime importance in trace work because the concentration aspects of the electrolysis permits the determination of minute amounts of a substance with reasonable accuracy. Thus, the analysis of solutions in the 10^{-6} to 10^{-10} M range becomes feasible by methods that are both simple and rapid.

The form of stripping analysis that has had the most widespread study involves the use of a micro mercury electrode for the deposition process; our discussion is largely confined to this particular application. For a more complete survey of stripping methods, the reader is referred to the review article by Shain.[13]

Electrodeposition. The electrodeposition step can be carried out in two ways. In the stoichiometric procedure the ion of interest is removed completely from a stirred solution at a constant cathode (or anode) potential. In the nonstoichiometric procedure only a fraction of the ion is deposited; here, not only must the electrode potential be controlled, but care must also be taken to reproduce the electrode size, the length of deposition, and the stirring rate for both the sample and the standard solutions employed for calibration. Nonstoichiometric procedures have the advantage of speed, since the electrolysis step need last only a few minutes.

Mercury electrodes of various forms have been most widely used for electrolysis, although platinum or other inert metals can also be employed. Generally it is desirable to minimize the volume of the mercury since this enhances the concentration of the deposited species. For the nonstoichiometric approach several methods have been devised to produce a microelectrode of reproducible dimensions. One of these employs the *hanging drop* electrode, which is illustrated in Figure 22-18. Here an ordinary dropping mercury capillary is employed to transfer a reproducible quantity of mercury (usually one to three drops) to a Teflon scoop. Note that the mercury capillary does *not* serve as an electrode in this application. The hanging drop electrode is then formed by rotating the scoop and bringing the mercury up to the platinum wire sealed in a glass tube. The drop adheres strongly enough so that the solution can be stirred. The drop can be replaced after an analysis by tapping the electrode to dislodge the mercury.

To use this apparatus for a nonstoichiometric deposition the drop is formed, stirring is begun, and a potential that is a few tenths of a volt more negative than the half-wave potential for the ion of interest is applied across the calomel-hanging drop system. Deposition is allowed to occur for a carefully measured period. Five minutes usually suffices for solutions that are 10^{-7} M or greater, 15 min for 10^{-8} M solutions, and 60 min for those that are 10^{-9} M. It should be emphasized that these times do not result in complete removal of the ion; the electrolysis period is determined by the sensitivity of the method ultimately employed for completion of the analysis.

[13]I. Shain, *Stripping Analysis* in I. M. Kolthoff and P. J. Elving, Eds., *Treatise on Analytical Chemistry,* Part 1, Vol. 4 (New York: Interscience Publishers, Inc., 1963), Chap. 50.

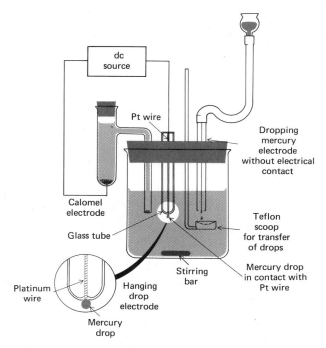

Figure 22-18. Apparatus for stripping analysis.

Completion of the analysis. A most convenient way for stripping and completing an analysis after deposition at a hanging drop electrode employs the same electrode for a voltammetric measurement. After completion of the carefully timed deposition, as described in the previous section, the stirring is stopped for perhaps 30 sec. The voltage is then decreased *at a fixed rate* from its original cathodic value toward the anodic, and the resulting anodic current is recorded as a function of the applied voltage. This technique, in which voltage is varied at a fixed rate, is termed *linear scan voltammetry* and produces a curve of the type shown in Figure 22-19.

In this experiment cadmium was first deposited from a $1 \times 10^{-8}\,M$ solution by application of a potential of about -0.9 V (vs. SCE), which is about 0.3 V more negative than the half-wave potential for this ion. After 15 min of electrolysis, stirring was discontinued; after an additional 30 sec, the potential was decreased at a rate of 21 mV per sec. At about -0.65 V, a rapid increase in anodic current occurred as a result of the reaction

$$Cd(Hg) \rightarrow Cd^{2+} + Hg + 2e$$

After reaching a maximum, the current decayed as a consequence of depletion of cadmium metal in the hanging drop. The peak current, after correction for the residual current, was found to be directly proportional to concentration of cadmium ions over a range of $10^{-6}\,M$ to $10^{-9}\,M$ and inversely proportional to deposition time. With reasonable care, an analytical precision of about 2 percent relative can be obtained. Analyses in this concentration

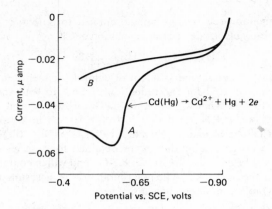

Figure 22-19. *A:* Current-voltage curve for anodic stripping of cadmium. *B:* Residual current curve for blank. Adapted from: R. D. DeMars and I. Shain, *Anal. Chem.*, **29**, 1825 (1957). (With permission of the American Chemical Society.)

range are not feasible by ordinary polarography. Mixtures can be resolved and analyzed by controlling the potential of deposition.

Many other variations of stripping analysis have been proposed. For example, a number of metals have been determined by electrodeposition on a platinum cathode either stoichiometrically or nonstoichiometrically. The quantity of electricity required to remove the deposit is then measured coulometrically. Here again, the method is particularly advantageous for trace analysis.

Problems

1. Calculate the lead concentration in a sample on the basis of the following data. Express the results in terms of milligrams of Pb per liter.

Solution	Observed current at -0.65 V (μamp)
25.0 of ml of 0.40 F KNO$_3$ diluted to 50.0 ml	12.4
25.0 ml of 0.40 F KNO$_3$ and 10.0 ml of sample diluted to 50.0 ml	58.9
25.0 ml of 0.40 F KNO$_3$, 10.0 ml of sample, and 5.0 ml of 1.7×10^{-3} F Pb^{2+} diluted to 50.0 ml	81.5

2. The standard addition method was employed for the polarographic determination of indium, the accompanying data being obtained.

Solution	Observed current at -0.70 V (μamp)
25.0 ml of 0.40 F KCl diluted to 100.0 ml	8.7
25.0 ml of 0.40 F KCl and 20.0 ml of sample diluted to 100.0 ml	49.1
25.0 ml of 0.40 F KCl, 20.0 ml of sample, and 10.0 ml of 2.0×10^{-4} F In(III) diluted to 100.0 ml	64.6

Calculate the milligrams of In(III) per liter of sample.

3. Derive an equation relating the anodic current to potential (vs. SCE) for a dropping mercury electrode in a dilute solution of Cl^-.

$$2Hg + 2Cl^- \rightarrow Hg_2Cl_2 + 2e$$

Is the half-wave potential independent of chloride concentration?

4. The half-wave potential for the reduction of Pb^{2+} to lead amalgam from 1 F KNO_3 is -0.405 V. What would be the half-wave potential if a 1.00×10^{-4} solution of Pb^{2+} in 1 F KNO_3 were made 1.00×10^{-2} M in Y^- ion as well? (Y^- is the anion of EDTA; formation constant for $PbY^{2-} = 1.1 \times 10^{18}$)

5. The following polarographic data were obtained for a series of solutions that were 1.00×10^{-4} F in Cd^{2+}, 0.100 F in KNO_3, and which contained various concentrations of the anion X^{2-}. The polarographic waves resulted from the reduction of Cd(II) to a cadmium amalgam.

Conc. X (F)	$E_{1/2}$ vs. SCE
0.00	-0.586
1.00×10^{-3}	-0.719
3.00×10^{-3}	-0.743
1.00×10^{-2}	-0.778
3.00×10^{-2}	-0.805

What is the formula and the formation constant for the complex formed between Cd^{2+} and X^-?

6. Copper(II) yields a reversible wave at the dropping mercury electrode as a result of reduction of Cu^{2+} to copper amalgam. In order to determine the composition and the formation constant for the complex between Cu^{2+} and the anion A^{3-}, a series of solutions that were 1.00×10^{-4} F in Cu^{2+}, 0.100 F in KNO_3, and which had the following concentrations of A^{3-}, were prepared.

Conc. A^{3-} (F)	$E_{1/2}$ vs. SCE
0.00	$+0.020$
1.00×10^{-3}	-0.382
3.80×10^{-3}	-0.404
8.30×10^{-3}	-0.413
1.20×10^{-2}	-0.416

What is the formula and formation constant for the complex?

7. Sketch the titration curves that would be expected in the following amperometric titrations with a dropping mercury electrode:
   (a) SO_4^{2-} with standard $Pb(NO_3)_2$ at a potential of -1.2 V (vs. SCE);
   (b) a mixture of Pb^{2+} and Ba^{2+} with a standard chromate solution at an applied potential of -1.2 V ($K_{sp} \sim 10^{-10}$ for $BaCrO_4$; $\sim 10^{-14}$ for $PbCrO_4$);
   (c) Ni^{2+} with dimethylglyoxime at an applied potential of -1.85 V (dimethylglyoxime is reduced at this potential).

23 Chromatographic and Other Fractionation Methods

The physical and chemical properties upon which analytical methods are based are seldom if ever entirely specific. Instead, these properties are shared by numerous species; as a consequence, a preliminary separation must often precede the final measurement step, particularly if the sample is complex.

Within the last two to three decades vast strides have been made in the development of separation techniques. Earlier, analytical separations were accomplished largely by chemical precipitation, batch distillation, and simple extraction. None of these methods, however, can isolate the components in many of the complicated mixtures that are of interest to the modern chemist — for example, the individual amino acids produced by hydrolysis of a protein, the members of a homologous series of hydrocarbons or alcohols, or the various ions of the rare-earth series. Many separations such as these are now practical as a consequence of the development of a group of fractionation methods. The tremendous impact of these methods on science is attested by the two Nobel prizes that have been awarded for discoveries in the field: the first to A. W. K. Tiselius in 1948 and the second to A. J. P. Martin and R. L. M. Synge in 1952.

NATURE OF THE SEPARATION PROCESS

All separation procedures have in common the distribution of the components in a mixture between two phases which subsequently can be mechanically separated. If the ratio of the amount of a given component present in each phase (the *distribution ratio*) differs significantly from that of another, a separation of the two is potentially feasible. To be sure, the complexity of the procedure depends upon how great the distribution ratios for the two components differ from one another. Where the difference is extreme, a single-stage process suffices. For example, a single precipitation with silver ion is adequate for the isolation of chloride from many other anions. Here, the ratio of the chloride ion in the solid phase to that in equilibrium in the aqueous phase is immense, while the comparable ratios for, say, nitrate or perchlorate ions, approaches zero.

A somewhat more complex situation is encountered when the distribution ratio for one component is essentially zero, as in the foregoing example, but the ratio for the other is not very large. Here a multistage process is required. For example, uranium(VI) can be extracted into ether from an aqueous nitric acid solution. Although its distribution ratio in the two phases is only about unity for a single extraction, uranium(VI) can nevertheless be separated from other metal ions which have no tendency for transfer by repeated or *exhaustive* extraction of the aqueous solution with fresh portions of ether. The repetitive extraction is readily automated; the well-known Soxhlet extraction apparatus for removal of soluble components from a solid phase is an example of such a device.

The most complex procedures are required when the distribution ratios of the species to be separated are greater than zero and approach one another in magnitude; here, multistage *fractionation* techniques are necessary. These techniques do not differ in principle from their simpler counterparts; both are based upon differences in the distribution ratios of solutes between two phases. However, two factors account for the gain in separation efficiency associated with fractionation. First, the number of times that partitioning occurs between phases is increased enormously; second, distribution is caused to occur between fresh portions of both phases. An exhaustive extraction differs from a fractionation process in this latter respect; although many contacts between phases are provided during an exhaustive extraction, fresh portions of only one of the phases are introduced.

Fractionation techniques have been applied to all combinations of phases. Most commonly, continuous contact between the two phases is provided by flowing the one countercurrent to the other, thus in effect increasing enormously the number of contacts. In most methods a *mobile phase* is forced through a fixed *immobile phase* in a column. In fractional distillation, however, the vapor phase is caused to move countercurrent to the condensed liquid phase to provide multiple contact stages.

The development of multistage fractionation techniques has led to the major advances in the field of separation mentioned earlier; it is with respect to these methods that we shall be concerned in this and the following chapter.

Errors Resulting from the Separation Process

In general, separations are based upon equilibrium processes; as a consequence, complete separation of an interference from the species of interest is never possible. At best, the separation lowers the concentration of the interference to a tolerable level. An additional requirement is that losses of the component sought during the separation must be smaller than the allowable error in the analysis. Thus, the two factors to be considered in any separation are (1) the completeness of the recovery of the species to be determined, and (2) the degree of separation from the unwanted constituent. These factors can be expressed algebraically in terms of *recovery ratios R*. For example, if x is the amount of the wanted constituent X that is recovered in a separation and x_0, its amount in the original sample, the recovery ratio is given by

$$R_X = \frac{x}{x_0} \tag{23-1}$$

Clearly, this ratio should be as close as possible to one. We may write a similar recovery ratio for the unwanted component

$$R_Y = \frac{y}{y_0} \tag{23-2}$$

where y_0 and y represent the initial and final amounts of the interference Y. As R_Y becomes smaller, the better the separation process becomes.

The incomplete recovery of X always results in a negative error. The error arising from the incomplete removal of Y will be positive if this species contributes to the analytically measured quantity and negative if it lowers the magnitude of this measurement. In order to examine the nature of these errors let us assume that the analysis is based upon the measurement of some quantity M that is proportional to the amount x and y present in the solution after the separation process (M may represent mass, volume, absorbance, diffusion current, and so forth). Such being the case, we may write that

$$M_X = k_X x \tag{23-3}$$
$$M_Y = k_Y y \tag{23-4}$$

where k_X and k_Y are constants that are proportional to the sensitivity of the measurement for each constituent (k_Y will be negative when Y interferes by lowering the sensitivity of the test for X).

If both X and Y are present, the measured value of M represents the sum of their contributions:

$$M = M_X + M_Y \tag{23-5}$$

$M = k_X X + k_Y Y \qquad NORMALLY.$

In the absence of Y, no separation would be required; if we indicate the measured quantity under these circumstances as M_0, we may write

$$M_0 = k_X x_0 \tag{23-6}$$

The error arising *from the separation process* is then $(M - M_0)$; expressed in relative terms, it is $\quad M - M_0 = k_X X + k_Y Y - k_X X_0$

$= k_Y Y + k_X (X - X_0) \qquad$ relative error due to separation $= \dfrac{(M - M_0)}{M_0} \tag{23-7}$

Substituting equations (23-3), (23-4), (23-5), and (23-6) into equation (23-7) gives

$R_X = \dfrac{X}{X_0}$

$$\text{relative error} = \frac{k_X x + k_Y y - k_X x_0}{k_X x_0}$$

$R_Y = \dfrac{Y}{Y_0}$ The further substitution of equations (23-1) and (23-2) yields

$$\text{relative error} = \frac{k_X R_X x_0 + k_Y R_Y y_0 - k_X x_0}{k_X x_0}$$

This expression may be rearranged to   ERROR DUE TO LOSS OF X

$$\text{relative error} = (R_X - 1) + \frac{k_Y y_0}{k_X x_0} R_Y \tag{23-8}$$

ERROR DUE TO INCOMP. REM. OF Y

The first term in equation (23-8) represents the error that is associated with the losses of X during the separation process. Thus, if 99 percent of X is recovered during the separation ($x/x_0 = R_X = 0.99$), then a relative error of -0.01 or -1 percent results.

ie DEP UPON ORIGINAL Y   ORIGINAL X

The second term in equation (23-8) takes account of the error resulting from the incomplete removal of Y by the separatory operation. The magnitude of this error is not only related to the recovery ratio R_Y, but is also dependent upon the ratio of y_0 to x_0 in the sample. Thus, a highly favorable recovery ratio is required for the separation of a wanted minor constituent X from a major component Y. This situation is often encountered in trace analysis where the ratio of y_0 to x_0 at the outset may be as great as 10^6 or 10^7.

The error incurred in a separation is also seen to depend upon the relative sensitivity of the measurement M for the two constituents (k_Y/k_X). If this measurement is not greatly affected by the presence of Y (that is, if k_Y is small), then a relatively incomplete separation may be adequate for an analysis. On the other hand, if the measurement system is equally sensitive to both X and Y, a more complete separation is likely to be required.

Classification of Fractionation Procedures

A number of methods for classifying separation procedures have been suggested.[1] The listing of multistage fractionation processes in Table 23-1

[1]P. J. Elving, *Anal. Chem.*, **23**, 1202 (1951); H. G. Cassidy, *J. Chem. Ed.*, **23**, 427 (1946); L. B. Rogers in I. M. Kolthoff and P. J. Elving, Eds., *Treatise on Analytical Chemistry*, Part I, Vol. 2 (New York: Interscience Publishers, Inc., 1961), p. 920.

Table 23-1   Classification of Multistage Fractionation Procedures

Phase Type	Process Name	Sample Phase	Second Phase
Solid–liquid	Adsorption chromatography	Solution	Solid adsorbent
	Thin-layer chromatography[a]	Solution	Fine powder supported on a glass plate
	Ion-exchange chromatography	Solution	Ion-exchange resin
Liquid–liquid	Countercurrent extraction	Solution	Immiscible solvent
	Partition chromatography	Solution	Immiscible solvent on solid matrix
	Paper chromatography	Solution	Immiscible solvent on paper matrix
	Thin-layer chromatography[a]	Solution	Immiscible solvent on fine powder supported on a glass plate
	Gel chromatography	Solution	Solvent held in the interstices of a polymeric solid
Liquid–gas	Fractional distillation	Gas	Condensed liquid
	Gas-liquid chromatography	Gas	Solvent held on solid matrix
Gas–solid	Gas–solid chromatography	Gas	Solid adsorbent

[a]Phase type appears to depend upon the pretreatment of the powder; see page 631.

is based on the nature of the two phases that are involved. The arrangement is arbitrary to the extent that the true identity of the second phase has not been established unambiguously for some of these processes.

Partition of a Component between Phases

With the exception of distillation, all of the fractionation techniques in Table 23-1 involve contact between a mechanically fixed phase and a second phase that flows past it. We will refer to the former as the *stationary* or *immobile* phase and to the latter as the *mobile* phase. The stationary phase may be either a solid, or a liquid held as a thin layer on a solid matrix. The mobile phase is either a gas or a liquid. Of interest, then, is the manner whereby solutes distribute or partition themselves between the stationary and the mobile phase.

We shall define the *partition ratio K* as

$$K = \frac{C_s}{C_m} \tag{23-9}$$

where C_s is the total analytical concentration of a solute in the stationary phase and C_m is its concentration in the mobile phase. In the ideal case the partition ratio is constant over a wide range of solute concentrations; that is, C_s is directly proportional to C_m. More often than not, however, nonlinear relationships occur. Some typical distribution curves are shown in Figure 23-1.

The ideal relationship shown by curve C in this figure is often approximated by distribution equilibria between two immiscible liquids, provided association or dissociation reactions do not occur in one of the solvents. Where such equilibria do exist, a distribution relationship similar to B or D is more likely. For example, if the stationary phase is water and the mobile phase is benzene, a curve of type B is observed for a solute consisting of a weak organic acid. Only the undissociated acid is soluble in benzene. In the aqueous solution, however, appreciable amounts of both the undissociated acid and its conjugate base are present; moreover, the ratio between these species is concentration-dependent. Thus, at low concentrations a smaller fraction of the total acid is available for distribution between the two solvents and the partition ratio is greater. If, on the other hand, water represented the mobile phase, a curve such as D would be expected. Curves of type B are also commonly encountered in vapor-liquid equilibria, where the vapor is the mobile phase.

Curve A is a typical *adsorption* isotherm which relates the amount of solute that is adsorbed on the surface of a solid to the solute concentration in the solution that contacts the solid. At high solute concentrations all adsorption sites on the solid surface are occupied; the extent of adsorption then becomes independent of the solute concentration. Curves of this type can be described by the relationship

$$C_s = kC_m^n$$

where k and n are constants.

The assumption of a constant K will frequently be made in the deriva-

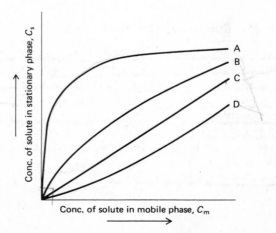

Figure 23-1. Some typical distribution curves.

tion of working equations for fractionation processes. The plots in Figure 23-1 suggest that this approximation will not introduce serious errors throughout the lower concentration ranges, where all curves are approximately linear. For wide concentration ranges, however, the derived expressions must be modified to take into account the variation in K with solute concentration.

Unfortunately, a uniform definition for the partition ratio is not employed in dealing with the various fractionation processes shown in Table 23-1. The difference lies in the assignment of phase concentrations to the numerator and denominator. Thus, in fractional distillation the partition coefficient K' is given by

$$K' = \frac{1}{K} = \frac{C_m}{C_s} \tag{23-10}$$

where the mobile phase is the gas. In liquid–liquid systems the concentration term for the nonaqueous phase ordinarily appears in the numerator; thus, depending upon the system, the partition ratio is defined by either (23-9) or (23-10). In the discussion that follows the definition that has been historically associated with each type of fractionation process is employed. We, however, use K to denote partition ratios as defined by (23-9) and K' for those defined by (23-10). The reader should be alert to the difference.

COUNTERCURRENT EXTRACTION[2]

All but one of the fractionation techniques shown in Table 23-1 involve a continuous flow of the mobile phase past the stationary phase. The exception is the countercurrent extraction process, in which contact between phases occurs in a large number (often several hundred) of discrete steps. The techniques for carrying out multiple extractions and the demonstration of their usefulness were pioneered by L. C. Craig of the Rockefeller Institute; as a consequence, the method and apparatus often bear his name.

The stepwise nature of the Craig countercurrent process permits ready development of fractionation theory. This theory is then easily extended to processes that employ continuous contact by introduction of the concept of the theoretical plate.

Description of the Craig Countercurrent Process

A schematic representation of five stages in the Craig process is shown in Figure 23-2. The separation process is begun by extracting a solution of the sample with an immiscible solvent. In Craig terminology this step is labeled "transfer zero" $(n = 0)$, and is depicted in the top row of the figure.

[2]For an extended discussion of this topic see L. C. Craig and D. Craig, *Laboratory Extraction and Countercurrent Distribution* in A. Weissberger, Ed., *Technique of Organic Chemistry*, 2d ed., Vol. III, Part I (New York: Interscience Publishers, Inc., 1956), pp. 149-332.

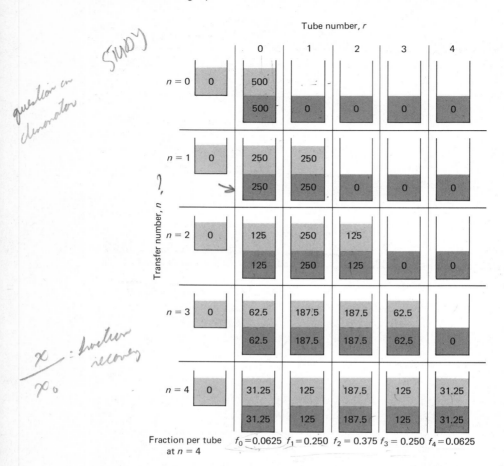

Figure 23-2. Schematic representation of the Craig countercurrent process for a solute ($K' = 1.00$).

This initial extraction takes place in a vessel that is labeled "tube zero" ($r = 0$). Also shown are four additional vessels, labeled tube 1, 2, 3, and 4, which are a part of a very large number r of identical containers; each of these vessels is filled with an equal volume of the more dense solvent (say, water). Tube zero also contains a known volume of a less dense organic solvent, and means are provided for introducing additional and identical volume increments of this solvent. For convenience, let us assume that the weight of solute is 1000 mg and that its partition ratio between the solvents is 1.00. Here, the aqueous layer is considered to be the stationary phase and the organic solvent is considered the mobile one. The partition ratio is defined as $K' = C_m/C_s$. Furthermore, although it is not a requirement of the process, it is convenient to stipulate that the volume of the organic solvent is identical to that of the aqueous phase; thus, when equilibrium is achieved in tube 0 after transfer 0, 500 mg of solute will be found in each phase.

The next step in the process (transfer number 1) involves transfer of the organic phase from tube 0 to tube 1, followed by introduction of an identical volume of fresh organic solvent into tube 0. Both vessels are then shaken, with the result that the solute is distributed among the four solutions as shown in line $n = 1$ of Figure 23-2.

This transfer and equilibration process is repeated, with an aliquot of fresh organic solvent being added to tube 0 at each transfer. The last row in the figure indicates the distribution of solute after four transfers ($n = 4$), where it will be seen that the solute has become distributed among all tubes, with the major portion moving progressively to the right in a wavelike fashion. The movement of the solute for a much larger number of transfers is shown in Figure 23-3; note that the distribution curves become broader and lower with an increased number of transfers and that the peak in the curve occurs farther and farther from the starting point as n becomes larger. In this figure the fraction of total solute contained in the two layers of each tube (that is, organic portion + aqueous portion) is plotted against the tube number as defined in Figure 23-2.

If we had chosen to use a solute with a less favorable partition ratio in Figure 23-2 (that is, $K' < 1.00$), the fraction transferred to the organic solvent at each step would have been smaller. Thus, the rate of movement of solute along the tubes would have been less. For a solute with partition ratio larger than unity, this movement would have been enhanced. These effects are depicted in Figure 23-4, where distribution plots after 50 transfers are shown for three solutes with different partition ratios. If, now, the process were performed on a single sample that contained all three of these solutes, and if there were no interactions among them, a distribution curve that is the sum of the three curves would be expected. For most purposes, the separation indicated would be adequate; if it were not, additional transfers could be performed to achieve greater resolution.

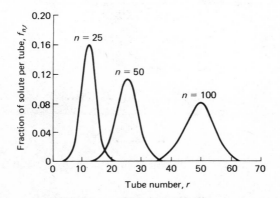

Figure 23-3. Effect of the number of transfers n upon the distribution of a solute with a distribution coefficient K' of 1.00. Equal volumes of the two phases were employed.

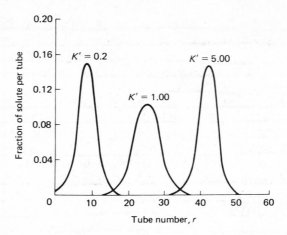

Figure 23-4. Effect of distribution coefficient on the distribution of solute after 50 transfers. Equal volumes of the two phases were employed.

Derivation of Distribution Curves

Distribution curves such as those shown in Figure 23-3 and 23-4 can be derived from partition ratio data. Such plots are useful for determining the number of transfers needed to achieve a given separation and for deciding from which tubes the various products should be harvested.

Let the quantity $f_{n,r}$ be the fraction of the initial solute A contained in tube r after n transfers have been performed. It is important to note that this fraction includes the solute in *both* layers in a given tube. Clearly, the sum of the fractions must add up to unity; that is,

$$1 = f_{n,0} + f_{n,1} + f_{n,2} + \cdots + f_{n,r}$$

Ideally, a constant ratio of solute concentrations will exist between the two phases in all tubes and will be given by

$$K' = \frac{C_m}{C_s} \tag{23-11}$$

Let us now designate x as the fraction of solute found in the mobile organic layer of any tube and y as the fraction in the stationary aqueous layer. Clearly,

$$x + y = 1$$

In any particular tube the weight of solute in each phase is given by

$$\text{wt in mobile phase} = C_m V_m$$
$$\text{wt in stationary phase} = C_s V_s$$

where C_m and C_s are expressed in grams per milliliter and V_m and V_s are the volumes of the two phases in milliliters. Then,

$$x = \frac{C_m V_m}{C_s V_s + C_m V_m}$$

and

$$y = \frac{C_s V_s}{C_s V_s + C_m V_m}$$

Substitution of equation (23-11) and rearrangement gives

$$x = \frac{K' V_m}{K' V_m + V_s} \qquad (23\text{-}12)$$

$$y = \frac{V_s}{K' V_m + V_s} \qquad (23\text{-}13)$$

Note that x and y are dependent only on K' when the volumes are fixed. Thus, if we are able to express $f_{n,r}$ in terms of x and y, we can generate data for distribution curves. To obtain such an expression let us analyze the first three steps of a countercurrent process.

Example.
1. *For $n = 0$:* At this stage the solute is contained in tube 0 only; thus,

$$f_{0.0} = x + y = 1.00$$

2. *For $n = 1$:* One transfer has been made and the solute is distributed between two tubes labeled 1 and 0 (that is, $r = 1$ and $r = 0$).
  (a) *For $r = 1$:* The solute in tube 1 comes from the transfer of the organic layer in tube 0, transfer 0, which contained the fraction x of the original solute. Thus,

$$f_{1,1} = x$$

Partition of the transferred solute occurs in tube 1, a fraction x remaining in the organic layer and a fraction y transferring to the aqueous phase. Since x and y retain the same values throughout,

$$\left(\begin{array}{c} \text{fraction of original sample} \\ \text{in organic layer, tube 1} \end{array} \right) = x \cdot x = x^2$$

$$\left(\begin{array}{c} \text{fraction of original sample} \\ \text{in aqueous layer, tube 1} \end{array} \right) = xy$$

Note that the sum of these quantities is $f_{1,1}$

$$f_{1,1} = x^2 + xy = x(x + y)$$

But $(x + y) = 1$; so $f_{1,1} = x$.

(b) *For r = 0:* The solute remaining in tube 0 came from the aqueous layer that contained the fraction y of the original. Thus,

$$f_{1,0} = y$$

After equilibration with fresh organic solvent, the distribution of the fraction in tube 0 will be

$$\left(\begin{array}{c}\text{fraction of original solute}\\\text{in organic phase, tube 0}\end{array}\right) = xy$$

$$\left(\begin{array}{c}\text{fraction of original solute}\\\text{in aqueous phase, tube 0}\end{array}\right) = y^2$$

3. *For n = 2:* The solute is now distributed among tubes 0, 1, and 2.
    (a) *For r = 2:* The solute in tube 2 comes from the organic layer of tube 1, transfer 1 containing the fraction x^2 of the original sample. Thus,

$$f_{2,2} = x^2$$

This fraction will be redistributed to give x^3 in the organic layer and x^2y in the aqueous phase of tube 2.
    (b) *For r = 1:* The solute now in tube 1 comes from two sources. A fraction xy remains in the aqueous layer. In addition, another fraction, also xy, will be introduced in the organic phase transferred from tube 0. Thus,

$$f_{2,1} = xy + xy = 2xy$$

    (c) *For r = 0:* The solute remaining in tube 0 consists of the fraction remaining in the aqueous phase after the first transfer

$$f_{2,0} = y^2$$

This analysis can be carried forward to demonstrate that, following transfer step 3, the solute will be partitioned among four tubes as follows:

tube 0	$f_{3,0} = y^3$
tube 1	$f_{3,1} = 3y^2x$
tube 2	$f_{3,2} = 3x^2y$
tube 3	$f_{3,3} = x^3$

These relationships are examples of the generalization that each of the fractions can be represented by one of the terms resulting from expansion of the binomial $(x + y)^n$, where n corresponds to the number of transfers. Any single term in the expansion of a binomial, and thus the corresponding value of $f_{n,r}$, is given by

$$f_{n,r} = \frac{n!}{r!\,(n-r)!}\, x^r y^{(n-r)} \qquad\qquad (23\text{-}14)$$

si Kcal

Example. Calculate the fraction of solute contained in tube 3 after five transfers have taken place, if the partition ratio for the solute is 4.00. Assume $V_m = V_s$.

From equations (23-12) and (23-13),

$$x = \frac{4}{4+1} = 0.8 \qquad x = \frac{K'V_n}{K'V_m + V_s}$$

$$y = \frac{1}{4+1} = 0.2$$

Substituting $n = 5$ and $r = 3$ into equation (23-14)

$$f_{5.3} = \frac{5 \cdot 4 \cdot 3 \cdot 2 \cdot 1}{(3 \cdot 2 \cdot 1)(2 \cdot 1)} (0.8)^3 (0.2)^2 = 0.205$$

Thus, 0.205 of the original solute would be found in tube 3.

The factorial quantities in equation (23-14) rapidly become huge as n is increased. Fortunately, most mathematical handbooks have tables listing factorial data: nevertheless, manipulation of the enormous numbers is cumbersome and often leads to errors.

Derivation of distribution data can be greatly simplified by assuming that curves such as those shown in Figures 23-3 and 23-4 approach the shape of the normal error curve (a Gaussian distribution). With this assumption, one can then express $f_{n,r}$ by means of an equation similar to that for the normal error curve. Thus, it has been shown that for values of n greater than 25, a good approximation of the solute distribution is given by

$$f_{n,r} = \frac{1}{\sqrt{2\pi nxy}} \exp\left[-\frac{(r_{max} - r)^2}{2nxy} \right] \tag{23-15}$$

where r_{max} is the number of the tube with the maximum solute content. If n is reasonably large, it can be shown that

$$r_{max} \cong nx \tag{23-16}$$

Furthermore, the fraction of solute in r_{max} is

$$f_{n,r_{max}} = \frac{1}{\sqrt{2\pi nxy}} \tag{23-17}$$

The arithmetic manipulations involved in the use of these equations are more easily performed than those for equation (23-12); for all practical purposes, the results obtained are the same, provided n is large.

Example. Verify the position of the peak in Figure 23-4 for the solute having a partition ratio of 5.00. Note that $V_m = V_s$ and $n = 50$. Calculate the fraction of solute contained in the tube corresponding to the peak and the fraction in tube ($r_{max} - 2$).

From equations (23-12) and (23-13),

$$x = \frac{5}{5 + 1} = 0.833$$

$$y = \frac{1}{5 + 1} = 0.167$$

According to equation (23-16)

$$r_{max} \cong 50 \times 0.833 = 41.7$$

Thus, the maximum amount of solute would be expected in tube 42. To obtain the fraction of solute in this tube we apply equation (23-17):

$$f_{50,42} = \frac{1}{\sqrt{2 \times 3.14 \times 50 \times 0.833 \times 0.167}} = \frac{1}{\sqrt{43.7}}$$

$$= 0.153$$

To obtain the fraction in tube 40 we employ equation (23-15):

$$f_{50,50} = \frac{1}{\sqrt{43.7}} \exp\left[-\frac{(42 - 40)^2}{2 \times 50 \times 0.833 \times 0.167}\right]$$

$$= 0.153 \; e^{-0.288}$$

$$\ln f_{50,40} = \ln 0.153 - 0.288$$

To convert the natural logarithm to base-ten terms we multiply both sides of the equation by 0.434. Thus,

$$\log f_{50,40} = \log 0.153 - 0.434 \times 0.288$$
$$= -0.815 - 0.125 = -0.940$$
$$f_{50,40} = 0.115$$

Application of the Craig Technique

The large number of successive extractions required by the counter-current technique is practical only if the equilibration and the transfer processes can be carried out automatically. Craig has designed an apparatus that permits more than 200 of these individual steps to be accomplished simultaneously. The apparatus consists of a series of interconnected extraction vessels constructed as shown in Figure 23-5. Initially, each cell is filled with the more dense liquid so that level a is attained with the cell in position C. When in this position, an upper layer of solvent can flow into, and can be contained in, arm d. When the cell is rotated in a counterclockwise manner to position A, the lighter solvent flows through C into the next cell. By rocking the cell between positions A and B, the two solvents are mixed. Note that no transfer can occur in these positions. After equilibration, the tube is held in position *B* to allow the layers to separate; then transfer is accomplished by returning the cell first to position *C* and then back to position *A*. Hundreds of cells can be mounted in a side-by-side array that is rotated and rocked by a motor that is programmed automatically to give the desired cycle.

The Craig countercurrent extraction process has been widely applied, particularly in the biochemical field, to the separation of closely related compounds. For example, Craig[3] has demonstrated its applicability by resolving a mixture of ten amino acids, despite the fact that the partition ratios for some of these compounds differed by less than 0.1. He also showed that the experimental distribution curves for the components of this mixture closely approximated the theoretical. Complex mixtures of fatty acids, polypeptides, nucleotides, aromatic amines, antibiotics, and many other organic substances have been resolved by the Craig technique. Few applications to inorganic systems exist in the literature, although there is no reason, in principle, why the technique should not be equally applicable to this type of sample.

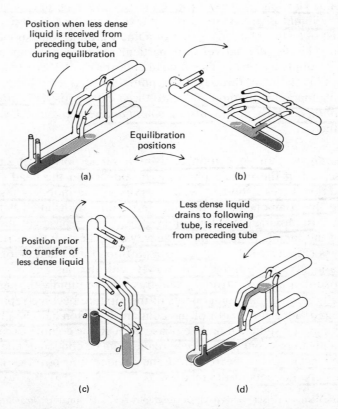

Position when less dense liquid is received from preceding tube, and during equilibration

Equilibration positions

(a)

(b)

Position prior to transfer of less dense liquid

Less dense liquid drains to following tube, is received from preceding tube

(c)

(d)

Figure 23-5. Tubes for Craig countercurrent distribution apparatus.

[3]L. C. Craig, *Anal. Chem.*, **22**, 1346 (1950).

CHROMATOGRAPHIC SEPARATIONS[4]

Chromatography encompasses a diverse group of separation methods (see Table 23-1) that are of great importance to the analytical chemist, for they often enable him to separate, isolate, and identify the components of mixtures that might otherwise be resolved with difficulty or not at all. The term chromatography is difficult to define rigorously, owing to the variety of systems and techniques to which it has been applied. In its broadest sense, however, chromatography refers to processes based on differences in rates at which the individual components of a mixture migrate through a stationary medium under the influence of a mobile phase.

General Description of the Chromatographic Process

The chromatographic separations of analytical chemistry are often carried out in tubular columns of glass or metal that have been packed with a porous solid. The solid serves as a stationary phase itself or as a support for a stationary liquid phase. Instead of a packed column, the stationary state may consist of a strip of coarse paper or a finely ground solid supported by a glass plate. These alternatives are particularly advantageous when the amount of sample is small. The fundamental principles are the same for separations involving all three types of immobile phase.

Chromatographic methods fall into three categories: (1) *elution analysis,* (2) *frontal analysis,* and (3) *displacement analysis.* Figure 23-6 contrasts these methods.

Elution analysis. In an elution analysis a single portion of sample, usually dissolved in the mobile phase, is introduced at the head of the column. Additional fresh solvent, which serves as the mobile phase, is then added and the components of the sample become distributed between the two phases. In moving down the column, the mobile phase carries the components of the sample in much the same way as in the countercurrent extraction process (Figure 23-2). Solutes with partition ratios that favor retention by the stationary phase move only slowly with the passage of the mobile phase; those that are less strongly held travel more rapidly through the column. Ideally, the various components of the mixture become separated into bands located along the length of the column. Isolation can then be accomplished by passing sufficient mobile phase through the column to cause these various bands to pass out the end, where they can be collected; alternatively, the column packing can be removed and divided into portions containing the various components of the mixture.

[4]For more complete discussions of chromatography see E. Heftmann, *Chromatography,* 2d ed, (New York: Reinhold Publishing Corp., 1967); E. Lederer and M. Lederer, *Chromatography,* 2d ed, (New York: American Elsevier Publishing Company, Inc., 1957); H. G. Cassidy, *Fundamentals of Chromatography* (New York: Interscience Publishers, Inc., 1957).

The process whereby the solute is washed through the column by addition of *fresh* solvent is called *elution,* and is by far the most popular way of performing a chromatographic separation. If a detector that responds to the solutes is placed at the end of the column and its signal is plotted as a function of time (or of volume of the added mobile phase), a series of peaks is obtained, as shown in the lower part of Figure 23-6. Such a plot is called a *chromatogram* and is useful for both qualitative and quantitative analysis. The positions of peaks serve to identify the components of the sample. Peak areas can be related to concentration.

Frontal analysis. In a frontal analysis a solution of the sample is fed *continuously* to the column. As shown in Figure 23-6, both solutes are initially retained on the column, but the less strongly held A moves faster than B. Thus, pure A is obtained for a period, followed by a mixture of A and B. The components can be identified from the positions of the discontinuities in the resulting chromatogram. The concentration of each component can also be estimated from the vertical displacements in the horizontal portions of the line; the relationships between detector response and time (or volume) for a frontal analysis are considerably more complex than those for elution analysis.

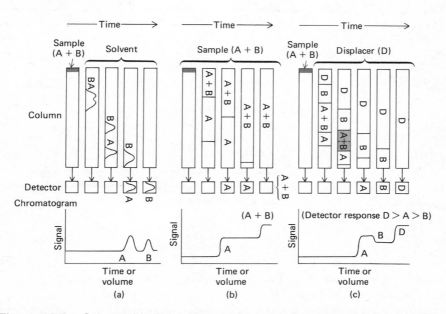

Figure 23-6. Schematic diagram showing three ways of column operation. (a) Elution analysis, (b) frontal analysis, (c) displacement analysis. Adapted from C. E. Bennett, S. Dal Nogare and L. W. Safranski, in I. M. Kolthoff and P. J. Elving, Eds., *Treatise on Analytical Chemistry,* Part I, Vol 3 (New York: Interscience Publishers, 1961), p. 1660. (With permission.)

Displacement analysis. In displacement analysis, as in the elution method, the sample is first introduced as a solution in a single portion. Here, however, the components are caused to move down the column by the continuous addition of a solution that contains some new substance (D in Figure 23-6) which is more strongly retained by the stationary phase than any of the components of the sample. This *displacer*, then, completely replaces the sample components on the stationary phase and forces them down the column. A similar sort of displacement occurs among the components of the sample, the least strongly held being displaced by the next most strongly held, and so forth. After being forced through a suitable length of column in this way, the sample components become dispersed into a series of contiguous zones, each containing one of the components in a pure form. The concentration of each component can be determined by the horizontal distance between the steps in the chromatogram.

A General Theory for Chromatographic Processes

We will first consider the classical theory of chromatography developed by Martin and Synge.[5] This theory was derived to describe liquid–liquid partition chromatography, in which the stationary phase is a liquid, coated on the surface of a finely divided solid support; the liquid is immiscible with the mobile phase. The assumption is made that the stationary phase behaves as a normal liquid and that the solid merely serves to hold it in place. With certain limitations and slight modification the treatment of Martin and Synge can be extended to all of the various chromatographic processes.

Partition ratio. In chromatographic theory the partition ratio is defined as

$$K = \frac{\text{concentration of solute in stationary phase}}{\text{concentration of solute in mobile phase}} = \frac{C_s}{C_m} \quad (23\text{-}18)$$

Note that K is the reciprocal of the K' employed in the earlier derivations (equations 23-11 to 23-13). We will, however, again let x be the fraction of solute in the mobile phase and y the fraction in the stationary phase. Substituting $1/K$ for K' converts equation (23-12) to

$$x = \frac{V_m}{V_m + KV_s} \quad (23\text{-}19)$$

Throughout this discussion, the assumption is made that the partition ratio is independent of solute concentration; that is, all systems behave ideally, as depicted by curve C in Figure 23-1. It should be noted that the expressions that will be developed become considerably more complicated if the theory must be modified to take account of nonlinear behavior.

[5] A. J. P. Martin and R. L. M. Synge, *Biochem. J.*, **35**, 1358 (1941).

The theoretical plate.   In developing the theory for chromatography, the countercurrent extraction process is used as a model, with the liquid that is held on the solid matrix corresponding to the lower stationary phase shown in Figure 23-2. The basic difference, of course, lies in the discrete stepwise contact between phases in the one process and the continuous contact in the other. As a consequence, true equilibrium can be expected at each step of the Craig process, whereas chromatography is a nonequilibrium process. To bridge the gap between the two, Martin and Synge introduced the concept of the theoretical plate, a construct that has long been used in fractional distillation theory. The theoretical plate is an imaginary layer of column of such thickness that the solution issuing from it has the composition that would obtain if equilibrium actually existed between the mobile and immobile phases midway in the layer. The theoretical plate then becomes analogous to one of the tubes in the Craig countercurrent apparatus.

A chromatographic column may be considered to contain a large number of theoretical plates with the following properties.

1. Since the column packing is assumed to be uniform, the thickness of each plate (measured along the path of the mobile phase) is identical; this parameter is called the *height equivalent of a theoretical plate* (HETP), and is given the symbol h.

2. A further consequence of uniform column packing is that the volume V_h of the mobile liquid contained in each plate is constant.

3. In each plate, partition occurs so that the fraction x of solute in the mobile phase is given by equation (23-19). The denominator of this relationship is termed the effective plate volume V_h. That is,

$$V_h = (V_m + K\, V_s) = h(A_m + K\, A_s) \qquad (23\text{-}20)$$

where A_m and A_s are the cross-sectional areas of the stationary and mobile phases. It should be noted that these areas can be determined from the length of the column packing and the total volumes of the two solvents contained therein; h, however, cannot be determined by direct measurement.

Movement of solute in a chromatographic column.   The treatment of Martin and Synge assumes that, at the outset, the sample is charged into and contained upon the first plate ($r = 0$). All subsequent plates, having identical volumes of mobile phase, are free of solute. An infinitely small volume dV of the pure mobile phase introduced at the top of the column then forces an equivalent volume from plate 0 to plate 1, from plate 1 to plate 2, and so forth. The increment passing from plate 0 to plate 1 carries with it the fraction dV/V_m of the solute originally contained in the mobile phase of plate 0. The fraction of total solute transferred is then

$$\frac{\text{fraction solute}}{\text{transferred}} = \frac{dV}{V_m} \cdot x$$

Substitution of (23-19) and (23-20) gives

$$\frac{\text{fraction solute}}{\text{transferred}} = \frac{dV}{V_h} = f_{1,1}$$

Clearly, the fraction of solute remaining in plate 0 will be

$$\frac{\text{fraction solute}}{\text{remaining}} = \left(1 - \frac{dV}{V_h}\right) = f_{1,0}$$

If we make the same kind of analysis as developed for the Craig process (p. 608), we find that, after three increments of dV ($n = 3$) have been introduced, the solute will be distributed among the first four plates as follows:

$$\text{plate 0} \qquad f_{3,0} = \left(1 - \frac{dV}{V_h}\right)^3$$

$$\text{plate 1} \qquad f_{3,1} = 3\left(1 - \frac{dV}{V_h}\right)^2 \left(\frac{dV}{V_h}\right)$$

$$\text{plate 2} \qquad f_{3,2} = 3\left(1 - \frac{dV}{V_h}\right) \left(\frac{dV}{V_h}\right)^2$$

$$\text{plate 3} \qquad f_{3,3} = \left(\frac{dV}{V_h}\right)^3$$

In general, the fractions in the various plates are given by the expansion of the binomial

$$\left[\left(1 - \frac{dV}{V_h}\right) + \frac{dV}{V_h}\right]^n$$

By analogy to equation (23-14)

$$f_{n,r} = \frac{n!}{r!\,(n-r)!} \cdot \left(\frac{dV}{V_h}\right)^r \left(1 - \frac{dV}{V_h}\right)^{(n-r)} \tag{23-21}$$

where $f_{n,r}$ is the fraction of the original solute in plate r after n transfers (or n additions of dV).

We have carried our treatment of the theory of partition chromatography at some length to demonstrate that the distribution of solutes along the length of a chromatographic column is similar to the distribution observed in the Craig apparatus. Without going into further detail, equation (23-21) can be modified to give

$$f_{n,r} = \frac{1}{\sqrt{2\pi r}} \left(\frac{n\,dV}{rV_h}\right)^r e^{(r\,-\,n\,dV/V_h)} \tag{23-22}$$

provided n is large.[6] Now $n\ dV$ is the total volume V of solvent introduced into the column. Thus,

$$f_{n,r} = \frac{1}{\sqrt{2\pi r}} \left(\frac{V}{rV_h}\right)^r e^{(r - V/V_h)} \tag{23-23}$$

A plot of $f_{n,r}$ versus V/V_h has the shape of the curves shown in Figure 23-3, the maximum occurring when $V/V_h = r$. Since r is the plate number, the linear distance the maximum has moved down the column is $rh = Vh/V_h$. Substituting equation (23-20) into this relationship gives

$$rh = \frac{hV}{V_h} = \frac{V}{A_m + K\ A_s} \tag{23-24}$$

where h is the thickness of one plate (HETP). Clearly, the position of maximum solute concentration will move farther and farther down the column as the volume V of added solvent is increased. Since V_h is a function of the partition coefficient of the solute, the positions of maxima for various solutes will appear at different locations on the column.

It is of interest to compare the rate of the solute migration peak with the rate of flow of the mobile phase. The position of the phase front is given by V/A_m, where A_m is the cross-sectional area of the mobile phase; thus, for any volume,

$$R_F = \frac{hV/V_h}{V/A_m} = \frac{A_m h}{V_h}$$

where R_F is called the *retardation factor*. Combining this expression with equation (23-20) gives

$$R_F = \frac{A_m}{A_m + K\ A_s} \tag{23-25}$$

where R_F is the rate of movement of the solute maximum compared with the rate of flow of the mobile solvent along the column. Note that all quantities on the right-hand side of equation (23-25) can be measured experimentally; therefore, R_F can be predicted.

If the sample contains two solutes M and N we may write

$$\frac{(R_F)_M}{(R_F)_N} = \frac{A_m + K_N\ A_s}{A_m + K_M\ A_s} \tag{23-26}$$

This equation then compares the rates of migration for the two solutes through the column in terms of experimentally measurable quantities.

[6]For a discussion of the approximations leading to this equation see H. A. Laitinen, *Chemical Analysis* (New York: McGraw-Hill Book Company, Inc., 1960), p. 495.

Example. A 14-mm I.D. glass tube was packed with 30.0 g of a finely divided solid having a density of 2.00 g/ml; the length of the packed portion was 18.0 cm. A mixture of two solutes was washed through the packing with a solvent for which the partition ratio (C_s/C_m) was 1.0 for solute A and 1.8 for solute B. Calculate the retardation factors for the two solutes and the volumes of solvent required to bring the concentration maxima to the end of the packing.

We first find A_s and A_m for the column as follows:

$$\text{volume of packing} = \frac{30.0 \text{ g}}{2.00 \text{ g/cm}^3} = 15 \text{ cm}^3$$

$$\begin{array}{l}\text{total volume of packed} \\ \text{portion of tube}\end{array} = \pi (0.70)^2 \times 18 = 28 \text{ cm}^3$$

$$\begin{array}{l}\text{volume of solvent} \\ \text{in packing}\end{array} = 28 \text{ cm}^3 - 15 \text{ cm}^3 = 13 \text{ cm}^3$$

Therefore,

$$A_m = 13/18.0 = 0.72 \text{ cm}^2$$
$$A_s = 15/18.0 = 0.83 \text{ cm}^2$$

From equation (23-25)

$$R_{F1} = \frac{0.72}{0.72 + 1.0 \times 0.83} = 0.46$$

$$R_{F2} = \frac{0.72}{0.72 + 1.8 \times 0.83} = 0.32$$

The volume required to bring the solute maxima to the end of the packing can be calculated with equation (23-24):

$$rh = 18.0 = \frac{V_1}{(A_m + K A_s)}$$

$$V_1 = 18(0.72 + 1.0 \times 0.83)$$
$$= 28 \text{ cm}^3$$

and
$$V_2 = 18(0.72 + 1.8 \times 0.83)$$
$$= 40 \text{ cm}^3$$

Limitations to the Plate Theory

The plate theory not only provides a reasonably satisfactory accounting for the shapes and rates of movement of bands upon a liquid-liquid chromatographic column, but can be extended to other types of chromatographic processes as well. It is therefore important to appreciate the limitations of the theory.

Throughout the derivation it was assumed that partition equilibrium is established within each plate during the flow of the mobile phase. The likelihood of this condition being realized is greatly enhanced if the layers of the two liquid phases are very thin. This consideration demands that the particle size of the solid support be as small as possible in order to increase its specific surface area and to minimize the space between particles (that is, to reduce the volume of the mobile phase). A reduction in particle size also has the effect of reducing the HETP, and consequently, increasing the number of plates in a column.

The probability of equilibrium being reached at each phase is also enhanced by a slow flow rate. The gain here, however, may be offset by the increased probability that the solute will diffuse from one plate to the next. Thus, there is an optimum flow rate for operation of any column; unfortunately, this rate cannot be predicted in advance. Diffusion and nonequilibrium effects are revealed by broadened bands and a lowered separation efficiency of the column.

A second assumption in the derivation was that the solute would behave ideally so that the partition coefficient would be independent of solute concentration. This assumption is frequently unjustified; the consequence is distortion or *tailing* of the partition curves, with the edges of the band becoming more drawn out. Decreased separation efficiency is the overall result.

In employing a column for a chromatographic separation, it is necessary to assume that the individual solutes in a mixture behave independently. Frequently, however, the partition coefficient of one solute will be altered by the presence of a second. Again, this effect will cause distorted bands.

A final limitation to the theory results from the assumption that the total sample is introduced to the first plate of the column only. In practice, the amount of sample is usually many times the capacity of one plate. Thus, the distribution process starts from several plates simultaneously and the resulting bands are somewhat broader than predicted by theory.

CHROMATOGRAPHIC METHODS EMPLOYING A LIQUID MOBILE PHASE

In this section we shall consider the techniques and applications of the five common chromatographic methods that are based upon a liquid mobile phase. These include partition, adsorption, ion exchange, paper, and thin-layer chromatography.

Partition Chromatography

Liquid–liquid partition chromatography was developed by Martin and Synge, who showed that the HETP of a properly prepared column can be as low as 0.002 cm.[7] Thus, a 10-cm column of this type may contain as many as

---

[7]A. J. P. Martin and R. L. M. Synge, *Biochem. J.,* **35,** 1358 (1941).

5000 plates; high separation efficiencies are to be expected even with relatively short columns.

Stationary phase. The most widely used solid support for partition chromatography has been silicic acid or silica gel. This material adsorbs water strongly; the stationary phase is thus aqueous. For some separations the inclusion of a buffer or a strong acid (or base) in the water film has proved helpful. Polar solvents, such as aliphatic alcohols, glycols, or nitromethane, have also been employed as the stationary phase on silica gel. Other support media include diatomaceous earth, starch, cellulose, and powdered glass; water and a variety of organic liquids have been used to coat these solids.

Mobile phase. The elution technique is generally favored over displacement or frontal analysis, and the mobile phase may be a pure solvent or a mixture of solvents which are at least partially immiscible with the stationary phase. Better separations sometimes are realized if the composition of a mixed solvent is changed continuously as elution progresses (*gradient elution*). Other separations are improved by elution with a series of different solvents. The choice of mobile phase is largely empirical.

Analysis of fractions. No single method has found general use for the detection of bands emerging from a partition chromatographic column. The composition of the eluent can be monitored continuously by comparing its refractive index, absorbance, conductivity, or other properties with those of the pure solvent. Alternatively, the fractions can be collected and analyzed by physical or chemical methods.

Applications. Partition chromatography has become a powerful tool for the separation of closely related substances. Typical examples include the resolution of the numerous amino acids formed in the hydrolysis of a protein, the separation and analysis of closely related aliphatic alcohols, and the separation of sugar derivatives. Figure 23-7 illustrates an application of the procedure to the separation of carboxylic acids. Here silica gel was employed as the support with 0.5 M sulfuric acid as the stationary phase. The column was eluted with a series of butanol-chloroform mixtures that varied from 15 to 50 percent in the alcohol. The eluted fractions were analyzed by titration with a standard solution of base. Where separation was incomplete, further treatment of the appropriate fractions on a longer column gave satisfactory isolation of the various components.

Adsorption Chromatography

Historically, the first chromatographic technique was based on adsorption. Its discoverer was a Russian botanist, M. Tswett, who in 1903 developed the technique to separate the various colored components of a plant

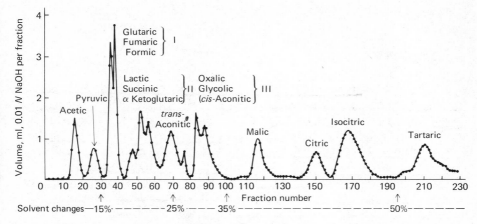

Figure 23-7. Partition chromatographic fractionation of acids on a silica gel column. Elution is by a chloroform solvent containing the indicated percentages of *n*-butyl alcohol. From W. A. Bulen, J. E. Varner and R. C. Burrell, *Anal. Chem.*, **24**, 187 (1952). (With permission of the American Chemical Society.)

extract. A quarter of a century passed before the significance of his discovery was fully appreciated by investigators engaged in the separation of biological and organic materials.

In all early applications the procedure was limited to the separation of colored substances (thus, the name *chromatography*) that could be identified by their appearance. A glass column was packed with a finely divided adsorbent such as silica, alumnia, calcium carbonate, or sucrose; the adsorbent was then wetted with a solvent, and a solution of the sample was introduced at the top of the column. The column was *developed* by washing with further portions of the solvent until colored bands of the solute appeared at various positions along its length. The various fractions were then recovered by pushing the packing out of the column, cutting it into sections, and treating each with a solvent that would cause the component to be desorbed.

The method was later simplified by washing the column with sufficient solvent until each of the adsorbed components had been eluted in turn and collected. Still later, such properties as refractive index or ultraviolet absorbance were employed to indicate when components of the mixture had been washed from the column; the scope of the procedure was thus broadened to include colorless materials.

Theory. The relationship between the concentration of the compound adsorbed to its concentration in a solvent is given by a curve called an adsorption isotherm (curve A in Figure 23-1). At low solute concentrations a rough linearity exists between concentrations in the solvent and the solid phases, and equations (23-18) to (23-26) can be employed to describe the process. Nonlinear behavior causes a distribution curve to depart from its bell-shape form, the leading edge being steeper and the trailing edge being more drawn out. These distortions are commonly encountered in adsorption chromatography and are responsible for separations that are not as complete as predicted by theory.

Applications. Separations based on adsorption chromatography rely on the equilibria that govern the distribution of the various solute species between the solvent and the surface of the solid. Large differences exist in the tendencies of compounds to be adsorbed. For example, a positive correlation can be discerned between adsorption properties and the number of hydroxyl groups in a molecule; a similar correlation exists with double bonds. Compounds containing certain functional groups are more strongly held than others. The tendency to be adsorbed decreases in the following order: acid > alcohol > carbonyl > ester > hydrocarbon. The nature of the adsorbent is also influential in determining the order of adsorption. Much of the available knowledge in this field is empirical; the choice of adsorbent and solvent for a given separation frequently must be made on a trial-and-error basis.

Adsorption chromatography has been used primarily for separation of organic compounds; several monographs concerned with applications have been published.[8]

Ion-exchange Chromatography

Ion exchange is a process in which an interchange of ions of like sign occurs between a solution and an essentially insoluble solid in contact with the solution. Many substances, both natural and synthetic, act as ion exchangers. Among the former are clays and zeolites; the ion-exchange properties of these materials have been recognized and studied for over a century. Synthetic ion-exchange resins were first produced in 1935 and have since found widespread laboratory and industrial application for water softening, water deionization, solution purification, and ion separation.

Synthetic ion-exchange resins are high molecular-weight, polymeric materials containing large numbers of ionic functional groups per molecule. For cation exchange there is a choice between strong acid type resins containing sulfonic acid groups (RSO_3H), or weak acid resins containing carboxylic acid ($RCOOH$) groups. The former have wider application. Anion-exchange resins contain basic functional groups attached to the polymer molecule. These are generally amines; strong base exchangers contain

---

[8]For example, see E. Lederer and M. Lederer, *Chromatography,* 2d ed. (New York: American Elsevier Publishing Company, Inc., 1957); H. G. Cassidy, *Fundamentals of Chromatography* (New York: Interscience Publishers, Inc., 1957); H. H. Strain, *Anal. Chem.,* **21**, 75 (1949); **22**, 41 (1950); **23**, 25 (1951); **24**, 50 (1952); **26**, 90 (1954); **30**, 620 (1958); **32**, 3R (1960); E. Heftmann, *ibid.,* **34**, 13R (1962); **36**, 14R (1964); **38**, 31R (1966); G. Zweig, *ibid.,* **40**, 490R (1968); G. Zweig and R. B. Moore, *ibid.,* **42**, 349R (1970).

quaternary amines ($RN(CH_3)_3^+OH^-$), and weak base types contain secondary or tertiary amines.

Ion-exchange equilibria. A cation-exchange process is illustrated by the equilibrium

$$x RSO_3^-H^+ + M^{x+} \rightleftharpoons (RSO_3^-)_x M^{x+} + xH^+$$

solid   solution   solid   solution

where M^{x+} represents a cation and R represents *a part* of a resin molecule. The analogous process involving a typical anion-exchange resin can be written as

$$x RN(CH_3)_3^+OH^- + A^{x-} \rightleftharpoons (RN(CH_3)_3^+)_x A^{x-} + xOH^-$$

solid   solution   solid   solution

where A^{x-} is an anion.

The identity of the mobile ion attached to an ion-exchange resin is determined by the electrolyte content of the solution with which the resin is in contact. For example, if a sulfonic acid resin is treated with a solution of sodium chloride, the following exchange reaction takes place:

$$RSO_3^-H^+ + Na^+ \rightleftharpoons RSO_3^-Na^+ + H^+$$

where R again represents one unit of the resin molecule. If the salt concentration is high, essentially all of the resin is converted to the sodium form. Similarly, if an anion-exchange resin is treated with strong sodium chloride, the reaction

$$RN(CH_3)_3^+OH^- + Cl^- \rightleftharpoons RN(CH_3)_3^+Cl^- + OH^-$$

is strongly favored, and the chloride form of the resin predominates. More than one functional unit of the resin is involved in exchange reactions with polyvalent ions. Thus, when a sulfonic acid resin is treated with calcium ions, the equilibrium can be expressed as

$$(RSO_3^-H^+)_2 + Ca^{2+} \rightleftharpoons (RSO_3^-)_2 Ca^{2+} + 2H^+$$

Ion-exchange equilibrium constants. The simplest approach to the theoretical treatment of ion-exchange equilibria is to employ the mass-action law. In general, the reaction can be expressed as

$$A + BR \rightleftharpoons B + AR$$

where A and B are ions in the aqueous phase and AR and BR represent the corresponding ions bound on the resin. The equilibrium constant for the system can be written as

$$K = \frac{a_B \, a_{AR}}{a_A \, a_{BR}} \qquad (23\text{-}27)$$

where a_B and a_A represent activities of the two ions in the aqueous solution and a_{AR} and a_{BR} represent their activities on the solid-resin phase.

The latter two terms can be replaced by products of activity coefficients and mole-fraction terms. That is,

$$K = \frac{a_B \, X_{AR} \, f_{AR}}{a_A \, X_{BR} \, f_{BR}} \tag{23-28}$$

where X_{AR} represents the mole fraction of the resin that is in the form of AR, and f_{AR} is an activity coefficient; X_{BR} and f_{BR} are analogously defined.

Equation (23-28) can be rearranged to give

$$\frac{a_B \, X_{AR}}{a_A \, X_{BR}} = K \frac{f_{BR}}{f_{AR}} = K_P \tag{23-29}$$

where K_P is called the *apparent, or practical equilibrium quotient.*

It is important to note that the activity ratio f_{BR}/f_{AR} cannot be directly measured by independent means, and the value of K is thus unknown. On the other hand, K_P can be determined experimentally by measuring the concentration of ions on a resin and their activities in the solution in equilibrium with the resin. Such experiments show that K_P is not entirely independent of changes in ionic concentrations on the resin or the activities of ions in the solution, particularly when the ions bear different charges. It has been found, however, that if one of the ions is in large excess, *both in the eluting solution and on the resin,* K_P is relatively constant and independent of changes in concentration of the other ion; that is, when $a_B \gg a_A$, $X_{BR} \gg X_{AR}$, and a_B is fixed, equation (23-29) becomes

$$\frac{X_{AR}}{a_A} = \frac{X_{BR} \, K_P}{a_B} \cong K_D \tag{23-30}$$

where K_D is a partition ratio analogous to K in equation (23-18). Thus, to the extent that K_D is constant, the general theory we have developed for chromatography also can be applied to column separations involving ion-exchange resins.

Note that values of K_D represent the affinity of a given resin for some ion A *relative* to another ion B. Where K_D is large, there is a strong tendency for the solid phase to retain ion A; where K_D is small, the reverse obtains. By selecting a common reference ion B, distribution ratios for different ions on a given type of resin can be compared. Such experiments reveal that polyvalent ions are much more strongly held than singly charged species. Within a given charge group, however, differences appear that are related to the size of the hydrated ion and other properties. Thus, for a typical sulfonated cation-exchange resin, values for K_D decrease in the order $Cs^+ > Rb^+ > K^+ > NH_4^+ > Na^+ > H^+ > Li^+$; for bivalent cations, the order is $Ba^{2+} > Pb^{2+} > Sr^{2+} > Ca^{2+} > Cd^{2+} > Cu^{2+} > Zn^{2+} > Mg^{2+}$.

Application of ion-exchange chromatography.[9] The techniques for fractionation of ions with K_D values that are relatively close to one another

[9]O. Samuelson, *Ion Exchangers in Analytical Chemistry* (New York: John Wiley & Sons, Inc., 1953).

are analogous to those described for adsorption and partition chromatography. For example, Beukenkamp and Rieman[10] report the separation of potassium and sodium ions by introducing the sample at the top of a column containing a sulfonic acid resin in its acidic form. The column is then eluted with a solution of hydrochloric acid. Sodium ions, being the less strongly held, move down the column more rapidly and can be collected before potassium ions appear in the effluent. In this application, partition of the sodium ions in each theoretical plate of the column involves

$$R^-H^+ + Na^+ \rightleftharpoons R^-Na^+ + H^+$$

and, since hydrogen ions are in excess in both solvent and resin, equation (23-29) applies. Thus,

$$K_D = \frac{X_{NaR}}{a_{Na}}$$

The potassium ion behaves in a similar fashion, but its partition coefficient is numerically larger. By a treatment similar to that described in the sections on partition chromatography and adsorption chromatography, equations can be developed that relate the migration rate of each ion to its partition ratio; from these relationships the length of column required to achieve a particular separation can be calculated.

A number of important separations are based upon ion exchange. Among these is the separation of the rare earths, primarily for preparation purposes. Here, fractionation is enhanced by elution with complexing agents which form complexes of differing stabilities with the various cations.[11] Ion-exchange resins have also been widely used to resolve mixtures of biological importance. For example, Figure 23-8 shows a partial

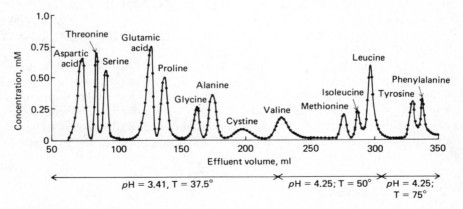

Figure 23-8. Separation of amino acids on a cation exchanger. (Dowex 50, Na form; 100 cm column). Total sample ~ 6 mg. Taken from S. Moore and W. H. Stein, *J. Biol. Chem.*, **192**, 663 (1951). (With permission of the copyright owners.)

[10]J. Beukenkamp and W. Rieman, *Anal. Chem.*, **22**, 582 (1950).
[11]See a series of papers in *J. Am. Chem. Soc.*, **69**, 2769-2881 (1947).

chromatogram of a synthetic mixture which simulates the composition of a protein hydrolysate. The amino acids, histidine, lysine, and arginine appear with higher volumes of eluent. Note that elutions were performed under varying conditions for different parts of the chromatogram. The positions of the several peaks are reported to be reproducible to better than five percent.

Separation of interfering ions of opposite charge. Ion-exchange resins are useful for removal of interfering ions, particularly where these ions have a charge opposite to the species being determined. For example, iron(III), aluminum(III), and other cations cause difficulty in the determination of sulfate by virtue of their tendency to coprecipitate with barium sulfate. Passage of a solution to be analyzed through a column containing a cation-exchange resin results in retention of all cations and the liberation of a corresponding number of protons. The sulfate ion, on the other hand, passes freely through the column, and the analysis can be performed on the effluent. In a similar manner, phosphate ion, which interferes in the analysis of barium or calcium ions, can be removed by passing the sample through an anion-exchange resin.

Concentration of traces of an electrolyte. A useful application of ion exchangers involves the concentration of traces of an ion from a very dilute solution. Cation-exchange resins, for example, have been employed to collect traces of metallic elements from large volumes of natural waters. The ions are then liberated from the resin by treatment with acid; the result is a considerably more concentrated solution for analysis.

Conversion of salts to acids or bases. An interesting application of ion-exchange resins is the determination of the total salt content of a sample. This analysis can be accomplished by passing the sample through the acid form of a cation-exchange resin; absorption of the cations causes the release of an equivalent quantity of hydrogen ion, which can be collected in the washings from the column and then titrated. Similarly, a standard acid solution can be prepared from a salt; for example, a cation-exchange column in the acid form can be treated with a weighed quantity of sodium chloride. The salt liberates an equivalent quantity of hydrochloric acid, which is then collected in the washings and diluted to a known volume.

Treatment of an anion-exchange resin with salts liberates the hydroxide ion by an analogous mechanism.

Paper Chromatography[12]

Paper chromatography is a remarkably simple form of the chromatographic method which has become an extremely valuable analytical tool

[12]For a more complete discussion see H. G. Cassidy, *Fundamentals of Chromatography* (New York: Interscience Publishers, Inc., 1957), Chap. 7; H. J. Pazdera and W. H. McMullen in I. M. Kolthoff and P. J. Elving, Eds., *Treatise on Analytical Chemistry*, Part I, Vol. 3 (New York: Interscience Publishers, Inc., 1961), Chap. 36.

to the organic chemist and the biochemist. In this application a coarse paper serves in lieu of a packed column. A drop of solution containing the sample is introduced at some point on the paper; migration then occurs as a result of flow by a mobile phase called the *developer*. Movement of the developer is caused by capillary forces. In some applications the flow is in a downward direction (*descending development*), in which case gravity also contributes to the motion. In *ascending development* the motion of the mobile phase is upward; in *radial development* it is outward from a central spot.

Theory. As ordinarily encountered, paper chromatography appears to be a type of partition chromatography in which the stationary phase is water absorbed on the hydrophilic surface of the paper. An organic solvent then serves as the mobile phase. By suitable treatment of the paper, the water can be replaced by a nonpolar, stationary liquid phase; aqueous solutions can then be used as developers. In still another variation, the paper is impregnated with anhydrous silica, alumina, or an ion-exchange resin; here, partition occurs as a consequence of solid–liquid or ion-exchange equilibria.

The general principles developed earlier for column partition chromatography appear to apply as well to paper media. Thus, the paper is conceived of as consisting of a series of theoretical plates within which equilibration occurs according to the relationship

$$K = C_s/C_m$$

It is customary to describe the travel of a particular solute in terms of its retardation factor R_F, which is defined as

$$R_F = \frac{\text{distance of solute motion}}{\text{distance of solvent motion}} \tag{23-31}$$

The retardation factor is also given by equation (23-25)

$$R_F = \frac{A_m}{A_m + K A_s}$$

The cross-sectional areas A_m and A_s are difficult to estimate, however, so that this equation is of little practical importance.

To compensate for uncontrolled variables, the distance traveled by a solute frequently is compared with that for a standard substance under identical conditions. The ratio of these distances is designated as R_{std}.

Techniques. Typical arrangements for paper chromatography are shown in Figure 23-9. The paper is usually 15 to 30 cm in length and one to several centimeters in width. A drop of the sample solution is placed a short distance from one end of the paper and its position is marked with a pencil. The original solvent is allowed to evaporate. The end of the paper nearest the sample is then brought in contact with the developer. Both paper and developer are sealed in a container to prevent evaporation losses (see

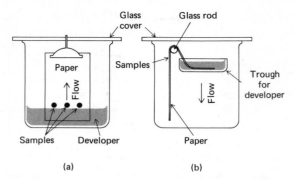

(a)   (b)

Figure 23-9. Apparatus for paper chromatography. (a) Ascending flow; (b) descending flow.

Figure 23-9). After the solvent has traversed the length of the strip, the paper is removed and dried. The positions of the various components are then determined in any of a variety of ways. Reagents that form colored compounds in the presence of certain functional groups are available for this purpose. For example, a ninhydrin solution causes the development of blue or purple stains that reveal the location of amines and amino acids on the paper. Detection has also been based upon the absorption of ultraviolet radiation, fluorescence, radioactivity, and so forth.

Development can be carried out in two directions (*two-dimensional paper chromatography*). Here, the sample is placed on one corner of a square sheet of paper. Development along one axis is then performed in the usual way. After evaporation of the solvent, the paper is rotated 90 deg and is again developed, but with a different solvent. Solutes that cannot be resolved by a single solvent are often separated by this technique. Figure 23-10 is a typical two-dimensional chromatogram.

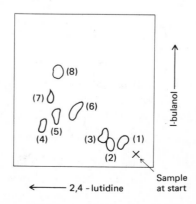

Figure 23-10. Two-dimensional chromatogram of an amino acid mixture. (1) Serine, (2) glycine, (3) alanine, (4) tryptophan, (5) methionine, (6) valine, (7) phenylalanine, and (8) leucine.

Applications. Paper chromatography has been widely employed for the qualitative analysis of inorganic, organic, and biochemical mixtures. Its greatest impact has been felt in the biochemical field where complicated mixtures of closely related compounds are so frequently encountered. One of the most important advantages of the method is its sensitivity; ordinarily, a few micrograms of sample are sufficient for an analysis.

Thin-layer Chromatography[13]

The techniques of thin-layer chromatography closely resemble those of paper chromatography. Here, however, partition occurs on a layer of finely divided adsorbent which is supported upon a glass plate.

Adsorbents. Adsorbent materials used for thin-layer chromatography include silica gel, alumina, diatomaceous earth, and powdered cellulose. A thin-layer plate is prepared by spreading an aqueous slurry of the finely ground adsorbent over the surface of a glass plate or a microscope slide. The plate is then allowed to stand until the layer has set up; for many purposes, it may be heated in an oven for several hours.

The type of equilibrium involved in thin-layer chromatography depends upon the composition of the layer and the way it has been prepared. Thus, if a silica gel film is deposited from aqueous solution and allowed to set up at room temperature, the particles undoubtedly remain coated with a thin film of water. If the sample is then distributed with an organic solvent, resolution will most likely be the result of a liquid-liquid partition. On the other hand, if the silica film is dried by heating, partition may involve solid–liquid adsorption equilibria.

Development and detection. The sample is spotted at one end of the plate and then developed by the ascending technique used in paper chromatography. Development is carried out in a closed container saturated with developer vapor. The plate is then dried and sprayed with a reagent for detection of the components or, more commonly, exposed to iodine vapor. Brown spots indicate the solute positions; identification is based upon R_F values.

Applications. Thin-layer chromatography is generally faster than paper chromatography and gives more reproducible R_F values. It is widely used for the identification of components in drugs, biochemical preparations, and natural products.

---

[13]See R. Maier and H. K. Mangold in C. N. Reilley, Ed., *Advances in Analytical Chemistry and Instrumentation.* Vol. 3 (New York: Interscience Publishers, Inc., 1964), p. 369.

Electrophoresis and Electrochromatography[14]

Electrophoresis is defined as the migration of particles through a solution under the influence of an electrical field. Historically, the particles referred to in this definition were colloidal and owed their charge to adsorbed ions. This definition has now become too restrictive; the term electrophoresis is presently applied both to the migration of individual ions as well as to colloidal aggregates. Electrophoretic methods provide a powerful means of fractionating the components of a mixture, be they aggregates or monodispersed.

Electrophoretic methods can be subdivided into two categories depending upon whether the separation is carried out in the absence or presence of a supporting or stabilizing medium. In the *free solution method* the sample solution is introduced as a band at the bottom of a U-tube filled with a buffered liquid. A field is applied by means of electrodes located at the tube ends; the differential movement of the charged particles toward one or the other electrode is observed. Separations occur as a result of differences in migration rates; these in turn are related to the charge-to-mass ratios and the inherent mobilities of the species in the medium. The free-solution method was perfected and applied to the separation of proteins by Tiselius; for this work he received the Nobel prize. The free-solution procedure has been of signal importance in the development of biochemistry; nevertheless, its widespread application has been hindered by such experimental problems as the tendency of the separated components to mix by convection as a consequence of thermal and density gradients as well as mechanical vibrations. In addition, elaborate optical systems are often needed to detect the bands of the separated species.

Many of the experimental difficulties associated with free-solution electrophoresis are avoided if the separations are carried out in a stabilizing medium such as a paper, a layer of finely divided solid, or a column packed with a suitable solid. Here, the experimental techniques closely resemble the various chromatographic methods discussed earlier, with the additional parameter of the superimposed electrical field. Depending upon the properties of the medium, the separations may result primarily from the electrophoretic effect or from a combination of electrophoresis and adsorption, ion exchange, or other distribution equilibria. Methods based upon electrophoresis in a stabilizing medium bear a variety of names, including *electrochromatography, zone electrophoresis, electromigration,* and *ionophoresis.* Our discussion is limited to this type of electrophoresis.

Experimental methods of electrochromatography. The solid media employed in electrochromatography are as numerous and varied as those

---

[14]For a detailed discussion see E. Heftman, *Chromatography,* 2d ed. (New York: Reinhold Publishing Corporation, 1967) Chaps. 10 and 11; J. R. Cann in I. M. Kolthoff and P. J. Elving, Eds., *Treatise on Analytical Chemistry,* Part I, Vol. 2 (New York: Interscience Publishers, Inc., 1961), Chap. 28.

found in the other chromatographic methods. Examples include paper, cellulose acetate membranes, cellulose powders, starch gels, ion-exchange resins, glass powders, and agar gels. Depending upon the physical nature of the solid, separations are performed on strips of paper or membranes, in columns, in trays, or in thin layers supported by glass or plastic. Despite certain disadvantages, filter paper is the most widely used stabilizing material; we will focus our attention on paper electrochromatography.

Figure 23-11 is a schematic diagram illustrating three of the many ways for performing a paper electrophoresis. In Figure 23-11(a), a strip of paper is held horizontally between two containers filled with a buffer mixture; the paper is well-soaked with buffer and evaporation is prevented by housing the apparatus in an air-tight container. The sample is introduced as a band at the center of the strip and a dc potential of 100 to 1000 V is applied across the two electrodes. The latter are sufficiently isolated from the paper to prevent the electrode reactions from altering the electrolyte composition of the buffer on the paper. Currents in the milliampere range ordinarily flow. After a suitable electrolysis period, the paper is removed, dried, and the component bands are detected by application of suitable colorimetric reagents.

The inverted V arrangement shown in Figure 23-11(b) is also widely used. Here, the sample is introduced at the apex of the V; cationic species migrate down one arm of the V and anionic constituents down the other.

Many other modifications of the apparatus shown in Figures 23-11(a) and 23-11(b) have been described, and several are available commercially. In some, the paper is supported (either horizontally or at some suitable angle) between two glass or plastic plates; the plates may be cooled to dissipate the heat generated by the current flow. In all of these devices, the rate at which components migrate is controlled primarily by their charge-to-mass ratios and their mobilities, with adsorption or other equilibria having only a secondary influence.

Figure 23-11(c) illustrates an apparatus for a type of two-dimensional electrochromatography. Here, the paper is held vertically and the sample components are carried down the sheet by the flow of a buffered solvent. Separation then occurs in the vertical direction as a consequence of differences in distribution ratios between the mobile and the fixed phases. In addition, however, a field is applied at right angles to the solution flow, and differential electromigration along the horizontal axis occurs as a consequence. Thus, the various species in the sample describe a radial path from the point at which the sample is introduced.

The technique illustrated in Figure 23-11(c) has been employed for both analytical and preparatory purposes. In the former, an amount of buffer just equal to that retained by the paper is allowed to pass before the experiment is discontinued; the paper is then removed, dried, and the detection reagents are applied. In preparatory applications the buffer flow is continued and the various fractions are collected as shown in Figure 23-11(c).

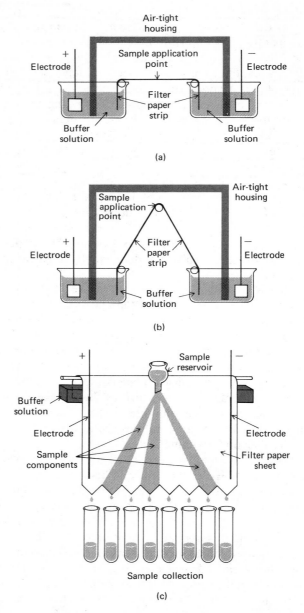

Figure 23-11. Some types of apparatus for paper electrochromatography.

Applications of electrochromatography.[15] Electrochromatographic methods are indispensible to the clinical chemist and the biochemist, who
use them for the fractionation of an amazing number of biological materials.
The widest application has perhaps been in clinical diagnosis for the separation of proteins and other large molecules contained in serum, urine, spinal
fluid, gastric juices, and other body fluids.

Electrochromatography has also been applied widely by biochemists
for the fractionation of smaller molecules such as alkaloids, antibiotics,
nucleic acids, vitamins, natural pigments, steroids, amino acids, carbohydrates, and organic acids.

It has also been shown that electrochromatography provides a convenient means for the separation of inorganic ions. Figure 23-12(a) illustrates
the separation of six metallic ions from a complexing medium. Note that
both anodic and cathodic migration has occurred depending upon the charge
of the complex ion. Note also that the movement of the nickel(II) ion has
been retarded by the precipitant dimethylglyoxime.

Figure 23-12(b) illustrates the path traveled by the components of a
sample when subjected to continuous electrophoresis and elution. By
collecting the eluate at appropriate places on the bottom edge of the paper,
solutions of the separated species can be obtained.

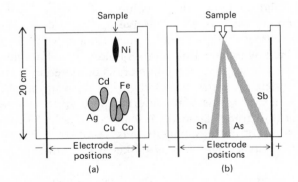

Figure 23-12. Electrochromatograms for inorganic ions. (a) Cations (0.05 M)
in 0.1 M tartaric acid; wash liquid was 0.01 M in ammonium tartrate, 0.005 M in
dimethylglyoxime, and 4 M in ammonia. Spots were detected with hydrogen
sulfide: 160 V, 100 mamp, 20 min. (b) Continuous separation. Solution 0.005
M each in As(III), Sb(III), and Sn(II); solution was also 0.02 M in tartaric acid,
0.04 M in lactic acid, and 0.04 M in *dl*-alanine. Paths were detected with hydrogen sulfide. 300 V, 95 mamp. Taken from H. H. Strain and J. C. Sullivan, *Anal.
Chem.*, **23**, 816 (1951). (With permission of the American Chemical Society.)

[15]For a review of applications see G. Zweig and J. R. Whitaker, *Paper Chromatography
and Electrophoresis,* Vol. 1 (New York: Academic Press, Inc., 1967).

Gel Chromatography[16]

Gel chromatography is a technique in which fractionation is based, at least in part, upon the molecular size and shape of the species in the sample. Several names have been given to this procedure, including gel-permeation chromatography, exclusion chromatography, and molecular-sieve chromatography.

Ordinarily, gel chromatography is performed on a column by the elution method. Here, the degree of retardation depends upon the extent to which the solute molecules or ions can penetrate that part of the solution phase which is held within the pores of the highly porous gel-like packing material. Molecules or ions larger than the pores of the gel are completely excluded from the interior, while smaller species are more or less free to enter these regions. Thus, larger molecules pass quickly through the column, while smaller ones are retarded to a greater or lesser extent depending upon their size, their shape, and sometimes upon their tendency to be adsorbed by the gel.

Column packing. The stationary phase in column chromatography consists of beads of a porous polymeric material that readily absorbs water (and in some instances other solvents) and swells as a consequence. The resulting solid contains a large volume of solvent fixed in the interstices of the polymeric network. The average size of the resulting pores or interstices is directly related to the quantity of solvent absorbed; this in turn is determined by the amount of cross linking in the polymer molecules.

One of the most widely used polymers is prepared by cross linking the polysaccharide dextran with epichlorhydrin. By varying the amount of the latter, a series of resins with differing pore sizes is obtained. These gels sell under the trade name of Sephadex®. Table 23-2 describes the properties of some of these Sephadex gels. Another group of commercial resins consists of polyacrylamide cross-linked with methylene bisacrylamide.

Theory of gel chromatography.[17]   The total volume of a column packed with a gel that has been swelled by water (or other solvent) is given by

$$V_t = V_g + V_i + V_o$$

where V_g is the volume occupied by the solid matrix of the gel; V_i is the volume of solvent held in its interstices and V_o is the free volume outside the gel particles. Assuming no mixing or diffusion, V_o also represents the volume of solvent required to carry through the column those components

---

[16]For further information concerning this technique see D. D. Bly, *Science*, **168**, 527 (1970).

[17]For further information see R. L. Pecsok and D. Saunders in R. A. Keller, Ed., *Separation Techniques in Chemistry and Biochemistry,* (New York: Marcel Dekker, Inc., 1967), pp. 83–85.

Table 23-2 Properties of Commercial Sephadex[a] Gels

Gel Designation	Volume Relationships[b] (ml/g original gel)			Approximate Exclusion Limit, mol wt
	V_g	V_i	V_o	
G-10	0.6	1.0	0.9	700
G-15	0.6	1.5	0.9	1,500
G-25	0.5	2.5	2	5,000
G-50	1	5	4	10,000
G-75	1	7	5	50,000
G-100	1	10	6	100,000
G-150	1	15	8	150,000
G-200	1	20	9	200,000

[a]Pharmacia Fine Chemicals, Inc., Piscataway, N.J.
[b]V_g = volume occupied by solid gel matrix.
V_i = volume of solvent held interstitially.
V_o = liquid volume between the gel beads.

that are too large to enter the pores of the gel. In fact, however, some mixing and diffusion will occur, and as a consequence the components will appear in a gaussian-shaped band with a concentration maximum at V_o. For those components that are small enough to enter freely into the interstices of the gel, the band maxima will appear at the end of the column at a volume corresponding to $(V_i + V_o)$. Generally, V_i, V_o, and V_g are of the same order of magnitude; thus, a gel column permits the separation of the large molecules of a sample from the small with a minimal volume of wash liquid.

Molecules of intermediate size range are able to penetrate some fraction K_d of the interstitially held solvent; for these, the elution volume V_e is

$$V_e = V_o + K_d V_i \qquad (23\text{-}32)$$

Equation (23-32) describes the behavior of a gel column with respect to all solutes. For molecules too large to enter the gel pores, $K_d = 0$ and $V_e = V_o$; for molecules that can enter the pores unhindered, $K_d = 1$ and $V_e = (V_o + V_i)$. In arriving at equation (23-31) the assumption was made that no interaction, such as adsorption, occurs between the solute molecules and the gel surfaces. If such interaction does occur, the amount of interstitially held solute will be increased; for interacting solutes that can freely enter the pores, K_d will be greater than unity.

Table 23-3 gives experimentally determined values of K_d for some dextran-type gels.

Applications of gel chromatography. Tightly cross-linked gels with small pore sizes, such as Sephadex G-25 and G-50, have found wide application to the *desalting* or removal of low molecular weight molecules from high molecular weight, natural product molecules. For example, it is evident

Table 23-3 K_d Values with Sephadex Gels[a]

		K_d			
Substance	*Approximate mol wt*	*Sephadex G-25*	*Sephadex G-75*	*Sephadex G-100*	*Sephadex G-200*
Ammonium sulfate	132	0.9	–	–	–
Potassium chloride	74	1.0	–	–	–
Tryptophan	204	2.2	1.2	–	–
Glycine	76	0.9	1.0	–	–
Ribonuclease	13,000	0	0.4	–	–
Trypsin	24,000	0	0.3	0.5	0.7
Serum albumin	75,000	0	0	0.2	0.4
Fibrinogen	330,000	0	0	0.0	0.0

[a]Pharmacia Fine Chemicals, Inc., Piscataway, N.J. From B. Gelotte, "Fractionation of Proteins, Peptides, and Amino Acids by Gel Filtration," in James and Morris, *New Biochemical Separations* (Princeton: Van Nostrand, 1964). See Table 23-2 for the properties of these gels.

from Table 23-3 that the G-25 gel will cleanly separate salts and amino acids from most proteins. Here the proteins, all with K_d values of zero, are completely excluded from the gel interior while the low molecular weight species have K_d values approaching unity. As a result, clean separation can be expected even with large sample volumes (as much as 20 to 30 percent of the total volume of the bed). This type of separation performs the same function as a dialysis through a cellophane membrane, but is more rapid and convenient to perform.

Sephadex G-25 and G-50 gels have also proved useful for the fractionation of peptides having a size range intermediate between the proteins shown in Table 23-3 and the low molecular weight species. These compounds have K_d values greater than zero but less than one. Elution techniques for small samples are similar to those used in other types of chromatography.

The more porous gels shown in Table 23-3 have found wide application to the fractionation and the purification of macromolecules such as proteins, nucleic acids, and polysaccharides. For example, Sephadex G-75 permits the clean separation of ribonuclease, trypsin, pepsin, and cytochrome c from higher molecular weight proteins such as serum albumin, hemoglobin, and fibrinogen (see Table 23-3).

Finally, it should be noted that gel permeation chromatography has been used by polymer chemists and biochemists for the estimation of molecular weights of large molecules. Here, the elution volume of the unknown is compared with elution volumes for a series of standard compounds having the same chemical characteristics.

Problems

1. (a) Perform a simple calculation to verify the positions of the maxima shown in Figures 23-3 and 23-4.

    (b) Calculate the fraction of the solute contained in the tube corresponding to each of the maxima above.

2. (a) If solutes A and B are found to have partition coefficients of 0.5 and 2.0 for an organic solvent-water mixture respectively, how many transfers in the Craig apparatus are required in order that their maxima in a distribution curve would differ by 20 tubes? Where would the two maxima occur? Assume that equal volumes of solvent are used in each tube.

    (b) Calculate the fraction of each solute present in each of the tubes corresponding to the maxima in (a).

    (c) If r'_{max} is the tube number corresponding to the first maximum in part (a), calculate the fraction of each solute in tubes $(r'_{max} + 5)$, $(r'_{max} + 10)$, and $(r'_{max} + 15)$.

    (d) Plot the data from (a) and (c) and estimate the tube number corresponding to the intersection of the lines for the two solutes.

    (e) From the curves in (d), calculate the fraction of B that would be found in tubes 30 and above. Calculate the approximate fraction of A in these same tubes.

3. Repeat the calculations in Problem 2 if the two solutes have partition coefficients 0.300 and 3.00.

4. Derive a distribution curve (fraction of solute versus tube number) after 80 transfers for a solute separated in a Craig apparatus if its partition coefficient between the phases were 4.0 and if equal volumes of the two solvents were employed.

5. A glass column having a cross-sectional area of 1.00 cm² was packed to a depth of 10.0 cm with a finely ground solid packing. The density of the solid was 2.5 g/cm³ and 15.0 g were required.

    (a) Calculate the retardation factors for solutes having partition ratios between the solid and a liquid phase (C_s/C_m) of 1.0, 2.0, and 4.0.

    (b) What volumes of the liquid phase would need to be added to bring the maxima for each of the solutes in (a) to the end of the column?

6. A column for partition chromatography was prepared by packing a 10-mm I.D. glass tube with 30 g of anhydrous silica gel (density = 2.3 g/cm³), which was subsequently treated so that it held 7.0 g of water on its surface. The gel packing occupied 32.0 cm of the tube. Two solutes were placed at the head of the column and eluted with chloroform. Calculate the volume of $CHCl_3$ required to bring the maxima for each of the solutes to the end of the packing if their partition ratios were 2.0 and 3.2, respectively.

24 Gas–liquid Chromatography

In gas chromatography the components of a vaporized sample are fractionated as a consequence of partition between a mobile gaseous phase and a stationary phase held in a column. *Gas-solid chromatography* (GSC) represents a subclassification in which the fixed phase is a solid; the partition process, then, involves gaseous adsorption equilibria. In *gas-liquid chromatography* (GLC) the stationary phase is a liquid, supported on an inert solid matrix; here gas-liquid equilibria are important.

Gas-solid chromatography has thus far had only limited application owing to tailing of elution peaks (a direct result of the nonlinear character of adsorption isotherms), difficulties in reproducing surface areas, and semipermanent retention of active gases on solid surfaces. Gas-liquid chromatography, on the other hand, has become the most important and widely used of all of the column chromatographic methods and is probably the fractionation process used most often by all types of chemists.

The concept of gas-liquid chromatography was first described in 1941 by Martin and Synge, who were also responsible for liquid-partition chromatography. Over a decade was to elapse, however, before the value of this method was demonstrated experimentally.[1] Since that time, the growth in applications of the procedure has been phenomenal.[2]

[1] A. J. James and A. J. P. Martin, *Analyst,* **77,** 915 (1952); *Biochem. J.,* **50,** 679 (1952).

[2] For detailed discussions of the technique, see A. I. M. Keulemans, *Gas Chromatography,* 2d ed. (New York: Reinhold Publishing Corporation, 1959); S. Dal Nogare and R. S. Juvet, Jr., *Gas-Liquid Chromatography* (New York: Interscience Publishers, Inc., 1962); H. P. Burchfield and E. Starrs, *Biochemical Applications of Gas Chromatography* (New York: Academic Press, Inc., 1962).

In principle, gas-liquid and liquid-liquid partition chromatography differ only in the detail that the mobile phase of the former is a gas rather than a liquid. The sample is introduced as a gas at the head of the column; those components that have a finite solubility in the stationary liquid phase distribute themselves between this phase and the gas according to the equilibrium law. Elution is then accomplished by forcing an inert gas, such as nitrogen or helium, through the column. The rate of movement of the various components along the column depends upon their tendency to dissolve in the stationary liquid phase. A distribution coefficient favoring the solvent results in a low rate; on the other hand, those components whose solubility in the liquid phase is negligible move rapidly through the column. Ideally, bell-shaped elution curves are obtained. Qualitative identification of the components is based upon the time required for the peak to appear at the end of the column; quantitative data are obtained from evaluation of peak areas. A typical gas chromatogram is shown in Figure 24-1.

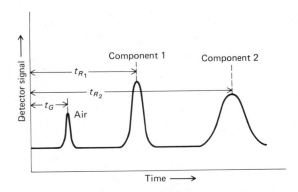

Figure 24-1.   Gas chromatogram for a two-component system.

PRINCIPLES OF GAS-LIQUID CHROMATOGRAPHY

Although frontal and displacement methods can be applied, only the elution method is of importance in gas chromatography. An inert gas (*carrier gas*) ordinarily serves as the eluent.

Definition of Some Terms

The time required for the maximum for a solute peak to reach the detector in a gas chromatographic column is called the *retention time* (see Figure 24-1). If the sample contains a component whose solubility in the liquid phase is extremely low (oxygen or air, for example), its rate of movement will approach the rate of flow of the carrier gas, and its retention time t_G will be shorter than the retention time t_R of a component that partitions between the gas and liquid phases; note the air peak in Figure 24-1.

The *retention volume* V_R of a component is defined as the volume of gas required to carry a component maximum through the column. This quantity is related to retention time by

$$V_R = t_R F_c$$

where F_c is the volume flow rate of the gas at the outlet corrected to the temperature of the column.[3] For a component that does not partition appreciably with the liquid phase,

$$V_G = t_G F_c$$

A more useful term is the *corrected retention volume* V_R^0, which is the volume corrected to the average pressure in the column. James and Martin[4] have shown that

$$V_R^0 = j\, t_R\, F_c \quad \text{and} \quad V_G^0 = j\, t_G\, F_c \tag{24-1}$$

where

$$j = \frac{3[(p_i/p_o)^2 - 1]}{2[(p_i/p_o)^3 - 1]} \tag{24-2}$$

where p_i and p_o are the gas pressures at the inlet and the outlet of the column.

The *retardation factor* R_F is the ratio between the rates of movement of sample and carrier gas through the column. If the column has a length l,

$$\frac{\text{rate of movement}}{\text{for component}} = \frac{l}{t_R}$$

Similarly,

$$\frac{\text{rate of movement}}{\text{for carrier gas}} = \frac{l}{t_G}$$

Thus,

$$R_F = \frac{l/t_R}{l/t_G} = \frac{t_G}{t_R} \tag{24-3}$$

Combination of equations (24-1) and (24-3) gives R_F in terms of corrected retention volumes. That is,

$$R_F = \frac{V_G^0}{V_R^0} \tag{24-4}$$

Theory of Gas-liquid Chromatography

The plate theory, employed for liquid-liquid partition chromatography, is readily adapted to the gas-liquid system. Ideally, the distribution of a

[3] The flow rate will frequently be measured when the gas is at room temperature. Then $F_c = F_c' \times T_c/T_r$ where F_c' is the observed rate and T_c and T_r are the column and room temperatures in degrees Kelvin.

[4] A. J. James and A. J. P. Martin, *Analyst*, **77**, 915 (1952); *Biochem. J.*, **50**, 679 (1952).

solute between a gas and a liquid phase can be described by means of a partition coefficient K_g, where

$$K_g = \frac{\text{conc. of solute in liquid phase (wt/ml)}}{\text{conc. of solute in gas phase (wt/ml)}} \qquad (24\text{-}5)$$

At the low concentrations used in gas-liquid chromatography, K_g is often independent of concentration.

With this definition equation (23-25) can be applied to a gas-liquid system; that is,

$$R_F = \frac{A_m}{A_m + K_g A_s} \qquad (24\text{-}6)$$

where A_m is the cross-sectional area of the gas phase and A_s is that for the stationary liquid phase. For purposes of gas-liquid chromatography it is convenient to express R_F in terms of volumes of the gas and liquid phases; letting V_L be the total volume of the liquid contained in the column, and remembering that V_G^0 is the corrected volume of the gas phase,

$$V_G^0 = A_m l$$

and

$$V_L = A_s l$$

where l is the length of the column. Substituting into equation (24-6), we find

$$R_F = \frac{V_G^0}{V_G^0 + K_g V_L} \qquad (24\text{-}7)$$

A comparison of equation (24-7) with equation (24-4) reveals that the corrected retention volume is given by

$$V_R^0 = (V_G^0 + K_g V_L) \qquad (24\text{-}8)$$

By substituting equation (24-1) into equation (24-7) we obtain

$$t_R = \frac{(V_G^0 + K_g V_L)}{j F_c} \qquad (24\text{-}9)$$

Thus, the retention time is dependent upon the volume of the liquid and gas phases (V_L and V_G^0) contained in the packed column as well as upon the partition coefficient. As might be expected, t_R becomes smaller as the flow F_c of the eluting gas is increased.

The relative retention times for two components in a mixture can be obtained from equation (24-9). Thus,

$$\frac{(t_R)_1}{(t_R)_2} = \frac{V_G^0 + (K_g)_1 V_L}{V_G^0 + (K_g)_2 V_L} \qquad (24\text{-}10)$$

where the subscripts 1 and 2 refer to the two components.

Example. The following data were obtained for a gas-liquid chromatographic column.

Column temperature:	50°C
Inlet pressure:	1250 torr
Exit pressure:	750 torr
Volume of liquid phase at 50°C:	2.80 cm³
Flow rate (gas at 25°C):	17.8 cm³/min
Retention time, air:	0.28 min
Retention time, methyl acetate:	4.30 min
Retention time, ethyl acetate:	4.80 min

Calculate: (a) the retention volumes for air, methyl acetate, and ethyl acetate; (b) the retardation factors for the two compounds; and (c) the partition coefficients at 50°C.

(a) In order to correct the retention volumes to the average column pressure we first employ equation (24-2):

$$j = \frac{3\left[\left(\frac{1250}{750}\right)^2 - 1\right]}{2\left[\left(\frac{1250}{750}\right)^3 - 1\right]} = 0.734$$

Then employing equation (24-1) and correcting the flow rate to column temperature,

$$V^0_{air} = V^0_G = 0.734 \times 0.28 \times 17.8 \times \frac{323}{298} = 4.0 \text{ cm}^3$$

$$V^0_{R1} = 0.734 \times 4.30 \times 17.8 \times \frac{323}{298} = 60.8 \text{ cm}^3$$

$$V^0_{R2} = 0.734 \times 4.80 \times 17.8 \times \frac{323}{298} = 67.9 \text{ cm}^3$$

(b) The retardation factors are obtained from equation (24-4):

$$R_{F1} = \frac{4.0}{60.8} = 0.066$$

$$R_{F2} = \frac{4.0}{67.9} = 0.059$$

(c) The partition coefficients can be obtained from equation (24-8):

$$K_{g1} = \frac{V^0_{R1} - V^0_G}{V_L} = \frac{60.8 - 4.0}{2.80} = 20.3$$

$$K_{g2} = \frac{67.9 - 4.0}{2.80} = 22.8$$

Measurement of column efficiency. The efficiency of a gas chromatographic column is conveniently described in terms of its theoretical plates n, a quantity that can be estimated from the width of an elution band and the retention time. Figure 24-2 illustrates how the required measurements are made.

The relationship shown in Figure 24-2 can be understood by consideration of equation (23-22), which applies equally well to gas-liquid chromatography. As we have noted (p. 619), the maximum in the plot of this equation occurs when $V/V_h = r$. When the maximum appears at the end of the column, $r = n$, where n is the number of theoretical plates in the column. The quantity V is the volume of gas required to bring the peak to the end of the column; if the flow rate is constant, V is proportional to d in Figure 24-2. Thus, we may write

$$n = \frac{V}{V_h} = kd \qquad (24\text{-}11)$$

where k is a proportionality constant.

The properties of equation (23-22) are such that the two intercepts of the tangents to the curve with the base line occur at $(n + 1) + 2(n)^{1/2}$ and $(n + 1) - 2(n)^{1/2}$; the peak width w in terms of V/V_h is then equal to the difference between these quantities, or $4(n)^{1/2}$. Thus,

$$4(n)^{1/2} = \left(\frac{V}{V_h}\right)_2 - \left(\frac{V}{V_h}\right)_1 = kw \qquad (24\text{-}12)$$

Combining equations (24-11) and (24-12) yields

$$n = 16 \left(\frac{d}{w}\right)^2$$

It should be pointed out that the magnitude of n is somewhat dependent upon the compound used in its determination.

Factors affecting column efficiency. The fractionation efficiency of any chromatographic column is determined by the number of its theoretical plates; with a gas-liquid column in particular, the number of plates depends

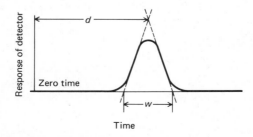

Figure 24-2. Theoretical plate determination; $n = 16(d/w)^2$.

in a complex way on a host of experimental variables. Investigations concerned with the kinetics of transport phenomena in a column have provided an understanding of some of the factors that affect the HETP, and hence column efficiency.[5] The *van Deemter* equation has evolved from such studies; this relationship takes the form

$$h = A + \frac{B}{v} + Cv \qquad (24\text{-}14)$$

where h is HETP and v is the linear gas velocity. The terms A, B, and C are all constants for a given column.

The A term (called the *eddy diffusion* term) is a function of nonideal column behavior that results from the multitude of pathways by which a molecule can find its way through the packing. These paths differ somewhat in length so that the residence times in the column for molecules of the same species are likewise variable. As a consequence, the molecules arrive at the end of the column over a time interval. This spread tends to broaden the elution band, and thus increase the HETP. Eddy diffusion is *independent* of gas velocity.

A second effect that leads to band broadening is *longitudinal diffusion* (B/v in equation 24-14). As solute molecules move through the column, molecular diffusion forces cause a migration from the concentrated center of the band toward the more dilute edges. The result is a broadening of the band and lower column efficiency. As the gas velocity increases, less time is available for this process to occur; thus, as shown in equation (24-14), longitudinal diffusion is diminished as the velocity v becomes larger. The effect of molecular diffusion in the liquid phase is so much smaller than in the gas that it can be neglected.

The so-called *mass transfer term Cv* is related to the rate of the partition of solute between the two phases. As the gas velocity becomes large, time is not available for attainment of true equilibrium. Thus, the full efficiency of the column is not achieved and an increase in HETP is observed.

Figure 24-3 shows the contribution of each of the three effects as a function of gas velocity (broken lines) as well as their net effect (solid line) on the HETP of a column. Clearly, the optimum efficiency is realized at the gas velocity corresponding to the minimum in the curve; because the loss in efficiency is greatest at low velocities, low flow rates are undesirable in gas-liquid chromatography.

Curves similar to the solid line in Figure 24-3 are readily obtained experimentally and have provided information concerning the variables that influence A, B, and C in equation (24-14). Band broadening due to eddy diffusion (A) is minimized by careful packing of the column with small, spherical particles. Particular care is needed to avoid open channels.

The longitudinal diffusion term (B) can be reduced somewhat if a denser

[5]See, for example, A. I. M. Keulemans, *Gas Chromatography*, 2d ed. (New York: Reinhold Publishing Corporation, 1959), pp. 129-138.

carrier gas is employed. Increases in pressure or decreases in temperature, of course, result in a greater gas density.

The most important variable that affects the rate of approach to equilibrium (C) appears to be the thickness of the liquid layer. Here, a marked improvement in column efficiency is obtained with very thin layers which enhance the probability of true equilibration. Equilibrium is also more likely at higher temperatures and with lower solvent viscosities.

Implicit in the foregoing treatment are the assumptions that the entire sample is introduced as an undiluted plug into the first plate and that the sample size is small enough so that the liquid phase in that plate is not entirely saturated with the solute. If these conditions are not met (and they seldom are), band broadening also results.

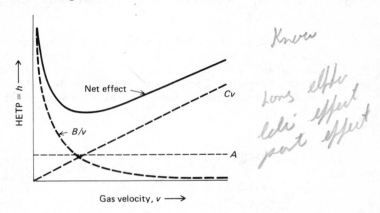

Figure 24-3. Schematic representation of the effect of factors in equation (24-14) as a function of gas flow rate.

Resolution of solutes. Of great practical interest in chromatography is the question of how many plates are required to resolve two solutes. The answer depends upon the relative sizes of the two partition ratios as well as upon the degree of separation that is considered satisfactory. In discussing this question we shall employ the term *separation factor* α, which is simply a ratio of the partition ratios for the two solutes to be separated; that is, $\alpha = K_{g2}/K_{g1}$. From equation (24-10) it is apparent that $\alpha = (t_R)_2/(t_R)_1$ when $K_g V_L \gg V_G^0$, a condition that is frequently satisfied.

Glueckauf[6] has established the relationships among the plate number **n**, the separation factor α, and the degree of band overlap; he has shown that the efficiency of separation depends upon the mole ratio of the solutes to be separated. His treatment reveals a linear relationship between the quantity

$$\eta \frac{(m_1^2 + m_2^2)}{2m_1 m_2} \tag{24-15}$$

[6]E. Glueckauf, *Trans. Faraday Soc.*, **51**, 34 (1955).

and the number of theoretical plates contained in a column; here, m_1 and m_2 are the number of moles of solutes 1 and 2 in the mixture and η_i is the fractional impurity in each band; that is,

$$\eta_1 = \frac{\Delta m_1}{m_2} \quad \text{and} \quad \eta_2 = \frac{\Delta m_2}{m_1}$$

where Δm_1 and Δm_2 are the moles of one solute contaminating the band of the other. Generally, in separating two bands the attempt is made to equalize the fractional impurity in each band (that is, $\eta_1 = \eta_2 = \eta$). Figure 24-4 shows the linear plots of the quantity given in equation (24-15) with respect to the number of theoretical plates for a series of different separation factors. The following example illustrates the application of this plot.

> **Example.** Two substances have a separation factor of 1.6. How many plates would be required to separate a mixture in which the molar ratio of the two substances was (a) 4:1 and (b) 1:1, if it is desired to have no more than 1 percent of each occurring in the band of the other? Repeat the calculations for substances having a separation factor of 1.1.

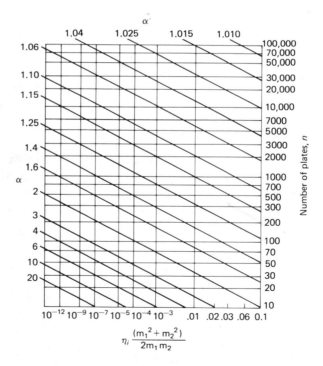

Figure 24-4. Relationship among fractional impurity η, separation factor α, and number of theoretical plates n. From E. Glueckauf, *Trans. Farady Soc.*, **51**, 34 (1955). (With permission.)

For mixture (a), $\eta = 0.01$ and

$$\eta\left(\frac{m_1^2 + m_2^2}{2m_1m_2}\right) = 0.01\left(\frac{4^2 + 1^2}{2 \times 4 \times 1}\right) \cong 0.02$$

Reference to Figure 24-4 shows that the required number of theoretical plates is approximately 60.

For a 1:1 mixture of the two solutes,

$$\eta\left(\frac{m_1^2 + m_2^2}{2m_1m_2}\right) = 0.01$$

and we see that about 80 plates would be necessary.

Figure 24-4 indicates that a 4:1 mixture of solutes with a separation factor of 1.10 would require about 1500 plates for an allowable fractional impurity of 1 percent. Approximately 2000 plates would be required for a 1:1 mixture.

APPARATUS

The apparatus for gas-liquid chromatography can be relatively simple and inexpensive; more than two dozen models are offered by various instrument manufacturers. The essential components are illustrated in Figure 24-5. A brief description of each component follows.

Carrier Gas Supply

Helium, nitrogen, carbon dioxide, and hydrogen from tank sources constitute the most commonly used carrier gases. Hydrogen has the obvious disadvantage of explosion danger. Suitable flow-regulating valves are required, and some means of reproducing the flow rate is desirable. A simple and accurate device for the latter purpose is a soap-bubble meter, shown

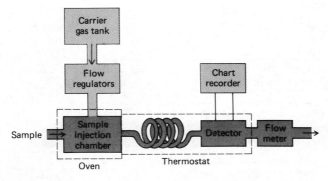

Figure 24-5. Block diagram of a gas-chromatographic apparatus.

in Figure 24-6. A soap film is formed in the path of the gas when a rubber bulb containing an aqueous solution of soap or detergent is squeezed; the time required for this film to move between two graduations on the buret is measured and can be converted to flow rate.

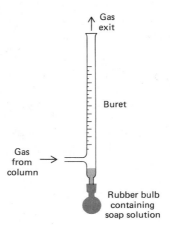

Figure 24-6.   Soap-bubble meter.

Sampling System

As mentioned earlier, good column efficiency requires that the sample be of suitable size and be introduced as a "plug" of vapor; slow injection and oversized samples cause band spreading and poor resolution. Liquid samples are introduced by injection through a silicone rubber diaphragm with a hypodermic syringe. To prevent overloading of the column, the volumes must be small (of the order of 1 to 20 microliters). The injection port ordinarily is heated to a temperature above the boiling point of the sample so that vaporization is rapid. Special apparatus has been developed for introduction of solids and gases.

Columns

Two types of columns are employed in gas-liquid chromatography. The capillary type is fabricated from capillary tubing, the bore of which is coated with a very thin film of the liquid phase. Capillary columns have a very low pressure drop and can thus be of great length; columns of several hundred thousand theoretical plates have been described. These columns, however, have very low sample capacities.

More convenient is the packed column, which consists of a glass or metal tube of roughly 0.25-inch diameter that ranges in length from 5 to 50 feet. Ordinarily, the tubes are folded or coiled so that they can be conveniently fitted into a thermostat. Typically, columns contain 100 to 1000 theoretical plates per foot. The best packed columns have a total of 30,000 and 60,000 theoretical plates.

Solid support. The ideal solid support would consist of small (20 to 40 μ), uniform, spherical particles with good mechanical strength. In addition, this material would be inert at elevated temperatures and be readily wetted by the liquid phase to give a uniform coating. No substance that meets all of these criteria perfectly is yet available.

The most widely used supports are made from diatomaceous earth and are marketed under such trade names as Celite, Dicalite, Chromosorb, C-22 Firebrick, and Sterchamol. Other supports have been fashioned from powdered Teflon, alumina, Carborundum, and micro glass beads.

Liquid phase. Desirable properties for the liquid phase in a gas-liquid chromatographic column include (1) *low volatility*: ideally, its boiling point should be at least 200° higher than the maximum operating temperature for the column; (2) *thermal stability*; (3) *chemical inertness*; and (4) *solvent characteristics* such that K_g values for the solutes to be resolved fall within a suitable range.

No single liquid meets all of the requirements, the last in particular. As a consequence, it is common practice to have available several interchangeable columns, each with a different stationary phase. Although some qualitative guidelines exist to aid in the choice of the liquid phase, the decision remains largely a matter of trial and error.

Equation (24-10), indicates that the retention time for a solute depends directly upon its partition ratio which, in turn, is related to the nature of the stationary phase. Clearly, to be useful in gas-liquid chromatography the solvent must generate different partition ratios among solutes; in addition, however, these ratios must be neither extremely large nor extremely small. The latter requirement is particularly important; if K_g values are too small, the solutes pass through the column so rapidly that no significant separation occurs. If, on the other hand, the ratios are large, the time required to remove solutes from the columns becomes inordinate.

In order to have a reasonable residence time in the column, a solute must show at least some degree of compatability (solubility) with the solvent. Thus, their polarities should be at least somewhat alike. A stationary liquid such as squalane (a high-molecular weight saturated hydrocarbon) or dinonylphthalate might be chosen for separation of members of a nonpolar homologous series such as hydrocarbons, ethers, or esters. On the other hand, a more polar liquid such as polyethyleneglycol would probably be more effective for the separation of alcohols or amines. For aromatic hydrocarbons, benzyldiphenyl might prove appropriate.

Among solutes of similar polarity, the elution order usually follows the order of boiling points; where these differ sufficiently, clean separations are feasible. Solutes with nearly identical boiling points but different polarities frequently require a liquid phase that will selectively retain one (or more) of the components by dipole interaction or by adduct formation. The data in Table 24-1 illustrate these effects. Here the retention times (relative to ethane) for a series of paraffinic and olefinic hydrocarbons are compared for three liquid phases. The first, triisobutylene, is nonpolar; the other two are polar. The third solvent, in addition, contains silver nitrate, which selectively forms loose adducts with specific olefins. Note that the retention times on triisobutylene closely correlate with the boiling points of the liquids, and that little or no separation is realized among the compounds with similar volatilities. Acetonyl acetone, on the other hand, is moderately polar and is capable of inducing polarization in the olefins; selective retention results. Note also that the retention times for the nonpolarizable paraffins are significantly less in the polar solvent than in the nonpolar one. The selectivity of adduct formation between olefins and silver nitrate permits some useful separation as well; for example, isobutene and butene-1 are not separated on the first two columns but are resolved in the presence of silver nitrate; improved separation of *cis*- and *trans*-butene-2 also occurs.

Another important interaction that often enhances selectivity is hydrogen bond formation. For this effect to operate the solute must have a polar hydrogen atom and the solvent must have an electronegative group (oxygen, fluorine, or nitrogen), or conversely.

Table 24-2 lists a few common solvents for gas–liquid chromatography arranged in order of decreasing polarity.

Column preparation. The support material is first screened to limit the particle size range. It is then made into a slurry with a volatile solvent that contains an amount of the stationary liquid calculated to produce a thin coating on all of the particles. After solvent evaporation, the particles appear dry and are free flowing.

Columns are fabricated from glass, stainless steel, copper, or aluminum.

Table 24-1 Separation of Hydrocarbons with Various Stationary Liquids[a]

Compound	Boiling Point, °C	Triiso-butylene	Acetonyl Acetone	$AgNO_3$ in Glycol
Ethane	−104	1.0	1.0	1.0
Isobutane	−11.7	5.6	2.2	0.75
Isobutene	−6.9	6.7	4.75	3.25
Butene-1	−6.3	6.7	4.75	6.25
Butadiene	−4.4	6.7	10.0	10.0
n-Butane	−0.5	7.5	3.0	0.75
trans-Butene-2	0.88	8.0	5.85	1.75
cis-Butene-2	3.72	8.9	6.8	5.5

Relative Retention Time On

[a]Data from B. W. Bradford, D. Harvey and D. E. Chalkley, *J. Inst. Petroleum*, **41**, 80 (1955). With permission.

Table 24-2 Some Common Solvents for Gas-liquid Chromatography

Solvent	Commercial Source	Maximum Temperature, °C	Typical Separations
Squalane ($C_{30}H_{62}$)		140	Hydrocarbons
Apiezon L (high-molecular-weight paraffin)	Metropolitan-Vickers Electrical Co., Ltd.	300	High boiling hydrocarbons, esters and ethers
Dinonylphthalate		140	Wide variety of compound types
Silicone oil (DC 550)	Dow-Corning	200	Wide variety of compound types
Silicone rubber gum (SE-30)	General Electric	350	High temperature separations of a variety of compounds
Carbowax 20M (polyethylene-glycol)	Union Carbide	250	Aromatics, alcohols, amines, and other nitrogen containing compounds

They are filled by slowly pouring the coated support into the straight tube with gentle tapping or shaking to provide a uniform packing. Care must be taken to avoid channeling. After the column has been packed, it is bent or coiled in an appropriate shape to fit the oven.

A properly prepared column may be employed for several hundred analyses.

Column thermostatting. Column temperature is an important variable that must be controlled to a few tenths of a degree for precise work. Control systems that have been employed include circulating air baths, electrically heated metal blocks, and jackets fed with vapor from a constant boiling liquid.

The optimum column temperature depends upon the boiling point of the sample and the degree of separation required. Roughly, a temperature equal to or slightly above the average boiling point of a sample results in an elution period of reasonable length (10 to 30 min). For samples with a broad boiling range it is often desirable to increase the column temperature, either continuously or in steps as the separation proceeds.

In general, optimum resolution is associated with minimal temperature; the cost of lowered temperature, however, is an increase in elution time and therefore the time required to complete an analysis. Figure 24-7 illustrates this principle.

Detection Systems

Detection devices for a gas-liquid chromatograph must respond rapidly and reproducibly to the low concentrations of the solutes emitted

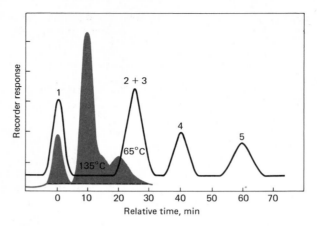

Figure 24-7. Effect of temperature on the separation of hexane isomers: (1) 2,2-Dimethylbutane; (2) 2,3-Dimethylbutane; (3) 2-Methylpentane; (4) 3-Methylpentane; (5) *n*-Hexane. From C. E. Bennett, S. Dal Nogare and L. W. Safranski, in I. M. Kolthoff and P. J. Elving, Eds., *Treatise on Analytical Chemistry*, Part I, Vol. 3 (New York: Interscience Publishers, 1961), p. 1690. (With permission.)

from the column. The solute concentration in a carrier gas at any instant is only a few parts in a thousand at most; frequently the detector is called upon to respond to concentrations that are smaller by one or two orders of magnitude (or more). In addition, the interval during which a peak passes the detector is usually a second or less; the detector must thus be capable of exhibiting its full response during this brief period.

Other desirable properties of the detector include linear response, good stability over extended periods, and uniform response for a wide variety of compounds. No single detector meets all of these requirements although more than a dozen different types have been proposed. We shall describe the most widely used of these.

Thermal conductivity detectors. A relatively simple and broadly applicable detection system is based upon changes in the thermal conductivity of the gas stream; an instrument employed for this purpose is sometimes called a *katharometer*. The sensing element of this device is an electrically heated source whose temperature at constant electrical power depends upon the thermal conductivity of the surrounding gas. The heated element may consist of a fine platinum or tungsten wire or alternatively a semiconducting thermistor. The resistance of the wire or thermistor gives a measure of the thermal conductivity of the gas; in contrast to the wire detector the thermistor has a negative temperature coefficient.

In chromatographic applications a double detector is always employed, one element being placed in the gas stream *ahead* of the sample injection chamber and the other immediately beyond the column. In this way the thermal conductivity of the carrier gas is canceled and the effects of variation of column temperature, pressure, and electrical power are minimized. The resistances of the twin detectors are usually compared by incorporating them into two arms of a simple bridge circuit such as that shown in Figure 24-8.

The thermal conductivities of hydrogen and helium are roughly six to ten times greater than most organic compounds. Thus, the presence of even small amounts of organic materials causes a relatively large decrease in thermal conductivity of the column effluent; the detector undergoes a marked rise in temperature as a result. The conductivities of nitrogen and carbon dioxide more closely resemble those of organic constituents; thus, detection by thermal conductivity is less sensitive when these substances are used as carrier gases.

Thermal conductivity detectors are simple, rugged, inexpensive, nonselective, accurate, and nondestructive of the sample. They are not as sensitive, however, as some of the other devices to be described.

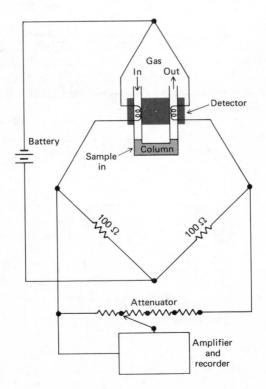

Figure 24-8. Schematic diagram of a thermal conductivity detector-recorder system.

Flame ionization detectors.   Most organic compounds, when pyrolyzed at the temperature of a hydrogen-air flame, produce ionic intermediates which provide a mechanism by which current can be carried through the flame. By employing an apparatus such as that shown in Figure 24-9, these ions can be collected and the resulting ion current measured. The electrical resistance of a flame is very high (perhaps 10^{12} ohms), and the resulting currents are therefore minuscule. As can be imagined, fairly complicated and expensive electronic circuitry is needed to measure these tiny currents.

The ionization of carbon compounds in a flame is poorly understood, although it is known that the number of ions produced is roughly proportional to the number of reduced carbon atoms in the flame. Oxidized carbon atoms (carbonyl, alcoholic, amino, for example) produce fewer ions or none at all.

The hydrogen flame detector currently is one of the most popular and most sensitive detectors. It is more complicated and more expensive than the thermal conductivity detector, but has the advantage of higher sensitivity. In addition, it has a wide range of linear response. It is, of course, destructive of the sample.

β-Ray detectors.   In a β-ray detector the effluent from the column is passed over a β-emitter such as strontium-90 or tritium; argon, which serves as the carrier gas, is excited to a metastable state by the radiation. The excited argon atoms cause ionization of the sample molecules by collision, and the resulting ion current can be measured in essentially the same way as in the hydrogen flame detector.

β-Ray detectors are highly sensitive and possess the advantage of not altering the sample significantly (in contrast to the flame detector). The requirement of argon as the carrier is a disadvantage.

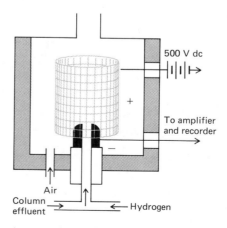

Figure 24-9.   Hydrogen flame ionization detector.

APPLICATIONS OF GAS-LIQUID CHROMATOGRAPHY

In evaluating the importance of gas-liquid chromatography it is important to distinguish between the two roles the method plays in its application to chemical problems. The first is as a tool for performing separations; in this capacity it is unsurpassed when applied to complex organic, metal-organic, and biochemical systems. The second and distinctly different function is that of providing the means for completion of an analysis. Here, retention times or volumes are employed for qualitative identification, while peak heights or areas provide quantitative information. In its analytical role gas-liquid chromatography is considerably more limited than some of the other methods considered in earlier chapters. As a consequence, an important trend in gas-liquid chromatography appears to be in the direction of combining the remarkable fractionation qualities of the method with the superior analytical properties of such instruments as mass, ultraviolet, infrared, and nmr spectrometers.

Qualitative Analysis

Gas chromatograms are widely used as criteria of purity for organic compounds. Contaminants, if present, are revealed by the appearance of additional peaks; the areas under these peaks provide rough estimates of the extent of contamination. The technique is also useful for evaluating the effectiveness of purification procedures.

In theory, retention-time data should be useful for the identification of components of mixtures. In fact, however, the number of variables that must be controlled in order to obtain reproducible retention times limits the applicability of the technique. On the other hand, gas-liquid chromatography provides an excellent means of confirming the presence of a suspected compound in a mixture, provided an authentic sample of the substance is available. No new peaks in the chromatogram of the mixture should appear upon addition of the pure compound, and enhancement of one of the existing peaks should be observed. The evidence is particularly convincing if the effect can be duplicated by using different columns and different temperatures.

Retention volume or time plots. It has long been recognized that, within a homologous series, a plot relating the logarithm of the retention time (or volume) to the number of carbon atoms is frequently linear, provided the lowest members of the series are excluded. This relationship can be useful for qualitative work since it is often possible to separate one homologous series from another by chromatography (if the two have different polarities) or by other physical or chemical means.

An even more useful relationship can sometimes be obtained by injecting portions of the sample on two columns with stationary phases that differ in polarity. For a homologous series a plot of the retention time on one column versus that on the other (employing a log scale) yields a straight line (see Figure 24-10); more important, the intercepts of the plots differ from one series to another. Thus, resolution of species with similar boiling points but differing polar characteristics frequently can be achieved by obtaining chromatograms for the sample on two columns. In addition, the identity of peaks may be considerably less ambiguous. This technique is quite analogous to two-dimensional paper chromatography.

Separation factors. We have seen (p. 647) that the separation factor $\alpha_{1,2}$ for compounds 1 and 2 is given by the relationship

$$\alpha_{1,2} = \frac{K_{g_1}}{K_{g_2}} \cong \frac{(t_R)_1}{(t_R)_2}$$

If a standard substance is chosen as compound 2, then $\alpha_{1,2}$ can provide a quantitative index for identification of compound 1 that is largely independent of column variables other than temperature; that is, numerical tabula-

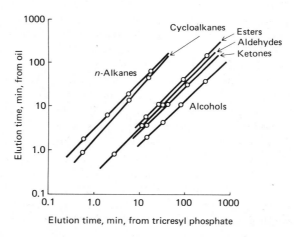

Figure 24-10. Behavior of several homologous series on two columns. Points from left to right correspond to the following:

n-alkanes: butane, pentane, hexane, heptane
cycloalkanes: cyclopropane, cyclopentane, cyclohexane
esters: methyl-, n-propyl-, n-butyl acetate
aldehydes: acet-, n-propion-, n-butyr-aldehyde
ketones: acetone, 2-butanone, 2-pentanone
alcohols: methyl, ethyl, n-propyl, n-butyl alcohol

From J. S. Lewis, H. W. Patton and W. L. Kaye, *Anal. Chem.*, **28**, 1370 (1956). (With permission of the American Chemical Society.)

tions of separation factors for pure compounds relative to a common standard can be prepared and then used for the characterization of samples on any column. The amount of such data available in the literature is presently limited, however.

Quantitative Analysis

The detector signal from a gas-liquid chromatographic column has had wide use for quantitative and semiquantitative analyses. Under carefully controlled conditions an accuracy of about 1 percent relative is attainable. As with most analytical tools, reliability is directly related to the amount of effort spent in calibration and control of variables as well as to the nature of the sample.

Analytical parameters. Quantitative analysis can be based either on peak heights or on peak areas. Peak heights, while more convenient to measure, are less satisfactory because of their greater dependence upon experimental variables. Flow rate is particularly critical, since peaks tend to become broader and lower as the residence time of the solute on the column increases. In addition, peak heights are more strongly affected by column temperatures, porosity of the packing, and column length. Despite these limitations, peak height is often a better analytical parameter than peak area for solutes with low retention times. The peaks for such solutes are narrow and tall; accurate determination of their areas is thus difficult.

Peak areas generally increase linearly with the reciprocal of flow rate, and a satisfactory correction for change in this variable can be applied. In contrast to peak heights the *relative* areas of the peaks for two compounds remain constant and independent of flow rate, a valuable characteristic that permits use of internal standards. Temperature fluctuations have only a small effect on absolute peak areas and none whatsoever on relative ones.

Area measurement. The area A of a chromatographic peak can be evaluated in any of several ways.

1. Cut out the peak; compare its mass with that for a portion of the recorder paper of known area.
2. Measure the area with a planimeter.
3. Multiply the height of the peak by its width at half-height; insofar as the curve has Gaussian characteristics this product is equal to $0.84\,A$.
4. Form a triangle by drawing tangents through the inflection points on either side of the peak, and connect these with a third line along the recorder base line; the resulting triangle will have an area that is approximately $0.96\,A$.
5. Use a mechanical or electronic integrator; such devices are included as accessories to many recorders.

Methods (4) and (5) appear to be most widely used.

Calibration. Calibration curves can be based upon weight, volume, or mole percent. A separate curve is needed for each solute to be determined since no detector responds in exactly the same way for each compound. The accuracy of the analysis depends upon how closely the standard samples approximate the unknown in composition, the care with which they are prepared, and the closeness with which the operational variables of the columns are controlled. The effect of column variables can be partially offset by adding an internal standard in known amount to both calibration standards and samples. The internal standard should have a retention time similar to the solute of interest; its peak, however, must be separated from those of all of the components of the sample—a requirement that is often difficult to realize.

Gas-liquid Chromatography with Selective Detectors

All of the detection systems for qualitative and quantitative analysis that we have considered thus far are nonselective in the sense that they respond in more or less the same way to the solutes emerging from a column (ideally, the nonselective detector would respond identically to every compound). The employment of selective or specific detectors, however, makes it possible in principle to combine the tremendous separatory power of gas-liquid chromatography with the superior analytical qualities of the instruments we have considered in earlier chapters.

Two general approaches are possible. In the first, the solute vapors are collected as separate fractions in a cold trap, a nondestructive and nonselective detector being employed to indicate their appearance. The fractions are then identified by infrared, mass, or other spectroscopic techniques. The main limitation to this approach lies in the very small (usually micromolar) quantities of solute contained in a fraction; nonetheless, the general procedure has proved useful in the qualitative analyses of complex mixtures.

A second general method involves the use of a selective detector to monitor the column effluent continuously. Some examples will illustrate this approach.

In their first publication on gas chromatography, James and Martin employed an automatic titration cell as a detector for volatile fatty acids; the same device was later applied to mixtures of aromatic and aliphatic amines. In this method the effluent from the column was led directly into a cell containing an acid-base indicator solution. The absorbance of the solution was monitored photometrically, the output from the cell being employed to control continuously the addition of titrant; the volume of the latter was recorded on a chart to produce an integral curve of volume versus time. Both qualitative and quantitative information could be retrieved from the recorded curve.

A coulometric method has been employed as the detector in the analysis of the various chlorinated compounds contained in pesticide residues.

After the halogenated compounds have been separated on the column, the effluent is mixed with oxygen and burned over a catalyst. The hydrochloric acid formed is absorbed in an automatic coulometric cell which generates silver ions on demand. An integral curve of quantity of electricity versus time is obtained.

Rapid scan techniques have also been employed to yield the entire spectrum of a fraction during the brief period of its appearance at the end of a column. For example, both infrared and mass spectral instruments that produce a spectrum in a few seconds have been designed for this purpose. These record a spectrum of each solute as it leaves the column.

Problems

1. A gas-liquid chromatographic column was operated under the following conditions:

Flow rate (50°C)	20 cm³/min
Column temperature	50°C
Inlet pressure	1000 torr
Outlet pressure	760 torr
Retention time air	0.50 min
Retention time n-hexane	3.50 min
Retention time n-heptane	4.10 min
Volume liquid phase (50°C)	3.12 cm³

    Calculate:
    (a) the corrected retention volumes for air (V_r^0), n-hexane (V_{R1}^0), and n-heptane (V_{R2}^0),
    (b) the retardation factor for the two organic compounds,
    (c) the partition coefficients (at 50°C) for the two compounds,
    (d) the separation factor for the two solutes.

2. The column described in Problem 1 was employed to separate an equimolar mixture of n-hexane and n-heptane. From the recorder data it was estimated that about 3 percent of each component appeared in the band of the other. How many theoretical plates did the column have?

3. For the column described in Problem 1, hexene-1 was found to have a retention time of 3.86 min.
    (a) Calculate the retention volume, the retardation factor, and the partition coefficient for hexene-1.
    (b) If the column had about 900 theoretical plates, what mole percent of hexene-1 would be expected in the band for hexane if a 1:1 mixture of the two components were passed through the column? A 10:1 mixture?

4. How many theoretical plates would be required for a gas-liquid column to separate two solutes having partition coefficients of 2.7 and 3.2 if
    (a) no more than 1 percent of each solute is to appear in the band of the other and the solutes are present in a 3:1 ratio,
    (b) no more than 1 percent of each solute is to appear in the band of the other and the solutes are present in a 1:1 ratio,

(c) no more than 5 percent of each solute is to appear in the band of the other and the solutes are present in about equimolar amounts.

5. A gas-liquid chromatographic column was operated under the following conditions:

Flow rate (room temperature)	40.0 cm³/min
Column temperature	122°C
Inlet pressure	1080 torr
Outlet pressure	770 torr
Retention times, minutes	air 0.24; benzene 1.41; toluene 2.67; cumene 5.34; ethylbenzene 4.18
Volume liquid phase (122°C)	14.1 cm³

Calculate:

(a) the corrected retention volumes for each species,
(b) the retardation factor for each substance,
(c) the partition coefficients for each compound,
(d) the separation factors for each pair of hydrocarbons,
(e) the minimum number of plates that the column would have to contain to permit separations such that no more than 0.10 percent of the lower boiling compound would appear in the band adjacent to it. Assume the mixture is equal molar in all of its constituents.

Answers
to Problems

CHAPTER 2

1. (a) $\nu = 3.3 \times 10^{17}$ Hz; $\sigma = 1.11 \times 10^7$ cm^{-1}
   (b) $\nu = 5.09 \times 10^{14}$ Hz; $\sigma = 1.70 \times 10^4$ cm^{-1}
   (c) $\nu = 2.38 \times 10^{13}$ Hz; $\sigma = 7.94 \times 10^2$ cm^{-1}
   (d) $\nu = 1.50 \times 10^8$ Hz; $\sigma = 5.00 \times 10^{-3}$ cm^{-1}
2. (a) 2.19×10^{-9} erg/photon; 3.14×10^4 kcal/mole; 1.37×10^3 eV
   (b) 3.37×10^{-12} erg/photon; 48.6 kcal/mole; 2.11 eV
   (c) 1.58×10^{-13} erg/photon; 2.27 kcal/mole; 9.85×10^{-2} eV
   (d) 9.94×10^{-19} erg/photon; 1.43×10^{-5} kcal/mole; 6.21×10^{-7} eV
3. (a) $\lambda = 588.8$ nm; $\nu = 5.093 \times 10^{14}$ Hz
   (b) $\lambda = 439.6$ nm; $\nu = 5.093 \times 10^{14}$ Hz
   (c) $\lambda = 284.5$ nm; $\nu = 5.093 \times 10^{14}$ Hz
4. Percent loss air to glass = 6.7
   Percent loss glass to solution = 1.4
   Percent loss solution to glass = 1.4
   Percent loss glass to air = 6.7
   Total percent original lost by reflection = 15.4

CHAPTER 3

1. (a) 5.85×10^3 liter cm^{-1} mole^{-1}
   (b) 3.70×10^{-2} cm^{-1}
   (c)   36.4 percent

2. (a) 6.00×10^2 liter cm^{-1} mole^{-1}
   (b)   0.048
   (c)   0.738
3. 0.716
4. 15.7 percent
5. (a) 0.740
   (b) 0.439
   (c) 0.370
   (d) 2.02×10^{-5} M
6. (a) 7.4×10^{-5} M
   (b) 25.0 percent
   (c) 1.54×10^{-5} M
7. 700 nm, 2nd order
   467 nm, 3rd order
   350 nm, 4th order
   280 nm, 5th order
   233 nm, 6th order
8. (a) 4.00×10^{-4} M, A = 1.028
      3.00×10^{-4} M, A = 0.753
      1.00×10^{-4} M, A = 0.221
      0.500×10^{-4} M, A = 0.097
   (b) 4.00×10^{-4} M, A = 1.094
      3.00×10^{-4} M, A = 0.821
      1.00×10^{-4} M, A = 0.274
      0.50×10^{-4} M, A = 0.137

CHAPTER 4

1. $C_M = \dfrac{\epsilon'_N A'' - \epsilon''_N A'}{b(\epsilon''_M \epsilon'_N - \epsilon'_M \epsilon''_N)}; \quad C_N = \dfrac{\epsilon''_M A' - \epsilon'_M A''}{b(\epsilon''_M \epsilon'_N - \epsilon'_M \epsilon''_N)}$

2. (a)

λ, nm	A	λ, nm	A	λ, nm	A
415	0.797	490	0.609	580	0.615
430	0.808	505	0.590	590	0.602
440	0.789	520	0.603	600	0.584
450	0.763	535	0.626	615	0.548
460	0.723	550	0.630	630	0.496
475	0.660	565	0.632	645	0.432

   (b) $A_{440} = 0.337$; $A_{590} = 0.627$
   (c) $A_{440} = 0.215$; $A_{590} = 0.491$
   (d) $A_{440} = 0.313$; $A_{590} = 0.229$

3. (a) [A] = $7.11 \times 10^{-4} M$ [B] = 3.85×10^{-5} M
   (b) [A] = 6.28×10^{-4} M   [B] = 1.48×10^{-5} M
   (c) [A] = 8.06×10^{-4} M   [B] = 2.30×10^{-5} M

4.

		Absorbance	
λ, nm	(a)	(b)	(c)
415	0.571	0.250	0.387
430	0.605	0.229	0.374
440	0.602	0.214	0.358
450	0.587	0.202	0.343
460	0.556	0.192	0.325
475	0.497	0.184	0.303
490	0.436	0.191	0.297
505	0.383	0.221	0.314
520	0.343	0.269	0.354
535	0.312	0.319	0.398
550	0.283	0.350	0.423
565	0.265	0.368	0.437
580	0.246	0.369	0.433
590	0.234	0.367	0.429
600	0.221	0.367	0.420
615	0.202	0.344	0.398
630	0.180	0.314	0.362
645	0.153	0.277	0.318

5. (a) $[X] = 1.46 \times 10^{-4} M$ $[Y] = 2.39 \times 10^{-4} M$
   (b) 0.703
   (c) $[X] = 1.57 \times 10^{-4} M$ $[Y] = 1.74 \times 10^{-4} M$
   (d) 0.704
6. (a) Orange
   (b) Orange
   (c) 485 nm
   (d) Approx. 0.28
   (e) 555 nm
7. (a) 6.04
   (b) 0.342
8. $A_{440} = 0.208$; $A_{680} = 0.381$
9. (a) 5.64
   (b) 2.32×10^7

10.

	(a)	(b)	(c)
		Absorbance	
λ, nm	$[HIn]/[In^-] = 3:8$	$[H^+] = 5.40 \times 10^{-7}$	$[H^+] = 1.44 \times 10^{-6}$
440	0.158	0.234	0.310
470	0.159	0.249	0.339
480	0.161	0.251	0.343
485	0.162	0.253	0.344
490	0.162	0.253	0.344
505	0.174	0.258	0.342
535	0.231	0.280	0.330
555	0.342	0.342	0.342
570	0.458	0.410	0.360
585	0.543	0.456	0.368

10. (continued)

	Absorbance		
	(a)	(b)	(c)
λ, nm	$[HIn]/[In^-] = 3:8$	$[H^+] = 5.40 \times 10^{-7}$	$[H^+] = 1.44 \times 10^{-6}$
600	0.618	0.495	0.372
615	0.646	0.506	0.365
625	0.646	0.500	0.353
635	0.637	0.488	0.339
650	0.606	0.450	0.308
680	0.454	0.340	0.231

11. (a) 5.44×10^{-7}
    (b) $A_{420} = 0.663$; $A_{550} = 0.363$; $A_{680} = 0.116$
    (c) 7.72
    (d) 8.10
12. In an unbuffered solution, the ratio of $[In^-]$ to $[HIn]$ will change with dilution. Thus, at the wavelengths (420 and 680 nm) where $\epsilon_{In^-} \neq \epsilon_{HIn}$, nonlinear behavior would result. At 550 nm, direct proportionality would be observed since the two ε's are identical here.
13. 2.36 percent
14. 21.6 mg Fe/liter
15. a. ML_2
    b. 6.8×10^8
16. 1.58×10^6
17. ThQ_2
18. $FeSCN^{2+}$
19. (a) Direct plot of data
    (b) 2.53×10^3
    (c) MnQ^+
    (d) 9×10^4
20. 3.7×10^6
21. 0.4×10^{-4} to 3×10^{-4}
22. (a) 1.23 percent
    (b) 1.40 percent
    (c) 1.18 percent
    (d) 7.37 percent
23. 1.81×10^{-4} *F*
24. (a) $A = 1.0$ to 2.0
    $T = 0.10$ to 0.010
    (b) 1.7 percent for $C = 0.50 \times 10^{-4}$ *M*
    58 percent for $C = 1.00 \times 10^{-4}$ *M*
    (c) 1.4 percent for $C = 0.10 \times 10^{-4}$ *M*
    1.2 percent for $C = 0.30 \times 10^{-4}$ *M*
    (d) $A_r = 0.10$ to 1.10
    $T_r = 0.794$ to 0.0794
    0.22 percent for $C = 0.50 \times 10^{-4}$ *M*
    7.3 percent for $C = 1.00 \times 10^{-4}$ *M*
25. (a) $A = 0.75$ to 1.5
    $T = 0.178$ to 0.0316
    (b) 1.95 percent at $C = 5.0 \times 10^{-5}$ *M*
    5.5 percent at $C = 10.0 \times 10^{-5}$ *M*

(c) 1.64 percent at $C = 2.5 \times 10^{-5} \ M$
   1.95 percent at $C = 5.0 \times 10^{-5} \ M$

(d) $A_r = 0.030$ to 0.78
   $T_r = 0.932$ to 0.165
   0.37 percent at $C = 5.0 \times 10^{-5} \ M$
   1.05 percent at $C = 10.0 \times 10^{-5} \ M$

26. (a) $A = 0.010$ to 0.050
   $T = 0.977$ to 0.891

  (b) 18 percent for $[Cd^{2+}] = 5 \times 10^{-7} \ M$
   3.9 percent for $[Cd^{2+}] = 25 \times 10^{-7} \ M$

  (c) For $[Cd^{2+}] = 5 \times 10^{-7} \ M$
   $A_r = 0.085$; $T_r = 0.822$; $\Delta C/C = 2.3$ percent
   For $[Cd^{2+}] = 25 \times 10^{-7} \ M$
   $A_r = 0.812$; $T_r = 0.154$; $\Delta C/C = 0.50$ percent

27. (a) $A = 0.015$ to 0.075
   $T = 0.966$ to 0.841

  (b) 18 percent for $[Fe^{3+}] = 1.0 \times 10^{-6} \ M$
   4.1 percent for $[Fe^{3+}] = 5.0 \times 10^{-6} \ M$

  (c) For $[Fe^{3+}] = 1.0 \times 10^{-6} \ M$
   $A_r = 0.093$; $T_r = 0.807$; $\Delta C/C = 3.1$ percent
   For $[Fe^{3+}] = 5.0 \times 10^{-6} \ M$
   $A_r = 1.013$; $T_r = 0.0966$; $\Delta C/C = 0.72$ percent

CHAPTER 6

1. (a) 1.9×10^6 dynes/cm
  (b) 2.08×10^3 cm^{-1}
2. (a) 3.99×10^3 cm^{-1}
  (b) 2.92×10^3 cm^{-1}
3. (a) Inactive
  (b) Active
  (c) Active
  (d) Active
  (e) Inactive
  (f) Active
  (g) Inactive
4. (a) 3.04×10^3 cm^{-1}
  (b) 2.24×10^3 cm^{-1}
5. Allyl alcohol, $CH_2{=}CH{-}CH_2OH$
6. Possible compounds are

   or

(The spectrum is that of *o*-methyl acetophenone.)

O
//
7. Acrolein, $CH_2=CH-C-H$ with H_2O contaminant
8. Propanenitrile, CH_3CH_2CN
9. Diphenylamine, $C_6H_5NHC_6H_5$
10. Probably an aliphatic alcohol (the spectrum is of 2-hexanol)
11. Benzaldehyde, C_6H_5CHO
12. Probably an aliphatic ester (the spectrum is of *t*-butylacetate)

CHAPTER 7

1. Singlet at $\delta > 11$
   Quartet at $\delta \sim 2.2$
   Triplet at $\delta = 0.9$
2. (a) Quartet at $\delta = 9.8$
       Doublet at $\delta \sim 2.4$
   (b) Singlet at $\delta > 11$
       Singlet at $\delta \sim 2.2$
   (c) Triplet at $\delta = 1.8$
       Quartet at $\delta = 3.4$
3. (a) Singlet at $\delta = 2.1$
   (b) Singlet at $\delta = 2.1$
       Triplet at $\delta = 1.1$
       Quartet at $\delta = 2.4$
   (c) Singlet at $\delta = 2.1$
       Quartet at $\delta = 2.6$
       Doublet at $\delta = 1.1$
4. (a) Singlet at $\delta > 11$
       Triplet at $\delta \sim 4$
       Octet at $\delta \sim 1.8$
       Triplet at $\delta = 0.9$
   (b) Singlet at $\delta > 11$
       Doublet at $\delta = 1.8$ to 2.2
       Series of 12 at $\delta \sim 4.0$
       Doublet at $\delta = 1.5$
   (c) Triplet at $\delta = 3.5$
       Series of 12 at $\delta \sim 1.8$
       Triplet at $\delta = 0.9$
   (d) Quartet at $\delta \sim 4$
       Doublet at $\delta = 1.5$
5. (a) Singlet at $\delta \sim 2.2$
       Singlet at $\delta = 6.5$ to 8
   (b) Triplet at $\delta = 1.1$
       Quartet at $\delta = 2.6$
       Singlet at $\delta = 6.5$ to 8
   (c) Quartet at $\delta \sim 1.5$
       Doublet at $\delta = 0.9$

6. Ethyl bromide
7. α-Bromobutyric acid
8. Methyl ethyl ketone
9. Ethyl acetate
10. (a) Ethylbenzene
    (b) Xylene
11. Cumene
12. Tetrahydrothiophene

CHAPTER 16

1. (a) 0.766
   (b) 0.452
   (c) 0.157
   (d) 0.042
2. (a) 0.375
   (b) 0.835
   (c) 0.192
   (d) 0.816
3. 15.4 days
4.

	σ_N	$(\sigma_N)_r$
(a)	5.0 counts	20 percent
(b)	15.8 counts	6.3 percent
(c)	50 counts	2.0 percent
(d)	158 counts	0.63 percent

5.

	σ_R	$(\sigma_R)_r$
(a)	15.9 cpm	12.6 percent
(b)	11.2 cpm	8.9 percent
(c)	5.0 cpm	4.0 percent
(d)	3.5 cpm	2.8 percent

6.

	$\pm k\,\sigma_N$	$\pm k\,(\sigma_N)_r$
(a)	14.5 counts	3.2 percent
(b)	35.2 counts	7.8 percent
(c)	55.0 counts	12.1 percent

7.

	$\pm k\,\sigma_N$	$\pm k\,(\sigma_N)_r$
(a)	11.4 counts	23.7 percent
(b)	21.4 counts	12.6 percent
(c)	32.8 counts	8.3 percent
(d)	50.1 counts	5.4 percent

8.

	$\pm k\,\sigma_R$	$\pm k\,(\sigma_R)_r$
(a)	9.6 cpm	3.2 percent
(b)	9.2 cpm	3.1 percent
(c)	10.1 cpm	3.5 percent
(d)	11.4 cpm	4.2 percent

9. (a) 1480 counts
   (b) 1290 counts
   (c) 1170 counts
10. 0.111 ppm
11. 2.37 μg I⁻/ml

CHAPTER 17

1. (a) 0.681 V
   (b) −0.254 V
   (c) −0.295 V
   (d) 0.594 V
   (e) 0.067 V
2. (a) −0.499 V
   (b) 0.183 V
   (c) 0.0178 V
   (d) −0.169 V
   (e) 0.912 V
   (f) 0.496 V
3. (a) 0.681 V
   (b) 0.727 V
   (c) −0.074 V
   (d) −0.11 V
   (e) 0.674 V
   (f) 0.217 V
4. (a) 0.420 V
   (b) 1.45 V
5. (a) 1.20 V
   (b) 1.46 V
6. (a) 0.43 V; galvanic
   (b) −0.41 V; electrolytic
   (c) −0.605 V; electrolytic
   (d) −0.88 V; electrolytic
   (e) 1.05 V; galvanic
7. (a) 0.54 V; galvanic
   (b) −0.203 V; electrolytic
   (c) −0.31 V; electrolytic
   (d) 1.26 V; galvanic
   (e) 0.510 V; galvanic
8. (a) 0.118 V; galvanic
   (b) 0.266 V; galvanic
9. −0.613 V
10. −0.75 V
11. 0.072 V
12. −0.980 V
13. 0.135 V
14. 1.8×10^{-4}
15. 1.1×10^{-12}
16. (a) 0.814 V
    (b) −0.443 V
    (c) −0.075 V
17. (a) −0.645 V
    (b) 1.20 V
    (c) −0.57 V

18. 0.573 V
19. (a) 0.553 V
    (b) 1.32×10^{-2} mg Cu^{2+}/sec
20. (a) 0.393 V
    (b) 0.371 V
21. $E_{theory} = 0.402$ V $\therefore$ cell is being affected by polarization.

CHAPTER 18

1. (a) 0.035 V
   (b) $Cu|CuBr$ (sat'd), Br^- $(xM)\|SCE$
   (c) $pBr = \dfrac{E_{SCE} - E^0_{CuBr} - E_{cell}}{0.059} = \dfrac{0.207 - E_{cell}}{0.059}$
   (d) 2.22
2. (a) $Ag|AgIO_3$(sat'd), IO_3^- $(xM)\|SCE$
   $$pIO_3 = -\frac{(0.114 + E_{cell})}{0.059}$$
   (b) $Ag|Ag_2C_2O_4$(sat'd), $C_2O_4^{2-}$ $(xM)\|SCE$
   $$pC_2O_4 = -\frac{2(0.233 + E_{cell})}{0.059}$$
   (c) $Ag|Ag(S_2O_3)_2^{3-}$ $(1.00 \times 10^{-4}\ M)$, $S_2O_3^{2-}$ $(xM) \| SCE$
   $$pS_2O_3 = \frac{0.466 - E_{cell}}{0.059}$$
   (d) $Pt|I_2(xM)$, I^- $(1.00 \times 10^{-4}\ M)\|SCE$
   $$pI_2 = \frac{2(0.614 + E_{cell})}{0.059}$$
   (e) $Pt \mid Fe^{3+}$ (xM), Fe^{2+} $(1.00 \times 10^{-4}\ M)$ $\| SCE$
   $$pFe(III) = \frac{0.766 + E_{cell}}{0.059}$$

3. (a) $pCrO_4 = -\dfrac{2(0.204 + E_{cell})}{0.059}$

   (b) $pCrO_4 = 2.76$
4. (a) 0.606 V
   (b) 6.3×10^{-4} mole/liter
   (c) 2.19
5. (a) 5.75
   (b) 1.95
   (c) 0.17
6. (a) 6.58
   (b) 2.24×10^{-7} to 3.09×10^{-7}

7. 6.3×10^{-18}
8. 7.7×10^{-8}
9. 2.3×10^{17}
10. $\pm 0.017\ pH$

11. Vol 0.100 F

Ag$^+$, ml	[Ag$^+$]	E_{Ag} vs. SCE
5.0	1.73×10^{-11}	-0.078
15.0	2.86×10^{-11}	-0.065
25.0	5.50×10^{-11}	-0.048
30.0	8.80×10^{-11}	-0.037
35.0	1.87×10^{-10}	-0.017
39.0	9.82×10^{-10}	$+0.026$
40.0	1.05×10^{-6}	$+0.204$
41.0	1.10×10^{-3}	$+0.382$
45.0	5.26×10^{-3}	$+0.422$
50.0	2.00×10^{-2}	$+0.439$

12. Vol of 0.080 F

Ce^{4+}, ml	E vs. SCE
5.0	-0.084
10.0	-0.045
15.0	-0.037
25.0	-0.026
40.0	-0.018
49.0	$+0.024$
50.0	$+0.469$
51.0	$+1.358$
55.0	$+1.399$
60.0	$+1.417$

13. (a) 0.365 V
    (b) 0.124 V
    (c) 0.083 V
    (d) 0.509 V

14. 3.77
15. 7×10^{18}
16. 1.25×10^9
17. 7.6×10^{-8}
18. 5.3×10^{-11}
19. 6.3×10^{18}
20. 5.4×10^{-5}
21. 1.0×10^{-6}

CHAPTER 19

Vol of reagent, ml	Problem 1 k	Problem 2 k	Problem 3 k	Problem 4 k
0.00	2.48×10^{-4}	1.31×10^{-7}	9.10×10^{-5}	3.91×10^{-5}
2.00	2.24×10^{-4}	2.01×10^{-5}	9.90×10^{-5}	3.10×10^{-5}
4.00	1.99×10^{-4}	4.09×10^{-5}	1.08×10^{-4}	4.18×10^{-5}
6.00	1.75×10^{-4}	6.48×10^{-5}	1.18×10^{-4}	5.71×10^{-5}
8.00	1.51×10^{-4}	9.27×10^{-5}	1.33×10^{-4}	7.37×10^{-5}
10.00	1.26×10^{-4}	1.26×10^{-4}	1.66×10^{-4}	9.12×10^{-5}
12.00	2.12×10^{-4}	1.33×10^{-4}	2.27×10^{-4}	1.41×10^{-4}
14.00	2.97×10^{-4}	2.05×10^{-4}	3.06×10^{-4}	1.90×10^{-4}
16.00	3.82×10^{-4}	2.49×10^{-4}	3.83×10^{-4}	2.40×10^{-4}

Vol of reagent, ml	Problem 5 k	Problem 6 k	Problem 7 k
0.00	3.91×10^{-5}	1.33×10^{-4}	1.33×10^{-4}
2.00	3.56×10^{-5}	1.31×10^{-4}	1.29×10^{-4}
4.00	5.14×10^{-5}	1.29×10^{-4}	1.24×10^{-4}
6.00	7.10×10^{-5}	1.26×10^{-4}	1.19×10^{-4}
8.00	9.15×10^{-5}	1.24×10^{-4}	1.15×10^{-4}
10.00	1.14×10^{-4}	1.22×10^{-4}	1.10×10^{-4}
12.00	1.15×10^{-4}	1.47×10^{-4}	1.33×10^{-4}
14.00	1.15×10^{-4}	1.72×10^{-4}	1.56×10^{-4}
16.00	1.16×10^{-4}	1.97×10^{-4}	1.79×10^{-4}

CHAPTER 20

1. (a) -1.09 V
   (b) 0.40 V
   (c) 0.77 V
   (d) -2.26 V
   (e) 122 min
   (f) -2.30 V
2. (a) -1.35 V
   (b) 0.50 V
   (c) 0.85 V
   (d) -2.70 V
   (e) 458 min
   (f) -2.74 V
3. (a) Co deposits first
   (b) -0.70 to -1.04 V (vs. SCE)
4. (a) Not feasible
   (b) —
   (c) Perform separation in the presence of CN^- to delay deposits of Ni

5. (a) Not feasible
   (b) Feasible
   (c) For (b), +0.039 to −0.057 V (vs. SCE)
6. (a) 0.262 V
   (b) −0.17 V
   (c) −0.61 V
   (d) −0.042 V
7. (a) 0.036 V
   (b) −0.61 V
   (c) −0.201 V
   (d) −0.430 V
8. (a) 1.159 V
   (b) 0.155 amp
   (c) 0.961 V
   (d) 0.128 amp; onset of concentration polarization is likely to prevent current being this large
9. (a) 0.419 V
   (b) 0.056 amp
   (c) 0.221 V
   (d) 0.029 amp; onset of concentration polarization is likely to prevent current being this large
10. (a) 0.539 V
    (b) 0.303 V
    (c) 0.0675 amp
    (d) 0.0379 amp

CHAPTER 21

1. 5.76 percent Ni
   1.35 percent Co
2. 19.3 percent BaI_2
   12.0 percent $BaCl_2$
3. 6.18 percent Fe_3O_4
4. (a) 1.03×10^{-3} F
   (b) 0.608 to 0.372 V
   (c) 14,570 ohms
   (d) 0.18 percent
   (e) 0.007 percent
5. 0.633 percent As_2O_3
6. 0.455 mg/cm^2
7. 142 mg H_2S/liter
8. 19.6 ppm phenol
9. (a) 2.90 ppm
   (b) 3.08 ppm
10. 0.0396 mg $CaCO_3$/ml
11. 130 g/eq
12. 0.352 g NaCN/liter
13. 103 mg

CHAPTER 22

1. 362 mg Pb/liter
2. 29.9 mg In(III)/liter
3. $E_{applied} = E_{SCE} - E^0_{Hg_2Cl_2 \rightarrow Hg} + 0.059 \log \dfrac{(i_d - i)}{k}$

   $E_{1/2} = E_{SCE} - E^0_{Hg_2Cl_2 \rightarrow Hg} + 0.059 \log \dfrac{[Cl^-]}{2}$
4. -0.878 V
5. CdX_2^{2-}
   $K_f = 3.2 \times 10^{10}$
6. CuA^-
   $K_f = 5.1 \times 10^{16}$
7.

Vol Pb(NO₃)₂
(a)

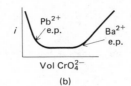

Vol CrO₄²⁻
(b)

Vol DMG
(c)

CHAPTER 23

1. (a) $r_{max} = nx$

$K = 1$	$x = 0.5$
$n = 25$	$r_{max} = 12.5$
$n = 50$	$r_{max} = 25$
$n = 100$	$r_{max} = 50$

$K = 0.20$	$x = 0.0167$	$r_{max} = 8$
$K = 1.00$	$x = 0.500$	$r_{max} = 25$
$K = 5.00$	$x = 0.833$	$r_{max} = 42$

   (b) $f_{25,12} = 0.16$
   $f_{50,25} = 0.11$
   $f_{100,50} = 0.080$

   $f_{50,8} = 0.15$
   $f_{50,25} = 0.11$
   $f_{50,42} = 0.15$

2. (a) $n = 60$; $r_{max}^A = 20$; $r_{max}^B = 40$
    (b) $f_{60,20}^A = 0.11$; $f_{60,20}^B = 3.5 \times 10^{-8}$
         $f_{60,40}^A = 3.5 \times 10^{-8}$; $f_{60,40}^B = 0.11$
    (c)

r	$f_{60,r}^A$	$f_{60,r}^B$
20	0.11	3.5×10^{-8}
25	0.0428	2.35×10^{-5}
30	0.0025	0.0025
35	2.35×10^{-5}	0.0428
40	3.5×10^{-8}	0.11

    (d) Curves intersect at tube 30
    (e) In tubes 30 and above, f^A is between 0.5 and 1.0 percent
         f^B is between 99 and 99.5 percent
3. (a) $n = 39$; $r_{max}^A = 9$; $r_{max}^B = 29$
    (b) $f_{39,9}^A = 0.152$ $f_{39,29}^B = 0.148$
    (c) $f_{39,14}^A = 0.025$ $f_{39,14}^B = 1.3 \times 10^{-8}$
         $f_{39,19}^A = 1.1 \times 10^{-4}$ $f_{39,19}^B = 1.0 \times 10^{-4}$
         $f_{39,24}^A = 1.3 \times 10^{-8}$ $f_{39,24}^B = 0.024$
    (d) Curves intersect at tube 19
    (e) Essentially complete separation is achieved

4. $r_{max} = 64$
    $f_{80,64} = 0.112$
    $f_{80,62} = 0.096 = f_{80,66}$
    $f_{80,60} = 0.060 = f_{80,68}$
    $f_{80,58} = 0.028 = f_{80,72}$
    $f_{80,56} = 0.0089 = f_{80,74}$
    $f_{80,55} = 0.0045 = f_{80,75}$
5. (a) $K = 1$; $R_F = 0.4$
         $K = 2$; $R_F = 0.25$
         $K = 4$; $R_F = 0.143$
    (b) $K = 1$; $V = 10$ cm³
         $K = 2$; $V = 16$ cm³
         $K = 4$; $V = 28$ cm³
6. $K = 2.0$; $V = 19.1$ ml
    $K = 3.2$; $V = 27.5$ ml

CHAPTER 24

1. (a) $V_G^0 = 8.6$ cm³
         $V_{R1}^0 = 60.3$ cm³
         $V_{R2}^0 = 70.6$ cm³
    (b) $R_{F1} = 0.143$
         $R_{F2} = 0.122$
    (c) $K_{g_1} = 16.6$
         $K_{g_2} = 19.9$
    (d) $\alpha = 1.20$

2. $n \cong 300$
3. (a) $V^0_{G3} = 68.2 \text{ cm}^3$
        $R_{F3} = 0.129$
        $K_{G3} = 19.1$
    (b) 1:1; 2 percent
         10:1; 0.4 percent
4. (a) $n \cong 500$
    (b) $n \cong 700$
    (c) $n \cong 250$
5. (a) Air $V^0_G = 10.5 \text{ cm}^3$
         Benzene $V^0_R = 61.7 \text{ cm}^3$
         Toluene $V^0_R = 117 \text{ cm}^3$
         Cumene $V^0_R = 233 \text{ cm}^3$
         Ethylbenzene $V^0_R = 183 \text{ cm}^3$
    (b) Benzene $R_F = 0.170$
         Toluene $R_F = 0.090$
         Cumene $R_F = 0.045$
         Ethylbenzene $R_F = 0.057$
    (c) Benzene $K_g = 3.63$
         Toluene $K_g = 7.53$
         Cumene $K_g = 15.8$
         Ethylbenzene $K_g = 12.2$
    (d) Toluene-benzene $\alpha = 2.07$
         Cumene-benzene $\alpha = 4.35$
         Ethylbenzene-benzene $\alpha = 3.36$
         Cumene-toluene $\alpha = 2.10$
         Cumene-ethylbenzene $\alpha = 1.29$
         Ethylbenzene-toluene $\alpha = 1.62$
    (e) $n \cong 600$

Appendix

Table A-1 Some Standard and Formal Electrode Potentials[a]

Half-reaction	E^0, volts	Formal Potential, volts
$F_2 + 2H^+ + 2e \rightleftharpoons 2HF$	3.06	
$O_3 + 2H^+ + 2e \rightleftharpoons O_2 + H_2O$	2.07	
$S_2O_8^{2-} + 2e \rightleftharpoons 2SO_4^{2-}$	2.01	
$Co^{3+} + e \rightleftharpoons Co^{2+}$	1.842	
$H_2O_2 + 2H^+ + 2e \rightleftharpoons 2H_2O$	1.77	
$MnO_4^- + 4H^+ + 3e \rightleftharpoons MnO_2 + 2H_2O$	1.695	
$Ce^{4+} + e \rightleftharpoons Ce^{3+}$		1.70, 1 F HClO$_4$; 1.61, 1 F HNO$_3$; 1.44, 1 F H$_2$SO$_4$
$HClO + H^+ + e \rightleftharpoons \frac{1}{2}Cl_2 + H_2O$	1.63	
$H_5IO_6 + H^+ + 2e \rightleftharpoons IO_3^- + 3H_2O$	1.6	
$BrO_3^- + 6H^+ + 5e \rightleftharpoons \frac{1}{2}Br_2 + 3H_2O$	1.52	
$MnO_4^- + 8H^+ + 5e \rightleftharpoons Mn^{2+} + 4H_2O$	1.51	
$Mn^{3+} + e \rightleftharpoons Mn^{2+}$		1.51, aq. H$_2$SO$_4$
$ClO_3^- + 6H^+ + 5e \rightleftharpoons \frac{1}{2}Cl_2 + 3H_2O$	1.47	
$PbO_2 + 4H^+ + 2e \rightleftharpoons Pb^{2+} + 2H_2O$	1.455	
$Cl_2 + 2e \rightleftharpoons 2Cl^-$	1.359	
$Cr_2O_7^{2-} + 14H^+ + 6e \rightleftharpoons 2Cr^{3+} + 7H_2O$	1.33	
$Tl^{3+} + 2e \rightleftharpoons Tl^+$	1.25	0.77, 1 F HCl
$IO_3^- + 2Cl^- + 6H^+ + 4e \rightleftharpoons ICl_2^- + 3H_2O$	1.24	
$MnO_2 + 4H^+ + 2e \rightleftharpoons Mn^{2+} + 2H_2O$	1.23	1.24, 1 F HClO$_4$
$O_2 + 4H^+ + 4e \rightleftharpoons 2H_2O$	1.229	
$IO_3^- + 6H^+ + 5e \rightleftharpoons \frac{1}{2}I_2 + 3H_2O$	1.195	
$SeO_4^{2-} + 4H^+ + 2e \rightleftharpoons H_2SeO_3 + H_2O$	1.15	
$Br_2(aq) + 2e \rightleftharpoons 2Br^-$	1.087[b]	
$Br_2(1) + 2e \rightleftharpoons 2Br^-$	1.065	1.05, 4 F HCl
$ICl_2^- + e \rightleftharpoons \frac{1}{2}I_2 + 2Cl^-$	1.06	

678

Table A-1 (continued)

Half-reaction	E^0, volts	Formal Potential, volts
$V(OH)_4^+ + 2H^+ + e \rightleftharpoons VO^{2+} + 3H_2O$	1.00	1.02, 1 F HCl, HClO$_4$
$HNO_2 + H^+ + e \rightleftharpoons NO + H_2O$	1.00	
$Pd^{2+} + 2e \rightleftharpoons Pd$	0.987	
$NO_3^- + 3H^+ + 2e \rightleftharpoons HNO_2 + H_2O$	0.94	0.92, 1 F HNO$_3$
$2Hg^{2+} + 2e \rightleftharpoons Hg_2^{2+}$	0.920	0.907, 1 F HClO$_4$
$H_2O_2 + 2e \rightleftharpoons 2OH^-$	0.88	
$Cu^{2+} + I^- + e \rightleftharpoons CuI$	0.86	
$Hg^{2+} + 2e \rightleftharpoons Hg$	0.854	
$Ag^+ + e \rightleftharpoons Ag$	0.799	0.228, 1 F HCl; 0.792, 1 F HClO$_4$; 0.77, 1 F H$_2$SO$_4$
$Hg_2^{2+} + 2e \rightleftharpoons 2Hg$	0.789	0.274, 1 F HCl; 0.776, 1 F HClO$_4$; 0.674, 1 F H$_2$SO$_4$
$Fe^{3+} + e \rightleftharpoons Fe^{2+}$	0.771	0.700, 1 F HCl; 0.732, 1 F HClO$_4$; 0.68, 1 F H$_2$SO$_4$
$H_2SeO_3 + 4H^+ + 4e \rightleftharpoons Se + 3H_2O$	0.740	
$PtCl_4^{2-} + 2e \rightleftharpoons Pt + 4Cl^-$	0.73	
$C_6H_4O_2$ (quinone) $+ 2H^+$ $+ 2e \rightleftharpoons C_6H_4(OH)_2$	0.699	0.696, 1 F HCl, H$_2$SO$_4$, HClO$_4$
$O_2 + 2H^+ + 2e \rightleftharpoons H_2O_2$	0.682	
$PtCl_6^{2-} + 2e \rightleftharpoons PtCl_4^{2-} + 2Cl^-$	0.68	
$I_2(aq) + 2e \rightleftharpoons 2I^-$	0.620^b	
$Hg_2SO_4 + 2e \rightleftharpoons 2Hg + SO_4^{2-}$	0.615	
$Sb_2O_5 + 6H^+ + 4e \rightleftharpoons 2SbO^+ + 3H_2O$	0.581	
$MnO_4^- + e \rightleftharpoons MnO_4^{2-}$	0.564	
$H_3AsO_4 + 2H^+ + 2e \rightleftharpoons H_3AsO_3 + H_2O$	0.559	0.577, 1 F HCl, HClO$_4$
$I_3^- + 2e \rightleftharpoons 3I^-$	0.536	
$I_2(s) + 2e \rightleftharpoons 2I^-$	0.5355	
$Cu^+ + e \rightleftharpoons Cu$	0.521	
$H_2SO_3 + 4H^+ + 4e \rightleftharpoons S + 3H_2O$	0.45	
$Ag_2CrO_4 + 2e \rightleftharpoons 2Ag + CrO_4^{2-}$	0.446	
$VO^{2+} + 2H^+ + e \rightleftharpoons V^{3+} + H_2O$	0.361	
$Fe(CN)_6^{3-} + e \rightleftharpoons Fe(CN)_6^{4-}$	0.36	0.71, 1 F HCl; 0.72, 1 F HClO$_4$, H$_2$SO$_4$
$Cu^{2+} + 2e \rightleftharpoons Cu$	0.337	
$UO_2^{2+} + 4H^+ + 2e \rightleftharpoons U^{4+} + 2H_2O$	0.334	
$BiO^+ + 2H^+ + 3e \rightleftharpoons Bi + H_2O$	0.32	
$Hg_2Cl_2(s) + 2e \rightleftharpoons 2Hg + 2Cl^-$	0.268	0.242, sat'd. KCl; 0.282, 1 F KCl
$AgCl + e \rightleftharpoons Ag + Cl^-$	0.222	0.228, 1 F KCl
$SO_4^{2-} + 4H^+ + 2e \rightleftharpoons H_2SO_3 + H_2O$	0.17	
$BiCl_4^- + 3e \rightleftharpoons Bi + 4Cl^-$	0.16	
$Sn^{4+} + 2e \rightleftharpoons Sn^{2+}$	0.154	0.14, 1 F HCl
$Cu^{2+} + e \rightleftharpoons Cu^+$	0.153	
$S + 2H^+ + 2e \rightleftharpoons H_2S$	0.141	
$TiO^{2+} + 2H^+ + e \rightleftharpoons Ti^{3+} + H_2O$	0.1	0.04, 1 F H$_2$SO$_4$
$AgBr + e \rightleftharpoons Ag + Br^-$	0.095	

Table A-1 (continued)

Half-reaction	E^0, volts	Formal Potential, volts
$S_4O_6^{2-} + 2e \rightleftharpoons 2S_2O_3^{2-}$	0.08	
$Ag(S_2O_3)_2^{3-} + e \rightleftharpoons Ag + 2S_2O_3^{2-}$	0.01	
$2H^+ + 2e \rightleftharpoons H_2$	0.000	-0.005, 1 F HCl, HClO$_4$
$Pb^{2+} + 2e \rightleftharpoons Pb$	-0.126	-0.14, 1 F HClO$_4$; -0.29, 1 F H$_2$SO$_4$
$Sn^{2+} + 2e \rightleftharpoons Sn$	-0.136	-0.16, 1 F HClO$_4$
$AgI + e \rightleftharpoons Ag + I^-$	-0.151	
$CuI + e \rightleftharpoons Cu + I^-$	-0.185	
$N_2 + 5H^+ + 4e \rightleftharpoons N_2H_5^+$	-0.23	
$Ni^{2+} + 2e \rightleftharpoons Ni$	-0.250	
$V^{3+} + e \rightleftharpoons V^{2+}$	-0.255	-0.21, 1 F HClO$_4$
$Co^{2+} + 2e \rightleftharpoons Co$	-0.277	
$Ag(CN)_2^- + e \rightleftharpoons Ag + 2CN^-$	-0.31	
$Tl^+ + e \rightleftharpoons Tl$	-0.336	-0.551, 1 F HCl; -0.33, 1 F HClO$_4$, H$_2$SO$_4$
$PbSO_4 + 2e \rightleftharpoons Pb + SO_4^{2-}$	-0.356	
$Ti^{3+} + e \rightleftharpoons Ti^{2+}$	-0.37	
$Cd^{2+} + 2e \rightleftharpoons Cd$	-0.403	
$Cr^{3+} + e \rightleftharpoons Cr^{2+}$	-0.41	
$Fe^{2+} + 2e \rightleftharpoons Fe$	-0.440	
$2CO_2(g) + 2H^+ + 2e \rightleftharpoons H_2C_2O_4$	-0.49	
$Cr^{3+} + 3e \rightleftharpoons Cr$	-0.74	
$Zn^{2+} + 2e \rightleftharpoons Zn$	-0.763	
$Mn^{2+} + 2e \rightleftharpoons Mn$	-1.18	
$Al^{3+} + 3e \rightleftharpoons Al$	-1.66	
$Mg^{2+} + 2e \rightleftharpoons Mg$	-2.37	
$Na^+ + e \rightleftharpoons Na$	-2.714	
$Ca^{2+} + 2e \rightleftharpoons Ca$	-2.87	
$Ba^{2+} + 2e \rightleftharpoons Ba$	-2.90	
$K^+ + e \rightleftharpoons K$	-2.925	
$Li^+ + e \rightleftharpoons Li$	-3.045	

"Source for the majority of E^0 values is Wendell M. Latimer, *The Oxidation States of the Elements and Their Potentials in Aqueous Solutions,* 2d ed., (Englewood Cliffs, N.J.: Prentice Hall, Inc., © 1952). The formal potentials are from Ernest H. Swift, *Introductory Quantitative Analysis,* (Englewood Cliffs, N.J.: Prentice Hall, Inc., 1950). With permission.

^{b}These potentials are hypothetical, since they correspond to solutions that are 1.00 M in Br$_2$ or I$_2$. The solubilities of these two compounds at 25°C are 0.21 M and 0.0133 M respectively. In saturated solutions containing an excess of Br$_2$(l) or I$_2$(s), the standard potentials for the half-reactions Br$_2$(l) + 2e $\rightleftharpoons$ 2Br$^-$ or I$_2$(s) + 2e $\rightleftharpoons$ 2I$^-$ should be used. On the other hand, at Br$_2$ and I$_2$ concentrations less than saturation, these hypothetical electrode potentials should be employed.

Table A-2 Solubility Product Constants

Substance	Formula	K_{sp}
Aluminum hydroxide	$Al(OH)_3$	2×10^{-32}
Barium carbonate	$BaCO_3$	4.9×10^{-9}
Barium chromate	$BaCrO_4$	1.2×10^{-10}

Table A-2 (continued)

Substance	Formula	K_{sp}
Barium iodate	$Ba(IO_3)_2$	1.57×10^{-9}
Barium manganate	$BaMnO_4$	2.5×10^{-10}
Barium oxalate	BaC_2O_4	2.3×10^{-8}
Barium sulfate	$BaSO_4$	1.0×10^{-10}
Bismuth oxide chloride	$BiOCl$	7×10^{-9}
Bismuth oxide hydroxide	$BiOOH$	4×10^{-10}
Cadmium carbonate	$CdCO_3$	2.5×10^{-14}
Cadmium oxalate	CdC_2O_4	9×10^{-8}
Cadmium sulfide	CdS	1×10^{-28}
Calcium carbonate	$CaCO_3$	4.8×10^{-9}
Calcium fluoride	CaF_2	4.9×10^{-11}
Calcium oxalate	CaC_2O_4	2.3×10^{-9}
Calcium sulfate	$CaSO_4$	6.1×10^{-5}
Copper(I) bromide	$CuBr$	5.9×10^{-9}
Copper(I) chloride	$CuCl$	3.2×10^{-7}
Copper(I) iodide	CuI	1.1×10^{-12}
Copper(I) thiocyanate	$CuSCN$	4×10^{-14}
Copper(II) hydroxide	$Cu(OH)_2$	1.6×10^{-19}
Copper(II) sulfide	CuS	8.5×10^{-45}
Iron(II) hydroxide	$Fe(OH)_2$	8×10^{-16}
Iron(III) hydroxide	$Fe(OH)_3$	1.5×10^{-36}
Lanthanum iodate	$La(IO_3)_3$	6×10^{-10}
Lead carbonate	$PbCO_3$	1.6×10^{-13}
Lead chloride	$PbCl_2$	1×10^{-4}
Lead chromate	$PbCrO_4$	1.8×10^{-14}
Lead hydroxide	$Pb(OH)_2$	2.5×10^{-16}
Lead iodide	PbI_2	7.1×10^{-9}
Lead oxalate	PbC_2O_4	3.0×10^{-11}
Lead sulfate	$PbSO_4$	1.9×10^{-8}
Lead sulfide	PbS	7×10^{-28}
Magnesium ammonium phosphate	$MgNH_4PO_4$	2.5×10^{-13}
Magnesium carbonate	$MgCO_3$	1×10^{-5}
Magnesium hydroxide	$Mg(OH)_2$	5.9×10^{-12}
Magnesium oxalate	MgC_2O_4	8.6×10^{-5}
Manganese(II) hydroxide	$Mn(OH)_2$	4×10^{-14}
Manganese(II) sulfide	MnS	1.4×10^{-15}
Mercury(I) bromide	Hg_2Br_2	5.8×10^{-23}
Mercury(I) chloride	Hg_2Cl_2	1.3×10^{-18}
Mercury(I) iodide	Hg_2I_2	4.5×10^{-29}
Silver arsenate	Ag_3AsO_4	1.0×10^{-22}
Silver bromide	$AgBr$	7.7×10^{-13}
Silver carbonate	Ag_2CO_3	8.2×10^{-12}
Silver chloride	$AgCl$	1.82×10^{-10}
Silver chromate	Ag_2CrO_4	1.1×10^{-12}
Silver cyanide	$AgCN$	2×10^{-16}
Silver iodate	$AgIO_3$	3.1×10^{-8}

Table A-2 (continued)

Substance	Formula	K_{sp}
Silver iodide	AgI	8.3×10^{-17}
Silver oxalate	$Ag_2C_2O_4$	1.1×10^{-11}
Silver sulfide	Ag_2S	1.6×10^{-49}
Silver thiocyanate	AgSCN	1.1×10^{-12}
Strontium oxalate	SrC_2O_4	5.6×10^{-8}
Strontium sulfate	$SrSO_4$	2.8×10^{-7}
Thallium(I) chloride	TlCl	2×10^{-4}
Thallium(I) sulfide	Tl_2S	1×10^{-22}
Zinc hydroxide	$Zn(OH)_2$	2×10^{-14}
Zinc oxalate	ZnC_2O_4	7.5×10^{-9}
Zinc sulfide	ZnS	4.5×10^{-24}

Table A-3 Dissociation Constants for Acids[a]

Name	Formula	Dissociation Constant, 25°C		
		K_1	K_2	K_3
Acetic	CH_3COOH	1.75×10^{-5}		
Arsenic	H_3AsO_4	6.0×10^{-3}	1.05×10^{-7}	3.0×10^{-12}
Arsenious	H_3AsO_3	6.0×10^{-10}	3.0×10^{-14}	
Benzoic	C_6H_5COOH	6.3×10^{-5}		
Boric	H_3BO_3	5.8×10^{-10}		
1-Butanoic	$CH_3CH_2CH_2COOH$	1.51×10^{-5}		
Carbonic	H_2CO_3	4.6×10^{-7}	4.4×10^{-11}	
Chloroacetic	$ClCH_2COOH$	1.51×10^{-3}		
Citric	$HOOC(OH)C(CH_2COOH)_2$	7.4×10^{-4}	1.74×10^{-5}	4.0×10^{-7}
Formic	$HCOOH$	1.74×10^{-4}		
Fumaric	$trans$-$HOOCCH:CHCOOH$	9.6×10^{-4}	4.1×10^{-5}	
Glycine	H_2NCH_2COOH	4.5×10^{-3}		
Glycolic	$HOCH_2COOH$	1.32×10^{-4}		
Hydrazoic	HN_3	1.9×10^{-5}		
Hydrogen cyanide	HCN	2.1×10^{-9}		
Hydrogen fluoride	H_2F_2	7.2×10^{-4}		
Hydrogen peroxide	H_2O_2	2.7×10^{-12}		
Hydrogen sulfide	H_2S	5.7×10^{-8}	1.2×10^{-15}	
Hypochlorous	HOCl	3.0×10^{-8}		
Iodic	HIO_3	1.58×10^{-1}		
Lactic	$CH_3CHOHCOOH$	1.38×10^{-4}		
Maleic	cis-$HOOCCH:CHCOOH$	1.5×10^{-2}	2.6×10^{-7}	
Malic	$HOOCCHOHCH_2COOH$	4.0×10^{-4}	8.9×10^{-6}	

Table A-3 (continued)

Name	Formula	K_1	K_2	K_3
		Dissociation Constant, 25°C		
Malonic	$HOOCCH_2COOH$	1.58×10^{-3}	8.0×10^{-7}	
Mandelic	$C_6H_5CHOHCOOH$	4.3×10^{-4}		
Nitrous	HNO_2	5.1×10^{-4}		
Oxalic	$HOOCCOOH$	6.2×10^{-2}	6.1×10^{-5}	
Periodic	H_5IO_6	2.4×10^{-2}	5.0×10^{-9}	
Phenol	C_6H_5OH	1.05×10^{-10}		
Phosphoric	H_3PO_4	7.5×10^{-3}	6.2×10^{-8}	4.8×10^{-13}
Phosphorous	H_3PO_3	1.00×10^{-2}	2.6×10^{-7}	
o-Phthalic	$C_6H_4(COOH)_2$	1.3×10^{-3}	3.9×10^{-6}	
Picric	$(NO_2)_3C_6H_2OH$	4.2×10^{-1}		
Propanoic	CH_3CH_2COOH	1.32×10^{-5}		
Salicylic	$C_6H_4(OH)COOH$	1.05×10^{-3}		
Sulfuric	H_2SO_4	strong	1.2×10^{-2}	
Sulfurous	H_2SO_3	1.74×10^{-2}	6.2×10^{-8}	
Tartaric	$HOOC(CHOH)_2COOH$	9.4×10^{-4}	2.9×10^{-5}	
Trichloroacetic	Cl_3CCOOH	1.29×10^{-1}		

[a]Taken in part from I. M. Kolthoff and P. J. Elving, Eds., *Treatise on Analytical Chemistry*, Part I, Vol. 1 (New York: Interscience Publishers, Inc., 1959), p. 432; and in part from I. M. Kolthoff and V. A. Stenger, *Volumetric Analysis*, 2d ed., Vol. I (New York: Interscience Publishers, Inc., 1942), p. 282. With permission.

Table A-4 Dissociation Constants for Bases[a]

Name	Formula	*Dissociation Constant, K, 25°C*
Ammonia	NH_3	1.86×10^{-5}
Aniline	$C_6H_5NH_2$	3.8×10^{-10}
1-Butylamine	$CH_3(CH_2)_2CH_2NH_2$	4.1×10^{-4}
Dimethylamine	$(CH_3)_2NH$	5.9×10^{-4}
Ethanolamine	$HOC_2H_4NH_2$	3.2×10^{-5}
Ethylamine	$CH_3CH_2NH_2$	5.6×10^{-4}
Ethylenediamine	$NH_2C_2H_4NH_2$	$K_1 = 8.5 \times 10^{-5}$
		$K_2 = 7.1 \times 10^{-8}$
Glycine	$HOOCCH_2NH_2$	2.3×10^{-12}
Hydrazine	H_2NNH_2	3.0×10^{-6}
Hydroxylamine	$HONH_2$	1.07×10^{-8}
Methylamine	CH_3NH_2	4.8×10^{-4}
Piperidine	$C_5H_{11}N$	1.6×10^{-3}
Pyridine	C_5H_5N	1.4×10^{-9}
Trimethylamine	$(CH_3)_3N$	6.3×10^{-5}
Zinc hydroxide	$Zn(OH)_2$	$K_2 = 4.4 \times 10^{-5}$

[a]Taken in part from I. M. Kolthoff and V. A. Stenger, *Volumetric Analysis*, 2d ed., Vol. I (New York: Interscience Publishers, 1942), p. 284. With permission.

Table A-5 Stepwise Formation Constants[a]

Name and Formula, Ligand	Metal Ion	Ionic Strength	K_1	K_2	K_3	K_4
Acetate, CH_3COO^-	Ag^+	0	5.4	0.8		
	Cd^{2+}	0	5.0×10^1			
	Cu^{2+}	3	2.0×10^1		1.4	0.38
	Hg^{2+}	0	1.4×10^2	1.1×10^1		
	Pb^{2+}	0	$K_1 K_2 = 2.7 \times 10^8$			
Ammonia, NH_3	Ag^+	0	2.7×10^2	3.3×10^1		
	Cd^{2+}	0	2.0×10^3	6.9×10^3		
	Co^{2+}	0	3.2×10^2	9.1×10^1	2.0×10^1	6.2
			9.8×10^1	3.2×10^1	8.5	4.4 ($K_5 = 1.1$)
	Cu^{2+}	0	9.8×10^3	2.2×10^3	5.4×10^2	9.3×10^1 ($K_6 = 0.18$)
	Hg^{2+}	2	6×10^8	5×10^8	1.0×10^1	6
	Ni^{2+}	0	4.7×10^2	1.3×10^2	4.1×10^1	1.2×10^1 ($K_6 = 0.11$)
Bromide, Br^-	Zn^{2+}	0	1.5×10^2	1.8×10^2	2.0×10^2	9.1×10^1
	Ag^+	0	$AgBr(s) + Br^- \rightleftharpoons AgBr_2^-$, $K_{S2} = 2.0 \times 10^{-5}$; $AgBr_2^- + Br^- \rightleftharpoons AgBr_3^{2-}$, $K_3 = 4.6$ ($K_5 = 0.43$)			
	Hg^{2+}	0.5	1.1×10^9	1.9×10^8	2.6×10^2	1.8×10^1
	Pb^{2+}	0	1.7×10^1			
Chloride, Cl^-	Ag^+	0	$AgCl(s) + Cl^- \rightleftharpoons AgCl_2^-$, $K_{S2} = 2.0 \times 10^{-5}$; $AgCl_2^- + Cl^- \rightleftharpoons AgCl_3^{2-}$, $K_3 = 1$			
	Bi^{3+}	2	1.5×10^2	3.6×10^1	1.3×10^1	1.1×10^2
	Cd^{2+}	0	1.0×10^2	5.0	0.26	
	Cu^+	0	$Cu^+ + 2Cl^- \rightleftharpoons CuCl_2^-$	1.1	$\beta_2 = 8.7 \times 10^4$	
	Fe^{2+}	2	2.3			
	Fe^{3+}	0	3.0×10^1	4.5	0.1	
	Hg^{2+}	0.5	5.5×10^6	3.0×10^6	8.9	1.1×10^1
	Pb^{2+}	0	4.0×10^1	$Pb^{2+} + 3Cl^- \rightleftharpoons PbCl_3^-$		$\beta_3 = 4.8 \times 10^1$
	Sn^{2+}	0	3.2×10^1	5.4	0.62	0.28
Cyanide, CN^-	Ag^+	0	$Ag^+ + 2CN^- \rightleftharpoons Ag(CN)_2^-$, $\beta_2 = 1 \times 10^{20}$			
	Cd^{2+}	3	5.0×10^5	1.3×10^5	4.3×10^4	3.5×10^3
	Hg^{2+}	0.1	1.0×10^{18}	5×10^{16}	7×10^3	1×10^3

Ligand	Cation	μ	K₁	K₂	K₃	K₄	Other equilibria / constants
Cyanide, CN^-	Ni^{2+}	0					$Ni^{2+} + 4CN^- \rightleftharpoons Ni(CN)_4^{2-}$, $\beta_4 = 1 \times 10^{22}$
Ethylenediaminetetraacetate, Y^{4-}	See Table A-6						
Fluoride, F^-	Al^{3+}	0.5	1.3×10^6	1.0×10^5	7×10^3	5×10^2	$K_5 = 4 \times 10^1$, $K_6 = 5$
Hydroxide, OH^-	Fe^{3+}	0.5	1.8×10^5	1.0×10^4	1.0×10^3		
	Al^{3+}	0	6×10^9				$Al(OH)_3(s) + OH^- \rightleftharpoons Al(OH)_4^-$, $K_{S4} \sim 10$
	Cd^{2+}	0.1	$\sim 10^4$				
	Cu^{2+}	0	$\sim 10^6$				
	Fe^{2+}	0	$\sim 10^6$				
	Fe^{3+}	0	1×10^{11}	5×10^{10}			
	Hg^{2+}	0	3×10^{11}				
	Ni^{2+}	0	$\sim 10^4$				
	Pb^{2+}	0	7×10^7				$Pb(OH)_2(s) + OH^- \rightleftharpoons Pb(OH)_3^-$, $K_{S3} = 5 \times 10^{-2}$
	Zn^{2+}	0	1×10^4				$Zn(OH)_2(s) + 2OH^- \rightleftharpoons Zn(OH)_4^{2-}$, $K_{S4} = 0.13$
Iodide, I^-	Cd^{2+}	0	1.9×10^2	4.4×10^1	1.2×10^1	1.3×10^1	
	Cu^+	0					$CuI(s) + I^- \rightleftharpoons CuI_2^-$, $K_{S2} = 8 \times 10^{-4}$
	Hg^{2+}	0.5	7.4×10^{12}	9.0×10^{10}	6.0×10^3	2.1×10^2	
	Pb^{2+}	0	1×10^2				$PbI_2(s) + I^- \rightleftharpoons PbI_3^-$, $K_{S3} = 2.2 \times 10^{-5}$; $PbI_3^- + I^- \rightleftharpoons PbI_4^{2-}$, $K_4 = 1.4 \times 10^{-4}$
Oxalate, $C_2O_4^{2-}$	Al^{3+}	0	$K_1K_2 = 1 \times 10^{13}$		2×10^3		
	Fe^{3+}	0	2.5×10^9	6×10^6	1×10^4		
	Mg^{2+}	0	6.6×10^3	2.7×10^1			
	Mn^{3+}	2	1.0×10^{10}	3.9×10^6	7×10^2		
	Pb^{2+}	0					$Pb^{2+} + 2C_2O_4^{2-} \rightleftharpoons Pb(C_2O_4)_2^{2-}$, $\beta_2 = 3.5 \times 10^6$
Sulfate, SO_4^{2-}	Al^{3+}	0	1.6×10^3	8×10^1			
	Cd^{2+}	0	2×10^2				
	Cu^{2+}	0	1.6×10^2				
	Fe^{3+}	0	1.1×10^4	2×10^1			
Thiocyanate, SCN^-	Ag^+	0					$AgSCN(s) + SCN^- \rightleftharpoons Ag(SCN)_2^-$, $K_{S2} = 6 \times 10^{-8}$
	Cu^+	5					$CuSCN(s) + SCN^- \rightleftharpoons Cu(SCN)_2^-$, $K_{S2} = 4 \times 10^{-4}$
	Fe^{3+}	0	1.4×10^2	2×10^1			
	Hg^{2+}	0	$K_1K_2 = 1.8 \times 10^{17}$		5.1×10^2	6×10^1	

[a]These data were taken from L. G. Sillén and A. E. Martell, *Stability Constants of Metal-Ion Complexes*. (London: The Chemical Society, 1964). With permission.

Table A-6 Formation Constants for EDTA Complexes[a]

Cation	K_{MY}	$log\ K_{MY}$	Cation	K_{MY}	$log\ K_{MY}$
Ag^+	2×10^7	7.3	Cu^{2+}	6.3×10^{18}	18.80
Mg^{2+}	4.9×10^8	8.69	Zn^{2+}	3.2×10^{16}	16.50
Ca^{2+}	5.0×10^{10}	10.70	Cd^{2+}	2.9×10^{16}	16.46
Sr^{2+}	4.3×10^8	8.63	Hg^{2+}	6.3×10^{21}	21.80
Ba^{2+}	5.8×10^7	7.76	Pb^{2+}	1.1×10^{18}	18.04
Mn^{2+}	6.2×10^{13}	13.79	Al^{3+}	1.3×10^{16}	16.13
Fe^{2+}	2.1×10^{14}	14.33	Fe^{3+}	1×10^{25}	25.1
Co^{2+}	2.0×10^{16}	16.31	V^{3+}	8×10^{25}	25.9
Ni^{2+}	4.2×10^{18}	18.62	Th^{4+}	2×10^{23}	23.2

[a]Data from G. Schwarzenbach, *Complexometric Titrations* (New York: Interscience Publishers, Inc., 1957), p. 8. With permission. (Constants valid at 20°C and ionic strength of 0.1.)

Index

Four-Place Logarithms of Numbers

n	0	1	2	3	4	5	6	7	8	9
10	0000	0043	0086	0128	0170	0212	0253	0294	0334	0374
11	0414	0453	0492	0531	0569	0607	0645	0682	0719	0755
12	0792	0828	0864	0899	0934	0969	1004	1038	1072	1106
13	1139	1173	1206	1239	1271	1303	1335	1367	1399	1430
14	1461	1492	1523	1553	1584	1614	1644	1673	1703	1732
15	1761	1790	1818	1847	1875	1903	1931	1959	1987	2014
16	2041	2068	2095	2122	2148	2175	2201	2227	2253	2279
17	2304	2330	2355	2380	2405	2430	2455	2480	2504	2529
18	2553	2577	2601	2625	2648	2672	2695	2718	2742	2765
19	2788	2810	2833	2856	2878	2900	2923	2945	2967	2989
20	3010	3032	3054	3075	3096	3118	3139	3160	3181	3201
21	3222	3243	3263	3284	3304	3324	3345	3365	3385	3404
22	3424	3444	3464	3483	3502	3522	3541	3560	3579	3598
23	3617	3636	3655	3674	3692	3711	3729	3747	3766	3784
24	3802	3820	3838	3856	3874	3892	3909	3927	3945	3962
25	3979	3997	4014	4031	4048	4065	4082	4099	4116	4133
26	4150	4166	4183	4200	4216	4232	4249	4265	4281	4298
27	4314	4330	4346	4362	4378	4393	4409	4425	4440	4456
28	4472	4487	4502	4518	4533	4548	4564	4579	4594	4609
29	4624	4639	4654	4669	4683	4698	4713	4728	4742	4757
30	4771	4786	4800	4814	4829	4843	4857	4871	4886	4900
31	4914	4928	4942	4955	4969	4983	4997	5011	5024	5038
32	5051	5065	5079	5092	5105	5119	5132	5145	5159	5172
33	5185	5198	5211	5224	5237	5250	5263	5276	5289	5302
34	5315	5328	5340	5353	5366	5378	5391	5403	5416	5428
35	5441	5453	5465	5478	5490	5502	5514	5527	5539	5551
36	5563	5575	5587	5599	5611	5623	5635	5647	5658	5670
37	5682	5694	5705	5717	5729	5740	5752	5763	5775	5786
38	5798	5809	5821	5832	5843	5855	5866	5877	5888	5899
39	5911	5922	5933	5944	5955	5966	5977	5988	5999	6010
40	6021	6031	6042	6053	6064	6075	6085	6096	6107	6117
41	6128	6138	6149	6160	6170	6180	6191	6201	6212	6222
42	6232	6243	6253	6263	6274	6284	6294	6304	6314	6325
43	6335	6345	6355	6365	6375	6385	6395	6405	6415	6425
44	6435	6444	6454	6464	6474	6484	6493	6503	6513	6522
45	6532	6542	6551	6561	6571	6580	6590	6599	6609	6618
46	6628	6637	6646	6656	6665	6675	6684	6693	6702	6712
47	6721	6730	6739	6749	6758	6767	6776	6785	6794	6803
48	6812	6821	6830	6839	6848	6857	6866	6875	6884	6893
49	6902	6911	6920	6928	6937	6946	6955	6964	6972	6981
50	6990	6998	7007	7016	7024	7033	7042	7050	7059	7067
51	7076	7084	7093	7101	7110	7118	7126	7135	7143	7152
52	7160	7168	7177	7185	7193	7202	7210	7218	7226	7235
53	7243	7251	7259	7267	7275	7284	7292	7300	7308	7316
54	7324	7332	7340	7348	7356	7364	7372	7380	7388	7396